Imaging the Eye from Front to Back

with RTVue Fourier-Domain Optical Coherence Tomography

Imaging the Eye from Front to Back
with RTVue Fourier-Domain Optical Coherence Tomography

David Huang, MD, PhD
Center for Ophthalmic Optics and Lasers
Doheny Eye Institute
Keck School of Medicine
University of Southern California
Los Angeles, CA

Jay S. Duker, MD
Director, New England Eye Center
Professor and Chair, Department of Ophthalmology
Tufts Medical Center
Tufts University School of Medicine
Boston, MA

Joel S. Schuman, MD, FACS
Eye and Ear Foundation Professor and Chair
Department of Ophthalmology
University of Pittsburgh School of Medicine
Director, UPMC Eye Center
Pittsburgh, PA

James G. Fujimoto, PhD
Department of Electrical Engineering
and Computer Science
Research Laboratory of Electronics
Massachusetts Institute of Technology
Cambridge, MA

Robert N. Weinreb, MD
Distinguished Professor of Ophthalmology
Director, Hamilton Glaucoma Center
University of California, San Diego
La Jolla, CA

Bruno Lumbroso, MD
Professor Libero Docente
University of Rome
Rome Eye Hospital
Rome, Italy

SLACK
INCORPORATED

Library of Congress Cataloging-in-Publication Data

Imaging the eye from front to back with RTVue Fourier-domain optical coherence tomography / [edited by] David Huang ... [et al.].
       p. ; cm.
    Includes bibliographical references and index.
    ISBN 978-1-55642-963-7 (alk. paper)
    1. Optical coherence tomography. 2. Eye--Tomography. I. Huang, David.
    [DNLM: 1. Tomography, Optical Coherence--methods. 2. Diagnostic Techniques, Ophthalmological. 3. Eye Diseases--diagnosis. WW 141 I314 2010]
    RE79.I42I54 2010
    616.07'545--dc22

                         2010000148

DEDICATION

Yasuo Tano, MD
1948-2009

We dedicate this book to the memory of Professor Yasuo Tano, who was one of the original editors when we started working on this project. Sadly, Dr. Tano passed away suddenly on January 31, 2009. We regret that he was not able to see the completion of the book with us.

Professor Tano was the Chairman of the Department of Ophthalmology at Osaka University Medical School in Osaka, Japan. He was an important national and international leader of the ophthalmology community, having served as the president of the Asia-Pacific Academy of Ophthalmology, past president of the Japanese Ophthalmological Society, treasurer of the International Council of Ophthalmology, and president of the Club Jules Gonin. Professor Tano was a well-known innovator and teacher of vitreoretinal surgery, and he originally trained with Dr. Robert Machemer. He was a prolific researcher and writer, having published more than 660 peer-reviewed scientific papers and 59 textbooks.

Professor Tano was interested in optical coherence tomography from the very early days of this new technology. This led to many important publications in the past decade by Professor Tano and his colleagues at Osaka University.

We thank members of the Osaka faculty—Drs. Fumi Gomi, Nobuyuki Ohguro, Kaori Sayanagi, and Motokazu Tsujikawa—for contributing their chapters to this book. We thank Dr. Ryotaro Tano (Dr. Yasuo Tano's son) and Dr. Naoyuki Maeda for providing the photograph for this dedication.

CONTENTS

ACKNOWLEDGMENTS

We thank the team at Optovue Inc for introducing Fourier-domain optical coherence tomography (OCT) to ophthalmology and pioneering the use of a single device for imaging both the front and back of the eye. These initiatives have accelerated the growth of the whole field. Mr. Jay Wei, the CEO and founder of Optovue, deserves credit for having the vision to be ahead of the pack. Some of us have known Jay since the earliest days of OCT, when he helped develop the first commercial OCT product as head of the advanced development team at Humphrey Instrument. We appreciate his long-term commitment to this technology. Optovue also supported collaborative research with many of the editors and authors, and aided in the development and planning of this book. For this we again thank Mr. Jay Wei, and also Dr. Michael Sinai, Senior Director of Clinical Affairs.

The tremendous advances in OCT technology over the past 20 years were, of course, not due to only one team or one company, but to the efforts of many teams at universities and companies around the world. We acknowledge the contributions of our many colleagues that made possible the current state of art and knowledge.

We are deeply grateful to our co-authors, who have delivered extraordinary OCT images and cases to enrich this book. Special thanks go to our colleagues at the University of Osaka, who honored Dr. Yasuo Tano's commitment to contribute to this book, and more generally, to advance the application of OCT in ophthalmology.

We thank the Health Care Books and Journals team at SLACK Incorporated for producing a beautiful design for the book and smoothing the development of this major project. In particular, John Bond, Senior Vice President, and Jennifer Briggs, Acquisitions Editor, provided invaluable support.

We are, of course, indebted to our patients who inspire us to explore and develop many new clinical applications. We hope this book helps doctors make better clinical decisions that make a difference to their patients.

Finally, we thank our spouses and families for their support during the many hours we put in to edit this book.

ABOUT THE EDITORS

David Huang, MD, PhD is the Manger Chair of Corneal Laser Surgery and Associate Professor of Ophthalmology and Biomedical Engineering at the University of Southern California and Director of the Doheny Laser Vision Center. He co-invented OCT when he was a PhD student with Professor James Fujimoto at the Massachusetts Institute of Technology. He heads the Center for Ophthalmic Optic and Lasers (www.COOLLab.net), a group that pioneered corneal and anterior segment OCT. He leads the Advanced Imaging for Glaucoma Bioengineering Partnership (www.AIGStudy.net), which developed OCT ganglion cell complex mapping and retinal blood flow measurement.

Jay S. Duker, MD is the Director of the New England Eye Center and Professor and Chair of Ophthalmology at Tufts Medical Center and the Tufts University School of Medicine in Boston, Massachusetts. His research interests include clinical applications of OCT technology for disease of the posterior segment of the eye. He is co-editor of one of the best-selling textbooks in ophthalmology, Yanoff and Duker's *Ophthalmology*.

James G. Fujimoto, PhD is Professor of Electrical Engineering and Computer Science at the Massachusetts Institute of Technology. His research interests include biomedical optical imaging using OCT, as well as femtosecond photonics. Dr. Fujimoto is a member of the National Academy of Sciences, National Academy of Engineering, and American Academy of Arts and Sciences.

Bruno Lumbroso, MD has been Director of the Department of Ophthalmology of the Rome Eye Hospital and Professor LD of Clinical Ophthalmology in the University of Rome La Sapienza for more than 35 years. He is involved in postgraduate teaching of Ocular Imaging in European and Mediterranean countries. His main interest is in logical methods of retinal imaging analysis and interpretation.

Joel S. Schuman, MD, FACS is the Eye and Ear Foundation Professor and Chairman of Ophthalmology and Director of the UPMC Eye Center at the University of Pittsburgh, where he is also Professor of Bioengineering. Dr. Schuman co-invented OCT while a fellow at Massachusetts Eye and Ear Infirmary, Harvard University, and pioneered the use of OCT in glaucoma during his tenure at New England Eye Center, Tufts University School of Medicine.

Robert N. Weinreb, MD is the Distinguished Professor of Ophthalmology and the Director of the Hamilton Glaucoma Center at the University of California, San Diego. Known for its unique cross-disciplinary investigative programs, the Hamilton Glaucoma Center is home for a world-renowned team of scientists and staff. Dr. Weinreb has trained more than 120 postdoctoral fellows in glaucoma, many of whom hold distinguished academic positions throughout the world.

CONTRIBUTING AUTHORS

Divya Aggarwal, MD (Chapter 28)
Center for Ophthalmic Optics and Lasers
Doheny Eye Institute and Department of
    Ophthalmology
Keck School of Medicine
University of Southern California
Los Angeles, CA

Perry S. Binder, MD (Chapter 12)
Gordon Binder & Weiss Vision Insititute
San Diego, CA

Leonardo C. Castro, MD (Chapters 14 and 15)
Retina and Vitreous Research Fellow
New England Eye Center
Tufts University School of Medicine
Boston, MA
Retina Specialist
Centro Brasileiro de Especialidades Oftalmológicas
Araraquara, SP, Brazil

William D. Dilworth (Chapter 7)
Clinical Applications Manager
Optovue Inc
Fremont, CA

Amani A. Fawzi, MD (Chapter 22)
Doheny Eye Institute and Department of
    Ophthalmology
Keck School of Medicine
University of Southern California
Los Angeles, CA

Lindsey S. Folio (Chapter 26)
Research Specialist
Eye and Ear Institute
University of Pittsburgh Medical Center
Pittsburgh, PA

Fumi Gomi, MD, PhD (Chapter 17)
Department of Ophthalmology
Osaka University Medical School
Osaka, Japan

Tarek S. Hemeida, MD (Chapter 22)
The Armed Forces Hospital
Department of Ophthalmology
Alexandria, Egypt

Martin Heur, MD, PhD (Chapter 13)
Doheny Eye Institute
Keck School of Medicine
University of Southern California
Los Angeles, CA

Yutaka Imamura, MD (Chapter 18)
The Vitreous, Retina, Macula Consultants of New York
LuEsther T. Mertz Retina Research Center
Manhattan Eye, Ear and Throat Hospital
New York, NY

Alessandro Invernizzi, MD (Chapter 19)
Eye Clinic
Department of Clinical Science "Luigi Sacco"
University of Milan
Ospedale Luigi Sacco
Milano, Italy

Hiroshi Ishikawa, MD (Chapters 25 and 26)
UPMC Eye Center and Department of Ophthalmology
University of Pittsburgh School of Medicine
Pittsburgh, PA

Larry Kagemann, MS BME (Chapter 26)
Research Instructor
University of Pittsburgh School of Medicine
Eye and Ear Institute
University of Pittsburgh School of Engineering
Department of Bioengineering
Pittsburgh, PA

Yan Li, PhD (Chapters 2, 4, 5, 11, and 12)
Center for Ophthalmic Optics and Lasers
Doheny Eye Institute and Department of
    Ophthalmology
Keck School of Medicine
University of Southern California
Los Angeles, CA

Sheila Mahdaviani, MD (Chapter 5)
Center for Ophthalmic Optics and Lasers
Doheny Eye Institute and Department of
    Ophthalmology
Keck School of Medicine
University of Southern California
Los Angeles, CA

Farnaz Memarzadeh, MD (Chapter 5)
Center for Ophthalmic Optics and Lasers
Doheny Eye Institute and Department of
    Ophthalmology
Keck School of Medicine
University of Southern California
Los Angeles, CA

Nobuyuki Ohguro, MD (Chapter 24)
Associate Professor
Department of Ophthalmology
Osaka University Medical School
Osaka, Japan

Marco Pellegrini, MD (Chapter 19)
Eye Clinic
Department of Clinical Science "Luigi Sacco"
University of Milan
Ospedale Luigi Sacco
Milano, Italy

Marco Rispoli, MD (Chapters 9 and 21)
Department of Ophthalmology
Ospedale Eastman
Rome, Italy

Alfredo A. Sadun, MD, PhD (Chapter 28)
Center for Ophthalmic Optics and Lasers
Doheny Eye Institute and Department of
    Ophthalmology
Keck School of Medicine
University of Southern California
Los Angeles, CA

Camila Haydée Rosas Salaroli, MD (Chapter 12)
Center for Ophthalmic Optics and Lasers
Doheny Eye Institute and Department of
    Ophthalmology
Keck School of Medicine
University of Southern California
Los Angeles, CA

Maria Cristina Savastano, MD (Chapter 21)
Retina Fellow
Department of Ophthalmology
Catholic University
Policlinico Gemelli
Rome, Italy

Kaori Sayanagi, MD (Chapter 16)
Department of Ophthalmology
Osaka University Medical School
Osaka, Japan

Giovanni Staurenghi, MD (Chapter 19)
Eye Clinic
Department of Clinical Science "Luigi Sacco"
University of Milan
Ospedale Luigi Sacco
Milano, Italy

Ou Tan, PhD (Chapters 6 and 28)
Center for Ophthalmic Optics and Lasers
Doheny Eye Institute and Department of
    Ophthalmology
Keck School of Medicine
University of Southern California
Los Angeles, CA

Maolong Tang, PhD (Chapters 3, 4, 11, and 13)
Center for Ophthalmic Optics and Lasers
Doheny Eye Institute and Department of
    Ophthalmology
Keck School of Medicine
University of Southern California
Los Angeles, CA

Stephen L. Trokel, MD (Chapters 12 and 13)
Edward S. Harkness Eye Institute
College of Physicians and Surgeons
Columbia University
New York, NY

Motokazu Tsujikawa, MD, PhD (Chapter 20)
Associate Professor
Department of Ophthalmology
Osaka University Medical School
Osaka, Japan

Gianmarco Vizzeri, MD (Chapter 27)
Assistant Professor of Ophthalmology
University of Texas—Galveston
Galveston, TX

Yimin Wang, PhD (Chapter 10)
Center for Ophthalmic Optics and Lasers
Doheny Eye Institute and Department of
    Ophthalmology
Keck School of Medicine
University of Southern California
Los Angeles, CA

Andre J. Witkin, MD (Chapters 8 and 23)
Tufts Medical Center
New England Eye Center
Boston, MA

Gadi Wollstein, MD (Chapter 26)
Department of Ophthalmology
University of Pittsburgh School of Medicine
Pittsburgh, PA

Linda M. Zangwill, PhD (Chapter 27)
Professor of Ophthalmology
Hamilton Glaucoma Center
University of California San Diego
La Jolla, CA

Introduction

Optical coherence tomography, or OCT, has come a long way since its initial introduction in the ophthalmic market in 1996. Initially, OCT was slow and had limited functions, just like first-generation computers. There was a lack of software to help the clinician interpret the images. The resolution and definition of the images were low, like an analog television. Only academic eye centers and large retina practices could afford them. Anterior segment OCT systems came onto the market much later as a separate system. It was not possible to scan both the front and back of the eye with one OCT system.

The RTVue was the first of a new generation of OCT systems that broke through these limitations. It employed the Fourier-domain technology that increased the speed of OCT almost 100-fold over the old time-domain technology. This means that high-definition tomographs could now be obtained in such a short time that motion artifacts were imperceptible. Three-dimensional imaging became possible. The resolution became high enough that it became possible to detect damage to the internal structures of the photoreceptor layer. The same OCT system now has been adapted to both the imaging of the cornea, angle, retina, and optic nerve. Disease-specific software now aids the clinician to map the ganglion cell complex for glaucoma diagnosis and map the cornea for keratoconus detection.

With the increased performance and functionality, the new generation of OCT costs no more than the old. This means that OCT is no longer confined to only a few highly specialized clinicians, but rather is widely available to all eye practitioners, including the glaucoma specialist, the cornea specialist, the refractive surgeon, and the comprehensive ophthalmologist. Our book addresses this greater audience and describes how to use the multifunctional OCT system to diagnose diseases and plan surgery for a variety of conditions from the front to the back of the eye.

The book is fortunate to have outstanding contributions from many leaders in the field. The operating principles of OCT are explained by some of the original developers of the technology. The anterior segment and glaucoma applications are explained by pioneers in the area. A substantial portion of the book is devoted to macular diseases and is written by well-known authors from around the world. We hope you will enjoy this timely book on this rapidly advancing technology that is changing and will continue to change the practice of ophthalmology.

1 Introduction to Optical Coherence Tomography

James G. Fujimoto, PhD and David Huang, MD, PhD

INTRODUCTION

Optical coherence tomography (OCT) is a fundamentally new type of medical diagnostic imaging technology that performs high-resolution, micron-scale, cross-sectional imaging of the internal microstructure in biological tissues by measuring the intensity and echo time delay of light.[1-6] OCT is a powerful imaging modality because it enables the real-time, *in situ* imaging of tissue structure or pathology with resolutions of 1 to 15 μm, which is one to two orders of magnitude finer than conventional clinical imaging technologies such as ultrasound, magnetic resonance imaging (MRI), or computed tomography (CT). Since its development in 1991, OCT has been explored in a wide range of clinical applications.[7] In ophthalmology, it is now considered a standard of care, enabling imaging of the retina and anterior eye at resolutions that were previously impossible to achieve with any other noninvasive imaging method.

Figure 1-1 illustrates the first demonstration of OCT imaging, reported by Huang, showing the human retina *ex vivo* and corresponding histology.[1] The OCT image in Figure 1-1 is an axial image resolution of ~15 μm, which is almost one order of magnitude finer than ultrasound. Imaging was performed with infrared light at ~800 nm wavelength. The images are displayed using a false color scale, where different magnitudes of backscattered light are displayed as different colors on a rainbow color scale. The light signals detected in these OCT images are extremely small, between 4×10^{-10} (0.4 billionths) to 10^{-6} (1 millionth) of the incident light. Figure 1-1 is an OCT image of the human retina in the region of the optic disc

with corresponding histology. The contour of the optic nerve head (ONH), as well as retinal vasculature, can be easily seen in the OCT image. The retinal nerve fiber layer (NFL) can also be visualized, and this suggested the application of OCT for glaucoma detection. This image was acquired *ex vivo,* and postmortem retinal detachment with subretinal fluid accumulation is evident.

Because of the ease of imaging the eye, the development of OCT in ophthalmology proceeded relatively rapidly. Figure 1-2 is one of the first examples of *in vivo* imaging of the normal human retina.[3] The axial image resolution was 10 μm in tissue, and imaging was performed with light at ~800 nm wavelength. The OCT image shows the normal contour of the foveal pit and the ONH. The retinal NFL is seen as a highly backscattering layer that decreases in thickness as it moves away from the optic disc. The retinal pigment epithelium (RPE) and choroid are evident as a thin, highly backscattering layer posterior to the retina. These early results demonstrated that OCT could image retinal structure with unprecedented resolution.[3-5]

Figure 1-3 shows the first demonstration of OCT imaging of the anterior eye, reported in 1994.[2] The axial image resolution was 10 μm in tissue, and imaging was performed at ~800 nm wavelength. Later OCT systems for imaging the anterior eye used light at longer ~1300 nm wavelengths, which reduced optical scattering and improved image penetration depth. The image spans a transverse dimension of 21 mm, thus allowing cross-sectional imaging of the entire anterior chamber. The OCT image shows the corneal thickness and the depth of the anterior chamber. The curvature of the anterior and posterior surfaces of the cornea can also be measured.

Huang D, Duker JS, Fujimoto JG, Lumbroso B, Schuman JS, Weinreb RN.
Imaging the Eye from Front to Back with RTVue Fourier-Domain Optical Coherence Tomography (pp 1-22).
© 2010 SLACK Incorporated.

Figure 1-1. The first demonstration of OCT imaging, reproduced from Huang.[1] OCT imaging is analogous to ultrasound B-mode imaging, except light is used instead of sound. A cross-sectional image is generated by scanning a light beam across the tissue and then measuring the echo time delay and intensity of backscattered or backreflected light. This figure is an OCT image of the human retina *in vitro* and corresponding histology. The axial image resolution is 10 µm, and imaging was performed with infrared light at ~800 nm wavelength. The scale bar is 300 µm.

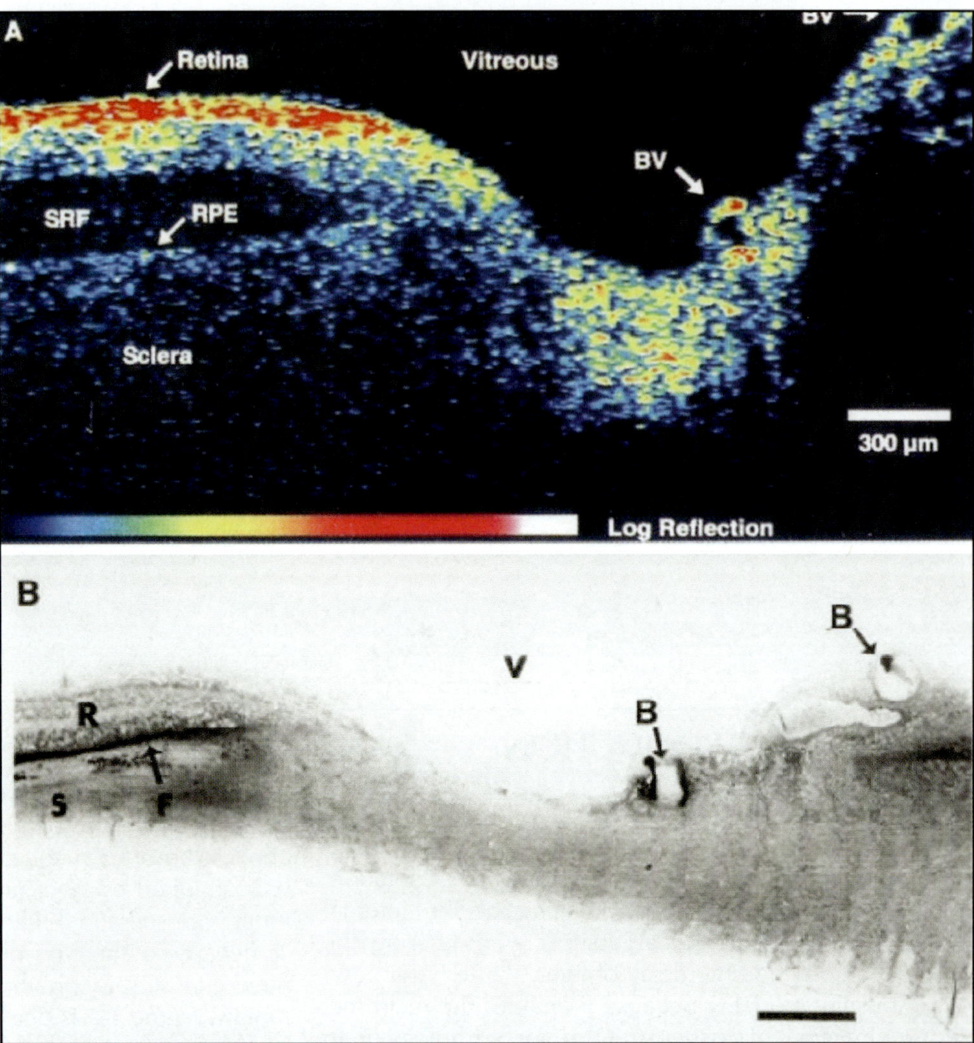

Figure 1-2. Early OCT image of the human retina *in vivo*. The image shows the normal retinal contour in the fovea and the optic disc region. The retinal NFL is evident as a highly backscattering layer that emanates from the optic disc and decreases in thickness as it approaches the fovea. The axial image resolution is 10 µm, and imaging was performed using light at ~800 nm wavelength. The image is displayed using a logarithmic false color scale, which maps the log of the backscattered, or backreflected, light intensity to a rainbow color scale. The maximum signal is approximately -50 dB of the incident signal, while the minimum detectable signal is approximately -95 dB. These early results demonstrated the feasibility of using OCT for retinal imaging and suggested its future application in ophthalmology. (Reprinted with permission from Hee MR, Izatt JA, Swanson EA, et al. Optical coherence tomography of the human retina. *Arch Ophthalmol.* 1995;113:325-332. Copyright © 1995 American Medical Association. All rights reserved.)

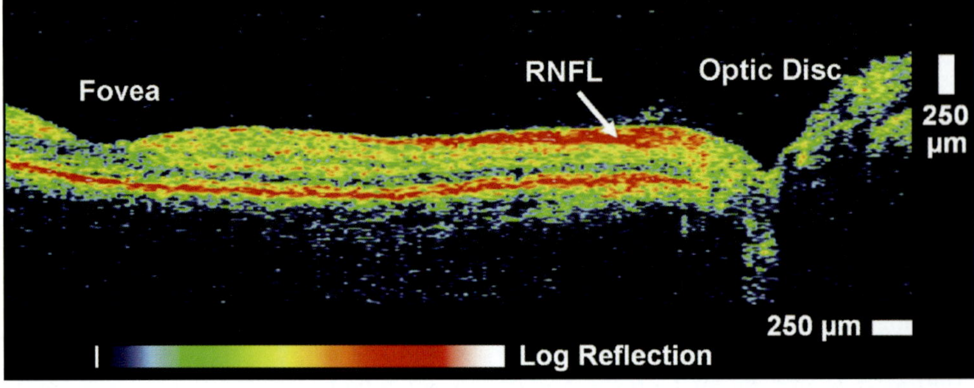

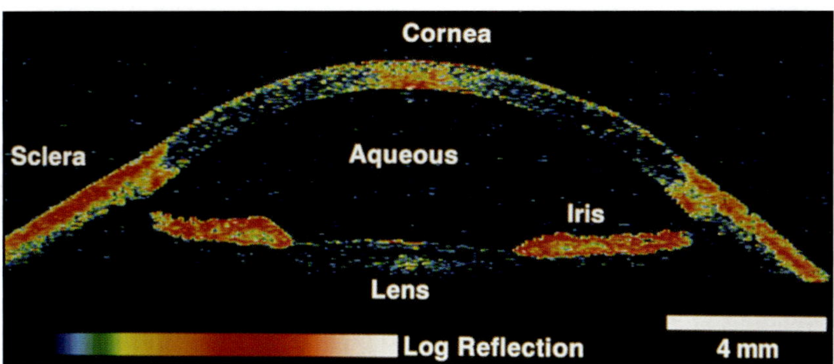

Figure 1-3. OCT image of the anterior chamber *in vivo*. This is the first demonstration of anterior eye imaging. The image shows the curvature of the anterior and posterior surfaces of the cornea, as well as the depth of the anterior chamber. The iris is visible, but it scatters light strongly, so deeper structures are shadowed. The axial image resolution is 10 μm, and imaging was performed with light at ~800 nm wavelength. (Reprinted with permission from Izatt JA, Hee MR, Swanson EA, et al. Micrometer-scale resolution imaging of the anterior eye *in vivo* with optical coherence tomography. *Arch Ophthalmol.* 1994;112:1584-1589. Copyright © 1994 American Medical Association. All rights reserved.)

However, it is necessary to compensate for image distortion effects from light refraction at the curved surface of the cornea. The sclera and iris are visible as highly optically scattering structures that produce shadowing of posterior features.

OCT is especially powerful in ophthalmology because it provides real-time, noncontact, cross-sectional imaging of the retina or the anterior eye with unprecedented resolution. Since OCT generates cross-sectional images of retinal morphology, it can provide vital diagnostic information that is complementary to conventional fundus photography and fluorescein angiography. OCT enables visualization of structural features of the retina, including the fovea and optic disc, as well as the internal architectural morphology of the retina, such as the NFL, ganglion cell layer, and photoreceptors.[3] OCT imaging of the anterior eye enables visualization of the cornea, iris, lens, and angle.[2] Early studies established the use of OCT imaging for detecting and monitoring various macular diseases, including macular edema, macular holes, central serous chorioretinopathy, age-related macular degeneration (AMD) and choroidal neovascularization, and epiretinal membranes.[4,8-14] OCT imaging can also be used to perform quantitative measurements or morphometry of the retina. OCT is especially powerful for the diagnosis and monitoring of diseases such as glaucoma or macular edema, because it can provide quantitative information of disease progression. Images can be analyzed quantitatively by using image-processing algorithms to automatically extract features such as retinal thickness or retinal NFL thickness.[5,15-17] The retinal NFL thickness, a diagnostic indicator for early glaucoma and disease progression, can be quantified and correlated with measurements of ONH structure or visual function.[17-24] Mapping and display techniques have been developed to represent OCT data in alternate forms, such as topographic thickness maps, in order to facilitate the interpretation and comparison to fundus images better.[14] Finally, since structural image information can be assessed quantitatively, OCT imaging can be used as a diagnostic tool to predict the probability of disease and to monitor disease progression and the effectiveness of treatment.

IMAGING WITH LIGHT VERSUS SOUND

OCT imaging is analogous to ultrasound B-mode imaging, except that it uses light instead of sound. OCT performs cross-sectional or volumetric imaging by measuring the echo time delay and intensity of backscattered or backreflected light from structures inside tissue. OCT images are 2D or 3D data sets representing variations in optical backscattering or backreflection in a cross-sectional plane or volume of tissue.

Because of the analogy between OCT and ultrasound, it is helpful to compare OCT imaging with ultrasound imaging. Ultrasound is widely used clinically for quantitative measurements of intraocular distances, as well as for imaging the anterior eye and globe.[25-29] Since ultrasound imaging depends on the reflection of sound waves from intraocular structures, it requires direct contact of the ultrasound probe to the cornea, or immersion of the eye in a liquid to transmit sound waves into the globe. The resolution of ultrasound depends on the frequency or wavelength of the sound waves. For typical ultrasound systems, sound-wave frequencies are in the 10 megahertz (MHz) regime, which yield spatial resolutions of ~150 μm. Ultrasound also has the advantage that sound waves are readily transmitted into most biological tissues and, therefore, it is possible to image structures deep within the body. High-resolution ultrasound imaging can be performed using higher frequency sound waves to achieve resolutions on the 20-μm scale.[27,28] However, these high-frequency sound waves are strongly attenuated in biological tissues, and imaging can be performed to depths of only 4 to 5 mm. Therefore, high-resolution ultrasound imaging of the retina is not possible.

OCT is an optical imaging technique that uses light rather than sound. The main limitation of optics is that light is highly scattered or absorbed in most biological tissues. Therefore, optical imaging is limited to tissues that are directly optically accessible or it requires the use of devices such as endoscopes or catheters. OCT is ideally suited for ophthalmology because of the ease of optical access to the eye. In addition, OCT can be performed without physical contact to the eye, thereby minimizing patient discomfort during examination. Imaging that uses light provides a significantly higher spatial resolution than what is possible with ultrasound. The first ophthalmic OCT images had axial resolutions of ~10 μm, ~10 to 20 times finer than standard ultrasound B-mode imaging.[3] Research OCT systems for ultrahigh-resolution ophthalmic imaging can achieve even finer resolutions of ~2 to 3 μm.[6,30] Current generation commercial OCT systems using spectral/Fourier-domain detection have axial image resolutions in the ~5 to 7 μm range. The inherently high resolution of OCT imaging permits visualization of individual retinal layers, thus facilitating the diagnosis of a wide range of retinal pathologies.

When light is directed onto the eye, it is backreflected from boundaries between different tissues and backscattered with different intensities from tissues with different optical properties. The distances and dimensions of different tissue structures can be determined by measuring the *echo* time delay of light backreflected or backscattered from the different structures at varying axial (longitudinal) distances.[31,32] However, it is important to note that the speed of light is almost a million times faster than the speed of sound. Since tissue dimensions are measured using the echo time delay, measurements involving light echoes require ultrafast time resolution.

Figure 1-4 illustrates the distance and time scales for light and sound. The speed of sound in water is approximately 1500 meters per second, while the speed of light is approximately 3×10^8 meters per second in air. Distance or axial range information may be determined from the echo time delay according to the formula: echo time delay = (distance the echo travels)/ (velocity of light or sound). Thus, the measurement of distances or structures with a resolution of ~100 μm, which would be typical for ultrasound, requires a time resolution of approximately 100 nanoseconds $(100 \times 10^{-9}$ sec). The measurement of structures with a resolution of ~10 μm, which can be achieved by OCT, requires a time resolution of approximately 30 femtoseconds $(30 \times 10^{-15}$ sec), approximately 1 million times faster. Fortunately, it is possible to perform ultrahigh-resolution time and distance measurements by using a simple optical technique known as interferometry.

MEASURING ECHOES OF LIGHT USING INTERFEROMETRY

OCT uses interferometry to perform ultrahigh-resolution time and distance measurements of light. The technique of low-coherence or white-light interferometry was first described by Sir Isaac Newton.[33] During the 1980s, low-coherence interferometry was used to per-

form high-resolution optical measurements in fiberoptic and optoelectronic components.[34-36] In order to perform distance measurements at tenths of a micron resolution (which corresponds to echo delay times of tens of femtoseconds), it is necessary to compare or correlate one optical beam or lightwave with another reference optical beam or lightwave. This type of measurement may be performed by using an optical device known as an interferometer.

Figure 1-5 is a schematic of an optical interferometer. The light source is a laser or other device that emits either short light pulses or short-coherence-length light. The optical beam from the light source is incident onto a partially reflecting mirror (beamsplitter) that splits the light into two paths. One light beam is sent to the patient's eye and is backreflected or backscattered from intraocular structures at different distances. The light beam returning from the eye consists of multiple echoes that give information about the distance and thickness of these intraocular structures. The second beam is reflected from a reference mirror at a known distance and acts as a time reference. This reflected reference light beam travels back to the partial mirror (beamsplitter), where it combines or interferes with the light from the eye. The detector measures the interference or correlation of light echoes from the eye with an echo that has travelled a known reference path delay. Different interferometric methods can be used to detect echo time delays of light. Previous-generation OCT systems used time-domain detection, while current-generation OCT is based on spectral or Fourier-domain detection.

AXIAL SCANS AND OPTICAL COHERENCE TOMOGRAPHY IMAGE GENERATION

The simplest measurement performed by OCT is analogous to an ultrasound A-mode scan, which gives information on tissue thickness. Figure 1-7 shows an early example of axial distance measurements or axial scans of the anterior chamber.[38] The plot shows the intensity of the backreflected or backscattered light from different structures within the anterior eye as a function of echo time delay or axial (longitudinal) distance. Echoes are generated from the anterior and posterior surfaces of the cornea, as well as from the anterior capsule of the lens. The intensity of the reflected light depends on the difference in refractive index between different tissues. The reflection from the anterior surface of the cornea is relatively large; however, reflections from internal surfaces (such as between the cornea and aqueous or different layers of the retina) are small. In addition, different tissues (such as the cornea, lens, and sclera

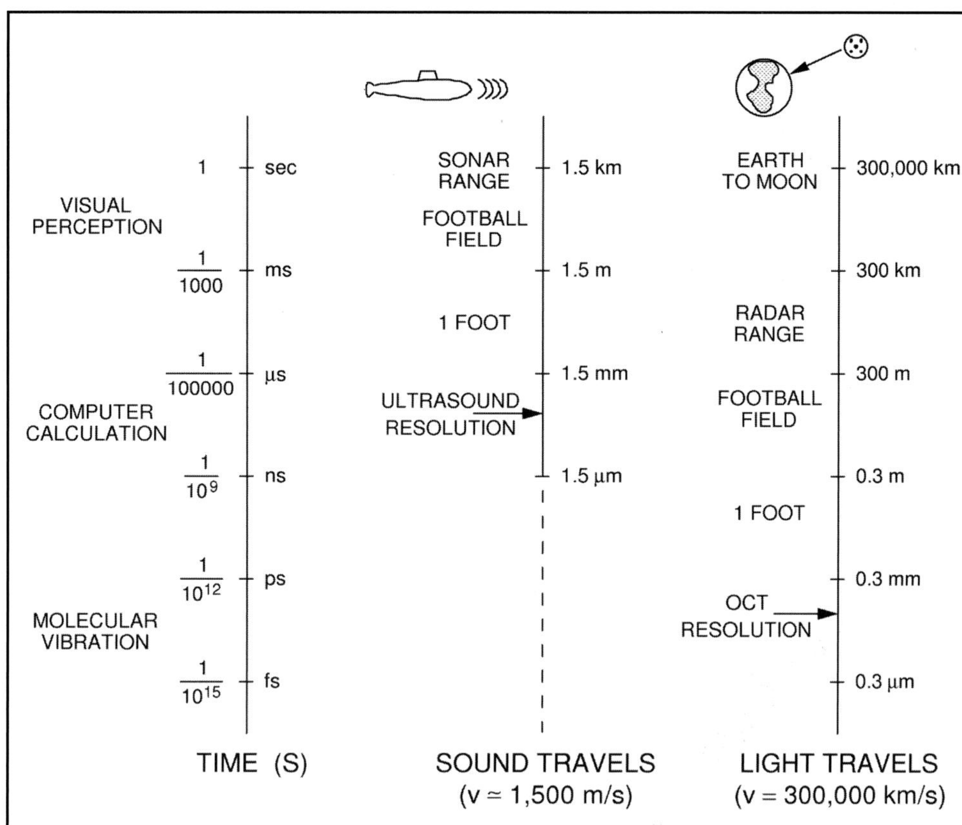

Figure 1-4. Distance and time scales for light and sound. The velocity of sound is approximately 1500 meters/sec, while the velocity of light is 3×10^8 (300,000,000) meters/sec. Standard ultrasound has a resolution limit of 150 µm, which corresponds to a time measurement of 100 nanoseconds (100×10^{-9} sec). In contrast, standard OCT imaging has a resolution of 10 µm, which corresponds to a time measurement of 30 femtoseconds (30×10^{-15} sec). In order to measure echoes of light from small structures, extremely fine time resolution is required. The scales are logarithmic; each vertical division represents a factor of 1000. Notice that light from the moon takes approximately 1.3 seconds to reach the earth.

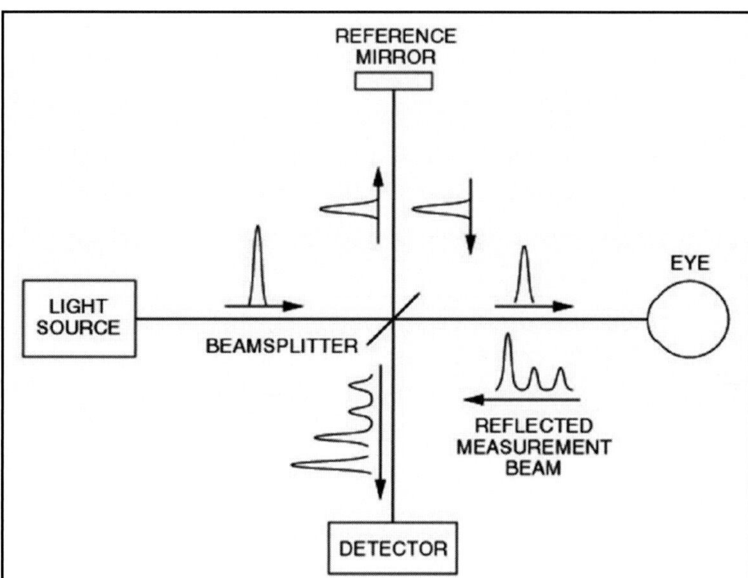

Figure 1-5. Measuring echo time delays of light. The echo delay time of light can be measured with high resolution by comparing or correlating one light beam with another. Light from a source is directed onto a partially reflecting mirror (beamsplitter) and is split into reference and OCT measurement beams. The OCT measurement beam is backreflected or backscattered from the tissue with different echo time delays, depending upon its internal microstructure. The light in the reference beam is reflected from a reference mirror at a known distance, which produces a known time delay. The light from the tissue, consisting of multiple echoes, and the light from the reference mirror, consisting of a single echo at a known delay, are combined by the interferometer and then detected. Previous-generation OCT systems used time-domain detection, while current-generation systems use spectral/Fourier-domain detection.

of the anterior eye or the different layers of the retina) will produce varying amounts of optical backscattering. This backscattering is detected in the OCT axial scan measurement.

The axial (longitudinal) distance measurement shown in Figure 1-7 permits a direct measurement of the corneal thickness as well as anterior chamber depth. The thickness of the tissue is calculated by measuring the optical echo delay and multiplying it by the speed of light in the tissue. The speed of light in the tissue is given by the speed in vacuum multiplied by the index of refraction of the tissue. Therefore, measuring the physical thickness in OCT relies upon knowledge of the tissue index of refraction. The intensity of the backreflected

How Time-Domain Optical Coherence Tomography Detection Works

Early OCT systems detected light echoes using *time-domain* detection. Figure 1-6 is a schematic of how time-domain OCT detection works. When the interferometer reference path is scanned, interference fringes are generated as a function of time at the output of the interferometer.[37-40] This process can also be understood by noting that the scanning reference mirror produces a Doppler shift of the light. If a narrow-line-width light source is used, the interference will occur over a wide range of path length differences. However, in order to detect light echoes, a low-coherence (broad-bandwidth) light source is required. Low-coherence light can be characterized as having statistical phase discontinuities over a distance known as the coherence length, which is inversely proportional to the frequency bandwidth of the light. When low-coherence light is used, interference is observed only when the reference path is matched to a light signal from the eye within the coherence length of the light. The interferometer measures the autocorrelation of the electric field of the light. The magnitude and time delay of light echoes are measured by scanning the reference arm and detecting the interference signal. The coherence length of the light source determines the axial image resolution. Since it is possible to have light sources with a few-micron coherence length, this enables the measurement of echoes of light with femtosecond timescale resolution.

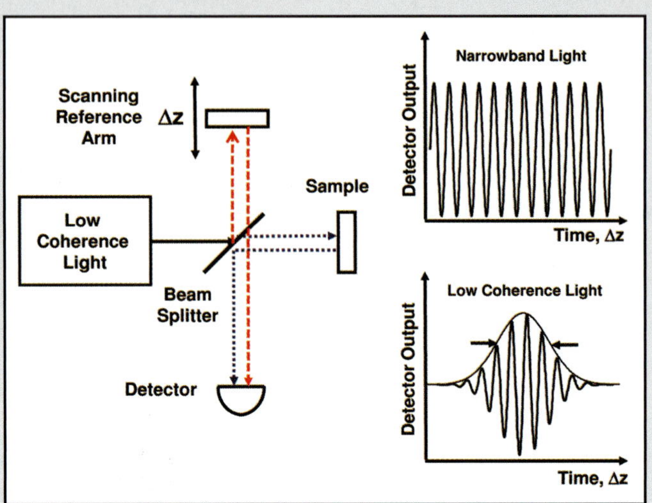

Figure 1-6. How time-domain OCT detection works. The echo time delay of a backreflected or backscattered light signal can be measured using low-coherence interferometry with a scanning reference arm. When a narrowband light source is used, the output of the interferometer consists of interference fringes. However, when a broadband or low-coherence light source is used, interference occurs only when the path lengths of the reference light and the light signal are equal. Detecting the interference signal as the reference path is scanned sequentially measures echos from different depth. This generates and axial scan which is analogous to ultrasound A-mode measurements.

Figure 1-7. Axial scan measurement of anterior chamber depth using low-coherence interferometery. The plots display the magnitude of the backreflected or backscattered optical intensity as a function of echo delay or distance. A large reflection is observed from the anterior surface of the cornea, while smaller reflections originate from the posterior corneal surface (cornea-aqueous boundary), and the lens or iris. Note the presence of scattered light that originates within the cornea. The graphs show the structural features that occur along the axis of the optical beam. If the transverse position of the optical beam is changed, then different features are measured (the depth of the anterior chamber versus the depth to the iris). (Reprinted with permission from Huang D, Wang J, Lin CP, Puliafito CA, Fujimoto JG. Micron-resolution ranging of cornea and anterior chamber by optical reflectometry. *Lasers Surg Med.* 1991;11:419-425.)

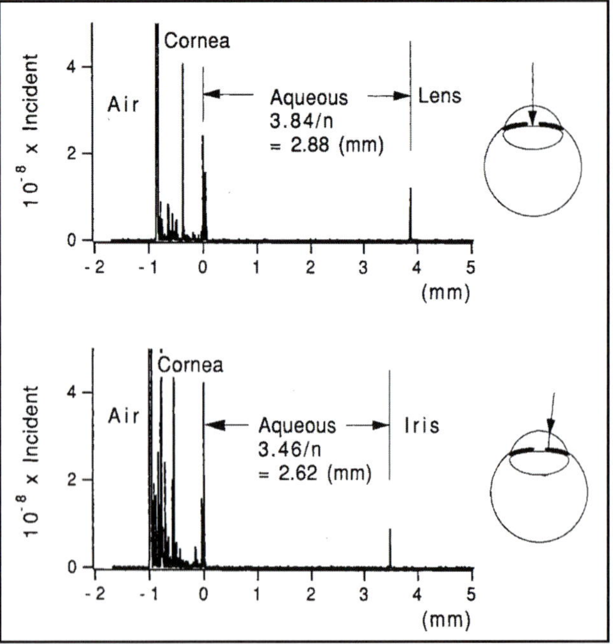

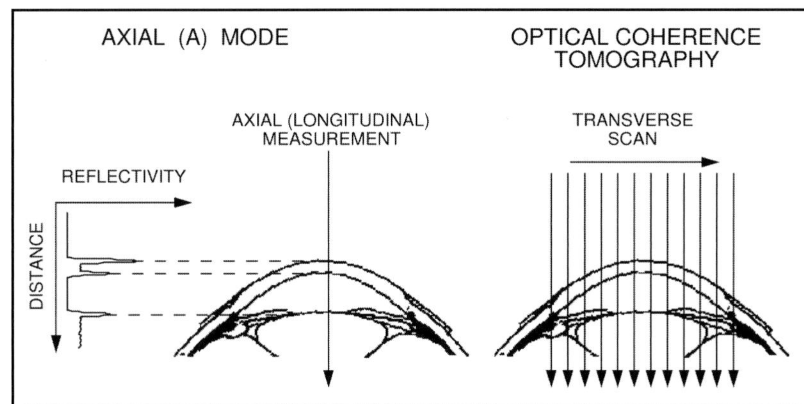

Figure 1-8. How OCT images are generated. OCT imaging is analogous to ultrasound B-mode imaging. OCT images are generated by performing rapid, successive axial (longitudinal) measurements at different transverse positions. Each axial measurement represents the echo delay of backreflected and backscattered light from microstructures inside tissue, thus giving a measurement of tissue dimensions and properties along the optical beam. By scanning the optical beam transversely while performing successive axial measurements, a 2D data set can be acquired. This data is a cross-sectional map of the backreflection and backscatter within the tissue, and it can be displayed as a cross-sectional image.

or backscattered light is extremely small, approximately 10^{-5} to 10^{-9} (-50 dB to -90 dB) of the incident intensity. Thus, very high-detection sensitivity to extremely weak, reflected light echoes is required in order to measure structures within the eye.

Once an axial scan or A-mode scan is obtained, the relative positions of different structures may be measured by scanning the transverse position of the optical beam. Figure 1-7 shows a second axial measurement with the optical beam aimed at the iris rather than at the anterior capsule of the lens. Because the light beam can be focused to a small spot size, the transverse position of the beam can be known with high precision. Thus, information on both the axial (or longitudinal) and the transverse microstructure of tissue can be measured. The transverse resolution is determined by the focused spot size of the light and it has a resolution analogous to the transverse resolution used in conventional microscopy.

OCT cross-sectional (B-mode) imaging is performed by acquiring successive axial (longitudinal) measurements, or A-scans, of the tissue at different transverse positions, as shown in Figure 1-8.[1-3,41] Successive, rapid axial measurements, or A-scans, are performed while scanning the optical beam in the transverse direction. The result is a set of axial scans where each scan represents the optical backreflection or backscattering as a function of depth in the tissue at a different transverse position. This 2D data set is displayed as a grayscale or false-color OCT image. The image in Figure 1-3 is an example of an OCT image of the anterior eye, which can be achieved by using the scan pattern in Figure 1-8.

Optical Coherence Tomography Instrumentation for Retinal Imaging

Retinal imaging requires that the OCT beam be focused and scanned on the retina. The instrument design for retinal imaging, which is similar to a fundus camera, is shown schematically in Figure 1-9. It is helpful to discuss briefly this instrument to understand how to achieve optimum quality in OCT images and to avoid artifacts. A high-power objective lens, similar to a handheld 78 D lens, is used to relay-image the retina onto an image plane inside the instrument. The instrument then relays the retinal image onto a video camera that enables real-time operator viewing of the fundus. The OCT imaging beam is coupled into the optical path of the instrument by using a partially reflective mirror (beamsplitter). It is then focused onto the retinal image plane by using another relay lens. The focused OCT beam is relay-imaged onto the retina by the objective lens and the patient's eye. The transverse spot size of the OCT beam on the retina is typically ~15 to 20 μm. Prior to OCT image acquisition, the objective lens position is adjusted in order to focus the OCT beam on the retina. This also focuses the video fundus image. The instrument also illuminates the fundus, which enables video viewing.

The transverse position of the OCT beam is scanned by two perpendicular x-y scanning mirrors inside the instrument. The optical system is designed so that the OCT beam pivots about the pupil of the eye when it is scanned, as shown in Figure 1-9. This prevents the OCT beam from being vignetted by the pupil, and it enables access to a wide field of view on the retina. In order to ensure that the OCT pivots about the pupil of the eye, the patient's eye must be located at a given distance from the ocular objective lens. If the instrument is too close or too far from this position, the OCT beam will change position transversely instead of pivoting about the pupil, and vignetting will occur. Vignetting causes reduction in the signal of OCT images near the edges of the scan and can lead to interpretation artifacts.

OCT imaging can be performed at different locations on the fundus by controlling the OCT beam scan. The location of the image also can be controlled by changing the patient's point of fixation. The instrument has an internal fixation target visible to the patient's eye that

Figure 1-9. Schematic of OCT instrument for retinal imaging. The OCT instrument is analogous to a fundus camera. An objective lens relays an image of the retina to a plane inside the instrument. Operator viewing of the fundus is performed with a video camera. A computer-controlled, 2-axis scanning mirror scans the OCT beam. A relay lens focuses the OCT beam onto the image plane, and the objective lens directs the OCT beam through the pupil onto the retina. The OCT beam is focused on the retina by adjusting the objective lens. When the OCT beam is scanned, the beam pivots about the pupil of the eye in order to minimize vignetting.

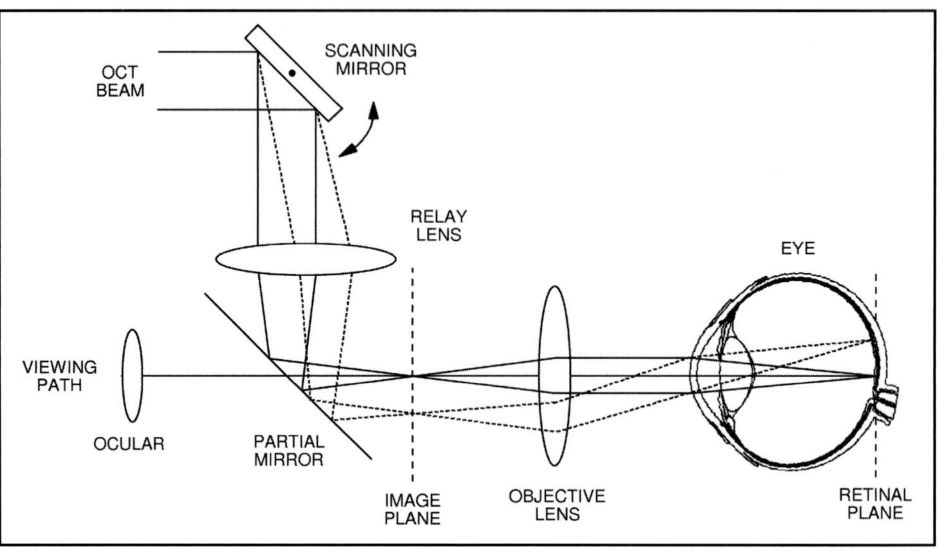

can be adjusted by computer control. The operator can view real-time video of the fundus image displayed in a window on the computer monitor. This enables precise aiming of the OCT imaging beam. When the OCT beam is scanned, images are generated and displayed in real time. The beam is also visible to the patient as a thin red line, whose position in the patient's visual field corresponds to points on the retina that are being scanned. Different scan patterns have been designed that are optimized for the diagnosis and monitoring of different retinal diseases. The different scan patterns used to image retinal features are described in detail in later chapters. Figure 1-10 shows an OCT image along the papillomacular axis. The figure is an example of an image acquired with the previous generation of OCT using time-domain detection. The axial image resolution is 8 to 10 µm.

Retinal imaging is performed with infrared light at 800 nm wavelengths. The allowable retinal limits for safe exposure have been thoroughly investigated and are governed by international standards such as the American National Standards Institute (ANSI standard). The maximum safe, permissible exposure depends upon the wavelength, the spot size, the duration of exposure, and repeated exposure. Because OCT instruments have extremely high-detection sensitivities, low power levels of ~750 µW can be used, which are well within safe exposure limits.

WHAT OPTICAL COHERENCE TOMOGRAPHY IMAGES SHOW

Image contrast in OCT depends upon differences in optical backreflection or backscattering between different tissue structures. Since light that reaches deeper tissue layers must pass through more superficial lay-

ers, shadowing effects (similar to those in ultrasound) can occur. Light incident onto tissue is transmitted, absorbed, or scattered.[42-45] Transmitted light remains unaffected and continues traveling into deeper tissue layers. Absorbed light is essentially removed from the incident beam. Absorption occurs because tissue chromophores, such as hemoglobin or melanin, absorb specific wavelengths in the incident light. Optical scattering is a property of a heterogenous medium. It occurs because of microscopic spatial variations in the refractive index within tissue. These refractive index variations can be caused by subcellular structures (such as nuclei, cytoplasm, or cell membranes) or bundles of smaller structures (such as nerve fibers or axons). Optical scattering causes incident light to be redirected in multiple directions. Light that completely reverses direction when it is scattered is called backscattered light. In contrast, backreflection of light can occur when light is incident on a boundary between two homogenous materials that have a different refractive index.

In tissues that are strongly absorbing or scattering, the intensity of the incident beam decreases exponentially with depth. In tissues other than the eye, attenuation of the optical beam from scattering restricts the OCT image penetration depth. However, the high sensitivity of OCT enables the detection of light from depths up to ~2 mm in most scattering tissues.[46,47] In ophthalmic imaging, the retina backscatters weakly, thus generating very small optical signal levels. In this case, the high sensitivity of OCT allows these weak optical signals to be detected so that retinal tissues can be imaged, even though they are virtually transparent.

OCT images are mostly comprised of "single backscattered" light. This light has propagated into tissue, backscattered once from the structure being imaged, and then propagated out of the tissue. The strength of the OCT signal from a tissue structure at a given depth

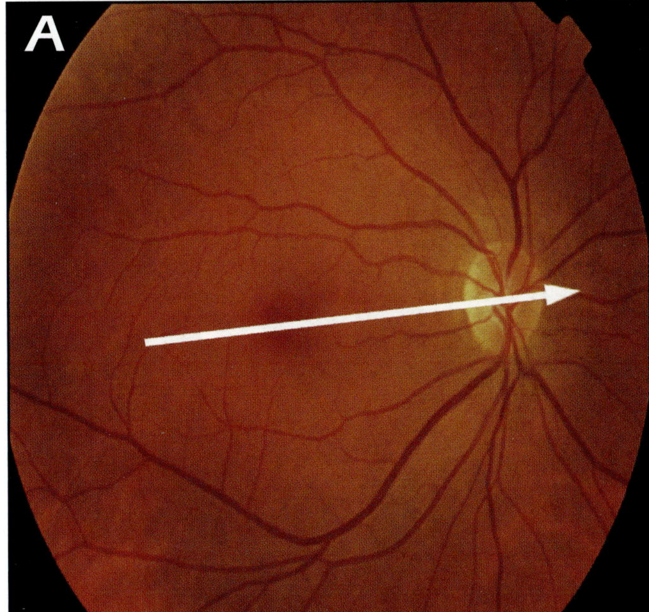
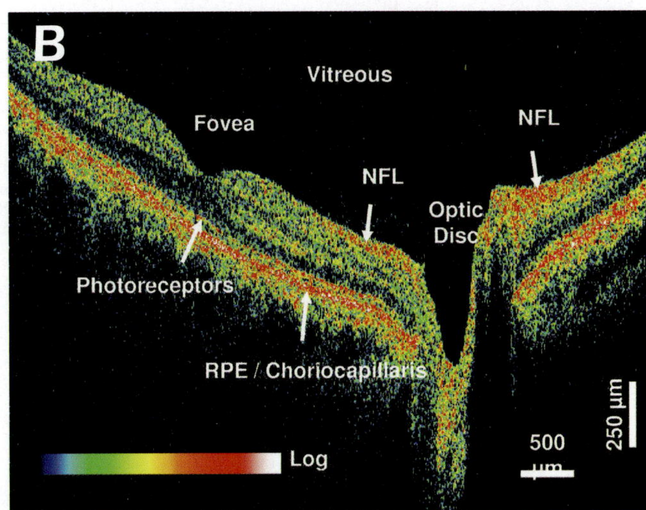

Figure 1-10. OCT image of a normal fovea and optic disc taken along the papillomacular axis (A). This image, acquired using the previous generation of OCT technology based on time-domain detection, has an axial resolution of 8 to 10 µm. The OCT scan is 10-mm long and its location is shown on a corresponding fundus photograph (B). The fovea and optic disc can be identified by their characteristic morphology, and the layered structure of the retina is apparent. The retinal NFL is highly reflective and increases in thickness toward the disc.

is defined by the amount of incident light that is transmitted without absorption or scattering to that depth, is directly backscattered, and propagates out of the tissue returning to the detector. Direct backreflection can also occur at the boundary of 2 materials that have different indices of refraction, such as the boundary between the cornea and the aqueous in the anterior eye (see Figure 1-3). When there is strong absorption or scattering, such as from blood vessels, hemorrhage, or the RPE, light is strongly attenuated and the shadowing of deeper structures can occur.

It is important to note that, although OCT images display the true dimensions of the structures being imaged (after correcting for index of refraction and refraction effects), contrast comes from different mechanisms than those in histology. In histology, selective stains are used to identify specific cellular or subcellar features. In OCT, image contrast occurs from intrinsic differences in tissue optical properties. Thus, care must be taken when interpreting OCT images, since they are not analogous to conventional histology. In addition, the false color scale used to display OCT images indicates differences in light-signal levels. While different tissues can have different signal levels because they have differences in backscattering or backreflection, the color in OCT images may not necessarily correspond to different types of tissue.

ULTRAHIGH-RESOLUTION OPTICAL COHERENCE TOMOGRAPHY

Since the introduction of commercial OCT instrumentation for retinal imaging in 1996, OCT technology has undergone several generations of improvement. Ultrahigh-resolution OCT retinal imaging was demonstrated in 2001, and it provided insight into the interpretation of retinal images.[30] Using specially designed, broadband light sources and OCT instruments, it was possible to achieve axial image resolutions of ~3 µm. This was significantly finer than the 10-µm axial resolution available in commercial instruments at that time.[30,48-50] Figure 1-11 is an ultrahigh 3-µm axial resolution image from Drexler et al.[30]

Ultrahigh-resolution OCT enables the visualization of individual retinal layers and it correlates with the well-known morphology of the retina.[51-54] Histologically, the retina can be seen as divided into 10 distinct layers, including 4 cell layers and 2 layers of neuronal interconnections. The interpretation of OCT imaging is supported by studies that compare OCT to histology and ultrahigh-resolution OCT imaging studies.[55-60] The interpretation of features in OCT images is also confirmed by imaging pathologies that produce known alterations of retinal architecture.[48-50]

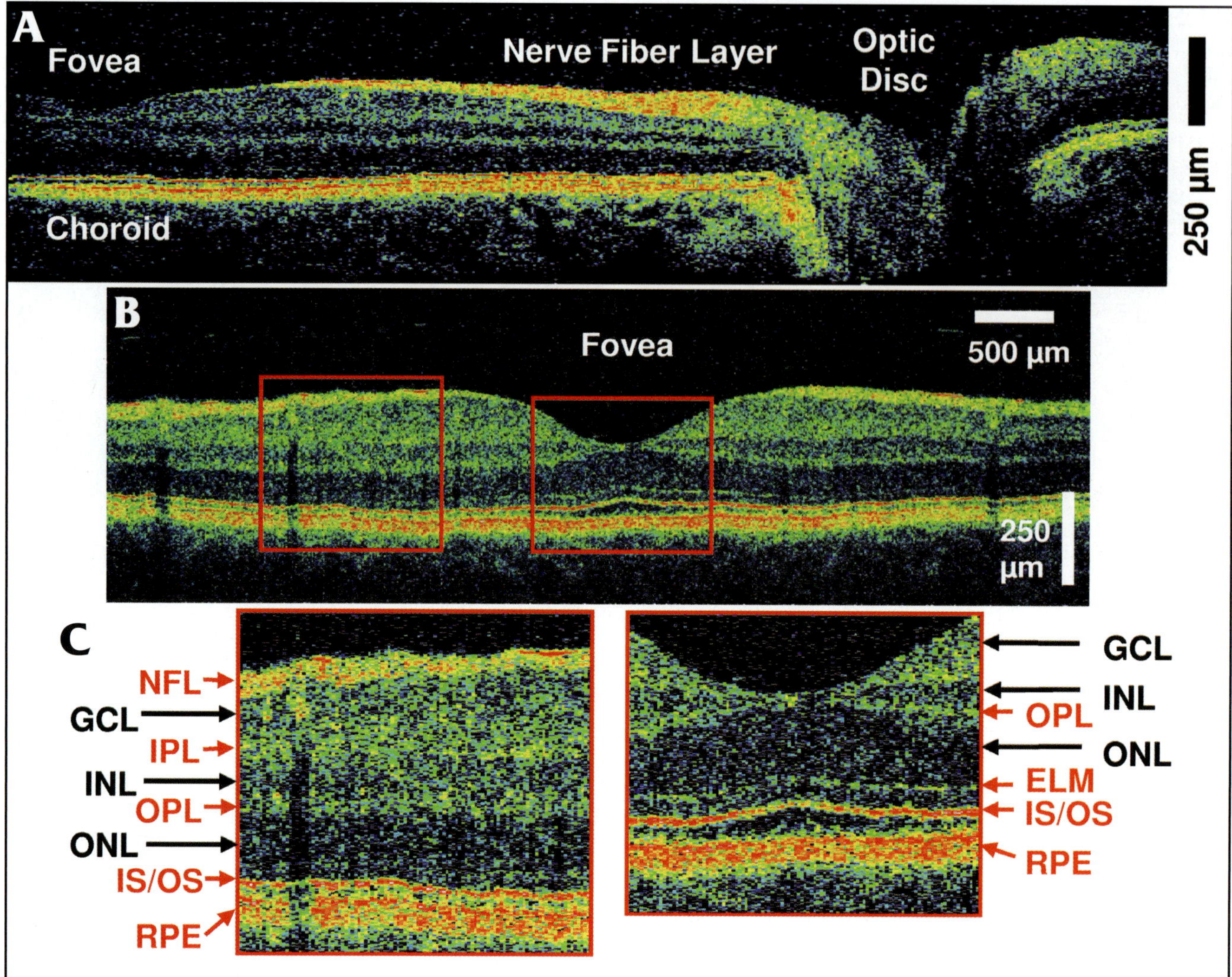

Figure 1-11. Ultrahigh-resolution OCT images of the macular region. Imaging is performed with 3-μm axial resolution. (A) OCT image with 6-mm transverse width and 512 transverse pixels. (B) High-magnification OCT image with 3-mm transverse width and 512 transverse pixels. (C) Enlargements of the fovea and parfoveal regions. The images are expanded in the vertical direction for better visibility of the retinal layers. The retina can be divided into 10 distinct layers, including 4 cell layers and 2 layers of neuronal interconnections. Proceeding from the inner retina to the outer retina, these layers are: the inner limiting membrane (not shown), the NFL, GCL, IPL, INL, OPL, ONL, ELM, IS and OS of the photoreceptor layer, and the RPE. The choriocapillaris and choroid are immediately posterior to the RPE. Ultrahigh-resolution imaging enables an excellent visualization of retinal microstructure, and it provides additional confirmation for the interpretation of standard-resolution images. (Data modified from Drexler W, Morgner U, Ghanta RK, et al. Ultrahigh resolution ophthalmic optical coherence tomography. *Nature Medicine.* 2001;7:502-507.)

The NFL and plexiform layers consist of axonal structures that are highly backscattering and they appear red in false color OCT images. In contrast, nuclear layers are weakly backscattering and appear blue-black. The first highly backscattering layer is the NFL, which is visible adjacent to the fovea. This layer is thin in the macular region. The 3 weakly reflective layers are the ganglion cell layer (GCL), inner nuclear layer (INL), and outer nuclear layer (ONL). The GCL increases in thickness in the parafoveal region. The moderately backscattering inner plexiform layer (IPL) is adjacent to the GCL and INL. The obliquely running photoreceptor axons, sometimes considered as a separate layer in the outer plexiform layer (OPL) that is known as Henle's fiber layer, are highly backscattering.

The boundary between the photoreceptor inner segment (IS) and outer segment (OS) is visible as a thin, highly backscattering band immediately anterior to the RPE and choroid. The reflection arising from this structure may be the result of a refractive index difference between the photoreceptor IS and the highly organized structure of the OS that contains stacks of membranous discs rich in the visual pigment rhodopsin.[54,61] The thickness of the photoreceptor IS and OS increases in the foveal region, thus corresponding to the well-known increase in the length of the OS of the cones in this region. The external limiting membrane (ELM) can sometimes be visualized as a thin backscattering layer posterior to the ONL and anterior to the boundary between the photoreceptor IS and OS. The ELM is not a physical membrane, but is an alignment of structures between the photoreceptors and the Müller cells.

The RPE, which contains melanin, is a strongly backscattering layer. The RPE can have banded structure, which is related to the rod and cone distribution, and interdigitation of the RPE cells between the photoreceptors.[60] The ability to image the photoreceptor morphology and its impairment is especially interesting as a marker of disease progression.[48,50,62]

Bruch's membrane, which is only 1 to 4 µm thick, cannot be visualized in the normal retina, because it is adjacent to the highly scattering RPE. In cases of RPE detachment or drusen, Bruch's membrane is sometimes visible. Finally, the choriocapillaris is vascular and strongly backscattering. The vascular structures of the choriocapillaris and choroid are highly scattering and they produce shadowing effects that limit the OCT imaging depth for deeper structures. In some cases, retinal blood vessels can be identified by their increased backscatter and by their shadowing of deeper structures.

High-Speed Spectral/ Fourier-Domain Optical Coherence Tomography

In addition to improvements in image resolution, new detection techniques now enable dramatic improvements in imaging speed.[63-69] Previous-generation OCT instruments used an interferometer with a scanning reference delay arm, which was known as time-domain detection. However, it is also possible to detect echoes of light in the Fourier domain by measuring the interference spectrum. This concept was first proposed by Fercher et al in 1995.[63] The first demonstration of retinal imaging was performed by Wojtkowski et al in 2002.[64] This result showed that sensitivities sufficient to image the retina could be achieved by using spectral/Fourier-domain OCT, and it encouraged a renewed interest in this technology. In 2003, 3 different research groups, working independently, demonstrated that Fourier-domain detection has a powerful sensitivity advantage over time-domain detection, since Fourier-domain detection essentially measures all echoes of light simultaneously.[70-72] The sensitivity enhancement is given approximately by the ratio of the axial resolution to the axial imaging depth. For most OCT systems, this is a sensitivity increase of 50 to 100 times, thus enabling a corresponding increase in imaging speeds. Studies by Nassif et al in 2004 demonstrated video-rate retinal imaging by using spectral/Fourier-domain OCT

to achieve 29,000 axial scans per second with 6 µm axial image resolution.[67] Wojtkowski et al demonstrated ultrahigh-resolution retinal imaging with ~2 µm resolution in tissue at 19,000 axial scans per second.[68] These and other advances stimulate growth in OCT research and commercial development.

Advantages of Spectral/ Fourier-Domain Optical Coherence Tomography

Spectral/Fourier-domain detection essentially measures all echoes of light simultaneously, while time-domain detection measures the echoes sequentially as a function of delay. This feature gives spectral/Fourier-domain OCT a dramatic increase in detection sensitivity, which enables OCT imaging ~50 to 100 times faster. High-speed spectral/Fourier-domain OCT has several advantages, including improved image quality, preservation of true retinal topography, improved retinal coverage, and accurate registration of the image set to fundus features. In general, for a given acquisition time, high-speed imaging can increase the number of axial scans or transverse pixels per image in order to yield high-definition images and to increase the number of cross-sectional images acquired sequentially in order to improve retinal coverage. Figure 1-13 shows a comparison of standard OCT using time-domain detection and high-speed, ultrahigh-resolution OCT using spectral/Fourier-domain detection, from Wojtkowski et al.[69] Figure 1-13A is a standard OCT image with time-domain detection, having 10 µm axial image resolution with 512 axial scans, acquired in ~1.3 seconds. Figure 1-13B is a high-speed, ultrahigh-resolution image using spectral/Fourier-domain detection, having ~2 µm axial resolution and 2048 axial scans, acquired in 0.13 seconds. The higher resolution and greater number of transverse pixels in the high-speed, ultrahigh-resolution OCT image improves the visualization of internal retinal structure. Increased imaging speeds minimize eye motion artifacts and enable the true topography of the ONH to be preserved.

In addition, high-speed OCT imaging enables the acquisition of complete 3D OCT data sets in a time comparable to that of previous OCT protocols that acquired several individual images. Figure 1-14 shows 3D OCT imaging of the optic disc using a raster scan pattern from Wojtkowski et al.[69] The 3D OCT volumetric data set contains comprehensive structural information. An *en face* view or OCT fundus images, which are identical to a standard retinal fundus view, can be generated by summing the data in the axial direction.[69,73] Individual cross-sectional OCT images (B-scans) can be precisely and reproducibly registered to *en face* features of the retina.

How Spectral/Fourier-Domain Optical Coherence Tomography Detection Works

Figure 1-12 shows how spectral/Fourier-domain OCT detection works. This detection technique uses a broadband light source with an interferometer and measures the interference spectrum by using a spectrometer and a high-speed, line scan camera. In spectral/Fourier-domain detection, different echo time delays of light are encoded as different frequencies in the spectrum of an interference signal. The echo time delays of light can be measured by Fourier transforming the interference spectrum. Spectral/Fourier-domain can be understood by noting that an interferometer acts like a periodic frequency filter, where the periodicity of the frequency filter is a function of the difference Δz between the sample and reference paths. The output from a broadband light source is split into 2 beams. One beam is directed onto the tissue to be imaged and is backreflected or backscattered from structures at different depths. The second beam is reflected from a fixed (not scanned) reference mirror. The signal beam and the reference beam have a relative time delay that is determined by the path length difference, Δz, which is related to the depth of the structure in the tissue. The interference of the 2 beams will have a spectral modulation that can be measured with a spectrometer. The periodicity of this modulation will be inversely related to the echo time delay Δz. Larger echo delays will produce higher frequency spectral modulation. The echo delays can be measured by Fourier transforming the interference signal. This produces an axial scan measurement (A-scan) of the magnitude and echo delay of the light signal from the tissue.

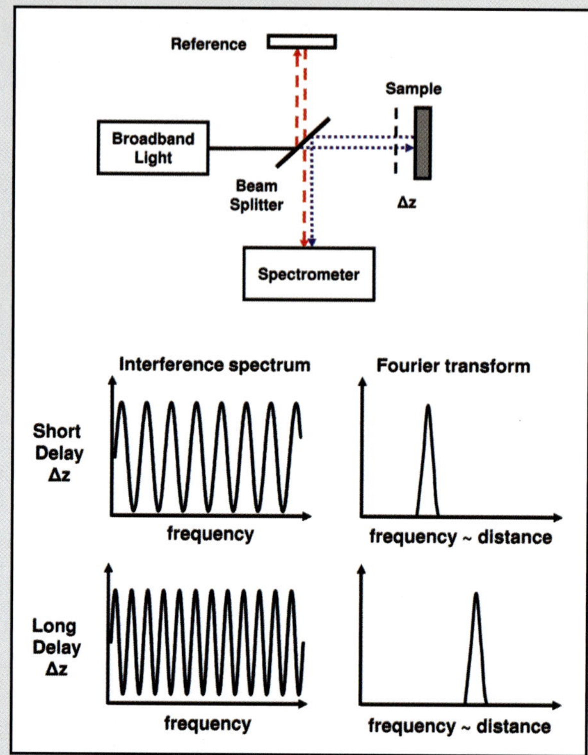

Figure 1-12. Spectral/Fourier OCT detection. Spectral/Fourier-domain OCT uses an interferometer with a broadband light source and measures the spectrum of the interference output with a spectrometer and a high-speed, line scan camera. Echo time delays of light are measured by noting that the interferometer acts like a spectral filter that has a periodic output spectrum dependent on the path length mismatch Δz (away from the zero delay position shown as dashed line). Larger path length differences generate higher frequency interference oscillations, where the frequency is proportional to the delay (bottom). Fourier transforming the spectral interference extracts the frequency and measures the delay (axial scan information). Spectral/Fourier-domain detection essentially measures all the echoes of light simultaneously and, therefore, has a significant sensitivity advantage when compared with time-domain detection.

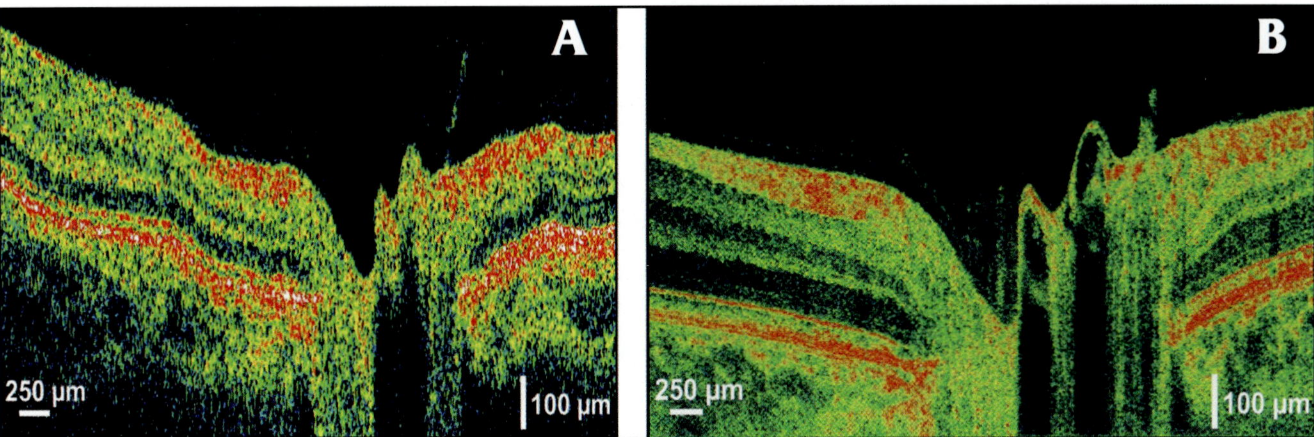

Figure 1-13. OCT with time-domain detection compared with high-speed, ultrahigh-resolution OCT using spectral/Fourier-domain detection. (A) Previous-generation OCT image of the optic disc with ~10 µm axial image resolution and 512 axial scans acquired in ~1.3 seconds. (B) High-speed, ultrahigh-resolution spectral/Fourier-domain OCT image with ~2 µm axial resolution and 2048 axial scans acquired in 0.13 seconds. The rapid acquisition speed preserves the true retinal contour and enables high-definition imaging with the large number of axial scans (transverse pixels) to be acquired. These images are from a research prototype system. Typical commercial instruments have 5 to 7 µm axial image resolution.

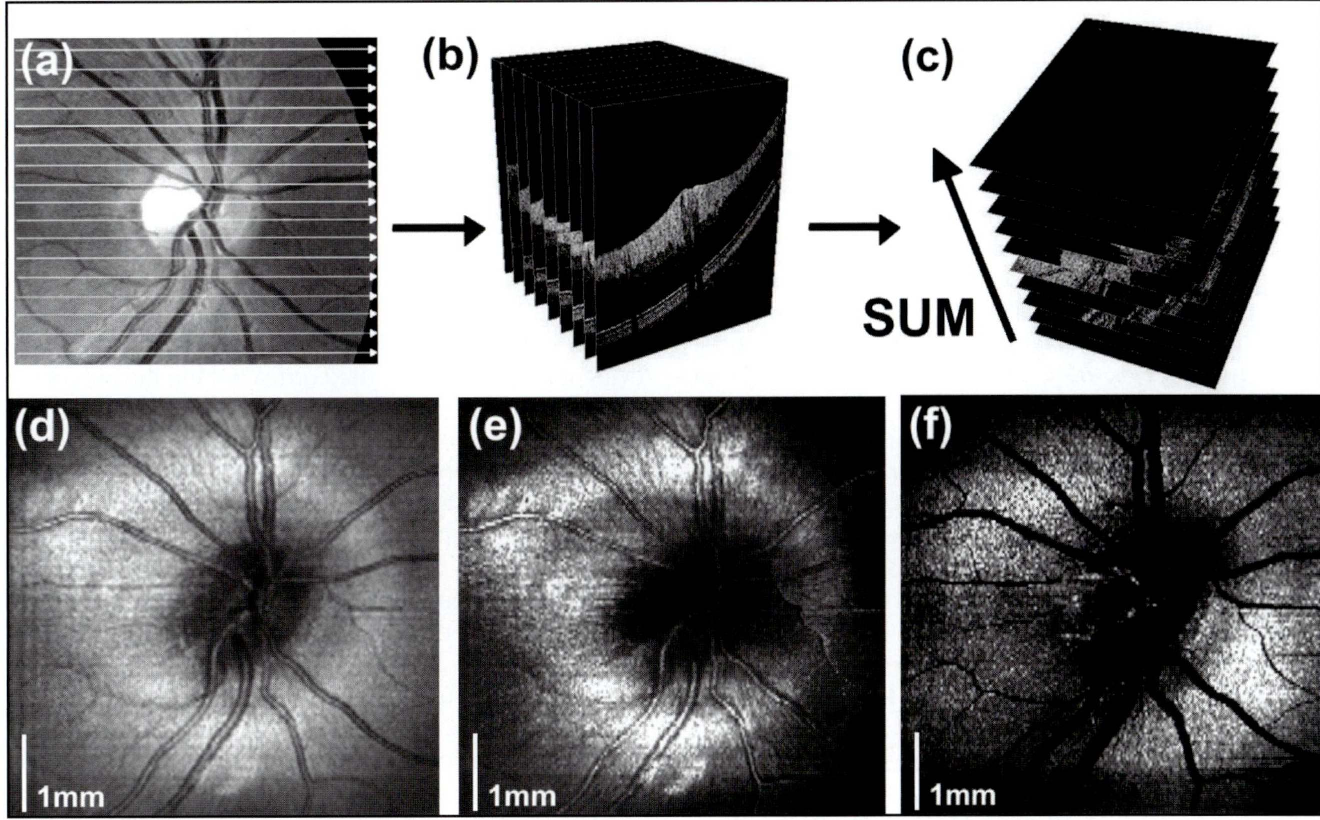

Figure 1-14. 3D OCT fundus image generated with raster scan. (A, B) High-speed OCT enables the acquisition of 3D OCT data that contain comprehensive volumetric information. (C, D) An OCT *en face* retinal fundus image can be generated directly from 3D OCT data by summing the signal along the axial direction. (E, F) OCT *en face* images can also be generated by displaying individual retinal layers, such as (E) the NFL or (F) RPE. Individual OCT cross-sectional images can be precisely and reproducibly registered with respect to features on the fundus image. (This figure was published in *Ophthalmology*, 112, Wojtkowski M, Srinivasan VJ, Fujimoto JG, et al, Three-dimensional retinal imaging with high-speed ultrahigh resolution optical coherence tomography, 1734-1746, Copyright Elsevier [2005].)

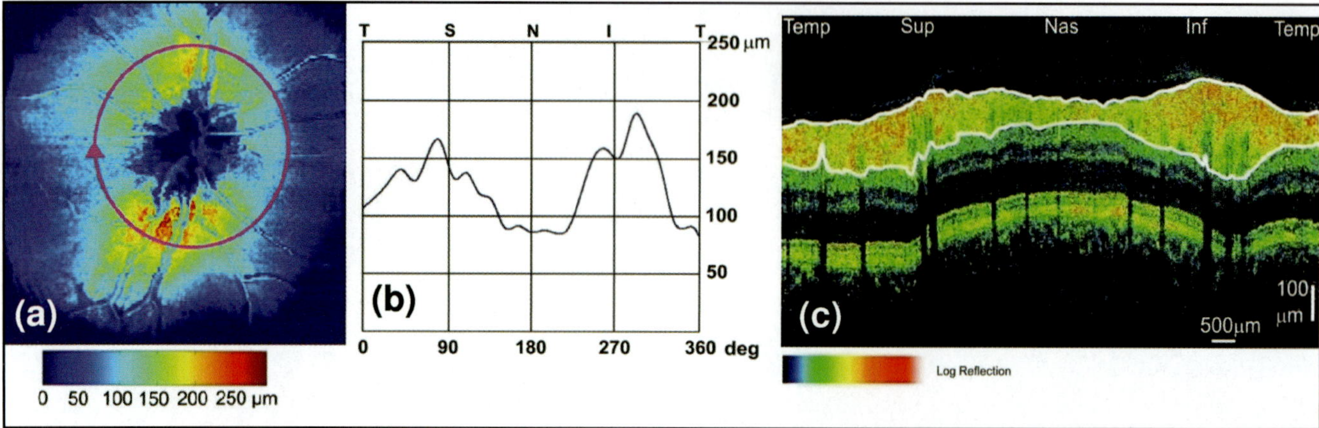

Figure 1-15. Circumpapillary measurement of the NFL from 3D OCT volumetric data. (A) False color map of NFL thickness generated from 3D OCT data. (B) Measurement of NFL thickness along a circumpapillary scan. (C) Virtual OCT circumpapillary image is extracted from the 3D OCT volumetric data set. The position of the circumpapillary image can be adjusted after processing to reduce errors from registration. (This figure was published in *Ophthalmology*, 112, Wojtkowski M, Srinivasan VJ, Fujimoto JG, et al, Three-dimensional retinal imaging with high-speed ultrahigh resolution optical coherence tomography, 1734-1746, Copyright Elsevier [2005].)

The 3D OCT data enable the generation of an NFL thickness map similar to that obtained by scanning laser polarimetry, except that OCT measures the NFL thickness by using cross-sectional image information, while scanning laser polarimetry measures the NFL thickness by using birefringence. This map can provide information on the radial and circumpapillary variations in the NFL thickness. Figure 1-15 is a plot of the NFL thickness variation measured along a 3.4-mm circle that is centered around the optic disc. Circumpapillary OCT images of any diameter and radial OCT images can be generated. Figure 1-15 also shows an example of a 3.4-mm-diameter circumpapillary OCT image generated from the 3D OCT data. OCT images and NFL maps can be precisely and consistently registered to the fundus by using the OCT fundus image generated from the same 3D OCT data.[69,73] This addresses a limitation in previous generations of OCT systems, where variations in the scan position can produce variations in measured NFL thickness values.

Anterior Eye Structure in RTVue Optical Coherence Tomography Images

The RTVue (Optovue Inc, Fremont, CA) performs imaging of the anterior eye by using a corneal adaptor module (CAM) that attaches to the objective lens of the instrument. Under normal operation, scanning the retina involves pivoting a collimated beam about the pupil plane of the eye (see Figure 1-9). The CAM focuses the OCT beam onto the cornea and produces a parallel (telecentric) displacement of the OCT beam as it is scanned, which is similar to the scan in Figure 1-8. Since the OCT beam is refracted when it is incident on the curved surface of the cornea, special software is required to *dewarp* the OCT image in order to display the true curvature of structures behind the cornea.

Figure 1-17 is an image of the cornea from the Optovue RTVue. Imaging is performed at 800 nm with 5 to 7 µm axial image resolution. The image has been dewarped to correct for refraction at the anterior corneal surface, and it shows the curvature of anterior and posterior surfaces of the cornea. A strong reflection is evident at the center of the cornea, where the air-cornea interface is exactly perpendicular to the incident OCT beam. This strong reflection produces echo artifacts in the OCT image that are visible in the A-scan lines at the center of the image. An enlargement of the OCT image shows a reflection from the anterior surface and the epithelial layer (ep), which is demarcated by low scattering. The corneal stroma has scattering in a striated pattern that results from collagen organization.

Figure 1-18 is an image of the normal anterior angle from the Optovue RTVue. Image averaging was performed to reduce speckle noise and improve sensitivity, thus increasing image penetration. The corneal epithelium (ep) joins with the sclera in region of the corneoscleral limbus (cl). Schwalbe's ring (sr) is on the posterior surface of the cornea at the termination of Descemet's membrane and is a useful landmark for measuring the anterior angle. Schlemm's canal (sc) is visible. The iris is highly pigmented and shadows deeper structures.

The scan patterns for imaging the anterior eye are discussed in Chapter 2 and scan procedures are described in Chapter 3, while interpretation of corneal and anterior angle images is discussed in Chapter 4.

ARTIFACTS IN SPECTRAL/FOURIER-DOMAIN IMAGING

While OCT that uses spectral/Fourier-domain detection has powerful advantages in terms of imaging speed, it also has important limitations that are not present in previous-generation OCT systems that use time-domain detection. Spectral/Fourier-domain OCT is subject to *mirror* artifacts in the images when the eye is positioned incorrectly. This can be understood by examining Figure 1-12. Spectral/Fourier-domain detection measures the echo time delays of light by comparing backreflected or backscattered light from the eye to light from a reference path delay that determines a *zero delay*. Spectral/Fourier-domain detection cannot distinguish between positive or negative delays when compared to this zero delay. Therefore, if the axial position of the eye is exactly at the reference zero delay position, the image appears folded about this zero delay as a mirror artifact.

Figure 1-16 shows a series of OCT retinal images that illustrate the mirror artifact. The different images, from the top to the bottom, are acquired by moving the OCT instrument toward the eye, so that the distance to the eye is decreasing. The echoes of light from the retina are measured with respect to a specific delay, the zero delay position from the instrument, which is determined by setting the reference path of the interferometer. The retina appears in a normal position when it is farther than the zero delay. When the instrument is moved toward the eye, so that the retina is exactly at the zero delay reference position, the portions of the retina that cross the zero delay appear folded over or mirror imaged. Finally, when the instrument is moved even closer to the eye, the retina is closer than the zero delay reference position and it appears inverted.

Instrument operators must take care to recognize the mirror artifact in OCT image data and to repeat image acquisition with the OCT instrument or reference delay position adjusted correctly so that the structure to be imaged does not overlap the zero delay position. Special care must also be exercised when interpreting OCT images, especially when the retina is tilted at an angle to the imaging frame or when there are structures, such as epiretinal membranes or in the case of pronounced edema, that have long axial lengths. These features often cross the zero delay position and appear as mirror images.

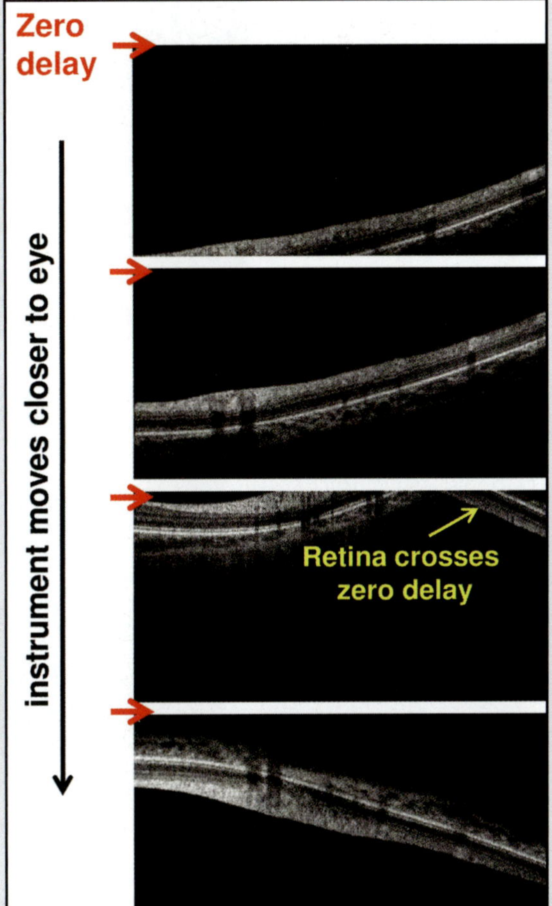

Figure 1-16. Artifacts in spectral/Fourier-domain OCT detection. Spectral/Fourier-domain detection cannot distinguish between echoes from positive or negative time delays, producing a "mirror image" artifact. Series of OCT images (top to bottom) as the instrument is moved closer to the eye. When the retina crosses the zero delay position of interferometer, from positive to negative delays with respect to the zero reference position, it appears as if it is in front of the zero delay. This produces a folded, or mirror image, appearance in the OCT image. When the instrument is moved even closer to the eye, such that the retina is in back of the zero delay, the image appears inverted. Care must be taken to avoid misinterpreting mirror image artifacts in OCT images.

In addition to the mirror image artifact, detection sensitivity in spectral/Fourier-domain detection varies with the axial position of the backscattering or backreflection signal. The instrument is most sensitive to echoes close to the zero reference delay position, and sensitivity decreases further from zero delay. This is seen in Figure 1-16, where the retina appears brighter when it is near zero delay. Conversely, features further from zero delay appear dimmer. Depending upon the application, it may be desirable to have increased sensitivity to signals that occur deeper in the retina in order to assess features below the RPE or to detect signals that are anterior to the retina, such as epiretinal membranes.

Figure 1-17. Imaging the cornea. RTVue image of the cornea; 2x enlargement shows details of the corneal epithelium (ep), stroma, and endothelium (en). The strong reflection from the central portion of the cornea, where the OCT beam is backreflected into the instrument, has an elevated background along the entire length of the axial scan, and shows multiple echoes from the strong reflection.

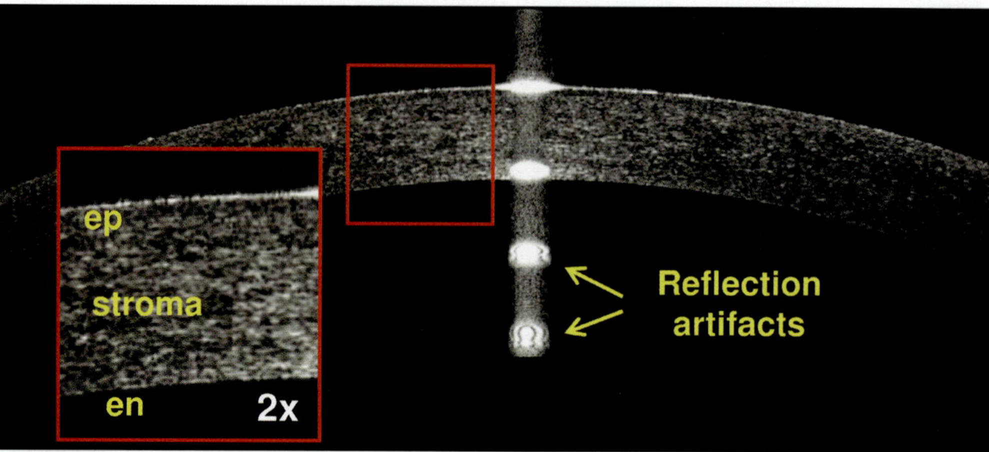

Figure 1-18. RTVue image of the normal anterior angle. The image is averaged 16 times to reduce speckle noise and enhance signal. The epithelium (ep), corneal limbus (cl), iris, Schwalbe's ring (sr) at the termination of Descemet's membrane, and Schlemm's canal (sc) can be seen. The trabecular meshwork is shadowed by the sclera.

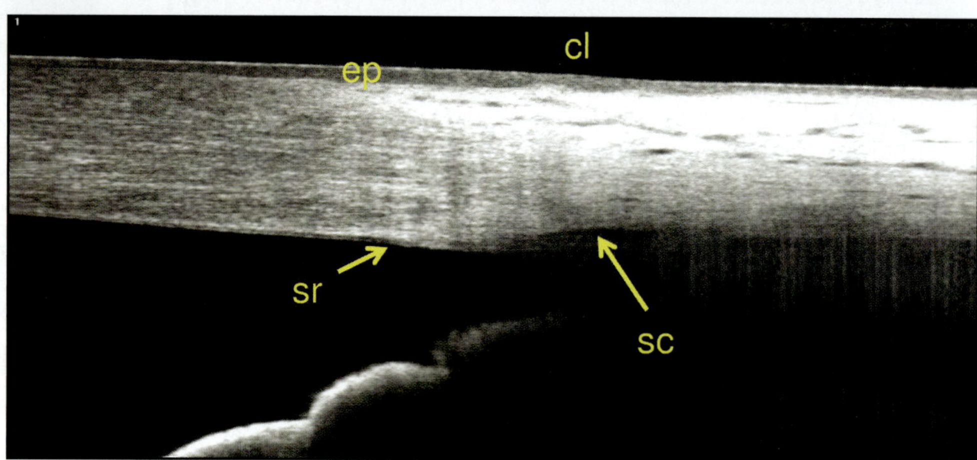

Retinal Structure in RTVue Optical Coherence Tomography Images

Figure 1-19 shows OCT images of the normal retina from the Optovue RTVue. The images have 5 to 7 µm axial image resolution and are acquired using different imaging protocols. The images are shown in grayscale in order to avoid artifacts from the false color scale and to better visualize differences in the images. The top image (Figure 1-19A) has 512 axial scans and is from a raster scan protocol. The middle image (Figure 1-19B) is a high-definition (HD line) with 4096 axial scans. The large number of axial scans improves the image quality by increasing the number of images pixels in the transverse direction. The bottom image (Figure 1-19C) is an average of 16 individual OCT images, each consisting of 1024 axial scans. Imaging averaging reduces speckle noise and gives a smoother appearance to the retinal layers.[74] Signal to noise is also increased, thus yielding a brighter image with improved image penetration behind the RPE and into the choroid. However, care

must be used because image averaging can blur fine image features.

Figure 1-20 shows 2x enlargements of the high-definition OCT image (4096 axial scans) and the averaged OCT image (16 images of 1024 axial scans). The images of the macula show characteristic features of the fovea with thinning of the overall retina and an increase in outer nuclear layer thickness. Increased thickness of the photoreceptor OS is also evident and can be seen from the reflection from the photoreceptor IS/OS junction. All major retinal layers can be visualized including the NFL, GCL, IPL, INL, OPL, ONL, ELM, the junction between the photoreceptor IS and OS, RPE, and the choroid. The averaged image, Figure 1-20B, shows a multilayer appearance of the RPE, which is thought to arise from variations in the cone-and-rod photoreceptor tips and the normal cone-and-rod distribution in the retina.[60] These structures are near the resolution limit of commercial OCT instruments and may not be visible in all subjects.

Generally, these results show that OCT images can provide detailed information about retinal pathology with resolutions approaching those of histopathology, on the level of architectural morphology. OCT images can, therefore, provide detailed information on retinal

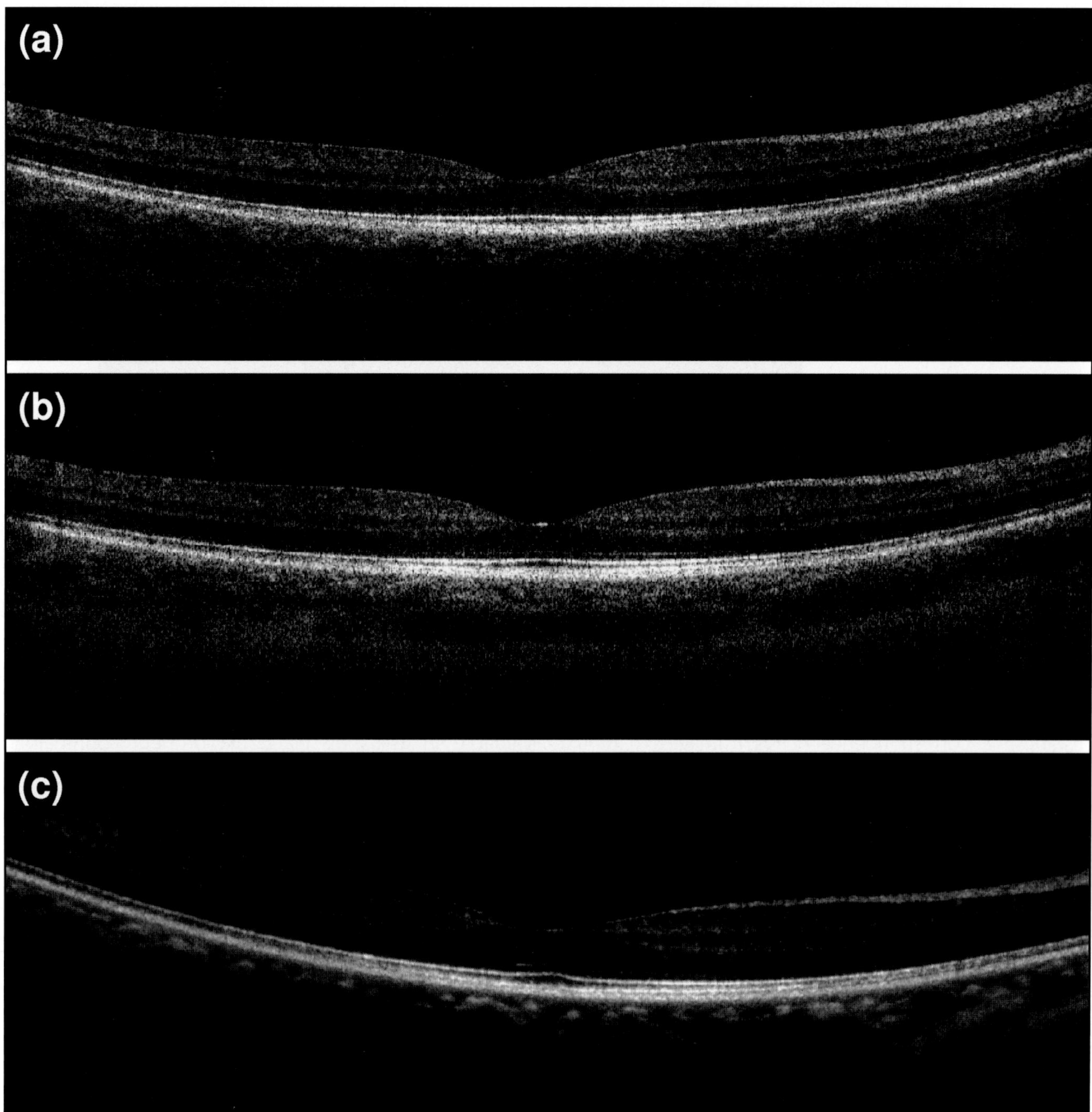

Figure 1-19. RTVue images of the normal retina. (A) Single OCT image from a raster scan with 512 axial scans (transverse pixels). (B) High-definition image (HD scan) consisting of 4096 axial scans (transverse pixels). Imaging with a higher density of axial scans improves continuity of retinal layers. (C) Averaged image consisting of 16 individual OCT images. Image averaging reduces speckle noise, thus giving the retinal layers a more homogenous appearance. Signal to noise is also improved, thus yielding increased image penetration into the choroid.

Figure 1-20. RTVue images of the normal retina. Enlargements of OCT images from Figure 1-19. (A) High-definition image and (B) averaged image. Retinal layers are visible, including the NFL, GCL, IPL, INL, OPL, ONL, ELM, IS/OS of the photoreceptor layer, and RPE.

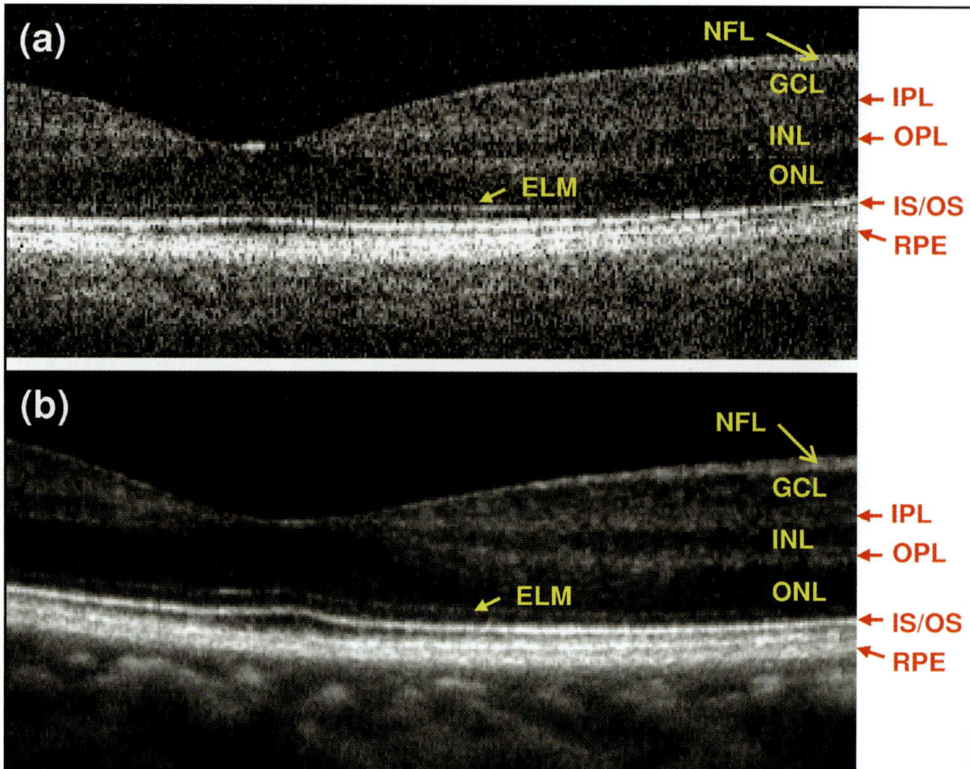

pathology. With the development of OCT atlases that establish a correspondence of OCT images to retinal disease, it should be possible to read OCT images in a manner similar to the way pathologists interpret histological micrographs.

Figure 1-21 is an OCT image of the normal optic disc. The OCT image shown is a high-definition (HD line) consisting of 4096 axial scans. The contour of the optic disc is demarcated by the boundary between the low-backscattering vitreous and the high-backscattering NFL. The normal cupping of the disc is evident. The OCT image shows an increase in NFL thickness as it approaches the neuroretinal rim, where the NFL is nearly the entire thickness of the retina. The retinal nerve fibers have a directional-dependent reflectance. The reflected light intensity from the nerve fibers decreases at the disc rim, where the nerve fibers are no longer perpendicular to the incident OCT optical beam, but bend into the ONH.

The RPE and choriocapillaris are visible as a highly backscattering layer that terminates at the lamina cribosa. The boundary between the photoreceptor IS and OS is also visible as a thin, highly backscattering feature immediately anterior to the RPE and choroid. The photoreceptor layer and RPE terminate in the region approaching the disc, and this can be used as a landmark to define the disc margin. The IPL and OPL are moderately reflective, while the INL and ONL are weakly reflective.

OCT imaging of the ONH and neuroretinal rim is valuable in assessing glaucoma or neuro-ophthalmic diseases. In evaluating the contour of the disc, it is important to note that OCT images are usually displayed with an expanded scale in the axial direction in order to allow better visualization of the thin retinal layers while still encompassing a transverse scan of several millimeters. Quantitative morphometry of the images that use the correct axial and transverse scales are, therefore, important.

Figure 1-22 is an example of 3D OCT data of the normal macula from the RTVue. The data are acquired by using a raster scan consisting of 101 frames of OCT images, each consisting of 513 axial scans. The total volume data set consists of 101 x 513, or 51,000 axial scans, and is acquired in ~2.2 seconds. Volumetric OCT data contains comprehensive information about retinal structure. An *en face* fundus image can be generated by summing the OCT signal in the axial direction. This image provides a view of the fundus that is similar to a scanning laser ophthalmoscope. Cross-sectional images (B-scans) can be extracted from the 3D OCT volumetric data set. The figure shows images along the temporal-nasal and superior-inferior direction. These individual images are registered to features on the fundus. The figure also shows a rendering of the 3D OCT volumetric data. The rendering can be manipulated to display the retina from a virtual perspective and cut away views can be generated.

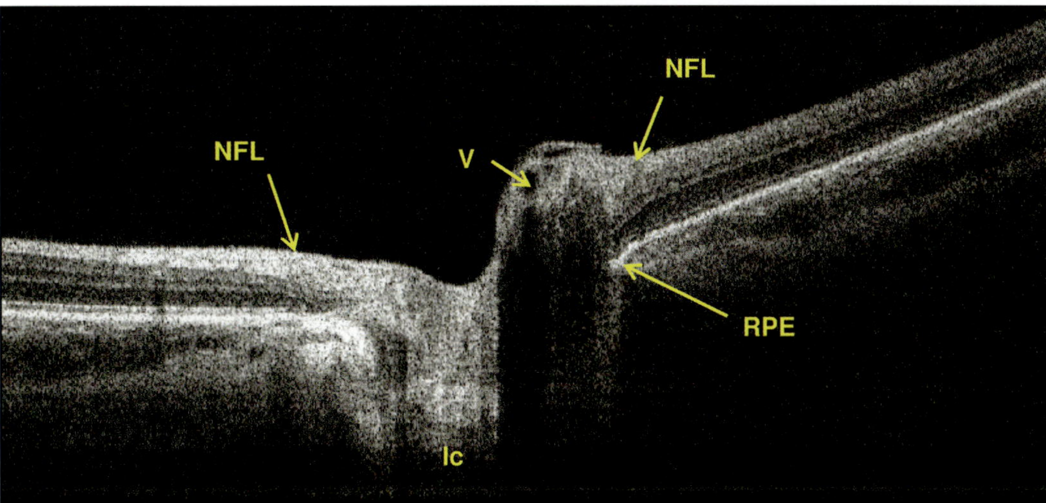

Figure 1-21. RTVue image of the normal optic disc. The NFL is visible and increases in thickness as it approaches the disc rim. The termination of the RPE/choriocapillaris and photoreceptors near the lamina cribrosa can be clearly visualized. Disc parameters such as cup and disc diameter, neuroretinal rim area, and cup-to-disc ratio may be measured by using this feature as a landmark. The lamina cribrosa (lc) and blood vessels (v) near the disc are also visible.

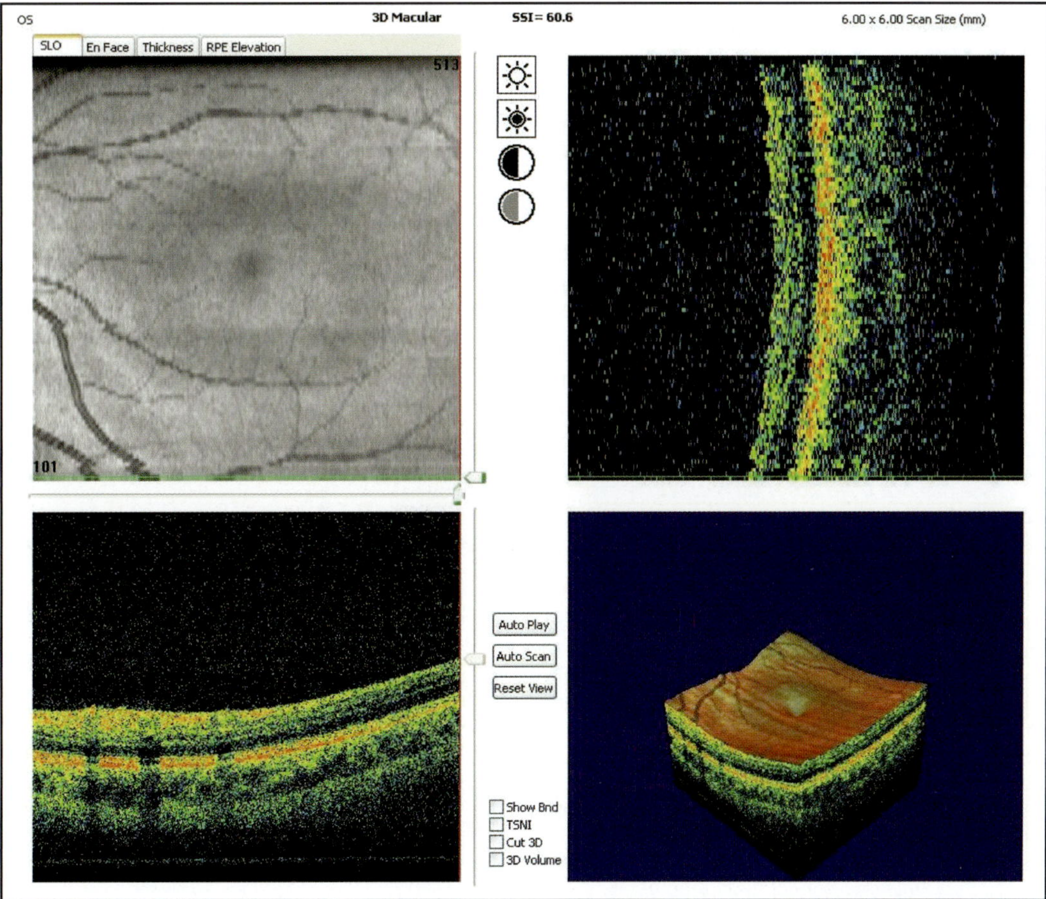

Figure 1-22. RTVue 3D OCT image of the normal retina. Multiple windows present the 3D OCT volumetric data. An *en face* OCT fundus image is shown in the upper left. Cross-sectional images (B-scans) in the horizontal (temporal nasal) or vertical (superior inferior) direction can be display (top right, bottom left), which are registered to features in the fundus. A 3D animation window (bottom right) shows a rendered view of the 3D OCT data set.

Chapters 5, 6, and 7 describe in more detail the scan protocols, procedures, and interpretation of retinal OCT images.

Conclusion

This chapter presents a summary of the history of OCT in ophthalmology as well as an overview of how OCT technology works, including time-domain detection and spectral/Fourier-domain detection. A discussion of what OCT images show and their general interpretation are also presented. The recent developments in OCT technology enable a dramatic improvement in OCT image quality, thus allowing the visualization of individual retinal layers. As additional clinical studies become available, OCT images could eventually be used to assess detailed retinal pathology on the level of architectural morphology, in a manner similar to that used by pathologists. The high imaging speeds now possible enable the acquisition of 3D OCT volumetric data sets that provide comprehensive information on retinal structure. OCT *en face* fundus images can also be generated that are analogous to images acquired by scanning laser ophthalmoscope. Cross-sectional (B-scan) images can be extracted from 3D OCT data sets and are precisely registered to features on the fundus. This capability will allow more accurate tracking of changes in retinal pathology from visit to visit. The high acquisition speeds also enable a more accurate assessment of true retinal topography. Quantitative measurement of retinal features, including retinal thickness, retinal NFL thickness, and other intraretinal layers, is also possible with improved accuracy and reproducibility. These capabilities promise improved sensitivity for disease diagnosis and the ability to more accurately monitor disease progression and treatment response.

References

1. Huang D, Swanson EA, Lin CP, et al. Optical coherence tomography. *Science.* 1991;254:1178-1181.
2. Izatt JA, Hee MR, Swanson EA, et al. Micrometer-scale resolution imaging of the anterior eye in vivo with optical coherence tomography. *Arch Ophthalmol.* 1994;112:1584-1589.
3. Hee MR, Izatt JA, Swanson EA, et al. Optical coherence tomography of the human retina. *Arch Ophthalmol.* 1995;113:325-332.
4. Puliafito CA, Hee, MR, Lin CP, et al. Imaging of macular diseases with OCT. *Ophthalmology.* 1995;102:217-229.
5. Schuman JS, Hee MR, Arya AV, et al. OCT: a new tool for glaucoma diagnosis. *Curr Opin Ophthalmol.* 1995;6:89-95.
6. Drexler W, Fujimoto JG. State-of-the-art retinal optical coherence tomography. *Progress in Retinal and Eye Research.* 2008;27:45-88.
7. Drexler W, Fujimoto JG. *Optical Coherence Tomography Technology and Applications.* New York, NY: Springer; 2008.
8. Puliafito CA, Hee MR, Schuman JS, Fujimoto JG. *Optical Coherence Tomography of Ocular Diseases.* Thorofare, NJ: SLACK Incorporated; 1996.
9. Hee MR, Puliafito CA, Wong C, et al. Quantitative assessment of macular edema with optical coherence tomography. *Arch Ophthalmol.* 1995;113:1019-1029.
10. Hee MR, Puliafito CA, Wong C, et al. Optical coherence tomography of macular holes. *Ophthalmology.* 1995;102:748-756.
11. Hee MR, Puliafito CA, Wong C, et al. Optical coherence tomography of central serous chorioretinopathy. *Am J Ophthalmol.* 1995;120:65-74.
12. Hee MR, Baumal CR, Puliafito CA, et al. Optical coherence tomography of age-related macular degeneration and choroidal neovascularization. *Ophthalmology.* 1996;103:1260-1270.
13. Wilkins JR, Puliafito CA, Hee MR, et al. Characterization of epiretinal membranes using optical coherence tomography. *Ophthalmology.* 1996;103:2142-2151.
14. Hee MR, Puliafito CA, Duker JS, et al. Topography of diabetic macular edema with optical coherence tomography. *Ophthalmology.* 1998;105:360-370.
15. Schuman JS, Hee MR, Puliafito. CA, et al. Quantification of nerve fiber layer thickness in normal and glaucomatous eyes using optical coherence tomography. *Arch Ophthalmol.* 1995;113:586-596.
16. Schuman JS, Pedut-Kloizman T, Hertzmark E, et al. Reproducibility of nerve fiber layer thickness measurements using optical coherence tomography. *Ophthalmology.* 1996;103:1889-1898.
17. Schuman JS, Pedut-Kloizman T, Pieroth L, et al. Quantitation of nerve fiber layer thickness loss over time in the glaucomatous monkey model using optical coherence tomography. *Invest Ophthalmol Vis Sci.* 1996;37:5255-5255.
18. Bowd C, Weinreb RN, Williams JM, Zangwill LM. The retinal nerve fiber layer thickness in ocular hypertensive, normal, and glaucomatous eyes with OCT. *Arch Ophthalmol.* 2000;118:22-26.
19. Zangwill LM, Williams J, Berry CC, Knauer S, Weinreb RN. A comparison of optical coherence tomography and retinal nerve fiber layer photography for detection of nerve fiber layer damage in glaucoma. *Ophthalmology.* 2000;107:1309-1315.
20. Bowd C, Zangwill LM, Berry CC, et al. Detecting early glaucoma by assessment of retinal nerve fiber layer thickness and visual function. *Invest Ophthalmol Vis Sci.* 200142:1993-2003.
21. Schuman JS, Wollstein G, Farra T, et al. Comparison of optic nerve head measurements obtained by optical coherence tomography and confocal scanning laser ophthalmoscopy. *Am J Ophthalmol.* 2003;135:504-512.
22. Guedes V, Schuman JS, Hertzmark E, et al. Optical coherence tomography measurement of macular and nerve fiber layer thickness in normal and glaucomatous human eyes. *Ophthalmology.* 2003;110:177-189.
23. Wollstein G, Schuman JS, Price LL, et al. Optical coherence tomography (OCT) macular and peripapillary retinal nerve fiber layer measurements and automated visual fields. *Am J Ophthalmol.* 2004;138:218-225.
24. Wollensak G, Aurich H, Wirbelauer C, Pham DT. Potential use of riboflavin/UVA cross-linking in bullous keratopathy. *Ophthalmic Research.* 2009;41:114-117.
25. Olsen T. Calculating axial length in the aphakic and the pseudophakic eye. *J Cataract Refract Surg.* 1988;14:413-416.
26. Olsen T. The accuracy of ultrasonic determination of axial length in pseudophakic eyes. *Acta Ophthalmologica.* 1989;67:141-144.
27. Pavlin CJ, Sherar MD, Foster FS. Subsurface ultrasound microscopic imaging of the intact eye. *Ophthalmology.* 1990;97:244-250.
28. Pavlin CJ, Harasiewicz K, Sherar MD, Foster FS. Clinical use of ultrasound biomicroscopy. *Ophthalmology.* 1991;98:287-295.
29. Pavlin CJ, Foster FS. Ultrasound biomicroscopy. High-frequency ultrasound imaging of the eye at microscopic resolution. *Radiologic Clinics of North America.* 1998;36:1047-1058.
30. Drexler W, Morgner U, Ghanta RK, et al. Ultrahigh-resolution ophthalmic optical coherence tomography. *Nature Medicine.* 2001;7:502-507.
31. Fujimoto JG, De Silvestri S, Ippen EP, et al. Femtosecond optical ranging in biological systems. *Optics Letters.* 1986;11:150-153.

32. Stern D, Lin WZ, Puliafito CA, Fujimoto JG. Femtosecond optical ranging of corneal incision depth. *Invest Ophthalmol Vis Sci.* 1989;30:99-104.

33. Born M, Wolf E, Bhatia AB. *Principles of optics: Electromagnetic theory of propagation, interference and diffraction of light.* 7th (expanded) ed. New York, NY: Cambridge University Press; 1999.

34. Youngquist R, Carr S, Davies D. Optical coherence-domain reflectometry: a new optical evaluation technique. *Optics Letters.* 1987;12:158-160.

35. Takada K, Yokohama I, Chida K, Noda J. New measurement system for fault location in optical waveguide devices based on an interferometric technique. *Applied Optics.* 1987;26:1603-1608.

36. Gilgen HH, Novak RP, Salathe RP, Hodel W, Beaud P. Submillimeter optical reflectometry. *IEEE Journal of Lightwave Technology.* 1989;7:1225-1233.

37. Fercher AF, Mengedoht K, Werner W. Eye-length measurement by interferometry with partially coherent light. *Optics Letters.* 1988;13:1867-1869.

38. Huang D, Wang J, Lin CP, Puliafito CA, Fujimoto JG. Micron-resolution ranging of cornea and anterior chamber by optical reflectometry. *Lasers Surg Med.* 1991;11:419-425.

39. Hitzenberger CK. Measurement of corneal thickness by low-coherence interferometry. *Applied Optics.* 1992;31:6637-6642.

40. Swanson EA, Huang D, Hee MR, et al. High-speed optical coherence domain reflectometry. *Optics Letters.* 1992;17:151-153.

41. Swanson EA, Izatt JA, Hee MR, et al. In vivo retinal imaging by optical coherence tomography. *Optics Letters.* 1993;18;1864-1866.

42. Patterson MS, Chance B, Wilson BC. Time resolved reflectance and transmittance for the non-invasive measurement of tissue optical properties. *Applied Optics.* 1989;28:2331-2336.

43. Flock ST, Patterson MS, Wilson BC, Wyman DR. Monte Carlo modeling of light propagation in highly scattering tissues. I. Model predictions and comparison with diffusion theory. *IEEE Transactions on Biomedical Engineering.* 1989;36:1162-1168.

44. Flock ST, Wilson BC, Patterson MS. Monte Carlo modeling of light propagation in highly scattering tissues. II. Comparison with measurements in phantoms. *IEEE Transactions on Biomedical Engineering.* 1989;36:1169-1173.

45. Cheong WF, Prahl SA, Welch AJ. A review of the optical properties of biological tissues. *IEEE Journal of Quantum Electronics.* 1990;26:2166-2185.

46. Fujimoto JG, Brezinski ME, Tearney GJ, et al. Optical biopsy and imaging using optical coherence tomography. *Nature Medicine.* 1995;1:970-972.

47. Schmitt JM, Knuttel A, Yadlowsky M, Eckhaus MA. OCTof a dense tissue: Statistics of attenuation and backscattering. *Physics in Medicine and Biology.* 1994;39:1705-1720.

48. Drexler W, Sattmann H, Hermann B, et al. Enhanced visualization of macular pathology with the use of ultrahigh-resolution optical coherence tomography. *Arch Ophthalmol.* 2003;121:695-706.

49. Ko TH, Fujimoto JG, Duker JS, et al. Comparison of ultrahigh- and standard-resolution OCT for imaging macular hole pathology and repair. *Ophthalmology.* 2004;111:2033-2043.

50. Ko TH, Fujimoto JG, Schuman JS, et al. Comparison of ultrahigh- and standard-resolution optical coherence tomography for imaging macular pathology. *Ophthalmology.* 2005;112:1922.e1-1992. e15.

51. Hogan H, Alvarado JA, Weddell JE. *Histology of the Human Eye: An Atlas and Textbook.* Philadelphia, PA: WB Saunders; 1971.

52. Gass JDM. *Stereoscopic Atlas of Macular Diseases: Diagnosis and Treatment.* 3rd ed., Vol 1. St. Louis, MO: CV Mosby; 1987:46-65.

53. Krebs W, Krebs I. *Primate Retina and Choroid—Atlas of Fine Structure in Man and Monkey.* New York, NY: Springer Verlag; 1991.

54. Spalton DJ, Hitchings RA, Hunter PA. Anatomy of the retina. In: *Atlas of Clinical Ophthalmology.* 2nd ed. St. Louis, MO: Mosby; 1994.

55. Toth CA, Narayan DG, Boppart SA, et al. A comparison of retinal morphology viewed by optical coherence tomography and by light microscopy. *Arch Ophthalmol.* 1997;115:1425-1428.

56. Huang Y, Cideciyan AV, Papastergiou GI, et al. Relation of optical coherence tomography to microanatomy in normal and rd chickens. *Invest Ophthalmol Vis Sci.* 1998;39:2405-2416.

57. Li Q, Timmers AM, Hunter K, et al. Noninvasive imaging by optical coherence tomography to monitor retinal degeneration in the mouse. *Invest Ophthalmol Vis Sci.* 2001;42:2981-2989.

58. Gloesmann M, Hermann B, Schubert C, et al. Histologic correlation of pig retina radial stratification with ultrahigh-resolution optical coherence tomography. *Invest Ophthalmol Vis Sci.* 2003;44:1696-1703.

59. Anger EM, Unterhuber A, Hermann B, et al. Ultrahigh resolution optical coherence tomography of the monkey fovea. Identification of retinal sublayers by correlation with semithin histology sections. *Exp Eye Res.* 2004;78:1117-1125.

60. Srinivasan VJ, Monson BK, Wojtkowski M, et al. Characterization of outer retinal morphology with high-speed, ultrahigh-resolution optical coherence tomography. *Invest Ophthalmol Vis Sci.* 2008;49:1571-1579.

61. Sidman R. The structure and concentration of solids in photoreceptor cells studied by refractometry and interference microscopy. *J Biophys Biochem Cytol.* 1957;3:1530.

62. Drexler W. Ultrahigh-resolution optical coherence tomography. *Journal of Biomedical Optics.* 2004;9:47-74.

63. Fercher AF, Hitzenberger CK, Kamp G, Elzaiat SY. Measurement of intraocular distances by backscattering spectral interferometry. *Optics Communications.* 1995;117:43-48.

64. Wojtkowski M, Leitgeb R, Kowalczyk A, Bajraszewski T, Fercher AF. In vivo human retinal imaging by Fourier domain optical coherence tomography. *Journal of Biomedical Optics.* 2002;7:457-463.

65. Wojtkowski M, Bajraszewski T, Gorczynska I, et al. Ophthalmic imaging by spectral optical coherence tomography. *Am J Ophthalmol.* 2004;138:412-419.

66. Cense B, Nassif N, Chen TC, et al. Ultrahigh-resolution high-speed retinal imaging using spectral-domain optical coherence tomography. *Optics Express.* 2004;12:2435-2447.

67. Nassif NA, Cense B, Park BH, et al. In vivo high-resolution video-rate spectral-domain optical coherence tomography of the human retina and optic nerve. *Optics Express.* 2004;12:367-376.

68. Wojtkowski M, Srinivasan VJ, Ko TH, et al. Ultrahigh-resolution, high-speed. Fourier domain optical coherence tomography and methods for dispersion compensation. *Optics Express.* 2004;12:2404-2422.

69. Wojtkowski M, Srinivasan VJ, Fujimoto JG, et al. Three-dimensional retinal imaging with high-speed ultrahigh-resolution optical coherence tomography. *Ophthalmology.* 2005;112:1734-1746.

70. Leitgeb R, Hitzenberger CK, Fercher AF. Performance of Fourier domain vs. time domain optical coherence tomography. *Optics Express.* 2003;11:889-894.

71. de Boer JF, Cense B, Park BH, et al. Improved signal-to-noise ratio in spectral-domain compared with time-domain optical coherence tomography. *Optics Letters.* 2003;28:2067-2069.

72. Choma MA, Sarunic MV, Yang CH, Izatt JA. Sensitivity advantage of swept source and Fourier domain optical coherence tomography. *Optics Express.* 2003;11:2183-2189.

73. Jiao S, Knighton R, Huang X, Gregori G, Puliafito CA. Simultaneous acquisition of sectional and fundus ophthalmic images with spectral-domain optical coherence tomography. *Optics Express.* 2005;13:444-452.

74. Sander B, Larsen M, Thrane L, Hougaard JL, Jorgensen TM. Enhanced optical coherence tomography imaging by multiple scan averaging. *Br J Ophthalmol.* 2005;89:207-212.

2 Anterior Segment Scan Patterns

Yan Li, PhD and David Huang, MD, PhD

INTRODUCTION

The RTVue is a unique Fourier-domain OCT system that can image both the anterior and posterior segments of the eye. It works at 830 nm wavelength and is capable of a scan speed of 26,000 axial scan per second. The RTVue has a depth resolution of 5 µm (full-width-half-maximum). The imaging depth of RTVue is limited to 2.3 mm.

CORNEAL ADAPTOR MODULE

The RTVue was originally designed for retinal imaging. To obtain high quality corneal and anterior segment images, a corneal adaptor module (CAM) was developed. The CAM consists of 2 adaptor lenses and a suite of scan patterns and analysis software. The adaptor lens is placed in front of the retinal objective lens to focus the OCT beam on the anterior segment (Figure 2-1). The CAM lenses were designed to provide telecentric scanning so that the OCT beam remains parallel to the central axis across the transverse scan range.

The wide angle lens (long lens, Figure 2-2) provides a scan width of up to 6 mm and a transverse resolution (focused spot size) of 15 µm. It is used for corneal mapping and wide angle cross-sectional imaging.

The high magnification lens (short lens, Figure 2-2) provides a scan width of up to 3 mm and a transverse resolution of 10 µm. It is useful for cross-sectional imaging of small features such as acanthamoeba.

The RTVue is capable of providing exquisite detail in its corneal and anterior segment cross-sectional images. In this book, we use CAM-L or CL to refer to the CAM long lens and use CAM-S or CS to stand for the CAM short lens. The RTVue CAM software provides both CAM-L and CAM-S scan patterns.

CAM-L SCAN PATTERNS

The CAM-L wide angle lens is the more commonly used of the 2 lenses. It is used to map the cornea and measure corneal features within the central 6 mm optical zone (eg, measure flap thickness after LASIK). CAM-L scan patterns can be selected by choosing CAM-L tab in examine mode (Figure 2-3). A summary of CAM-L scan patterns is listed in Table 2-1.

Pachymetry

The corneal mapping pattern on the RTVue-CAM is called pachymetry. It consists of 8 high-definition meridional scans acquired in only 0.32 seconds (Figure 2-4). The corneal thickness profile is measured by an automated algorithm that detects the anterior and posterior corneal boundaries on the cross-sectional images. A 6-mm diameter pachymetry map is formed by interpolation of the thickness profiles on the 8 meridians (see Figure 2-4).

Huang D, Duker JS, Fujimoto JG, Lumbroso B, Schuman JS, Weinreb RN.
Imaging the Eye from Front to Back with RTVue Fourier-Domain Optical Coherence Tomography (pp 23-30).
© 2010 SLACK Incorporated.

Figure 2-1. The corneal adaptor lens is placed in front of the retinal objective lens of the RTVue to focus the OCT beam on the anterior segment. A set of gooseneck lights is used to illuminate the anterior segment for concurrent video imaging, and can also be used for fixation of the contralateral eye.

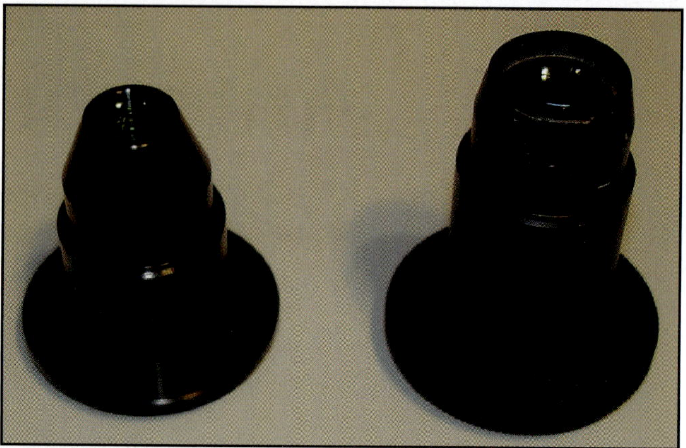

Figure 2-2. The corneal adaptor lenses. Left: the high magnification lens (short lens). Right: the wide angle lens (long lens).

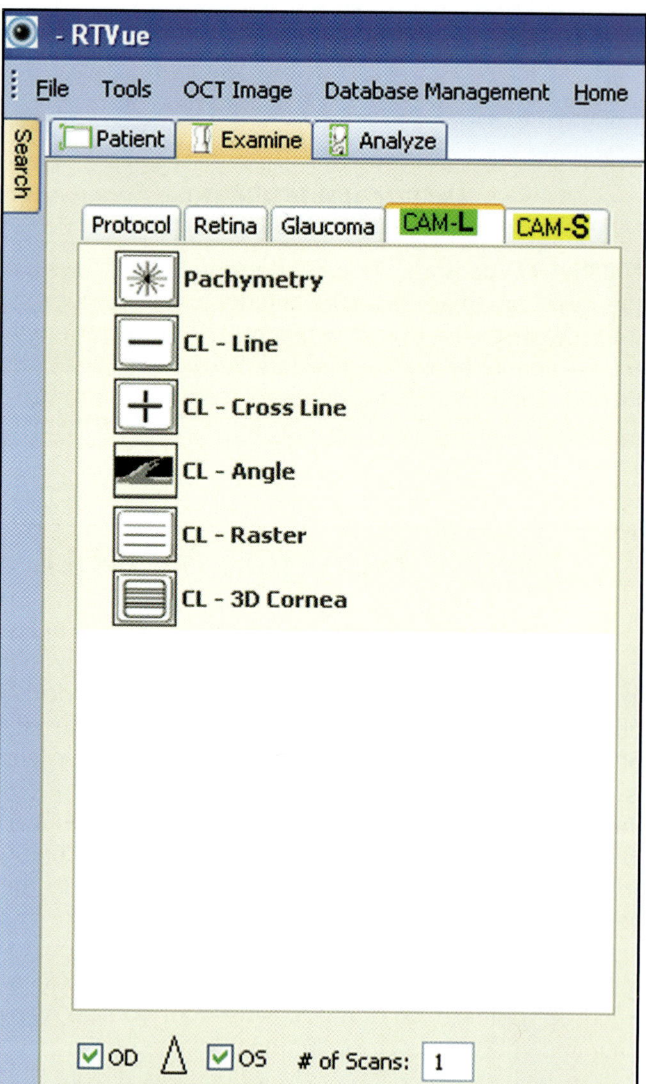

Figure 2-3. CAM-L scan list.

Table 2-1

CAM-L Scan Patterns

Scan Patterns	Scan Length (mm) Default	Scan Length (mm) Adjustable	Width (mm) Default	Width (mm) Adjustable	Axial Scan × Frame Numbers
1 Pachymetry	6	Fixed	N/A	N/A	1024 x 8 radial scans
2 CL-Line	6	2 ~ 6	N/A	N/A	1024 x 1 scan
3 CL-Cross Line	6	2 ~ 6	N/A	N/A	1024 x 2 cross scans
4 CL-Angle	3	2 ~ 6	N/A	N/A	1024 x 1 scan
5 CL-Raster	6	1 ~ 6	4	0 ~ 6	1024 x 17 raster scans
6 CL-3D Cornea	6	2 ~ 6	6	0 ~ 6	512 x 101 raster scans

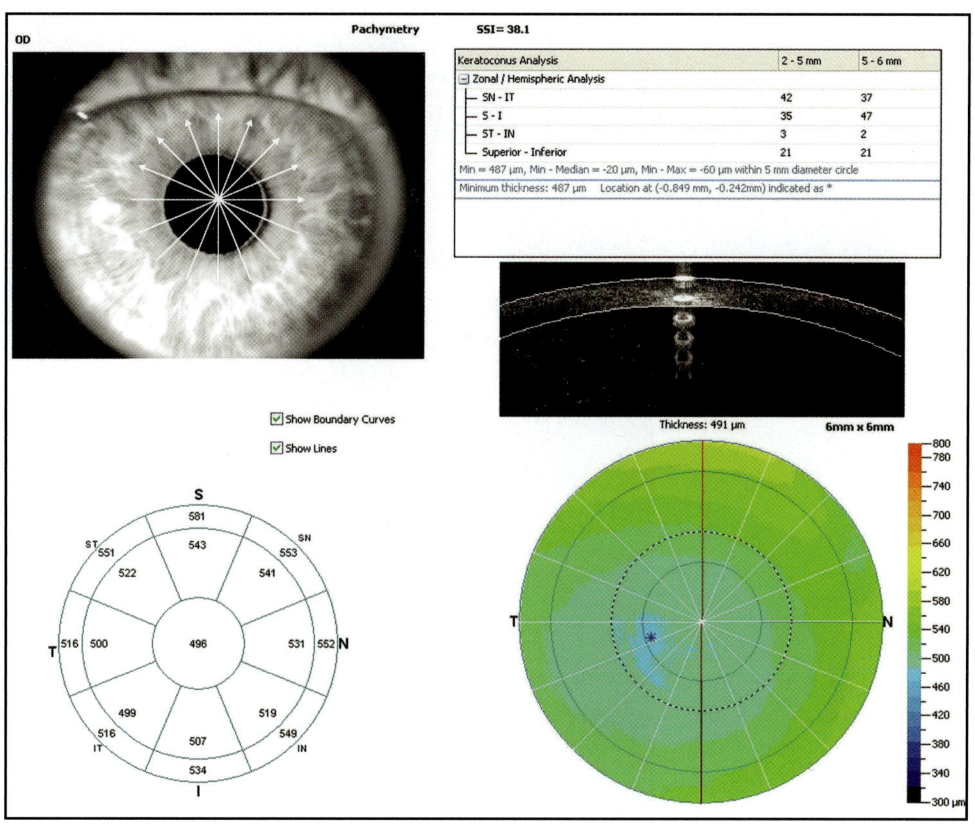

Figure 2-4. Pachymetry scan pattern (upper left) consists of 6 mm lines on 8 meridians. Each line consists of 1024 axial scans acquired in 0.04 seconds. The set of 8 meridians are acquired in 0.32 seconds. Pachymetry maps were calculated from the corneal mapping OCT scan. The map is shown both on a color scale (lower right) and in sector average (lower left).

Figure 2-5. CL-Line scan of a normal human cornea. (A) Single frame. (B) Average of 16 frames after registration. (C) Enlarged section of the frame-averaged image showing epithelium (Epi), Bowman's layer (BL), Descemet's membrane (DM), and endothelium (Endo).

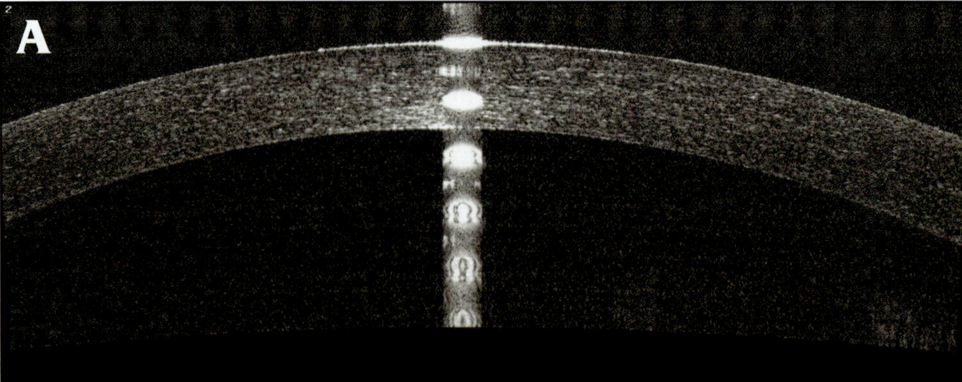

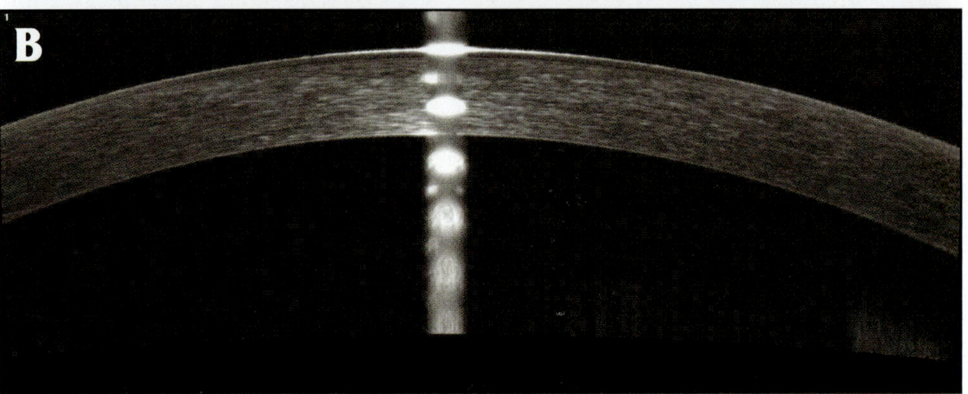

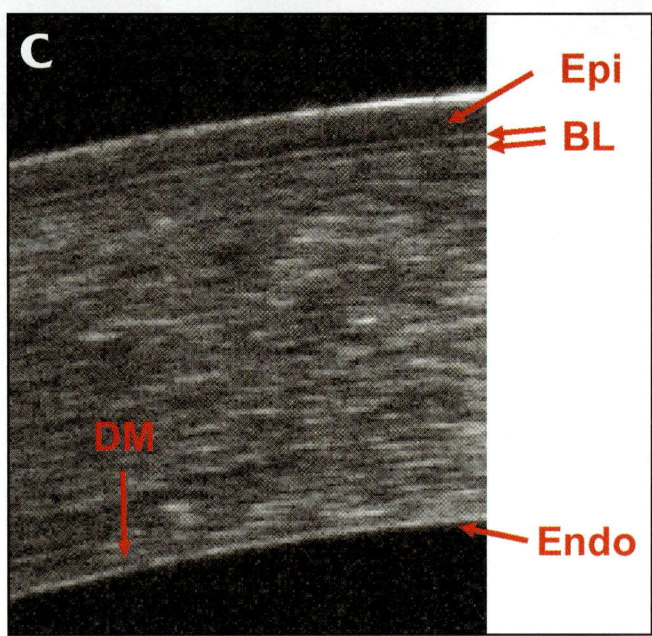

CL-Line

The CL-Line pattern is a line scan with a default setting of 6 mm length (2 to 6 mm, adjustable) and horizontal orientation. The high speed of the RTVue allows the acquisition of a high-definition image frame of 1024 lines (axial scans) in 0.04 seconds. The high depth resolution allows delineation of fine corneal layers. Since the image frames contain little motion artifacts, many frames, typically 8 to 16, can be accurately registered and averaged to reduce the speckle and background noise. Comparison of a single frame image (Figure 2-5A) and frame-averaged image (Figure 2-5B) shows improved delineation of corneal layers in the averaged image. In the enlarged section of the averaged image (Figure 2-5C), one can clearly distinguish the epithelium, the front and back of the Bowman's layer, Descemet's membrane, and the endothelium.

CL-Cross Line

The CL-Cross Line scan is a combination of 2 perpendicular CL-Line scans crossing at the center (Figure 2-6). The default line orientations are horizontal and vertical. Frame-averaged images are available as in the line scan.

CL-Angle

The CL-Angle scan is similar to CL-Line scan but scans deeper (2.3 mm vs 2 mm) (Figure 2-7). The default scan length is 3 mm.

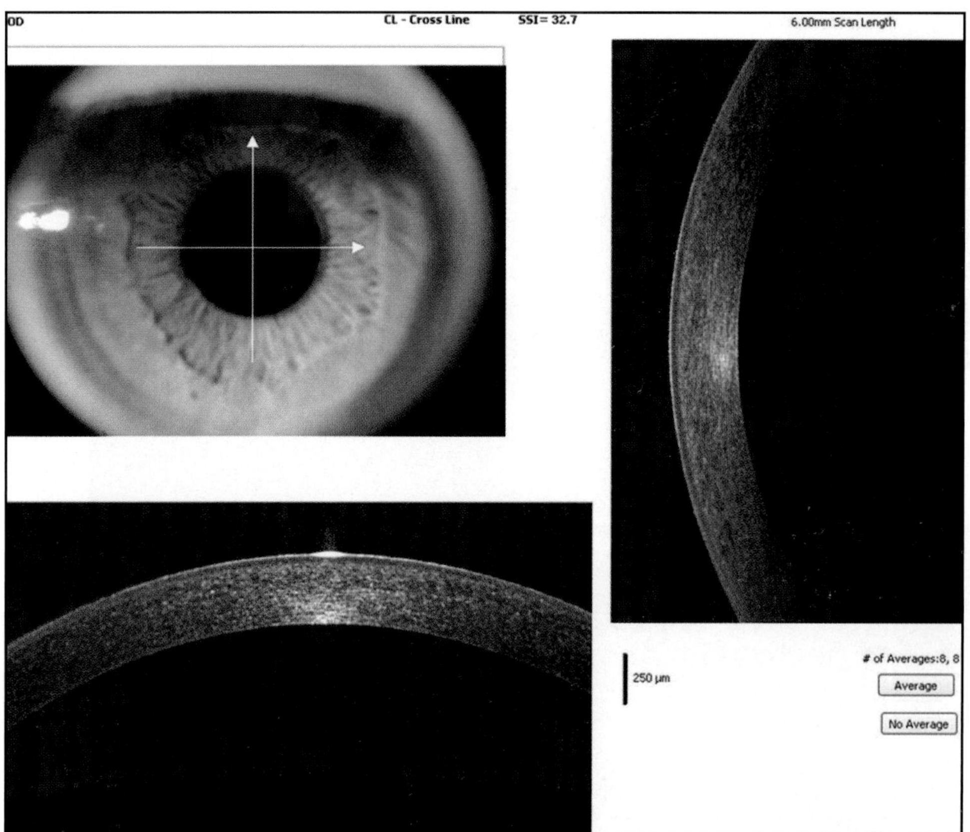

Figure 2-6. CL-Cross Line scan of a normal cornea. Both horizontal and vertical line scan was taken from the same shoot.

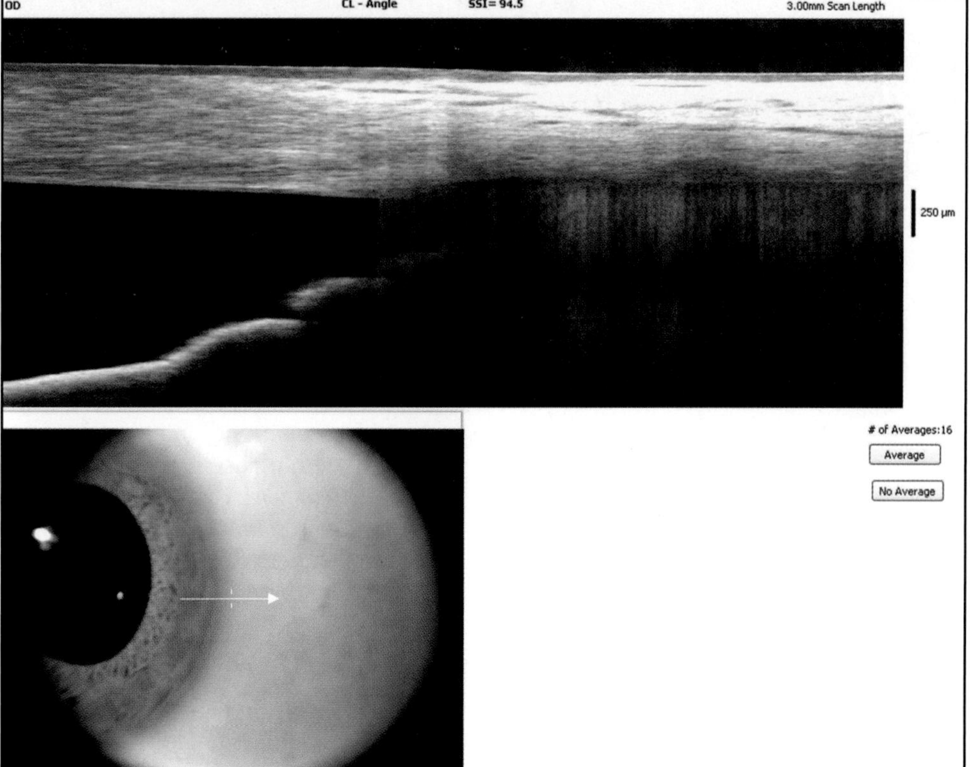

Figure 2-7. CL-Angle image of a normal subject.

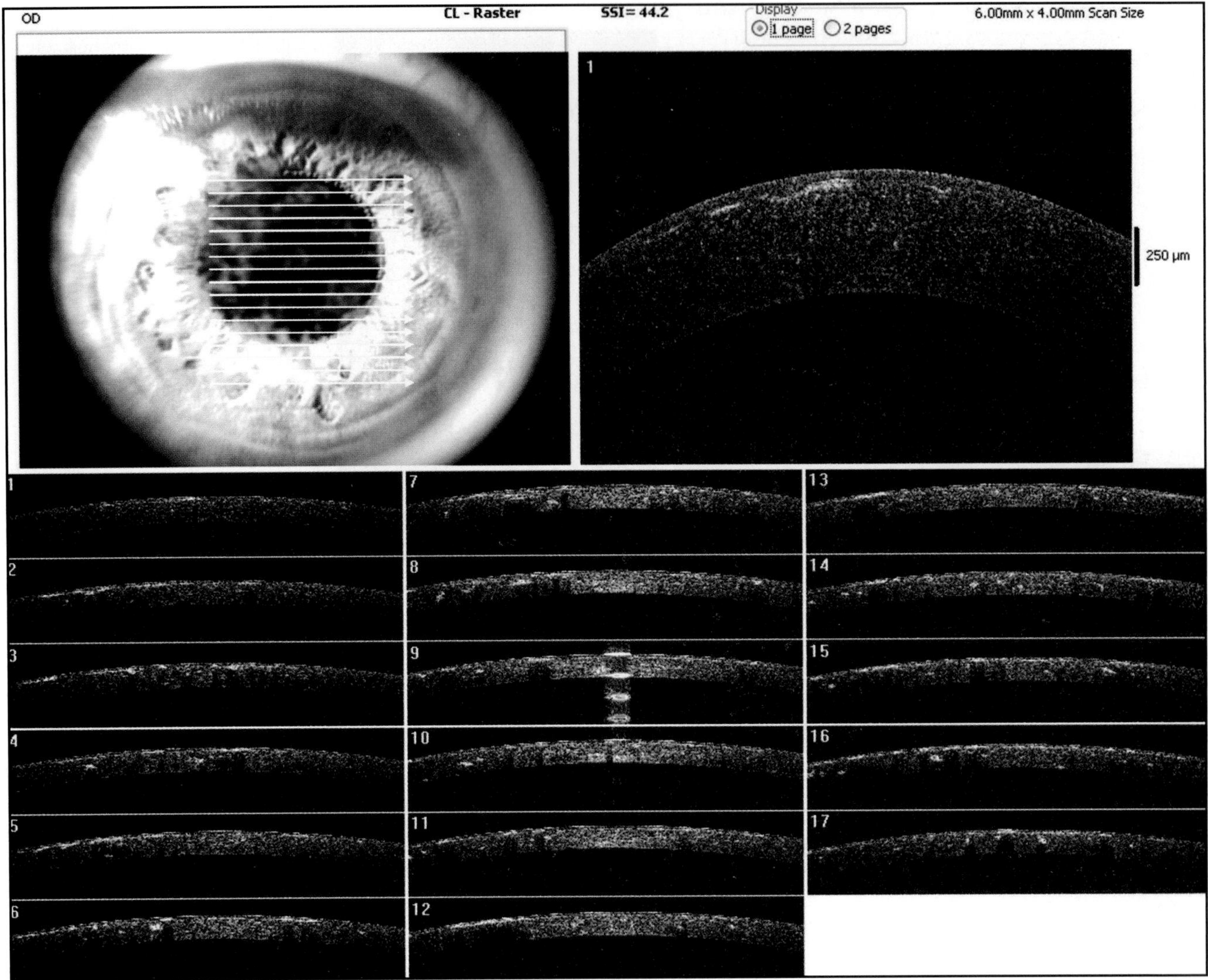

Figure 2-8. CL-Raster scan of a cornea with granular dystrophy.

CL-Raster

The CL-Raster scan consists of 17 line scans with 1024 axial scans in each cross-sectional image (Figure 2-8). The scan length is 6 mm by default and is adjustable between 1 and 6 mm. The raster spans 4 mm by default and is adjustable between 0 and 6 mm.

CL-3D Cornea

The CL-3D scan is composed of 101 line scans with 512 axial scans in each cross-sectional image. It generates a volume image of the tissue of interest. The scan length is 6 mm by default and is adjustable between 1 and 3 mm. The line scans span 6 mm by default and can be adjustable between 0 and 6 mm.

CAM-S SCAN PATTERNS

The CAM-S high-resolution lens is used to obtain highly detailed images of corneal and anterior segment structures. For example, one may use it to examine the gap at the edge of a LASIK flap in a case of suspected flap displacement. CAM-S scan patterns are summarized in Table 2-2. CAM-S scan patterns can be selected by choosing CAM-S tab in examine mode (Figure 2-9).

CS-Line

The CS-Line scan is similar to the CL-Line scan. The scan length is 2 mm by default and can be adjusted between 1 and 3 mm.

Table 2-2

CAM-S SCAN PATTERNS

| SCAN PATTERNS | SCAN LENGTH (MM) | | WIDTH (MM) | | AXIAL SCAN × FRAME NUMBERS |
	DEFAULT	ADJUSTABLE	DEFAULT	ADJUSTABLE	
1 CS-Line	2	1 ~ 3	N/A	N/A	1024 x 1 scan
2 CS-Cross Line	2	1 ~ 3	N/A	N/A	1024 x 2 cross scans
3 CS-Angle	3	1 ~ 3	N/A	N/A	1024 x 1 scan
4 CS-Raster	2	1 ~ 3	0.5	0 ~ 3	1024 x 17 raster scans
5 CS-3D Cornea	2	1 ~ 3	2	0 ~ 3	512 x 101 raster scans

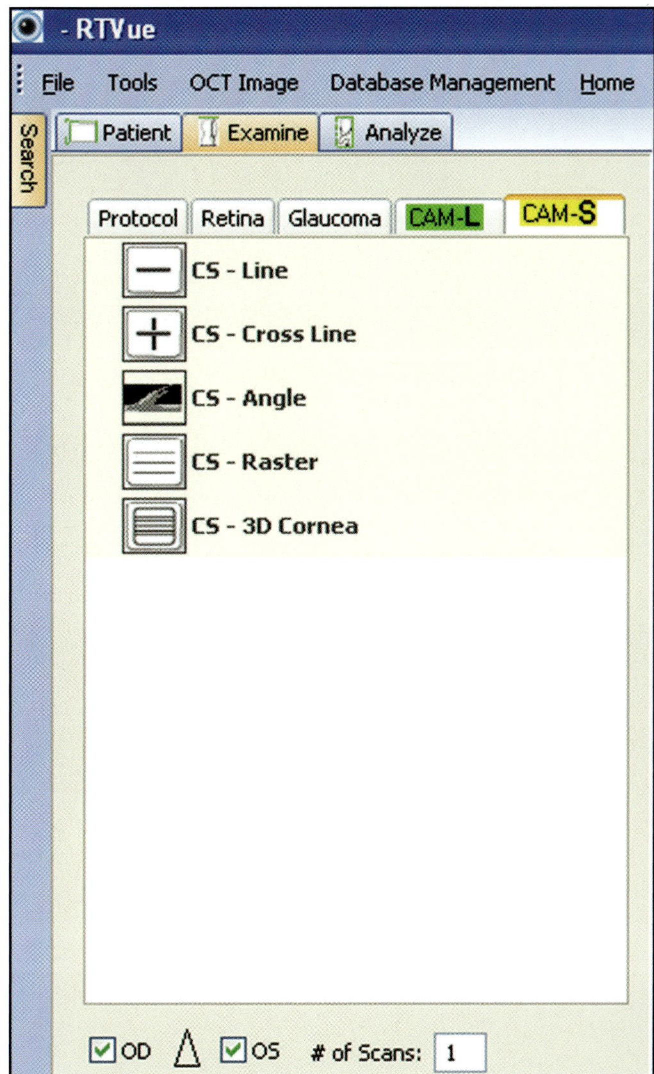

Figure 2-9. CAM-S scan list.

CS-Cross Line

The CS-Cross Line scan is similar to the CL-Cross Line scan. The scan length is 2 mm by default and is adjustable between 1 and 3 mm.

CS-Angle

The CS-Angle scan is similar to the CL-Angle scan. The scan length is 3 mm by default and is adjustable between 1 and 3 mm.

CS-Raster

The CS-Raster scan is similar to the CL-Raster scan. The scan length is 2 mm by default and is adjustable between 1 and 3 mm. The raster scans 0.5 mm by default and can be adjustable between 0 and 3 mm.

CS-3D Cornea

The CS-3D Cornea scan is similar to the 3D Cornea scan. The scan length is 2 mm by default and is adjustable between 1 and 3 mm. The line scans span 2 mm by default and can be adjustable between 0 and 3 mm.

SCAN DEPTH AND ORIENTATION

In addition, the imaging depth of CAM-L Angle and CAM-S Angle scans is 2.3 mm (in tissue). All other CAM patterns scan 2 mm deep in tissue. All patterns, except Pachymetry, can be rotated between -90 and 90 degrees.

Anterior Segment Scan Procedures

Maolong Tang, PhD and David Huang, MD, PhD

The CAM, a corneal adaptor module lens, when used in conjunction with RTVue, is for the *in vivo* imaging and measurement of the cornea and anterior chamber of the eye. Two CAM lenses are available for RTVue: CAM-L and CAM-S. CAM-L has lower magnification but wider field of view than CAM-S. For RTVue-CAM, the anterior segment scan procedure consists of 3 main steps:

1.   Mounting CAM lens

2.   Entering patient information in the patient menu

3.   Acquiring OCT scans in the examine menu

CORNEAL ADAPTOR MODULE

Attach the CAM lens onto the front of the objective lens used for retinal imaging. Pull the scanner all the way back, align using the live video image of the patient's eye, then gradually move the scanner forward until the cornea or other eye structure of interest is in the target zone (between 2 dashed red lines) as shown on the real-time OCT display on the monitor. This will also bring the iris and pupil into focus in the live video image. Note that the working distances between the lens adapter and cornea are *13 mm* on CAM-L model and *10 mm* on the CAM-S model. Do not move the scanner quickly and monitor the proximity to the patient in order to avoid incidentally hitting the patient's eye with the CAM lens surface.

PATIENT MENU

In the patient menu, the user can search, add, and edit patient information. It is designed to help you schedule a patient's visit in advance, preview today's or this week's scheduled patients, and search for a patient's history. The patient menu has 3 components (Figure 3-1):

1.   Patient list

2.   Patient information

3.   Visit information

Patient List

To create a new patient entry, click the [New Patient] button, fill out the information fields (fields marked with an asterisk (*) are required—if the manifest refraction spherical equivalent is 0 diopter then leave blank), and click the [Save] button at the very bottom of the window. If you would like to cancel this operation, then simply click the [Cancel] button to exit the New Patient screen.

Patient Information

The patient information is shown when a patient is selected.

1.   Required information fields are marked by an asterisk (*).

2.   Refraction value (patient's current refraction – spherical equivalent with add) when used, sets the

Huang D, Duker JS, Fujimoto JG, Lumbroso B, Schuman JS, Weinreb RN.
Imaging the Eye from Front to Back with RTVue Fourier-Domain Optical Coherence Tomography (pp 31-38).
© 2010 SLACK Incorporated.

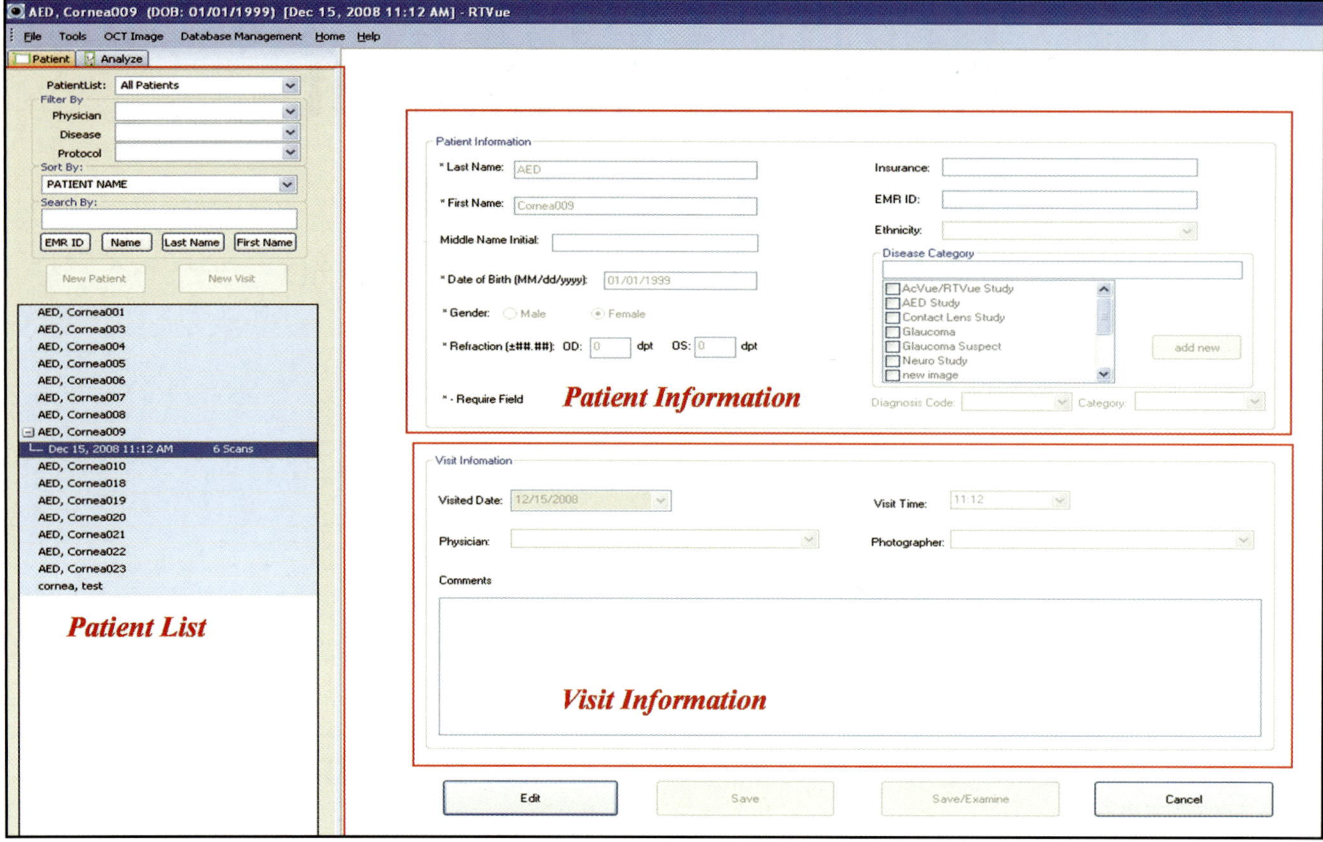

Figure 3-1. Patient menu on RTVue.

initial focus for scanner. The focus (scan beam and video fundus image) can be adjusted or fine-tuned in the Clinical Tab during the scan operation.

3. Disease Category is a "user defined" list of diseases. The disease category (multiple choices) can also be used as a filter for the patient list.

4. Diseases can be entered one at a time by clicking the [Add New] button. The selected Patient Name, DOB, and EMR ID will be displayed on the title bar of the Report window.

5. If the [Save] button is grayed-out, check to see that all required fields are filled.

Visit Information

To create a new visit entry, click the [New Visit] button, fill out the information fields, and click the [Save] button. If you would like to cancel this operation, then simply click the [Cancel] button to exit the New Visit screen. You can also create a new visit for an existing patient by right clicking on the patient name in the list, and selecting "Add New Visit."

Patient List Short-Cuts

Click on the patient's name to view patient visits. Right click on the patient name or visit date to view options:

- Add New Visit: Automatically create a new visit using the current date for the selected patient, which takes the system directly to the Examination screen

- Delete Current Visit: Deletes the selected visit (any visit in the patient visit history)

- Delete Current Patient: Deletes the selected patient (a warning message appears first to verify that this is the intended action)

Click [OK] to confirm delete or [Cancel] to withdraw the operation.

EXAMINE MENU

Choose CAM-L TAB to select scan patterns when using CAM-L model lens (Figure 3-2, highlighted in green) or choose CAM-S TAB to select scan patterns when using CAM-S model lens (Figure 3-2, highlighted in yellow).

Acquiring Cornea Optical Coherence Tomography Images

The following is a general procedure to acquire anterior segment images using RTVue:

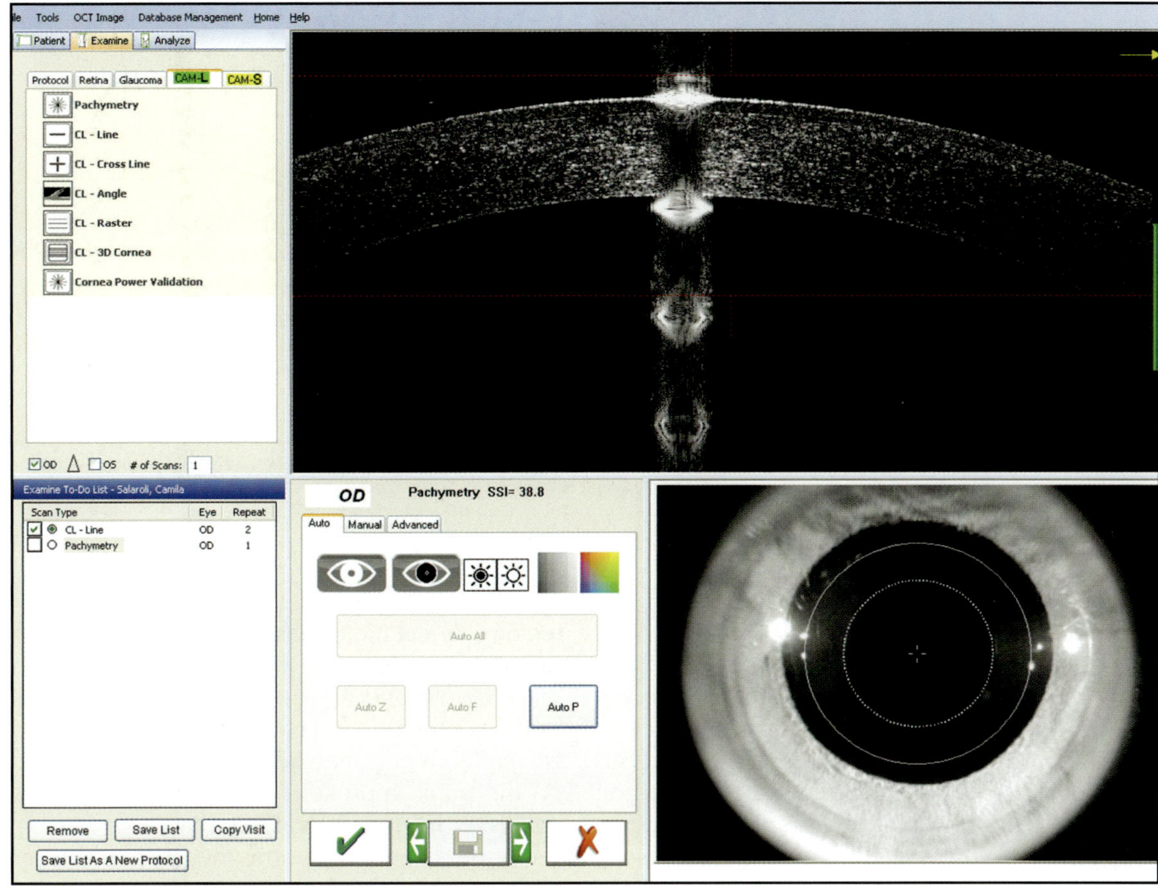

Figure 3-2. Examine menu on RTVue.

1. It is recommended to turn off the examine room light for scanning the angle. For corneal imaging, however, it is better to have a small pupil with room light on so alignment (Step #12) can be easier.

2. Attach the selected CAM lens (CAM-L or CAM-S).

3. Use the 2 red LEDs on headrest to illuminate the cornea and obtain live video image (one for each eye, these are NOT fixation lights). Figure 3-3 is a photo showing the location of the external LED lights.

4. Select a patient to be examined or create a new patient.

5. Select a visit or create a new visit.

6. Click the proper cornea examine tab (CAM-L or CAM-S).

7. Select and add scan pattern(s) or scan protocol.

8. Save the scan list.

9. Ask the patient to look straight into the CAM lens and fixate on the blue dot, or use the yellow external fixation LED on headrest to guide patient fixation if necessary.

10. Start the scan process.

11. Move the scanning module forward until the iris is in focus in the live video image like using a slit lamp. The image of the eye structure of interest should fall in the target zone (between 2 dashed red lines) as shown in Figure 3-2 in the upper right panel.

12. Align (center) the scan on the structure of interest by moving the joystick. Difference scan patterns require slight different alignment. This step is critical for proper dewarping and corneal measurements and is detailed later. Figure 3-2 shows centration on the pupil for the pachymetry scan in the lower right panel.

13. Adjust scan beam to desired length and orientation.

14. Adjust polarization using the [Auto-P] button (see Figure 3-2) if needed to improve signal strength.

15. Stop (capture) scan.

16. Review and process the OCT images.

17. Save the scan.

Aligning the Scan

Because the cornea is a dome, aligning the OCT cross-section along a meridional plane ensures that the

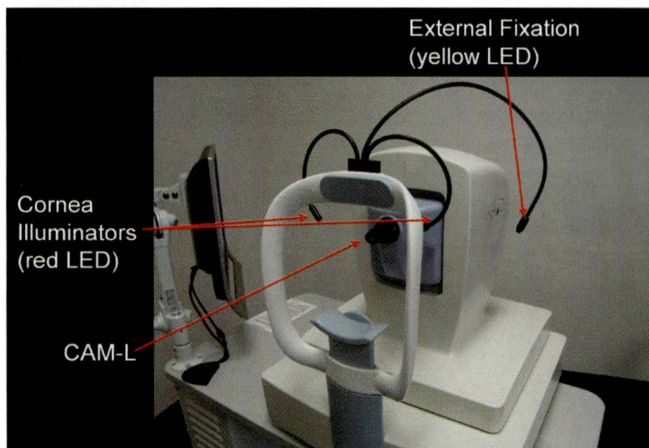

Figure 3-3. Proper position of external LED lights: 2 red corneal illuminators (short neck) and 1 yellow external fixation target (long neck).

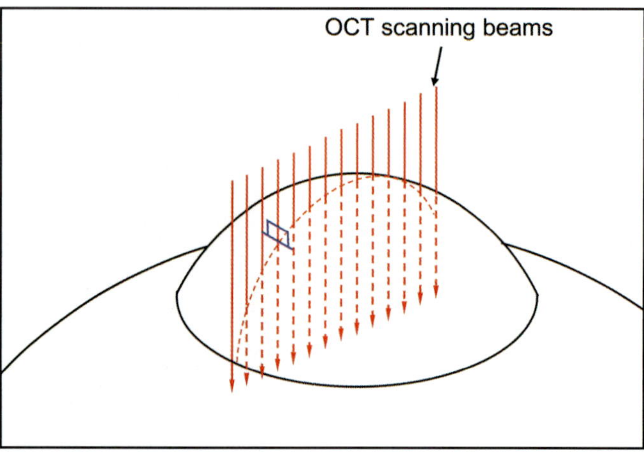

Figure 3-4. Put the scans approximately along a corneal meridian.

scan beams are perpendicular to the corneal surface, in the plane perpendicular to the cross-section (Figure 3-4). This means that refraction of the OCT beam (bending of the ray) at the corneal surface only occurs within the plane of the OCT cross-sectional plane. This makes it possible to correct for the image distortion caused by refraction using relatively simple 2D dewarping software. Therefore it is important to align the scan approximately along a corneal meridian in order to obtain proper dewarping and corneal measurements (see Figure 3-4). The alignment can be accomplished using several landmarks, depending on the scan pattern and structure of interest.

For pachymetry scan pattern, it is recommended to be centered on the pupil using the real-time video image and the circular overlay (see Figure 3-2, bottom right). An alternative landmark is the corneal vertex, the point on the cornea that is perpendicular to the optical (fixation) axis. Vertex centration is obtained by centering the bright vertical flare line on the real-time OCT image (see Figure 3-2, top right). Often the 2 landmarks are nearly at the same location. But we prefer to center at the pupil rather than the vertex, because the vertex position may be altered by corneal surgeries such as LASIK, photorefractive keratectomy, and phototherapeutic kera-

tectomy or ectatic diseases such as keratoconus. Thus if one wants to compare the difference in the pachymetry maps before and after corneal surgery, using the pupil as the centration landmark allows the 2 maps to be registered correctly. For keratoconus diagnosis, centering on the pupil improves the ability to detect the typical pattern of inferotemporal thinning.

For the line scan pattern, the pupil or the vertex flare both could be used as the landmark to ensure proper alignment (Figure 3-5). If the structure of interest is too far from corneal vertex to be included in a single frame, other landmarks can be used. For example, to scan the angle, put the line scan (or angle scan pattern) perpendicular to limbus (Figure 3-6). Use the external fixation (yellow light) to guide the patient's fixation (of the fellow eye) until the cornea/sclera edge is horizontal. To scan certain peripheral features on the cornea, such as an intracorneal ring implant (Intacs, Addition Technology Inc, Des Plaines, IL), LASIK flap edge, and corneal scar, it is recommended to put the line scan perpendicular to either the structure of interest or the pupil boundary (Figure 3-7A and B). To measure the lower tear meniscus, align a vertical scan at the 6 o'clock position, where the cornea and lower lid meet (Figure 3-8).

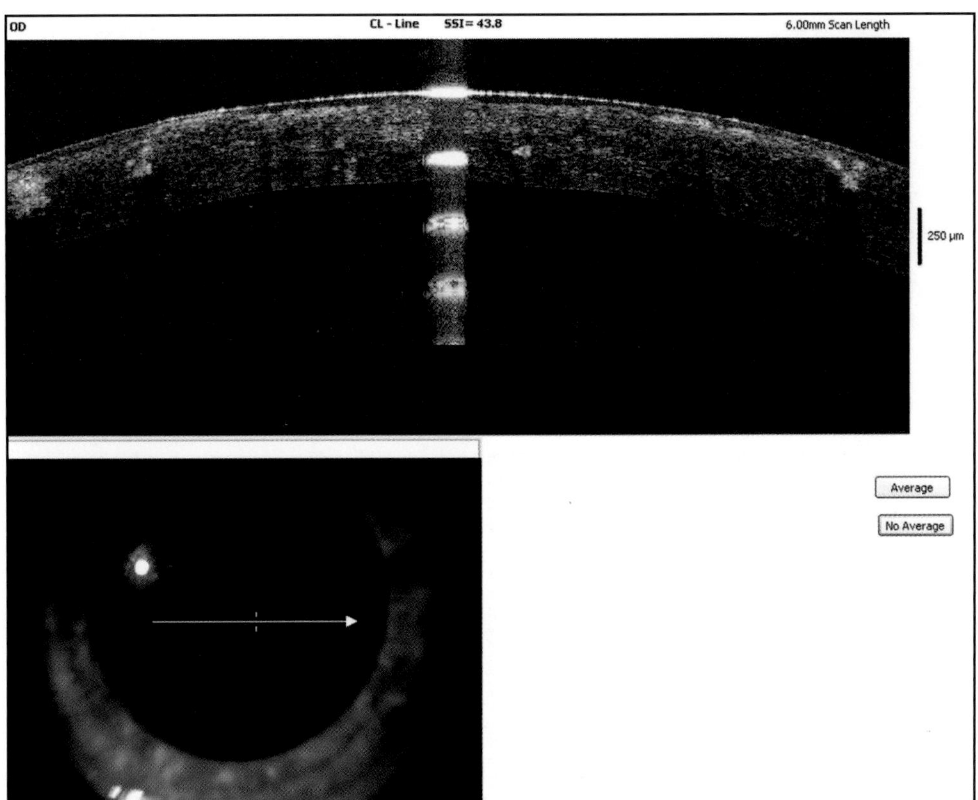

Figure 3-5. Include the corneal vertex in line scans, if possible.

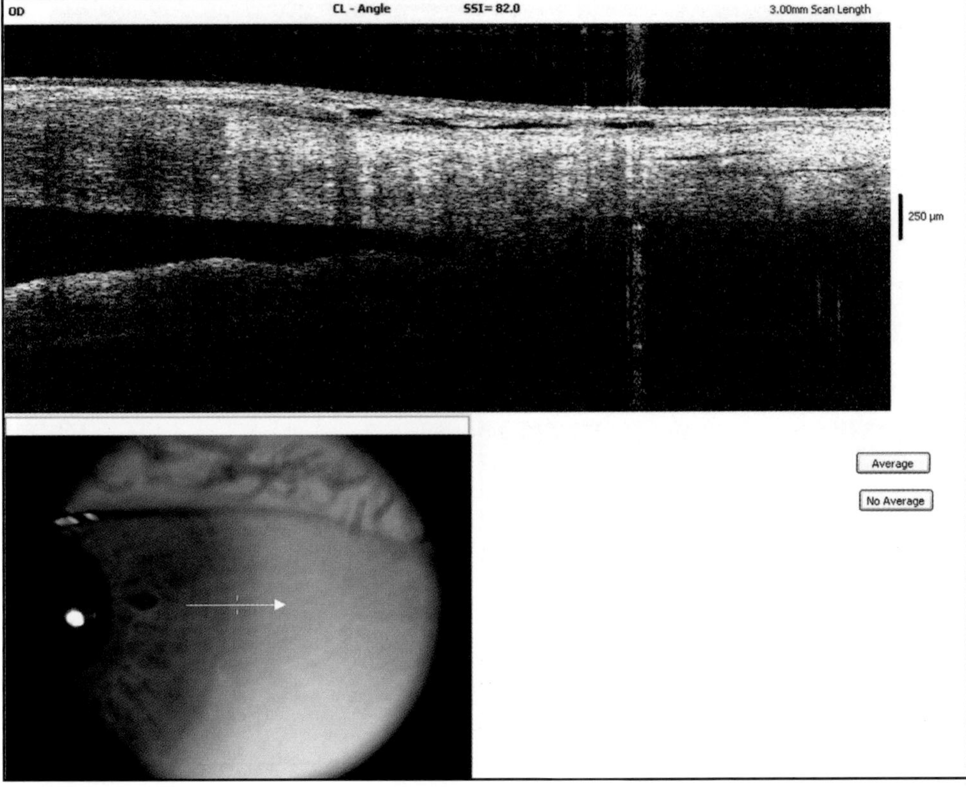

Figure 3-6. Put the line scan perpendicular to the limbus for scanning the angle.

Figure 3-7. (A) Put the line scan perpendicular to the Intacs segment. (B) Put the line scan perpendicular to the flap edge.

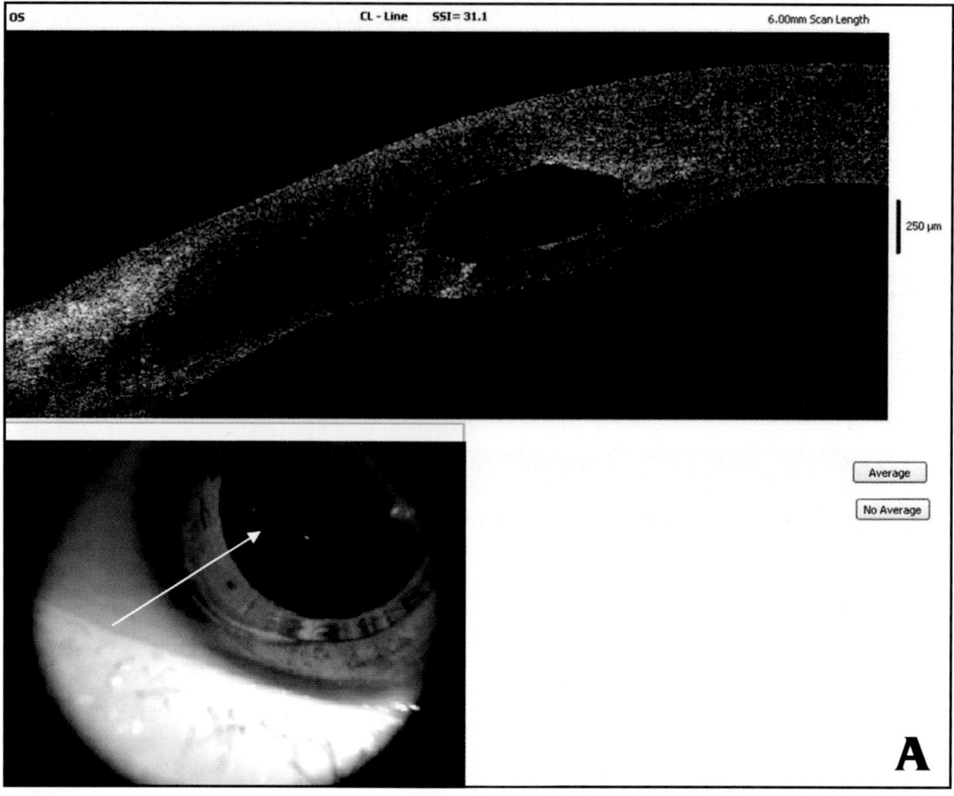

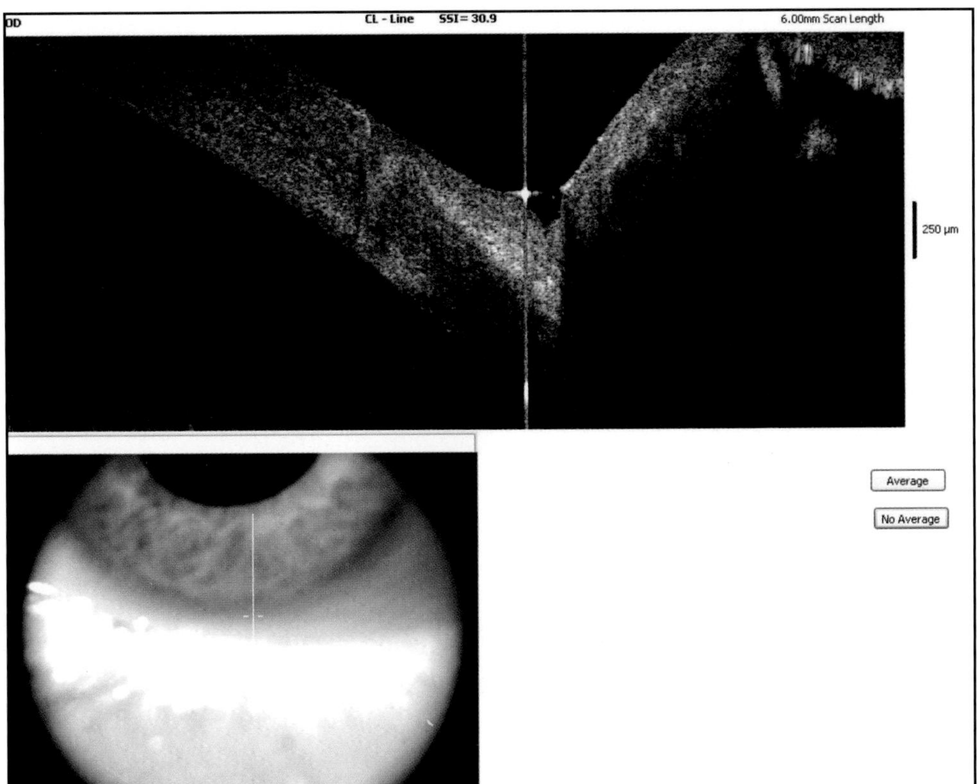

Figure 3-8. Use the line scan to measure lower tear meniscus. Put the line scan vertically at the 6 o'clock position, where the cornea and lower lid meet.

Interpretation of Corneal Images

David Huang, MD, PhD; Yan Li, PhD; and Maolong Tang, PhD

DEWARPING OF CORNEAL IMAGES

Anterior segment OCT images are distorted by refraction at several boundaries, where the refractive index undergoes major changes. Refraction changes the direction of the OCT probe beam. The RTVue corneal adaptor lenses (both the wide-field and high resolution options) provide telecentric scanning, where the beams remain parallel to the optical axis of the OCT system within the transverse scanning range (Figure 4-1). But at the air-tear interface, refraction causes the rays to converge. Refraction at the tear-cornea interface and cornea-aqueous interface also affect the OCT probe beam.

Another source of distortion is the change in image dimensions with the change in group index. The group index is inversely proportional to the group velocity of light while the refractive index is inversely proportional to the phase velocity of light. The group index and the refractive index are very close for most materials, including eye structures (Table 4-1).[1-3] The optical delay that OCT measures is proportional to the group index. Materials with high group indices, such as the cornea, appear thicker than its actual dimension (Figure 4-2A).

Both types of distortion could be corrected by a dewarping algorithm performed in a computer (Figure 4-2).[4] Exact dewarping requires perfect knowledge of every index transition in 3D. This presents a problem because most current OCT corneal imaging consists of a single or a few 2D cross-sections. Luckily, because of our knowledge of corneal structure, we can simplify the

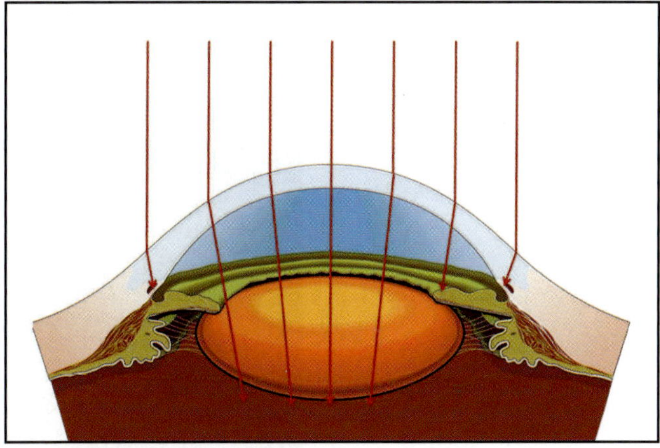

Figure 4-1. Illustration of refractive warping in corneal OCT. The probe beam is scanned in a telecentric fashion across a corneal meridian. The beam is parallel to the optical axis until it reaches the anterior corneal surface, at which point the beam direction is bent due to refraction.

dewarping of corneal images with minimal loss of accuracy in anatomic measurements.

Firstly, because the cornea is a dome structure, a cross-sectional scan along a meridional plane can be dewarped in 2D, without regard to the third dimension. Figure 4-3 shows that OCT beams in the meridional plane are perpendicular to the cornea in the off-meridian plane (the plane formed by the azimuth and the beam axis). Therefore, only the angle of the beam with the cornea within the meridional plane needs to be considered in the dewarping calculation.

Corneal meridians cross at the center of the cornea and are perpendicular to the azimuthal lines (similar to

Huang D, Duker JS, Fujimoto JG, Lumbroso B, Schuman JS, Weinreb RN.
Imaging the Eye from Front to Back with RTVue Fourier-Domain Optical Coherence Tomography (pp 39-46).
© 2010 SLACK Incorporated.

REFRACTIVE AND GROUP INDICES OF AIR, WATER, AND CORNEA AT 830 nm WAVELENGTH		Table 4-1
MEDIUM	**REFRACTIVE INDEX**	**GROUP INDEX**
Air	1.0003	1.0003
Water	1.331	1.341
Cornea	1.370	1.382

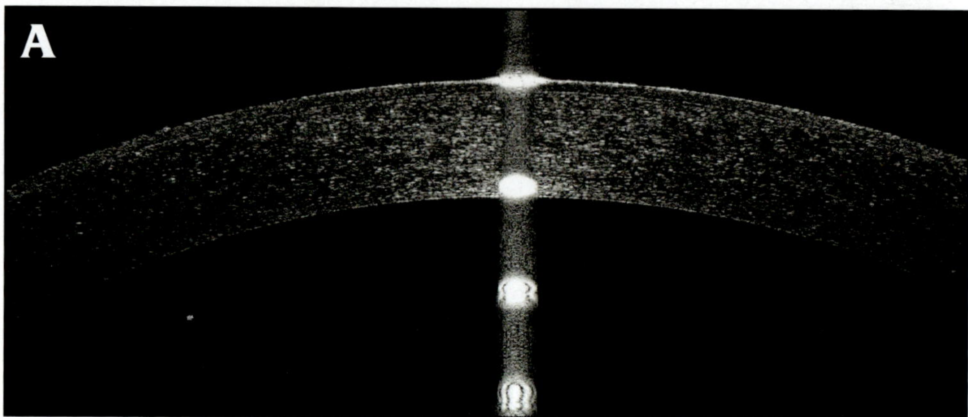
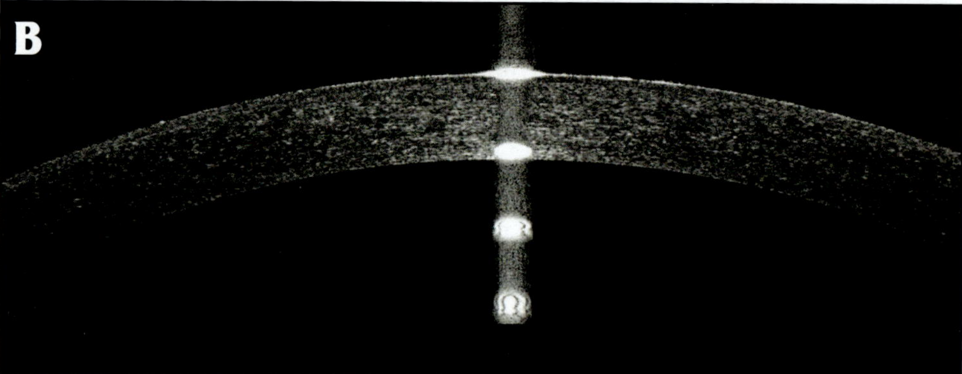

Figure 4-2. (A) Raw corneal OCT image without dewarping. (B) Dewarped corneal image.

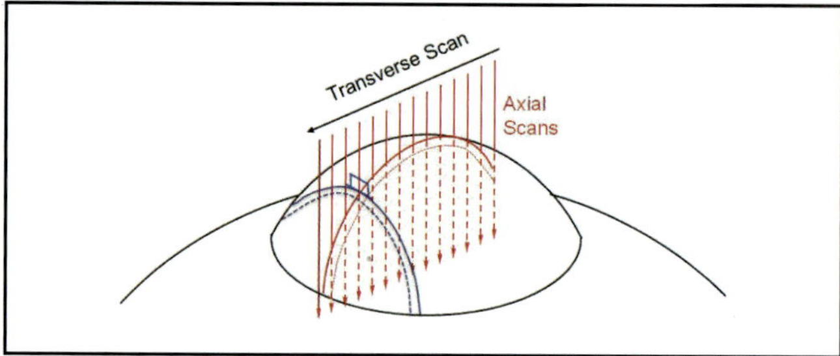

Figure 4-3. A meridional OCT scan (red) is made with the transverse scan along a corneal meridian. The OCT axial scans are perpendicular to the perpendicular plane (blue) formed by the beam axis and the azimuthal line.

latitude lines for the Earth). Therefore, we know that a OCT scan is meridional if it either crosses the center of the cornea or is perpendicular to the limbus. The center of the corneal dome can be identified by the vertex, which produces a strong axial flare. Alternatively, the center of the cornea can be located approximately by the center of the pupil or limbal circle. If a corneal section is not cut approximately along a meridian, then 2D dewarping cannot be accurately performed and

measurements of dimensions (eg, corneal thickness) and angles cannot be considered accurate.

Secondly, the anterior corneal boundary is the strongest index transition by far, and the posterior corneal boundary is approximately parallel. Therefore, for the purpose of dewarping, it is sufficient to identify the anterior corneal boundary. For measurement of corneal thickness and intracorneal dimensions, the corneal index should be used in dewarping. For the measurement of anterior chamber

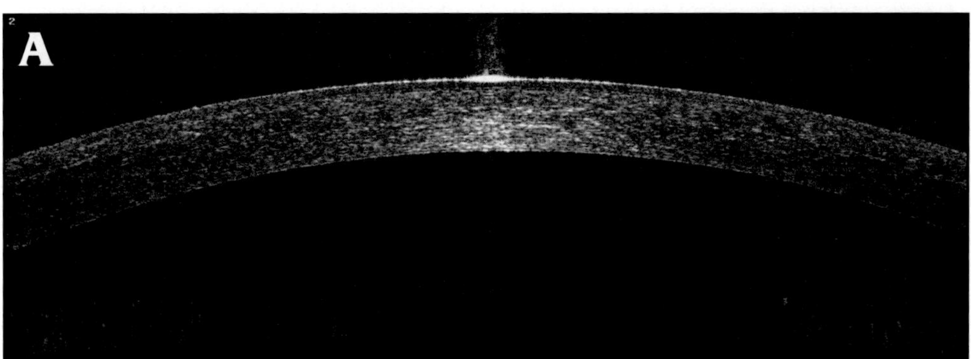

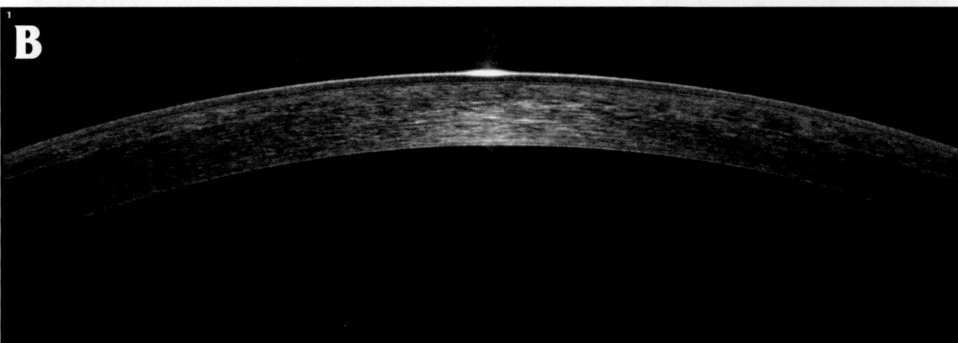

Figure 4-4. OCT image of a cornea 1 week after LASIK. (A) Single frame. (B) Average of 16 frames.

(or tear film) dimensions, the index of physiologic saline (which is the same as aqueous, tears) should be used. If the image does not include the anterior corneal boundary, then the dewarping cannot be performed and measurements of dimensions cannot be made accurately.

In summary, the reader of corneal and anterior segment OCT images should make sure that cross-sectional images are made approximately along meridional planes and include the anterior corneal boundary. If not, accurate quantitative measurements cannot be produced, though qualitative interpretations can still be made.

FRAME AVERAGING

Line scans in RTVue are automatically frame averaged. Typically, 16 consecutive frames of images are registered (lined up) and averaged (Figure 4-4). The averaged frame and a single frame (in case the averaging was unsuccessful) are saved on disk. Frame averaging suppresses speckle noise (snow-like texture) and improves the signal-to-noise ratio of the image. This is useful in identification of thin structures with low contrast, such as the LASIK flap interface.

REFLECTANCE OF CORNEAL STRUCTURES

The recognition of tissue boundaries in OCT images depends on the contrast between the backscattered or reflected signal strength. This contrast varies according to the angle between the incident OCT beam and the tissue of interest. For example, in the OCT image of a recently post-LASIK cornea, one can recognize a great variation in the contrast going from the center to the periphery (see Figure 4-4). In the very center, the very strong specular reflection from the air-tear interface dominates, saturating the (white out) adjacent region above and below, producing a vertical line of flare from the top to the bottom of the image. Slightly off center, the corneal stroma reflects strongly, making the relative recognition of the LASIK flap interface more difficult (Figure 4-5). The epithelium appears dark relative to the brighter stroma near the center. Further off the center, reflection from the stroma weakens and the flap interface and the epithelium appear relatively brighter. The variation in contrast from the center to the periphery is due to the variation in incidence angle. This can be understood by analyzing the relationship between the structure and the angular distribution of backscattering (Figure 4-6, Table 4-2). The air-tear interface is a specular reflection like a mirror, which preserves the directional nature of the incident light. Since the OCT beam usually has a cone angle of only a few degrees of arc, this specular reflection is very strong at the vertex (exact perpendicular incidence), but fades very quickly when the angle is off center by a few degrees. The corneal stroma is a collection of collagen fibers that are cylindrical in shape. Cylindrical reflectors (see Figure 4-6) backscatter light in a fan shape, with the narrow dimension in the plane parallel to the fiber. Since the collagen fibers in the corneal stroma are organized into lamellar sheets, the collective backscattering is somewhat mirror-like, though the directional selectivity is not as narrow as the specular

Figure 4-5. Enlarged peripheral (left) and paracentral (right) corneal images showing epithelium, Bowman's layer, flap interface, stroma, Descemet's membrane (DM), and endothelium.

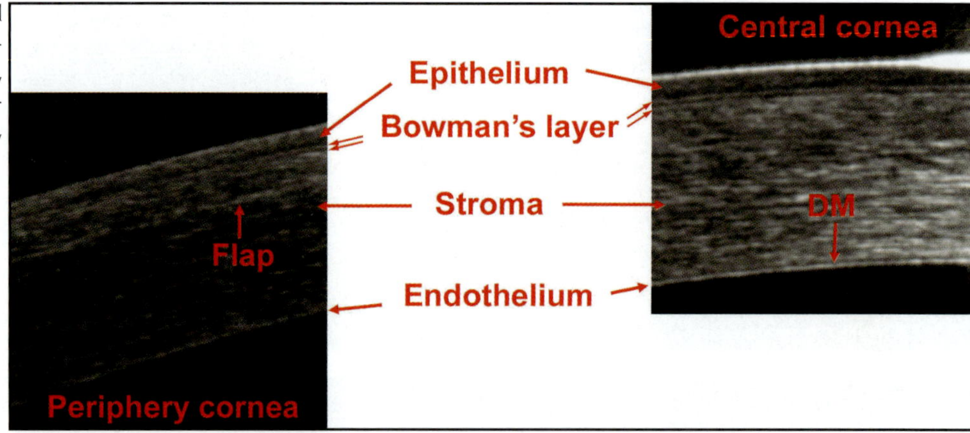

Figure 4-6. Illustration of 3 types of reflecting structures. The specular reflecting surface preserves the directional nature of the incident light (red arrow going toward the surface). A narrow cylinder reflects light into a fan that is narrow in the plane parallel to the cylinder's long axis but wide outside of that plane. A small particle scatters light into a wide cone.

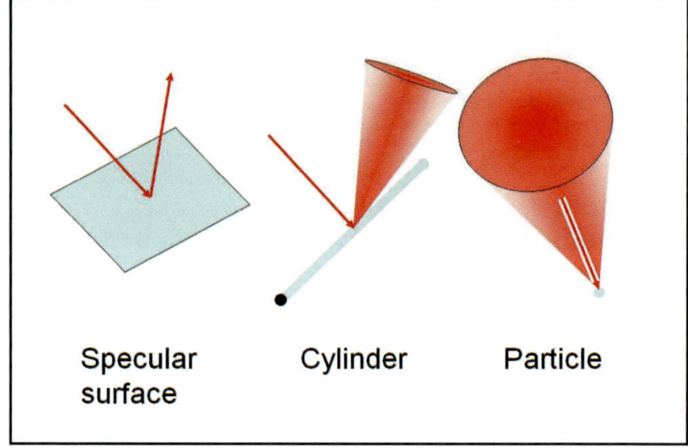

CHARACTERISTIC REFLECTANCE OF TISSUE AT PERPENDICULAR AND OBLIQUE INCIDENCE ANGLES — Table 4-2

TISSUE	PERPENDICULAR INCIDENCE	OBLIQUE INCIDENCE	TYPE OF BACKSCATTERING
Air-tear interface	Very high	Very low	Specular
Tear	Low	Low	Rare particles
Corneal epithelium	Medium	Medium	Particles
Corneal stroma	High	Low	Oriented cylinders
Cornea-aqueous interface	High	Low	Specular (primary), particles and cylinders
LASIK flap interface	Medium	Medium	Mixed, fades with time
Sclera	High	High	Unoriented cylinders
Iris stroma	Medium	Medium	Particles and cylinders
Iris pigment	High	High	Particles

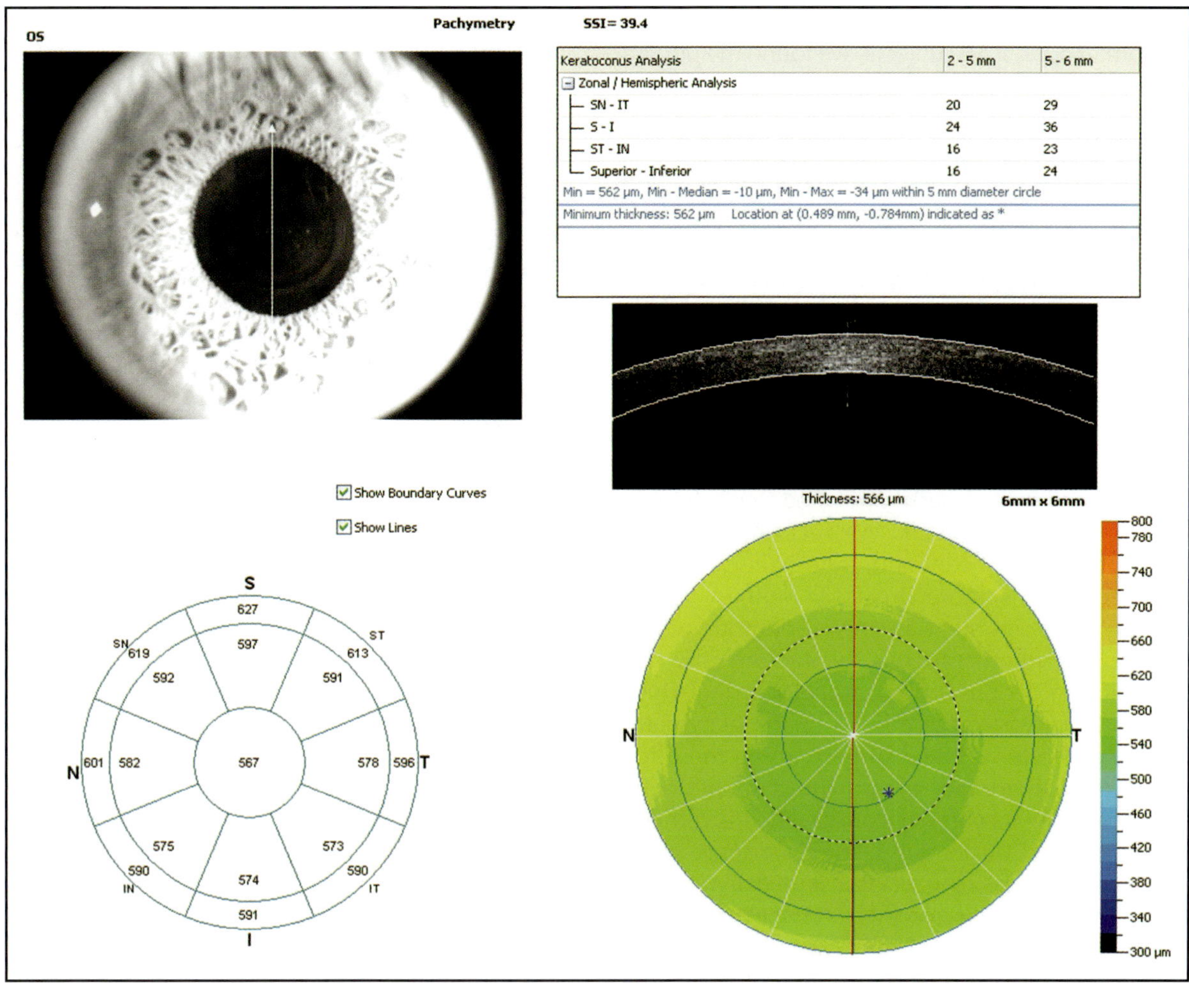

Figure 4-7. RTVue CAM Fourier-domain OCT pachymetry map printout.

reflection. The posterior stroma appears to reflect more directionally than the anterior stroma (brighter centrally compared to the periphery). This might be explained by the presence of interlamellar interweaving fibers in the anterior stroma.[5,6] The scleral collagen is less organized, therefore its backscattering is much less directional. Of course, the cornea is much more transparent than the sclera and backscatters much less because of the regular tight spacing between lamellae.[7] Some structures are theoretically mixed. For example, the cornea-aqueous interface includes a specular index transition, as well as backscattering from the endothelial cells (particles) and Descemet's membrane (cylinders of collagen fibers). The directional nature of the cornea-aqueous interface indicates that it is primarily specular. The LASIK flap interface is lamellar, but the nondirectional nature of the elevated reflectivity in the early postoperative period indicates that it is particle-like, possibly due to small scale irregularities of the flap cut.

THE PACHYMETRY SCAN

The high speed of the RTVue system reduces motion error and enables more accurate mapping of the cornea. The corneal mapping pattern on the RTVue CAM (corneal adaptor module) is called pachymetry. It consists of 8 high-definition meridional scans acquired in only 0.32 seconds (refer to Chapter 2). The corneal thickness profile is measured by an automated algorithm detecting the anterior and posterior corneal boundaries on the cross-sectional images. The software has been improved over successive versions and now works well for a wide variety of normal and abnormal (post-LASIK, keratoconus, scarred) corneas. However, if there is any error in the automated boundary drawing, the operator can manually correct it. A 6-mm diameter pachymetry map was then formed by interpolating the thickness profiles on the 8 meridians. The maps (Figure 4-7, bottom) were divided into zones by octants and annular rings (2, 5, and 6 mm). The average pachymetry of each zone was displayed in the sector map (Figure 4-7, bottom left). The repeatability of the pachymetry map is excellent. In a study on normal eyes of 50 subjects, we found the pooled standard deviation of repeated measurements to be 1.3 to 1.7 μm for central (<2 mm) thickness and 2.8 to 3.9 μm for pericentral (2 to 5 mm) thickness measurements. The reliability of the map is best in the paracentral region where the corneal reflection is strong, and it decreases peripherally as the signal fades. Thus, the measurement error can occur in the peripheral zones of steep corneas and decentered scans. Therefore abnormal thickness in the peripheral cornea, outside the 5 mm diameter, should be interpreted with caution, and the boundary line on the relevant meridional OCT image should be checked.

There are 2 landmarks to center the corneal mapping scans: the corneal vertex and the pupil (refer to Chapter 3). These 2 landmarks are often very close to each other in normal subjects. We recommend using the pupil as the primary centration reference. Although the vertex position is more precisely defined, it can be altered by surgeries such as LASIK, photorefractive keratectomy, or phototherapeutic keratectomy; by diseases such as keratoconus; or by corneal scarring. Thus, if one wants to compare the difference in pachymetry maps before and after corneal surgery, using the pupil as the centration landmark allows the 2 maps to be registered correctly. In corneal scar cases, it is difficult to locate the corneal vertex due to the distorted corneal surface. In keratoconus, the ectasia is usually located inferotemporally, and therefore the corneal vertex is also shifted inferotemporally. Thus, using the pupil as the center might better reveal the asymmetric thinning in keratoconus.

Corneal Power Measurement

The RTVue system is able to directly measure the power of both the anterior and posterior corneal surfaces. It is a logical improvement to conventional approaches of measuring corneal power, like keratometry, because it bypasses the theoretical keratometric index and uses actual refractive indices.

The corneal power measurement on RTVue is based on the "Pachymetry" scan pattern. This pattern consists of 8 high-definition meridional scans. Within each scan, the anterior and posterior corneal boundaries are automatically detected. The best-fit spheres are fitted to the central 3 mm of anterior and posterior boundaries to obtain the radius of curvature. Then the anterior corneal power for each meridian is calculated by:

$$P_a = (n_1 - n_o)/R_a$$

where P_a is anterior corneal power along each meridian, n_1 is the refractive index of the cornea, n_o is the refrac-

tive index of the air, and R_a is the radius of curvature of the anterior surface.

The posterior corneal power for each meridian is calculated by:

$$P_p = (n_2 - n_1)/R_p$$

where Pp is posterior corneal power along each meridian, n2 is the refractive index of the aqueous, and Rp is the radius of curvature of the anterior surface.

The corneal powers along the 8 meridians are averaged to get the anterior and posterior corneal power. The total corneal power can be calculated by the thick lens formula:

$$K = K_a + K_p - D * K_a * K_p/n_1$$

where K is total corneal power, K_a is anterior corneal power, K_p is posterior corneal power, and D is central corneal thickness.

In order to improve the repeatability of corneal power measurements, the RTVue software automatically recorded 5 consecutive sets of pachymetry scans. After discarding scans with large motion artifacts or of poor quality, the 5 sets are averaged to provide the final results of total corneal power.

The corneal power measured by OCT is lower than that of conventional keratometry using the conventional keratometric index of 1.3375.[8] Based on data collected in our research, the RTVue OCT-derived corneal power was 1.2 D lower than that measured by the IOL-Master (Carl Zeiss Meditec Inc, Dublin, CA) automated keratometry. The OCT measurement is probably more accurate, since the keratometric index was adapted by convenience and never adequately validated.[8,9] However, since most intraocular lens (IOL) power formulas are calibrated to conventional keratometry, OCT measurements need to be converted to the conventional keratometric equivalent before use in conventional IOL power formulae. A dedicated OCT-based IOL power formula is under development.

References

1. Ciddor PE, Hill RJ. Refractive index of air. 2. Group index. *Applied Optics.* 1999;38(9):1663-1667.
2. Drexler W, Hitzenberger CK, Baumgartner A, et al. Investigation of dispersion effects in ocular media by multiple wavelength partial coherence interferometry. *Exp Eye Res.* 1998;66(1):25-33.
3. Lin RC, Shure MA, Rollins AM, et al. Group index of the human cornea at 1.3-micron wavelength obtained in vitro by optical coherence domain reflectometry. *Optics Letters.* 2004;29(1):83-85.
4. Westphal V, Rollins AM, Radhakrishnan S, Izatt JA. Correction of geometric and refractive image distortions in optical coherence tomography applying Fermat's principle. *Optics Express.* 2002;10(9):397-404.

5. Morishige N, Wahlert AJ, Kenney MC, et al. Second-harmonic imaging microscopy of normal human and keratoconus cornea. *Invest Ophthalmol Vis Sci.* 2007;48(3):1087-1094.

6. Teng SW, Tan HY, Peng JL, et al. Multiphoton autofluorescence and second-harmonic generation imaging of the ex vivo porcine eye. *Invest Ophthalmol Vis Sci.* 2006;47(3):1216-1224.

7. Smith TB. Modeling corneal transparency. *Am J Phys.* 2007;75(7):588-596.

8. Bennett AG, Rabbetts RB. *Clinical Visual Optics.* 3rd ed. Boston, MA: Butterworth-Heinemann; 1998:387-388.

9. Gobbi PG, Carones F, Brancato R. Keratometric index, videokeratography, and refractive surgery. *J Cataract Refract Surg.* 1998;24(2):202-211.

5 Interpretation of Angle Images

Farnaz Memarzadeh, MD; Sheila Mahdaviani, MD; and Yan Li, PhD

Using the newer Fourier-domain OCT technology, the RTVue achieves high scan speeds that allow acquisition of high-definition images in a small fraction of a second. For angle imaging, several consecutive frames of each scan can be registered and averaged to increase the signal-to-noise ratio and sharpen anatomic features (Figure 5-1). The shorter wavelength of 830 nm and broader bandwidth produce an excellent axial resolution of 5 µm in tissue. This is greater than a 3-fold improvement in resolution compared to the Visante (Carl Zeiss Meditec Inc, Dublin, CA), which has roughly a 17 µm resolution. This high resolution allows for visualization of very small anatomic details such as the Schwalbe's line, Schlemm's canal, and the trabecular meshwork (Figures 5-2 and 5-3), which were not possible with previous generations of anterior segment OCT systems.[1,2]

The shorter wavelength of 830 nm used in RTVue, however, encounters signal loss due to scattering at the limbus and angle. This causes poorer visualization of the scleral spur and iris root (see Figure 5-2), compared to longer wavelength (1310 nm) OCT systems such as the Visante. In anterior segment OCT imaging using longer wavelength, the scleral spur has been an important anatomic landmark to define quantitative angle measurement parameters used to assess the magnitude of angle opening, such as angle opening distance (AOD) and trabecular iris space area (TISA).[2,3] In the RTVue images, the scleral spur cannot be clearly visualized on some images due to the greater scattering of light at the

limbus. The Schwalbe's line, which represents the termination of Descemet's membrane, is an anatomic feature that is consistently visible on all high quality images (see Figures 5-2 through 5-5). The internal portion of the trabecular meshwork measures approximately 500 to 750 µm in length and starts immediately posterior to Schwalbe's line and extends posterior toward the scleral spur. Measurement of the distance between Schwalbe's line and the anterior surface of the iris defines the AOD at Schwalbe's line (AOD-SL), a method of quantifying the angle width on RTVue images (see Figures 5-2 and 5-3).

The high resolution of RTVue allows for visualization of very fine detail in the anterior segment such as dilated iris vessels, Schlemm's canal, and aqueous collector channels and veins (see Figures 5-3 and 5-4) that we had not previously been able to see with any noninvasive imaging technology. We have used this technology to image (see Figure 5-5) the changes that occur in the anterior chamber angle after selective removal of the trabecular meshwork and inner wall of Schlemm's canal via Trabectome (NeoMedix Corp, San Juan Capistrano, CA) surgery. This technology may also enable imaging of other glaucoma angle surgeries such as canaloplasty and trabecular stenting.

The RTVue system is also very useful clinically for evaluating the anterior segment in the presence of corneal edema or opacity (refer to Chapter 12). We have used this technology (Figures 5-6 and 5-7) to help guide our clinical decision making and surgical management.

Huang D, Duker JS, Fujimoto JG, Lumbroso B, Schuman JS, Weinreb RN.
Imaging the Eye from Front to Back with RTVue Fourier-Domain Optical Coherence Tomography (pp 47-52).
© 2010 SLACK Incorporated.

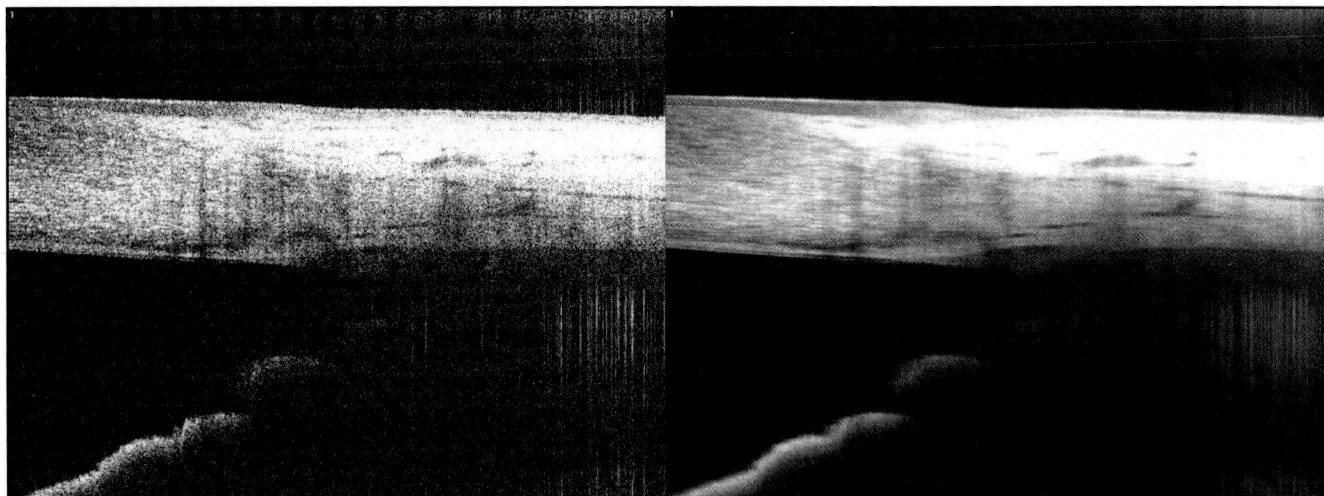

Figure 5-1. Comparison of a single-frame image (left) vs a 16-frame averaged image (right) of the anterior chamber angle. Frame averaging removes speckle and background noises, thus improving delineation of small anatomic features.

Figure 5-2. Cross-sectional OCT image of the nasal angle in a normal subject. The high resolution helps visualize the termination of the endothelium and Descemet's membrane (Schwalbe's line [SL]), a useful landmark on these images. Also visible are the external limbus (EL) and trabecular meshwork (TM). The scleral spur (SS) is faintly visible. The angle recess, iris root, and ciliary body are not visible due to blocking by the sclera. The AOD between SL and the anterior surface of the iris (AOD-SL, yellow caliper line) was large, measuring 473 μm, indicating that the angle is open.

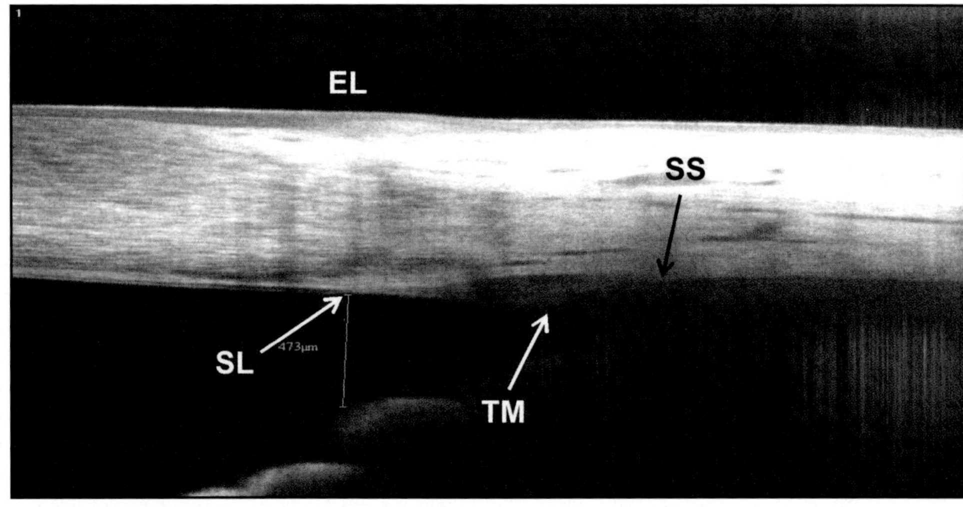

Figure 5-3A. Cross-sectional OCT image of the nasal angle in an eye with narrow angles. Note the small distance between the trabecular meshwork (TM) and the iris, indicating a narrow, potentially occluded angle. The AOD-SL (dotted line) was narrow, measuring 120 μm. Also visible are the Schlemm's canal (SC) and an aqueous collector vein.

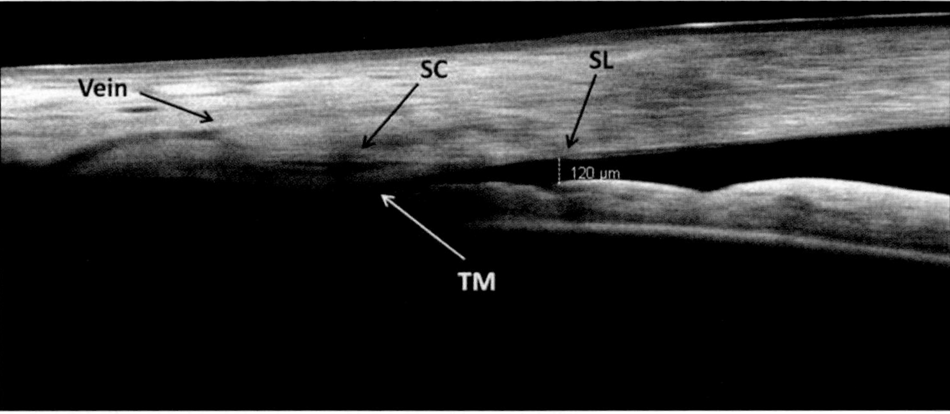

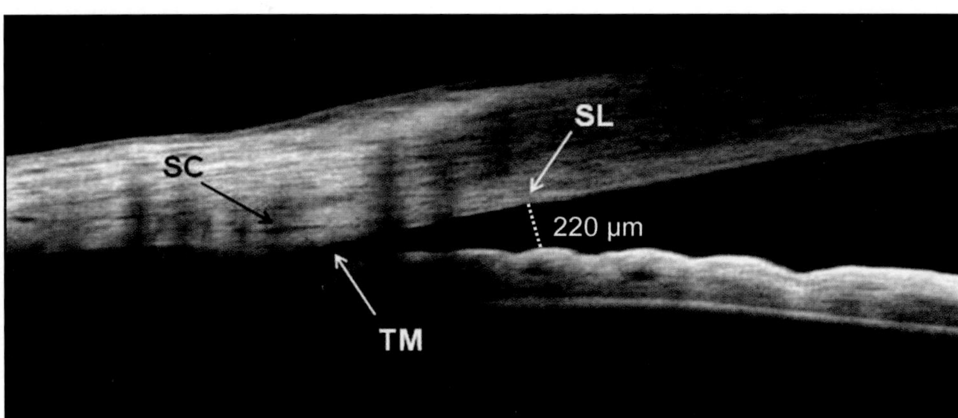

Figure 5-3B. Cross-sectional OCT image of the same patient after laser peripheral iridotomy. Note the widening of the angle after the procedure. The AOD-SL (dotted line, 220 µm) nearly doubled after the procedure.

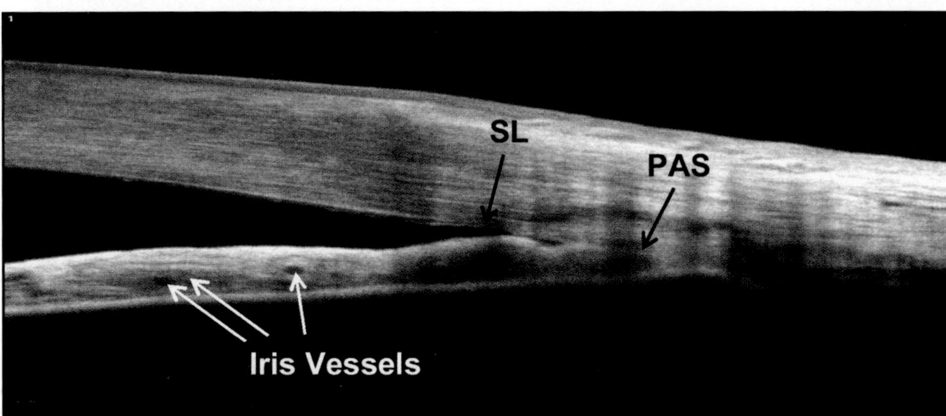

Figure 5-4. Cross-sectional OCT image of a patient with neovascular glaucoma and secondary angle closure. Peripheral anterior synechiae (PAS) extending just posterior to the Schwalbe's line (SL) and occluding the trabecular meshwork are visualized. Also notable are the dilated circumferential and longitudinal iris blood vessels.

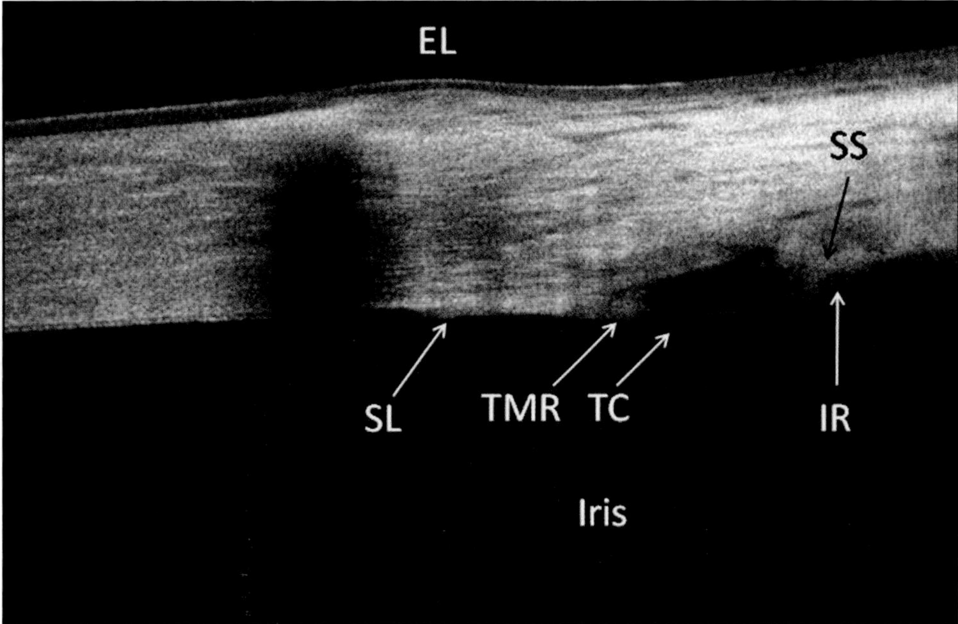

Figure 5-5. Cross-sectional OCT image of the nasal angle after Trabectome surgery. The shadowing in the peripheral cornea is caused by the limbal girdle of Vogt. The Trabectome removed the posterior trabecular meshwork, leaving a trabecular cleft (TC) of 374 µm in width and an anterior trabecular meshwork remnant (TMR). The iris root (IR) is very faint due to scleral shadowing but can be traced by contiguity with the iris. (Courtesy of Brian A. Francis, MD.)

Figure 5-6. OCT images of a 27-year-old aphakic male with a history of congenital glaucoma. The patient presented with 2 previous aqueous shunt implants, elevated intraocular pressure, and corneal edema. View of the tube in the anterior chamber was obscured due to corneal edema. Anterior segment OCT line scans (A and B) were along the course of the tube shunt, as shown in the video frame (C). Anterior segment OCT demonstrates that the tube was in contact with a retrocorneal membrane but not the corneal endothelium. The lumen and opening of the tube were not obstructed. The mirror image of the anterior cornea in the lower OCT frame (B) should be ignored—it was due to the complex conjugate ambiguity artifact of Fourier-domain OCT.

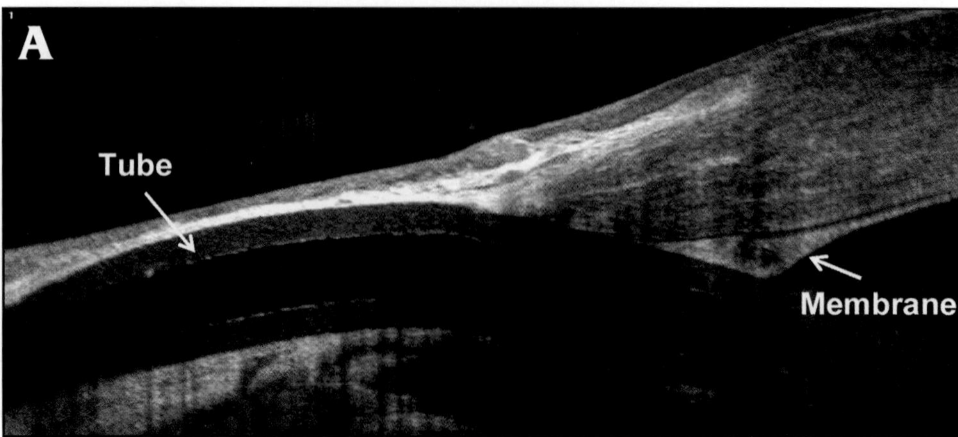

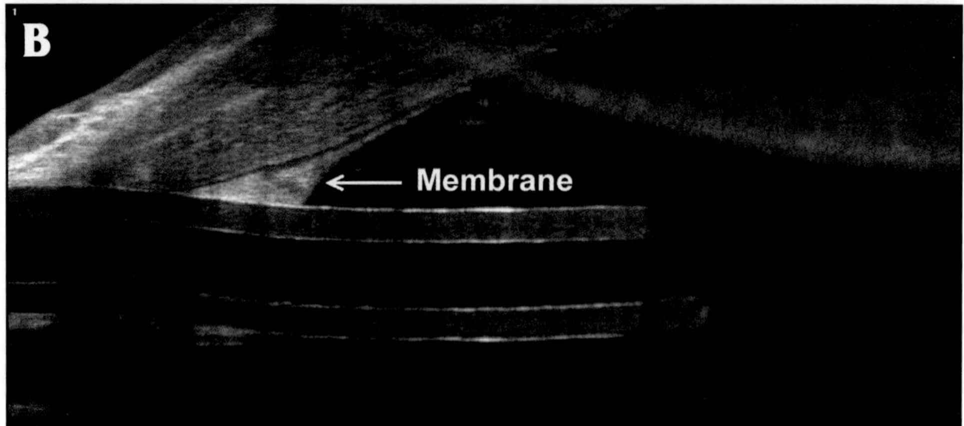

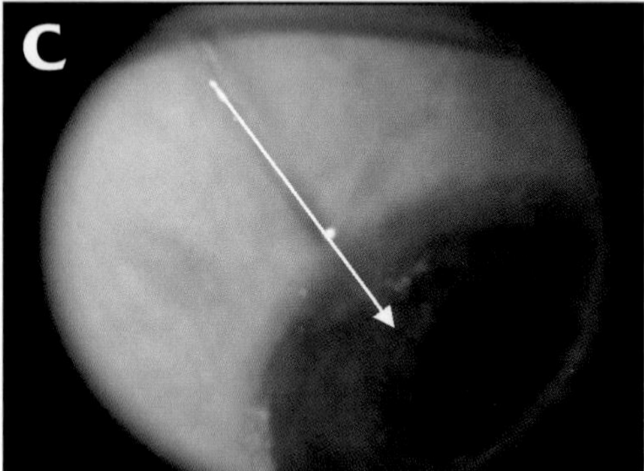

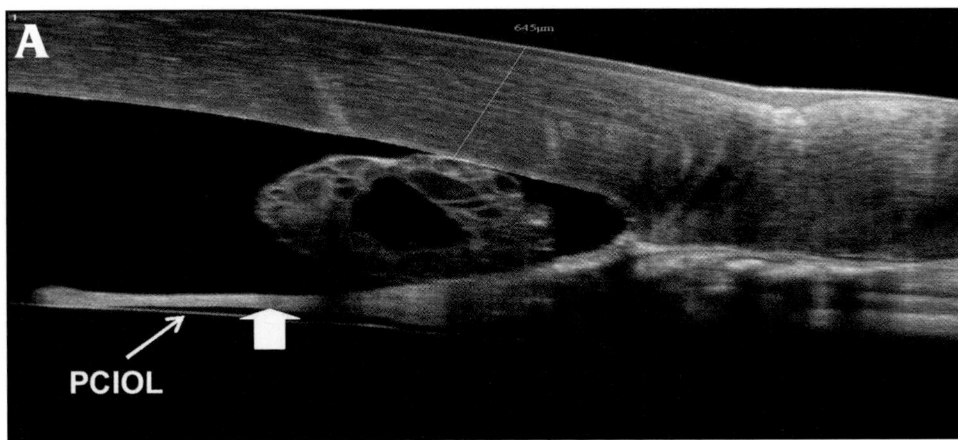

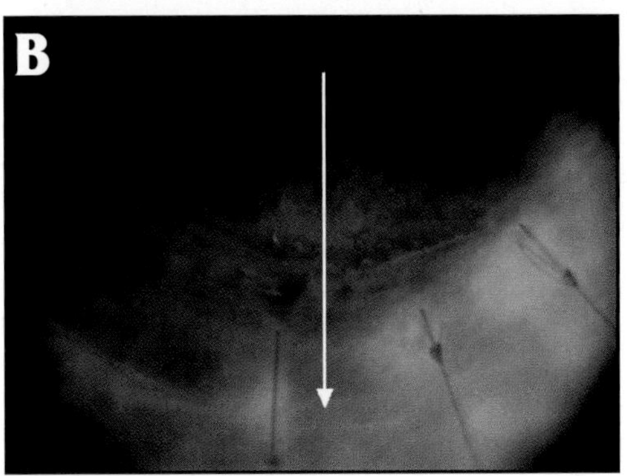

Figure 5-7. Cross-sectional OCT image of the inferior angle in a pseudophakic patient with secondary angle closure glaucoma, status is after multiple penetrating keratoplasties (PK). Clinical examination 15 years after cataract extraction and few months after PK demonstrated an edematous graft and probable old cortical material in the anterior chamber. After slit lamp examination, it was not possible to determine if there was any corneal endothelial touch with the iris because of scarring at the graft-host junction. The OCT image shows a hydrated cortical remnant lodged in the space between the cornea and a thickened anterior capsule (thick arrow). The posterior chamber intraocular lens (PCIOL) was visible behind the capsule. The cortex was in contact with the endothelium, indicating the need for its surgical removal. The angle was closed by a membrane that fused the cornea, iris, and anterior lens capsule. (A) is the OCT image and (B) is the video image showing direction of OCT scan (thin arrow line).

REFERENCES

1. Huang D, Li Y, Radhakrishnan S. Optical coherence tomography of the anterior segment of the eye. *Ophthalmol Clin North Am.* 2004;17(1):1-6.

2. Radhakrishnan S, Goldsmith J, Huang D, et al. Comparison of optical coherence tomography and ultrasound biomicroscopy for detection of narrow anterior chamber angles. *Arch Ophthalmol.* 2005;123(8):1053-1059.

3. Pavlin CJ, Harasiewicz K, Foster FS. Ultrasound biomicroscopy of anterior segment structures in normal and glaucomatous eyes. *Am J Ophthalmol.* 1992;113:381-389.

Posterior Segment Scan Patterns

6

Ou Tan, PhD and David Huang, MD, PhD

INTRODUCTION

The RTVue software allows a wide variety of 2D and 3D scan patterns to aid in the evaluation of retinal diseases and glaucoma. The 2D scan (also called B-scan, Figure 6-1) is composed of many axial scans (A-scan) arranged in a line or circle (Figure 6-2). The more densely the A-scans are packed (A-scan/mm), the higher the image "definition." Because Fourier-domain OCT systems like the RTVue are much faster than older time-domain OCT systems, it is now possible to produce high-definition B-scans in time frames (0.02 to 0.04 seconds) that allow very little motion artifacts.

The 3D scan patterns (Figure 6-3) are composed of many 2D scans. The most common pattern is the *raster* scan, which is composed of many parallel lines. The lines could also be arranged in a radial wheel spoke pattern called a *radial* scan. *Circular* scans are generally arranged in a concentric pattern. There are also composite patterns that include both radial and concentric scans and both vertical and horizontal lines.

A drawback of Fourier-domain OCT technology is that the signal level is not uniform across the image—it varies with depth. The signal is stronger at depths close to the position of the reference mirror. In order to maximize the signal in the vitreous and inner retinal layers, it is best to place the reference plane just anterior to the retina. In the RTVue software, this is called the *vitreoretinal* imaging mode. To maximize the signal in the choroid and outer retina, it is best to place the reference plane posterior to the choroids. This is called the *chorioretinal* imaging mode.

Depend on the purpose of scan pattern, we group the scan patterns in to two classes: retinal scan patterns and glaucoma scan patterns. The scan pattern parameter in this chapter is based on RTVue software version 4.0.

RETINAL SCAN PATTERNS

Retinal scan patterns were originally designed for retinal disease. Usually they are centered at the fovea for default. The scan depth is 1.92 mm. Except EMM5, the center of scan patterns can be manually edited or moved with the cursor. The default imaging model of retinal scan patterns is the chorioretinal mode. A summary of retinal scan patterns is listed in Table 6-1.

Line Scan

Line scan is used to provide a high-definition B-scan at the retinal location selected by the user. It allows for frame averaging to boost signal, reduce speckle, and enhance contrast. Thus, line scan is particular useful in eyes with low signal due to media opacity.

Line scan has a default scan length 6-mm long (adjustable from 2 to 12 mm). The default orientation angle is zero, which means a horizontal line from left to right on the retina. A B-scan includes 1024 A-scans and takes 0.039 seconds. The default fixation target is the fovea.

Huang D, Duker JS, Fujimoto JG, Lumbroso B, Schuman JS, Weinreb RN.
Imaging the Eye from Front to Back with RTVue Fourier-Domain Optical Coherence Tomography (pp 53-60).

Figure 6-1. A cross-sectional OCT image is also called a B-scan. It is composed of many axial scans (A-scans) that correspond to vertical lines (see the dotted line) in the image.

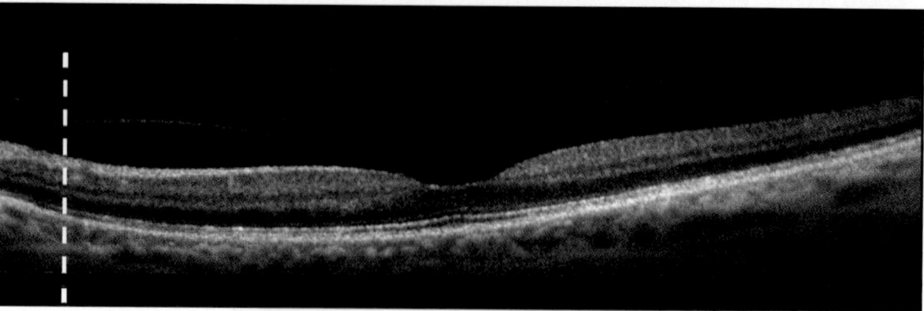

Figure 6-2. Two type of B-scans: (left) line scan and (right) circular scan.

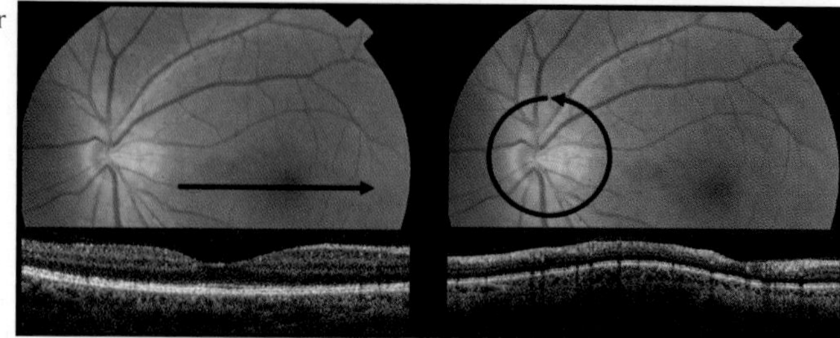

Figure 6-3. Three types of patterns to combine B-scans in a scan pattern: (left) raster pattern, (middle) radial pattern, and (right) concentric pattern.

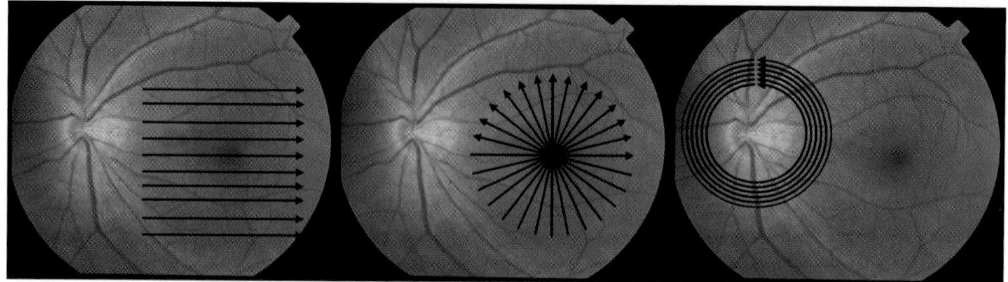

RTVue provides frame averaging for line scans. The B-scan is repeated at same location 16 times (16 frames). Then RTVue registers and averages the frames (Figure 6-4). The averaging process is fully automatic. If the eye moved in the scanning, only frames with good correlation are averaged to prevent blurring due to misregistration. RTVue also allows the user to switch between an averaged image and the original B-scan. A line scan averages 16 frames and actually takes 0.63 seconds.

Cross Hair Scan

The cross hair scan (Figure 6-5) is composed of a horizontal and a vertical line. It has a default scan length of 6 mm (adjustable from 2 to 12 mm). The orientation is fixed to horizontal and vertical directions. A cross hair scan takes 0.63 seconds. It also uses the frame averaging technique to enhance the image—8 horizontal and 8 vertical B-scans are averaged. The default fixation target is the fovea center. The cross hair scan is useful to obtain high quality B-scans of the macula centered on the fovea.

High Density Scans (High Density Line and High Density Cross Hair)

High density (HD) scans (Figure 6-6) are designed to improve the transverse definition of B-scans. Its B-scans consist of 4096 A-scans, 4 times of the normal line scan. HD scans include HD line scan and HD cross hair scan. The parameter is same as line scan and cross hair scan except HD scans do not use frame averaging. Thus HD scan is actually faster. HD line scan takes 0.16 second and HD cross hair scan takes 0.31 second.

Macular Map Scan

The macular map (MM5) scan (Figure 6-7) is a composite pattern consisting of both vertical and horizontal lines. It is optimized for mapping the macula. The line spacing is 0.25 mm in the central 3 mm square and 0.5 mm in the remainder of the 5 mm square area of the macula. The scan is centered on the fovea. It provides data for automated mapping of retinal thickness and a retinal pigment epithelium (RPE) elevation map. Another advantage of the MM5 scan pattern is that it provides both vertical and horizontal B-scans that intersect at 241 points in the macula. Thus, the user can

Table 6-1

Scan Pattern	Time (sec)	# A-Scan	Adjustability	Default
Line	0.63	1 x 1024	Scan length: 2 to 12 mm Angle: 0 to 180 Center: any	6 mm 0 fovea
Cross Hair	0.63	2 x 1024	Scan length: 2 to 12 mm Angle: fixed Center: any	6 mm 0/90 fovea
HD Line	0.16	1 x 4096	Scan length: 2 to 12 mm Angle: 0 to 180 Center: any	6 mm 0 fovea
HD Cross Hair	0.31	2 x 4096	Scan length: 2 to 12 mm Angle: fixed Center: any	6 mm 0/90 fovea
3D Macular	2.2	101 x 513	Scan length (H/W): 3 to 8 Angle: 0 to 180 Center: fixed	(H x W) 6 x 6 mm 0 fovea
EMM5	1.5	2 x 13 x 807 plus 2 x 8 x 512	Scan length (H/W): fixed Angle: fixed Center: fixed	(H x W) 6 x 6 mm 0/90 fovea
3D Reference	2.4	141 x 385	Scan length (H/W): fixed Angle: fixed Center: fixed	(H x W) 7 x 7 mm 0 fovea
Raster	0.34	17 x 512	Scan length (H): 2 to 10 mm (W) 0 to 6 mm Angle: 0 to 180 Center: any	(H x W) 6 x 4 mm 0 fovea
Grid	1.8	2 x 5 x 960	Scan length (H/W): 2 to 10 mm Angle: fixed Center: any	(H x W) 8 x 2 mm 0/90 fovea
MM6/Radial Slicer	0.47	12 * 1024	Scan length: 2 to 10 mm Center: any	6 mm fovea

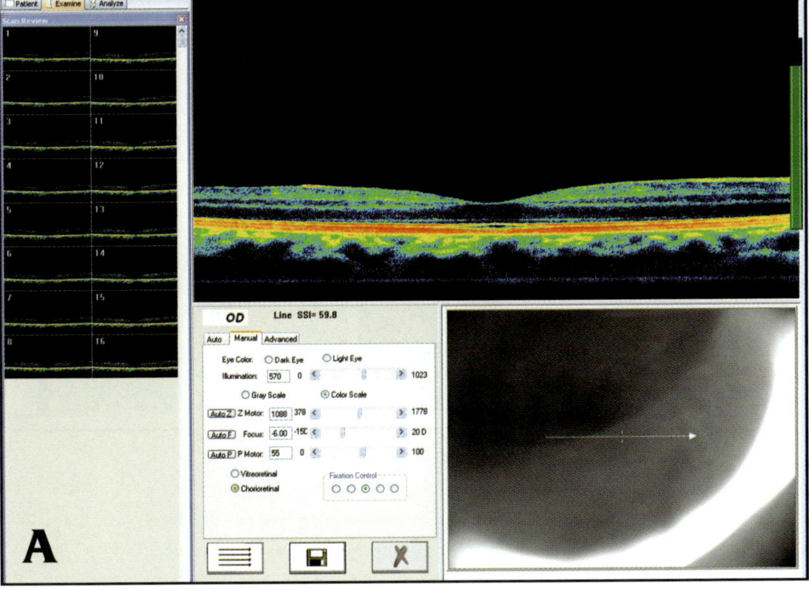

Figure 6-4. Line scan allows 16 B-scans at the same location and averages register frames to boost signal. (A) 16 frames and averaged image.

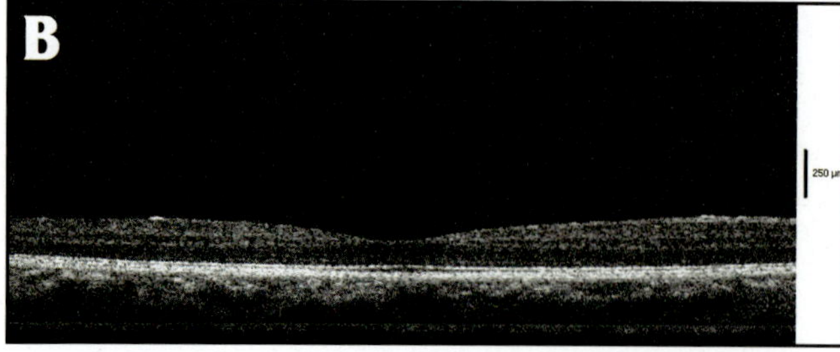

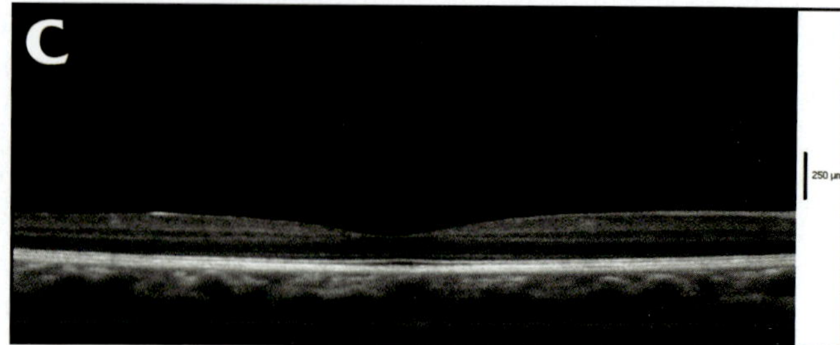

Figure 6-4. Line scan allows 16 B-scans at the same location and averages register frames to boost signal. (B) Single B-scan before average. (C) Averaged image (gray scale).

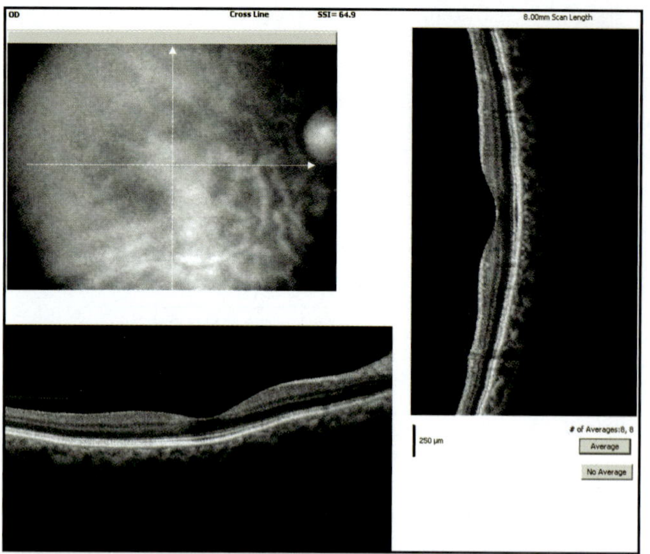

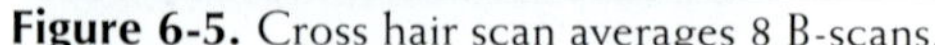

Figure 6-5. Cross hair scan averages 8 B-scans.

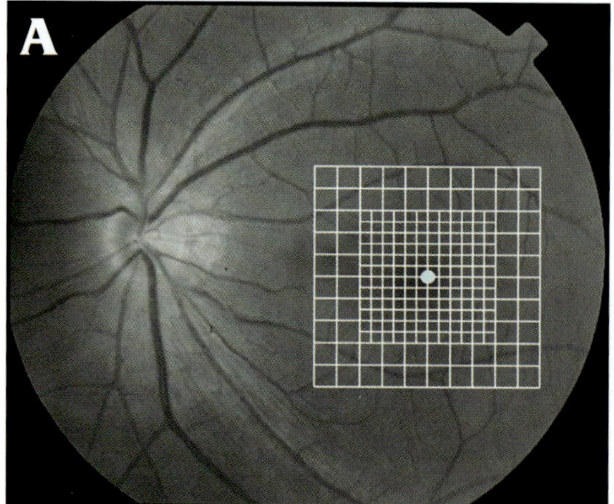

Figure 6-6. OCT image from high-definition line scan.

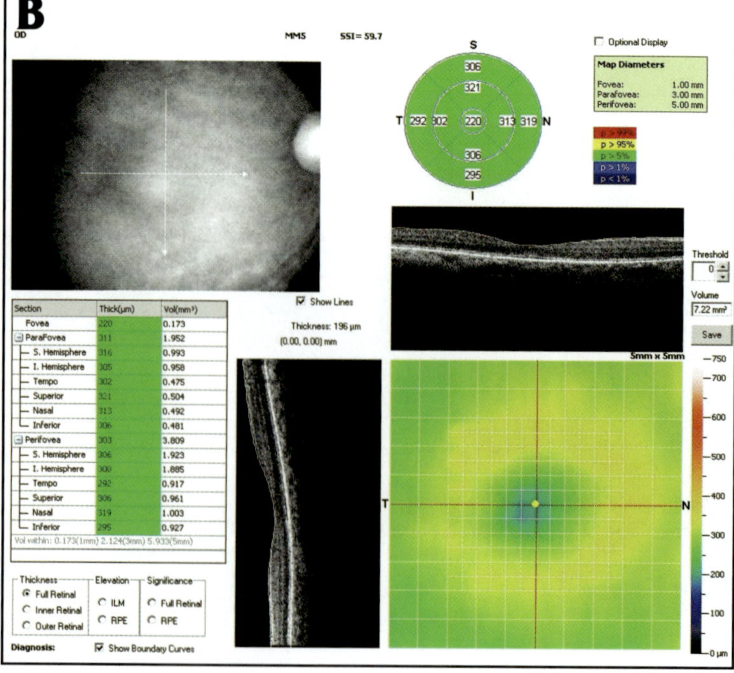

Section	Thick(µm)	Vol(mm³)
Fovea	220	0.173
ParaFovea	311	1.952
— S. Hemisphere	316	0.993
— I. Hemisphere	305	0.958
— Tempo	302	0.475
— Superior	321	0.504
— Nasal	313	0.492
— Inferior	306	0.481
Perifovea	303	3.809
— S. Hemisphere	306	1.923
— I. Hemisphere	300	1.885
— Tempo	292	0.917
— Superior	306	0.961
— Nasal	319	1.003
— Inferior	295	0.927

Vol within: 0.173(1mm) 2.124(3mm) 5.933(5mm)

Figure 6-7. MM5 scan. (A) Illustration of MM5 scan pattern. (B) Analysis of MM5 scan.

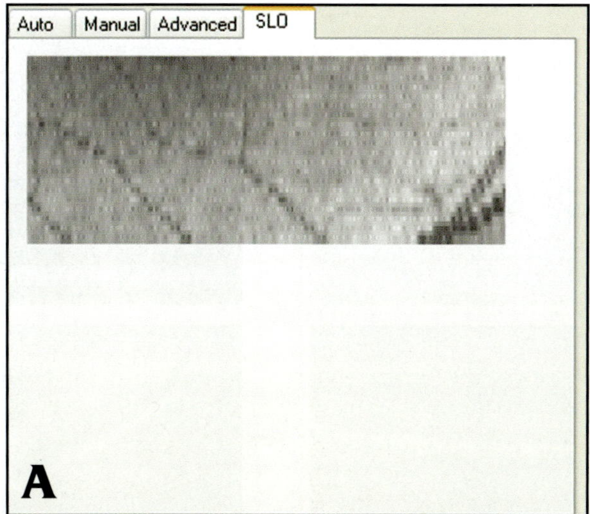

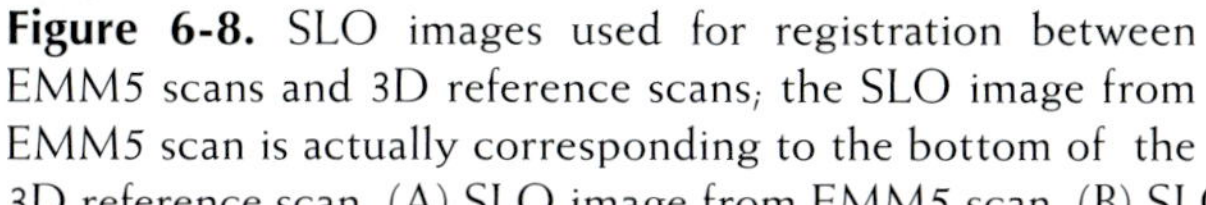

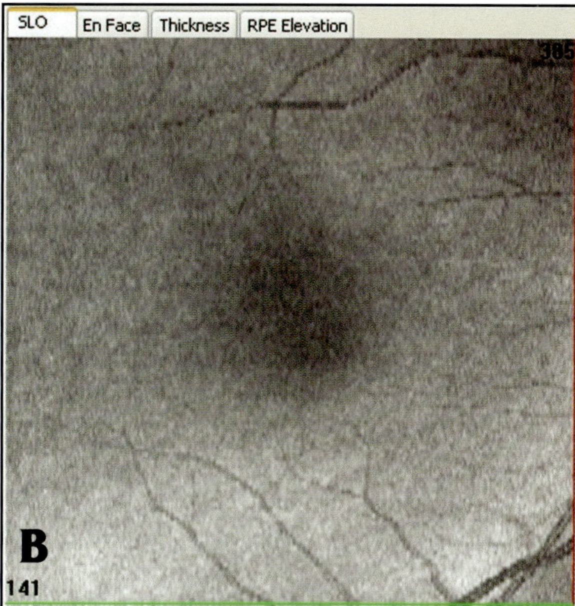

Figure 6-8. SLO images used for registration between EMM5 scans and 3D reference scans; the SLO image from EMM5 scan is actually corresponding to the bottom of the 3D reference scan. (A) SLO image from EMM5 scan. (B) SLO image from 3D reference scan.

choose to display high quality vertical and horizontal B-scans near a pathology of interest any where in the macula. MM5 scan has 19,500 A-scans and takes 0.78 seconds. The scan time is relatively short compared to a dense 3D scan.

Enhanced Macular Map Scan and Three-Dimensional Reference Scan

For patients with low vision, the MM5 scan may be off-center when eyes cannot be fixed at the target well. In order to compare the map with normative database or do progression analysis, it is important to register the MM5 scan to fovea center. RTVue provides an enhanced macular map (EMM5) scan which is the MM5 scan plus an extra raster scan in the inferior macula taken for the purpose of registration with the 3D reference scan. A summation (also known as *en face* projection) image is obtained from the raster scan by summing the OCT signal intensity of each A-scan. The summation image is also called the OCT SLO image. The summation image from the EMM5 scan is registered to the summation image obtained from the baseline 3D reference scan (Figure 6-8). In case the baseline OCT SLO image itself is not centered at the fovea, the user can manually adjust the location of the fovea center.

In order to keep the 5-mm square map around the fovea after the center correction, B-scans are added and it enlarges the scan region of the EMM5 scan to 6 mm. EMM5 takes 1.35 seconds with the extra raster scan.

A 3D reference scan is a special scan designed for registration of the EMM5 scan. It follows a raster pattern and covers a 7-mm square around the fovea. It is composed of 141 B-scans, each B-scan has 385 A-scans. It will create 7-mm square of a OCT SLO image. The 3D reference scan takes 2.4 seconds. It only needs to be done once for each patient at the first (baseline) visit.

Grid Scan

For the cross hair scan, it is difficult to compare 2 visits when the scan center has a small shift due to the eye movement or low vision. RTVue provides the grid scan for a solution. As showed in Figure 6-9, a grid scan has 5 parallel lines in a horizontal direction and 5 lines in a vertical direction. It could be used to register scans from 2 separate visits and extract a matched line scan for comparison.

Grid scans also apply frame averaging to enhance the image. Each line is repeated 5 times. Totally, the grid scan has 50 B-scans and takes 1.8 seconds. The default scan length is 8 mm (adjustable 2 to 10 mm). The default width of the 5 lines is 2 mm (adjustable 2 to 10 mm). The default fixation is fovea center.

Raster Scan

A raster scan is used to give a 3D view of the retina in a short time. Raster scans consist of 17 parallel lines with the same scan length. The lines are evenly distributed in the scan area. The default scan length is 6 mm (adjustable 2 to 10 mm). The default width of the raster scan is 4 mm (adjustable 1 to 6 mm). The user can rotate the lines and set to any angle. A raster scan takes 0.34 seconds.

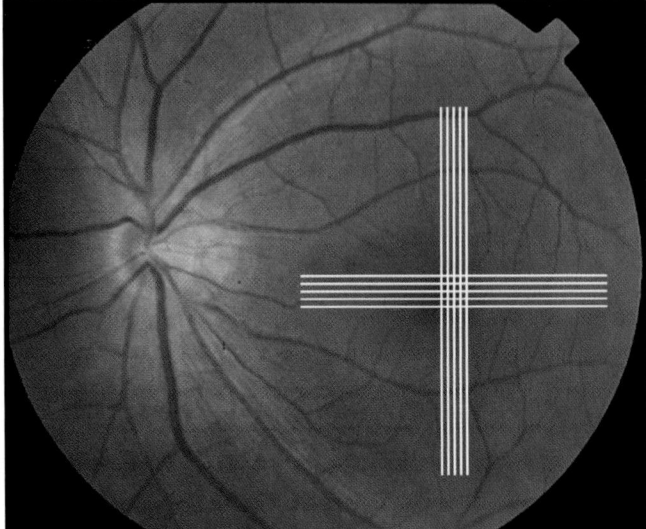

Figure 6-9. Illustration of grid scan pattern.

Radial Slicer Scan and Macular Mapping 6 mm

Radial slicer scan follows a radial pattern. It has the same function as a raster scan, but changes the rotation of the B-scans instead of the shift. The radial slicer scan has 12 radial scans, the user can change the scan length. The default fixation is fovea.

Macular mapping 6 mm (MM6) is a special radial slicer scan pattern with a fixed 6-mm scan length that provides analysis for the retinal thickness map. It accomplishes the same function as the MM5, but with fewer data points. MM6 scan has 12,288 A-scans and takes 0.47 seconds.

Three-Dimensional Macular Scan

A 3D macular scan provides a high-definition OCT image in both horizontal and vertical directions of the macula. The 3D macular scan covers a square region with an adjustable scan length from 3 to 8 mm. The default size is 6 mm. It contains 101 B-scans. Each B-scan has 513 A-scans. The 3D macular scan contains a total of 51,813 A-scans and takes 2.2 seconds. The analysis includes B-scan fly through, 3D display, *en face* summation of intensity in a retinal sub-layer or band, retinal thickness map, and RPE elevation map (Figure 6-10).

GLAUCOMA SCAN PATTERNS

Glaucoma scan patterns are designed for diagnosis and progression analysis of glaucoma and other neuropathy diseases. Usually these scan patterns center at the

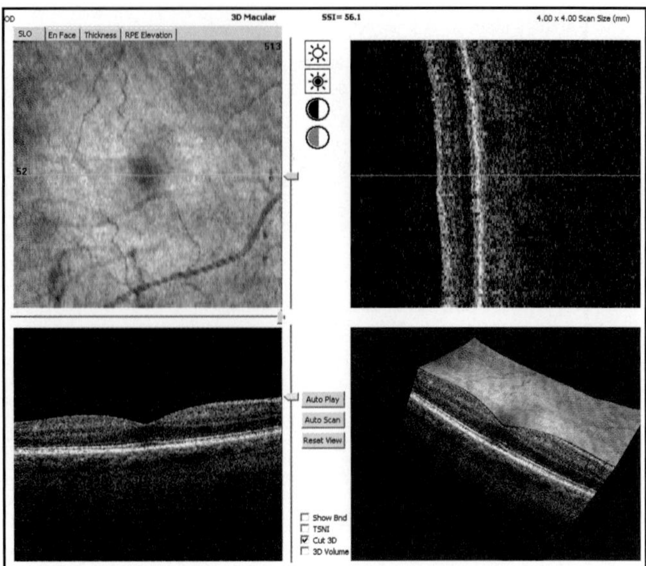

Figure 6-10. Analysis of a 3D macular scan with 4-mm scan length. Top left: SLO-like image from summation of intensity of OCT A-scans. Top right: A vertical section. Bottom left: A horizontal section. Bottom right: 3D display allows user to rotate, enlarge, and cut through 3D data. RTVue also provides *en face* summation and retinal thickness analysis function for 3D macular scan.

optic nerve, except for the ganglion cell complex (GCC) scan. The scan depth is 2.31 mm, which is wider than the retina scans. The default imaging model of the glaucoma scan patterns is vitreoretinal mode. A summary of glaucoma scan patterns is listed in Table 6-2.

Retinal Nerve Fiber Layer 3.45 Scan

The retinal NFL 3.45 scan repeats a *circular* scan with a 3.45-mm diameter for 4 revolutions. It provides retinal NFL thickness averaged from the 4 B-scans. It allows significant loss analysis of retinal NFL thickness profile or averages based on normal reference.

Ganglion Cell Complex Scan

The GCC scan is specially designed to measure the macular GCC, which is composed of the inner 3 layers of the retina (NFL, ganglion cell layer, and inner plexiform layer). Glaucoma selectively damages the GCC, and GCC thickness was found to be sensitive and repeatable for glaucoma diagnosis.[1,2] GCC scan consists of 15 vertical line scans covering a 7-mm square region. GCC scan centers at 1 mm temporal to fovea center for better coverage of the temporal region. GCC scan also includes a horizontal line scan at the middle. GCC scan takes 0.56 seconds. It provides analysis for GCC thickness map, GCC deviation map, and significant loss map based on normal reference (Figure 6-11).

Table 6-2

SUMMARY OF GLAUCOMA SCAN PATTERNS

SCAN PATTERN	TIME (SEC)	# A-SCAN	ADJUSTABILITY	DEFAULT
Retinal NFL 3.45	0.15	4 x 1024	Scan length: fixed	3.45 mm
			Center: fixed	Optic disc
GCC	0.25	1 x 467	Scan length (H/W): fixed	(HXW) 7 x 7 mm
		plus 15 x 400	Angle: fixed	0/90
			Center: 1 mm temporal to fovea	
ONH	0.55	3 x 965	Scan length: fixed	4.9 mm
		plus 3 x 775		
		plus 3 x 587	Center: fixed	Optic disc
		plus 4 x 425		
		plus 12 x 455		
3D Disk	2.2	101 x 513	Scan length (H/W): 3 to 8	(HXW) 6 x 6 mm
			Angle: 0 to 180	0
			Center: fixed	Optic disc

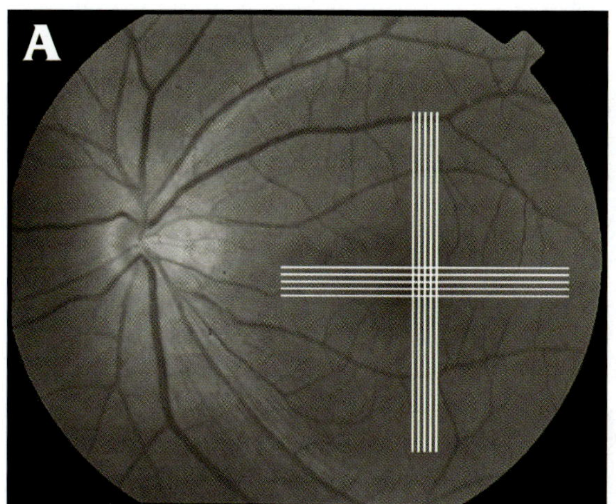

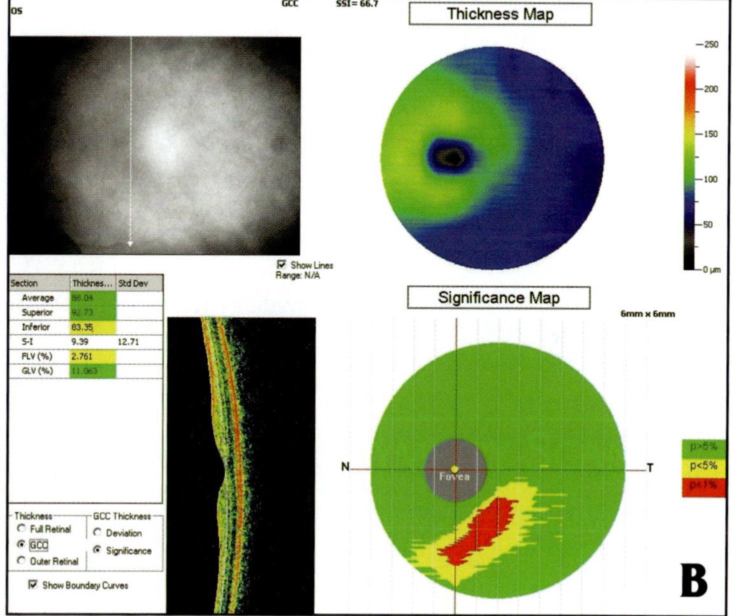

Figure 6-11. GCC scan and analysis. (A) Illustration of GCC scan pattern. (B) GCC analysis of a glaucoma patient.

Optic Nerve Head Scan

The optic nerve head (ONH) scan is designed to include retinal NFL thickness analysis and nerve head shape analysis, both found sensitive in glaucoma diagnosis and progression. It combines 2 different scan patterns (Figure 6-12). First, it scanned 13 circles with diameters of 1.3 to 4.9 mm, which are analyzed to provide the peripapillary nerve fiber thickness map. Then it makes 12 radial scans with a length of 3.7 mm, which are used to calculate the disc margin and nerve head shape parameters. All B-scans are centered at the optic disc. The ONH scan takes 0.55 seconds. As it is difficult to locate the disc center, the disc center is automatically detected and used to recenter the whole scan. For progression analysis purpose, the ONH scan is registered to a baseline OCT SLO image, which is obtained from a 3D disk scan. If a 3D disk scan is not available, the alternative way is to register the OCT to the disc margin drawn on the baseline infrared video image.

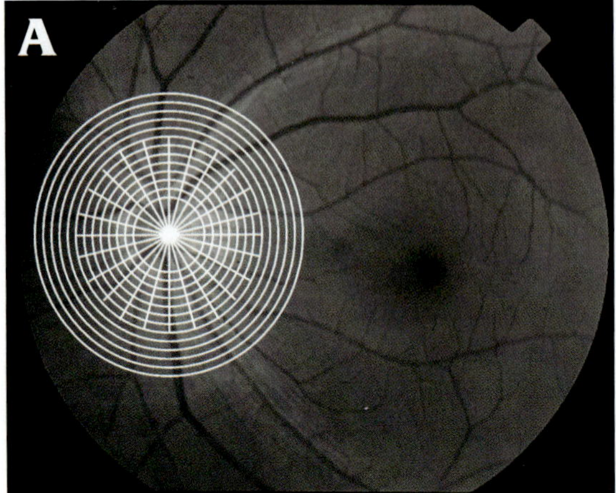

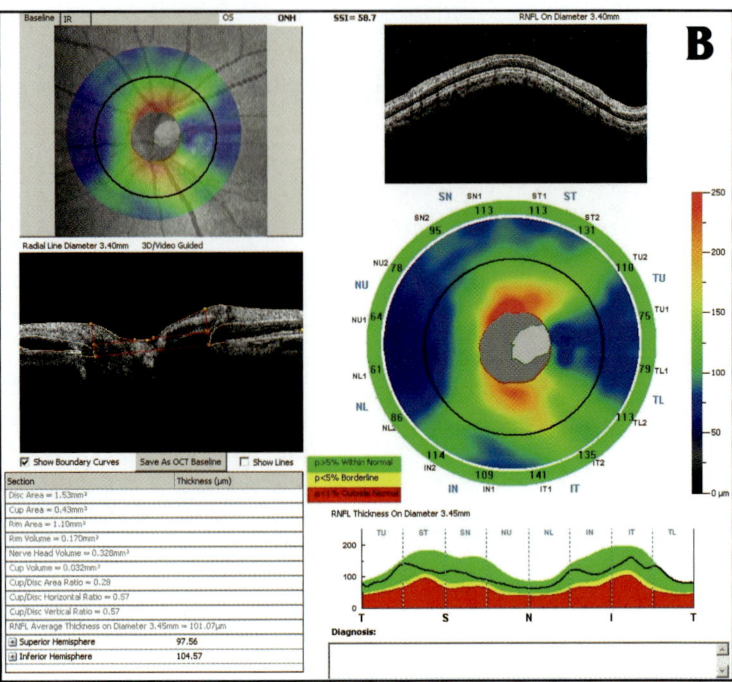

Figure 6-12. ONH scan and analysis. (A) Illustration of ONH scan pattern. (B) Retinal NFL and nerve head shape analysis; Note that the scan is registered to the OCT SLO image, which is shown at the top left of the figure.

Figure 6-13. Analysis of 3D disk scan with 6-mm scan length. Top left: SLO image with detected disc boundary, which will be used as baseline for ONH scan. RTVue allows user to edit the disc margin to match tip points on B-scans. RTVue also provides *en face* summation and retinal thickness analysis function for 3D disk scan.

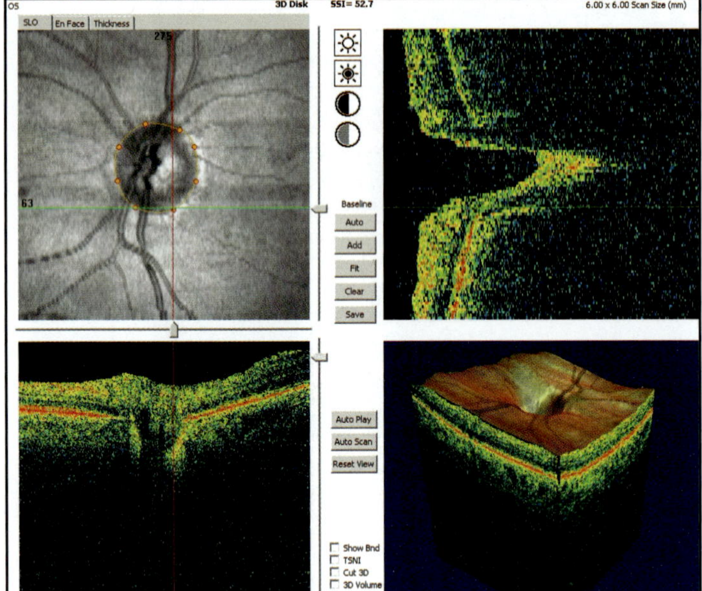

Three-Dimensional Disk Scan

A 3D disk scan (Figure 6-13) provides a high-definition 3D image of the nerve head and peripapillary region. The parameter and analysis is similar to a 3D macular scan, except a 3D disk centers at the ONH. It also provides a baseline for an ONH scan.

REFERENCES

1. Tan O, Chopra V, Lu A, et al. Detection of macular ganglion cell loss in glaucoma by Fourier-domain optical coherence tomography. *Ophthalmology*. In press.
2. Tan O, Li G, Lu AT, et al. Mapping of macular substructures with optical coherence tomography for glaucoma diagnosis. *Ophthalmology*. 2008;115(6):949-56.

7

Posterior Segment Scan Procedures

William D. Dilworth

OVERVIEW AND CONCEPTS

Understanding how glaucoma affects the eye and how it is measured helps when capturing the scans. This will be covered in other chapters and in a dedicated section of this chapter.

The RTVue-100 device is a light-based imaging system that uses a technology referred to by several names:

- Fourier-domain OCT
- Spectral-domain OCT
- High-speed OCT
- High-definition OCT

These are different names for the same thing. Each of these names refers to a different aspect of the exact same technology.

The RTVue-100 uses a coherent band of light-centered at approximately 840 nm (which is in the near-infrared range of the visible spectrum). Because the unit is light-based, the clarity of the media is an important consideration. Anything that blocks light will decrease the quantity of light being reflected from the eye structures and lower the signal-to-noise ratio—in effect weakening the signal. This does not mean that eyes with opacifications cannot be scanned accurately; it just means the structures may be harder to visualize.

The RTVue software is designed to allow for a wide range of scanning situations, from screening through comprehensive study-based examinations. Consideration should be given to the type of scanning that is needed in the practice prior to seeing patients.

SET-UP CONSIDERATIONS

During the initial set-up of the device, there are a few physical environment considerations that should be taken into account.

The uncrating and assembly of the unit should be done only by an Optovue representative.

Power

A grounded power supply is required. The unit may be configured to either 110v or 220v power supply, depending on the country it is being shipped to. In areas where the power is unstable or power loss is possible, it is advisable to install an uninterruptable power supply and line conditioner.

Space

There should be enough room for the operator and patient to maneuver into their scanning positions without difficulties. The area should be free of tripping hazards such as wires and loose carpets. As with all medical measurement devices, movement of the base unit should be kept to a minimum.

Lighting

The room should not be sunlit and should be able to have lighting conditions dimmed if needed. Ideally, the room lights should be controllable from the operator's position via a dimmer switch.

Huang D, Duker JS, Fujimoto JG, Lumbroso B, Schuman JS, Weinreb RN.
Imaging the Eye from Front to Back with RTVue Fourier-Domain Optical Coherence Tomography (pp 61-74).

Stability

The flooring of the room should be relatively solid to minimize vibration and motion effects.

Temperature

Extreme room temperature changes may have an adverse effect on the internal settings. The unit should be kept in a fairly stable environment. The unit's warm-up time may be significantly extended as a result of low overnight room temperatures.

Protocol Considerations

Also during set-up, the practice should decide on the scan protocols to be used. Scan protocols are simply collections of the available scan patterns. These are usually tailored to the practice by the installing representative, but are easily created should the needs of the practice change. The protocols may be set-up to span the range of the glaucoma evaluation the practice will be ordering.

Data Integrity Considerations

The RTVue device is shipped with 2 independent hard drives. These internal drives provide storage for the original and a backup copy of the scan data. In most practices the drive capacity should be adequate for approximately 2 years worth of scanning without the need to archive data. The software allows for an additional redundant backup location, if the user desires. The redundant backup location may be on a network server, a USB hard drive, or an added internal SATA archive drive location. Ideally, this archive location should be on a network server. As long as the archive location remains accessible to the RTVue's computer, it can be retrieved without user intervention.

Training Considerations

Operator training is available in many forms.

- On-site training sessions are available by contacting the Optovue main office.

  45531 Northport Loop West
  Fremont, CA 94538
  510-623-8868
  www.optovue.com/contact#subject_9

- On-line sessions (called webinars) are available by visiting the Optovue Web site. These can be accessed from any computer with an Internet connection and a reservation confirmation.
  www.optovue.com/support/training-schedule

- In-booth sessions are also available at most of the trade shows/conventions. Contact the main office for details on any specific show
  www.optovue.com/news/events

BEGINNING OF THE DAY

Turning on the RTVue-100

The RTVue-100 table power should be turned on at least 20 minutes prior to the first patient being scanned. This warm-up period allows the internal components to reach optimal operating temperature. Scans done before this time may have lower signal-to-noise ratios, resulting in poor scan quality. When the table is powered on, the scan head receives power as indicated by a small blue light on the operator side of the scan head.

The computer may be started at any time after the table has been turned on by pressing the large blue button on the front panel of the computer. The computer is a high-end standard PC with a few modifications. The Windows XP Pro operating system (Microsoft Inc, Redmond, WA) will boot, followed by the RTVue software. During this process the scan head will cycle some of the adjustment motors.

Do not install any third party software on the computer as this may void your warranty.

Cleaning

The main lens of the RTVue should be checked periodically for dust, fingerprints, or smudges. The lens should also be quickly checked if the signal strength is ever less than expected during an exam. This can be done by shining a small penlight directly onto the lens. If the lens is not clean, use simple isopropyl (rubbing) alcohol and a lint-free cloth (such as a microfiber lens cloth) to clean the lens. Avoid using excess fluids that may get into the lens housing or drip into electronic components.

All places and surfaces that the patient comes in contact with should be cleaned between each exam. These are the chin rest, the forehead rest, and any place where the patient touched. All of these parts may be cleaned with simple isopropyl (rubbing) alcohol and a cloth.

The exterior surfaces of the RTVue unit may be cleaned with a mildly damp cloth and any mild cleaning solution. Ensure that no excess fluid is allowed to enter the electronic portions of the device. Do not use any solvents as part of the cleaning procedure.

The monitor should be cleaned with a soft damp (not wet) cloth in a gentle circular fashion to remove fingerprints. Avoid chemicals or abrasives that could scratch or cloud the plastic screen surface and blur the monitor image.

BEGINNING A PATIENT EXAM

Positioning the Patient

- Seat the patient in front of the RTVue. It is recommended to use a sturdy chair with a back and locking (or no) wheels.

- Pull the scan head all the way to the back position before asking the patient to place his or her chin in the chin rest.

- Adjust the table height and chin rest height to be comfortable for the patient. Also, make sure the patient's canthus (outer corner of the eye) is approximately even with the marks on the forehead rest supports. To adjust the table height, use the switch located just under the tabletop on the side of the computer housing. To adjust the chin rest height, use the 2 small white buttons on the operator side of the RTVue scan head.

CLINICAL TIP

Patients may place their hands beside the unit or hold onto the upright sections of the headrest. In cases of patients with tremors, having the patient hold the uprights often allows them to remain still during the scan process.

GETTING FAMILIAR WITH THE SCREENS

Home Screen

The home screen (Figure 7-1) is a cover page that is used to protect patient information. It is completely HIPAA compliant and should be left on the screen whenever the RTVue is unattended.

This page can be called up from any screen and at any time by simply selecting the Home menu item located at the top of the screen.

Patient Tab Screen

Patient Information

There are 3 major screen modes in the RTVue device: Patient Information (Figure 7-2), Exam, and Analysis. These tabs are visited, in order, for routine scanning in clinical settings.

To search for an existing patient, select All Patients from the Patient List box, type some (or all) of the known information into the search box, and click on what you want the software to search.

- Press [Last Name] to search only the last name fields for the text entered.

- Press [First Name] to search only the first name fields in the patient database.

- Press [Name] to search anywhere in any of the name fields.

- You may also search based on the patient's EMR ID (which stands for electronic medical record identification) by entering the information in the search field and pressing the [EMR ID] button.

A short list of patients that match the search criteria will be displayed. Note that the information typed into the box may appear anywhere in the field, not just at the beginning.

You may also manually scroll down the list of patients and select the patient when the name is shown. The list may be sorted alphabetically by last name or in the order they were last scanned.

Decision point:

- If you find the patient name you are looking for, verify the date of birth and select the patient from this list and click on [New Visit] to create a session for today. You can skip over the "Adding a Patient" step below and go to the "Visit Information" section.

- If the name is not present, the patient does not exist in the database. You will need to click on [New Patient] to add this patient. A new visit will automatically be created when a new patient is added.

RECOMMENDED TECHNIQUE

Although the RTVue unit does allow visits to be moved if they have been captured under the wrong or duplicate patient, every effort should be made to capture the exams accurately under the correct patient. The best technique for avoiding this is to always search for the patient from the existing database.

Adding a Patient

If you are adding a new patient, you will need to provide some minimal patient information, as shown below:

- Last name
- First name

Figure 7-1. Home screen.

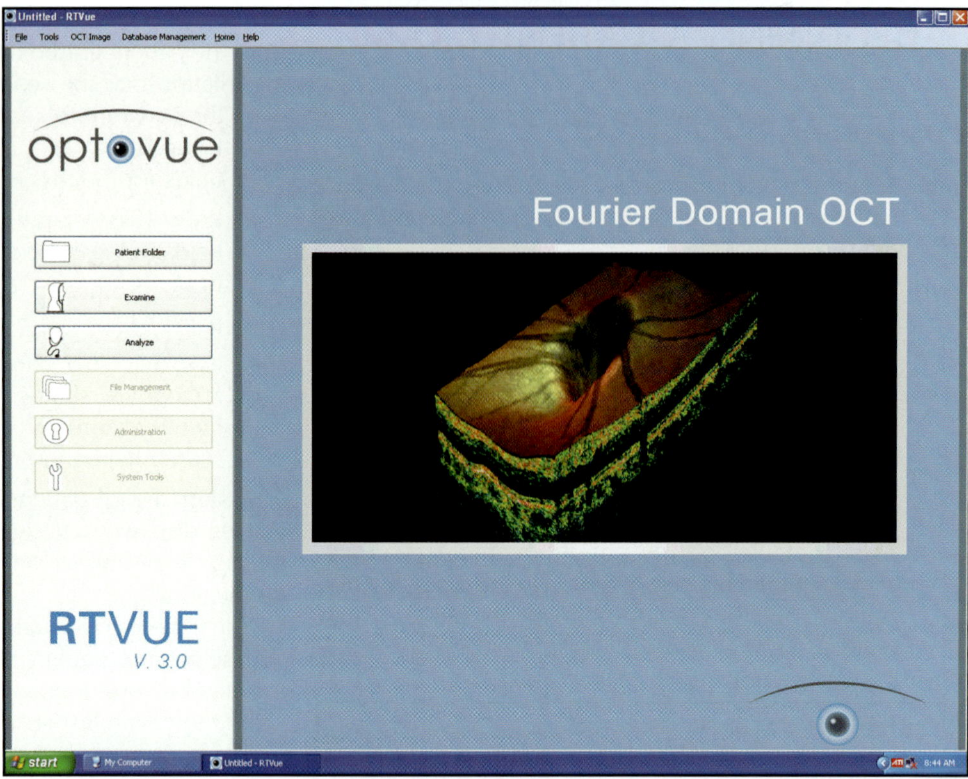

Figure 7-2. Patient information screen.

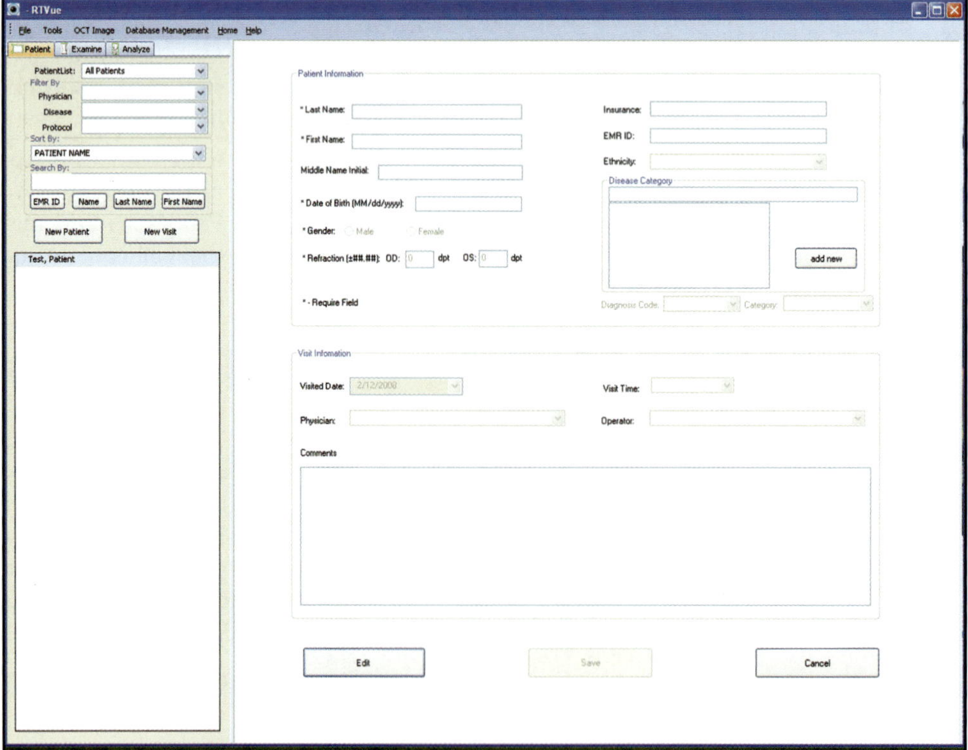

- Middle name (optional): It is a good practice to standardize the way a patient name is used in the practice. In cases where punctuation exists in the name, most practices elect to not enter it. Many practices also enter hyphenated names as a single last name. The use of upper and lower case helps to make the name more readable.

- Date of birth (selectable format, see User's Manual for formatting options): Patients over 18 years of age will have age-based normative data displayed.

- Gender: The male or female option must be selected. This has no effect on the normative data but is useful in helping to identify the patient.

- Ethnicity (optional, but strongly recommended): The normative database includes ethnicity as one of the factors. Specifying an ethnicity at this point allows for a narrower range of normal and allows the analysis to be more sensitive to variations from normal. If the patient is of mixed ethnicity, the Other/All category should be selected.

- Disease category (optional): Entering categories in this field allows for quick retrieval of similar patients for comparison or locating patients based on incomplete information

- Refraction (defaults to plano): The RTVue, as part of the automatic processing, determines the best refraction to use for the scan process. The refraction setting does not need to be set on this screen if the patient is between -10.00 and +10.00 D. If they are outside this range, a rough spherical refraction should be entered.

Visit Information

Whether the patient is new or existing, you will now be able to enter some optional visit information. This optional information will help you identify whom the patient is seeing and who performed a specific scan:

- Physician name: If the physician is not listed on the pull down menu, simply select the Add Physician item from the list. A quick dialog box will allow you to enter information to create a new physician on this list.

- Operator name: Similar to the physician list, operators may be added by selecting the Add Operator item from the list.

- Comments

To move on to the next phase, simply press the [Save & Exam] button, near the bottom of the screen.

Data Check

If any required information is missing or incorrectly formatted, you will be given an error message that will detail the problem. Click OK to clear the error and correct the information and press [Save & Exam] again.

Tips to make finding existing patients easier:
1. Be consistent in your naming practices within the office.

2. Use the disease categories to label the patients.

    a. New disease categories can be added easily by simply clicking on the [New] button in the Disease Category section. Type the new category into the text box and select OK. The new category will be added to the alphabetically sorted list.

    b. Any existing category can be selected by clicking on the checkbox next to the name of the disease category (note: do not just select the name of the disease, place a check in the box).

    c. Multiple categories may be selected for the same patient.

3. Use the filters to search and select patients by Physician, Disease, or Scan protocols.

    a. Use the pull-down menus to select any of these and show only the patients that match the items shown in the 3 filter sections.

    b. To clear the filter, select the blank line at the top of each of the pull-down menus.

4. Search by EMR numbers, since these are unique to each patient.

5. Use the Comments field in the Visit section to document operator comments. These comments are not printed on any reports. Therefore they are an excellent place to document patient specific notes, ie, "patient blinks a lot and has dry eyes, use artificial tears prior to scanning."

EDITING A PATIENT

Should you find that you need to edit any part of a patient's information, first select the patient from the patient list and then click the [Edit] button. This will allow you to change or add information. Once you are finished, click the [Save] button to save changes.

Exam Tab

General Scan Information and Set-Up

The second phase of normal scanning is the exam mode (Figure 7-3), where the scans are selected and preformed. Systems that are review stations, or have calibration settings that are not correctly synchronized,

Figure 7-3. Exam tab screen.

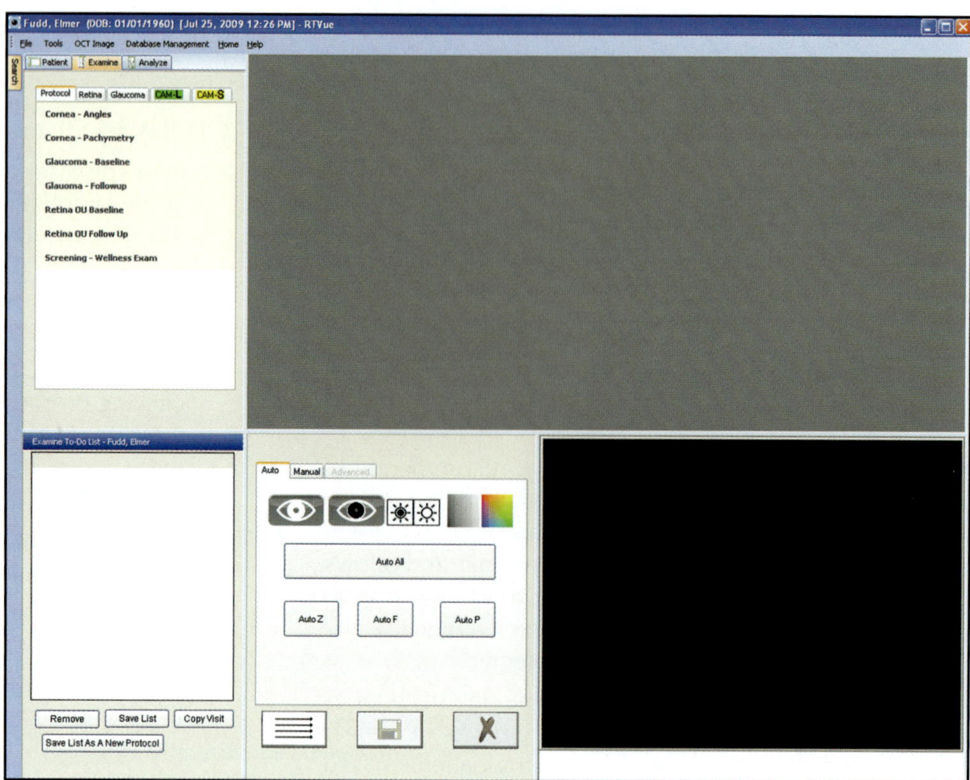

will not be shown in the exam tab. Refer to the User's Manual to see the correct calibration synchronization procedure.

Protocols

Protocols are set up in advance, usually as part of the machine set-up process. The RTVue has many scan pattern options available, many more than are required for most scanning needs. We use predefined groups to automatically select which scans are commonly requested. These groupings are called protocols. Each protocol contains a selection of scan patterns that are disease- or area-specific. These can be modified easily, as described in the User's Manual.

To select a protocol, simply select the Protocol tab and click on the protocol item from the list. The protocol will appear in the To Do list below. A typical protocol selection list and chosen selection may look like Figure 7-4.

Individual Scans

In most cases, the protocol scans will be the only scans required to obtain the information required for diagnosis. However, if the operator should decide he or she needs to add an additional scan to highlight a pathology for the physician, he or she can select from the lists of individual scans shown on the Retina, Glaucoma, or CAM tabs (if the unit is equipped with the CAM module). These are selected in the same manner as the protocol: select the tab and select the scan from the list (Table 7-1). By default, both eyes are selected for individual scans, but either eye may be unchecked prior to selecting the scan.

The To Do List

This is the list of all the scans you have selected to do. Typically in a clinical setting, this will do 3 or less scans per eye for new patients and only 1 or 2 for return patients. The RTVue will cue up each of the scans on the list, in order. Selecting any of the scans on the To Do list will make it the next scan to be taken.

Start the scan by clicking on the icon with the 4 arrows (see this button in Figure 7-3). The next scan for the current eye will be started. If no scan is selected in the To Do list, the top-most scan will be started. Continuing to scan the same eye after all scans have been completed will cause an error message to pop up on the screen.

Starting the scan will activate the infrared viewer screen in the lower right corner of the screen. Holding the joystick with one hand and placing the other hand on the base of the unit is the preferred manner of moving the scan head. While the scan head is pulled all the way back toward the operator, align the patient's pupil to the center of the scan. **Only then,** start moving toward the patient. The joystick controls are comparable to the controls for a slit lamp.

- Moving the joystick to the left and right adjusts the scan head to the left and right.
- Moving the joystick toward the patient moves the scan head toward the patient.

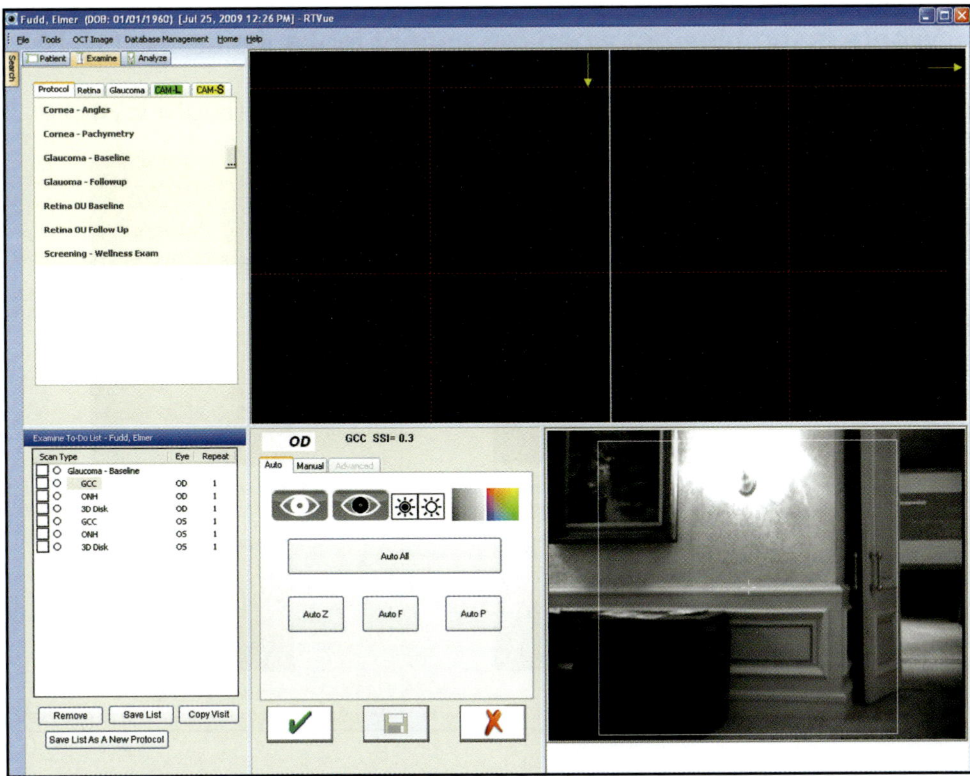

Figure 7-4. Capture screen.

<table>
<tr><td colspan="4" align="right">**Table 7-1**</td></tr>
<tr><td colspan="4" align="center">## SCAN PATTERNS USED FOR GLAUCOMA ANALYSIS</td></tr>
</table>

SCAN NAME	ICON	DESCRIPTION	DEFAULT SIZE
GCC (MM7)		1 x 467 – 7 mm horizontal line with 15 x 400 – 7 mm vertical lines at 0.5 mm interval, centered 1 mm temporal to fovea	6 mm x 6 mm
ONH (NHM4)		13 concentric rings (1.3 to 4.9 mm diameter) All centered on disk 3 x 965/ring (4.9, 4.6,4.3) 3 x 775/ring (4.0, 3.7, 3.4) 3 x 587 (3.1, 2.8, 2.5) 4 x 425/ring (2.2, 1.9,1.6, 1.3) 12 x 452/radial 3.4 mm lines	4.9 mm
3D Disk		101 x 513 frames (51,813 data points)	6 mm x 6 mm
Retinal NFL 3.45		4 x 3.45 mm diameter circular scans centered over disk	3.45 mm

- Twisting the joystick moves the scan head up and down.

Instruct the patient to look into the main lens and to focus on the blue fixation light inside the machine. The blue fixation light will be in different locations inside the device depending on the type of scan selected. The patient will also see the background red lighting and the actual scan pattern being taken as a slightly brighter red light (Figure 7-5).

Glide the joystick toward the patient's pupil and move through the pupil until the image of the retina fills the screen. Once the retina is visible, use very small movements with the joystick to reduce any lens glare artifacts and maximize the retinal image. It is very important to

Figure 7-5. Fixation.

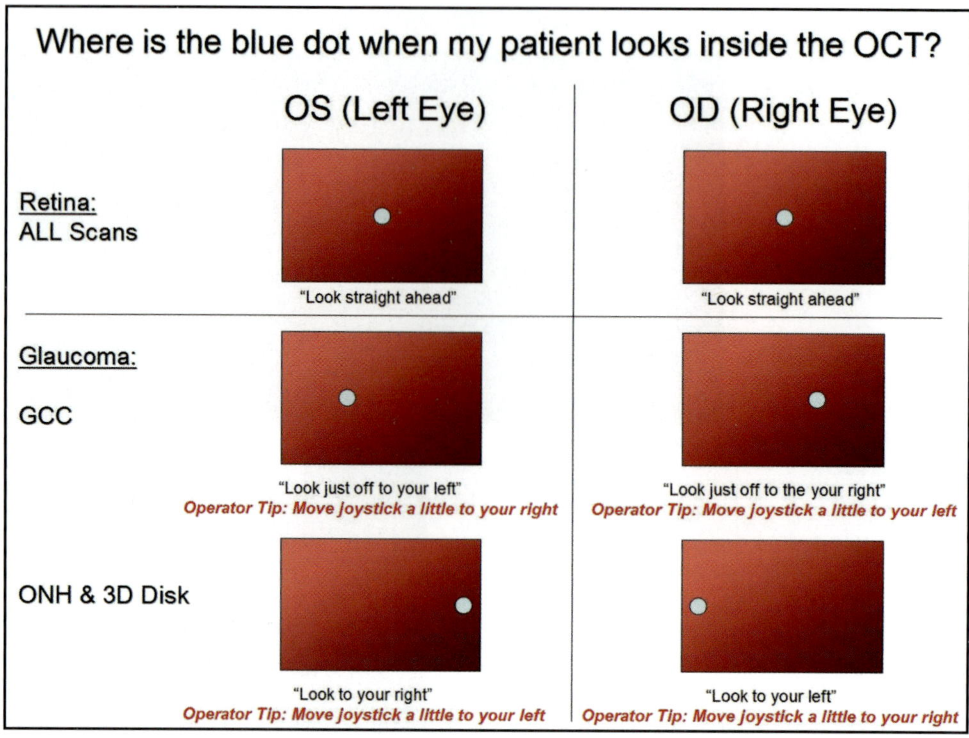

ignore the OCT window at this point and concentrate only on the infrared viewer image.

Once the infrared image has been maximized, click on the [Auto All] button. A good technique is to hold the side of the scan head housing with your left hand while this process is being run. The RTVue runs through 3 automatic routines to optimize the strength and location of the OCT signal.

The [Auto All] function consists of 3 automatic optimization routines:

- [Auto Z] finds the correct height within the OCT windows to display the scan
- [Auto P] finds the best Polarization setting
- [Auto F] finds the best Focus setting

Any of these may be run independently or together in the [Auto All] sequence.

CLINICAL TIP

The Auto All should only be used once for the first scan of each eye during the scanning session. Excessive use of the function causes unnecessary wear on the unit and slows the scanning process.

The OCT signal should now be evaluated for position and clarity. The optimum range is highlighted in all the OCT windows between the red horizontal lines.

However, it is critical that no scan be higher than the top of the OCT window. In order to measure the thickness of the retina, the OCT signal must show the top of the retina. The height of the OCT in the window may be controlled by a single left click and then using the scroll wheel on the mouse to adjust the height of the OCT signal within the top window.

If the edges or a section of the OCT scan are dimmer, it may indicate the light is being blocked by the edge of the iris, a vitreous floater, or cloudiness of the lens. If you see this, try moving the joystick very slightly to the right or left, up or down. Do not move the joystick toward or away from the patient. Often the difference between an adequate scan and an excellent one is just a 1/16th of an inch.

There are also vertical center lines shown on the OCT windows. These are displayed to help ensure that the midpoint of what you are scanning is in the center of your scan line. Patients may have eccentric fixation and this will help line up the scans correctly.

The strength of the OCT signal is shown by the height of the color bar on the far right edge of the OCT window. Strong signals will be shown as green and tall, poor signals will be yellow/red and short. The actual strength number is displayed as the SSI number just below the OCT window, but the color code on the side bar is easier to see during a live scan.

When performing the optic nerve head (ONH), or 3D disk scan, the actual scan pattern in the live infrared window will have to be centered on the optic nerve by dragging and dropping the scan pattern or by double clicking on the center of the optic nerve. Double clicking will cause the scan pattern to snap to the center of the optic nerve.

When the scan quality is good, the location correct, and the height correct, click on the check box or click on the joystick button. The green check box will only show when the scanner has acquired enough scans for the operator to capture an image (this check box can be seen in Figure 7-4).

It is important to remember that when you hit the capture button, you are not starting the scan, but ending it. This means that you can watch the patient and when everything is right, click the button.

The scan will show miniature renderings of each of the B-scans captured in the current scan pattern. This may be as few as 4 or as many as 146 depending on the scan taken; these will be arranged on the far left of the screen for review. Any of the miniatures, where the scan strength falls below 30, will have the colors inverted for easy identification. If the operator wants to review any scan prior to saving, they can simply click on that scan's miniature and it will be displayed in the large OCT window.

On some scan patterns (ie, ganglion cell complex [GCC] and ONH) a pair of green arrows may be seen on either side of the [Save] button (this can be seen in Figure 7-6). If these arrows are shown, there may be an alternate set of images available. This is useful when one or several of the individual scans are of poor quality or contains a blink.

Examples of good scans can be seen in Figures 7-6 through 7-8. Figures 7-9 and 7-10 are examples of bad scans and why they are considered bad.

If the scan is considered acceptable, it can be saved and the next scan pattern can be started. If the scan is not acceptable, pressing the [Scan] button will let the user re-acquire the same scan pattern. If the [Scan] button is pressed when no more scans are available on the To Do list, an error message will be shown. Just click [OK] to clear the error.

IMPORTANT NOTE

The operator has the advantage of the "live view." He or she will often see small abnormalities in the retina that may not be shown in a simple line scan. If the operator notes any retinal structure variations, he or she should evaluate his or her results to ensure these are clearly shown to the physician. If they are not, then the operator should add any additional scans needed to show the abnormality.

When all the scans have been captured, the user is ready to proceed to the Analysis Mode to print out (or electronically capture) the results.

Analysis Mode

The final section of the scanning process is analysis of the scans and printing out the results (Figure 7-11). Most physicians will have a preferred set of printouts based on each protocol used.

It is the operator's role to present the physician with an accurate visualization of pathologies to aid in the diagnostic process. To properly accomplish this, the operator should strive to advance his or her own knowledge of the retinal anatomy and the way the various diseases appear on the OCT scans. Operators who strive for better education and continuing education credits in workshops will recognize abnormal retina appearance quickly and anticipate the views that will best assist the physician in the diagnostic process. If the operator observes abnormal retina appearance, he or she should actively seek the physician's permission to perform additional scans in these areas. Often this takes the form of a standing order from the physician.

With this in mind, the operator should discuss, with the physician, the way he or she would like to be presented with data. The printouts suggested in this chapter are a guideline and not necessarily your doctor's preference.

Significance Maps—Deviations From Normal

When a glaucoma patient is first seen by the office, the doctor must assess if he or she has any glaucomatous damage to the retina. This can be measured in 3 locations: the ONH, the retinal NFL circle, and the GCC. In order to quickly assess each of these areas, the RTVue utilizes a very large normative database that allows comparison of the patient's measurements to the range of measurements obtained from normal eyes.

However, there are several factors that can affect what is considered normal for any given patient, including their ethnic heritage, their age, and their disk size. All of these are factored into the expected (statistically likely) normal values used in the normative database. A simple stop-light color coding system is used to report the patient's findings compared to normal: green, yellow, and red. The significance maps use these 3 colors to show if and where the patient may have damage to his or her retinal structures.

IMPORTANT TIP

The RTVue uses both thickness and significance color mapping. The colors shown in thickness maps always have a micron scale located near the map. These thickness maps are **not** based on the red, yellow, and green significance map scale.

Figure 7-6. Good GCC scan example.

- Good SSI number (above 60 in a patient with clear media) as shown by the number under the OCT image and the green bar beside the OCT image
- No inverted color miniatures (all scans above SSI 30)
- Scan is well positioned inside each of the windows
- OCT image is strong across entire image (no weak spots)
- Macula seen in the infrared window

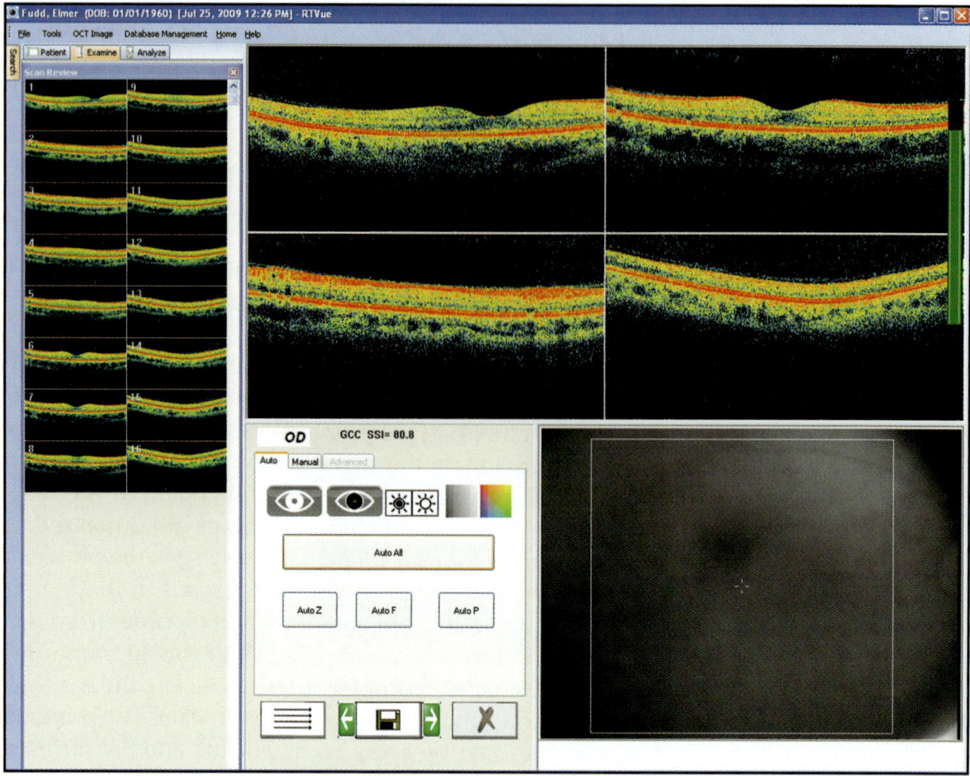

Figure 7-7. Good ONH scan example.

- Good SSI (above 60 in patients with clear media)
- No inverted color miniatures on left
- Scans are well positioned in the OCT windows
- Nerve head is well centered in the infrared window
- Nerve head is well centered in the SLO window
- No shears (patient movement shown as breaks in the blood vessels) in the SLO window

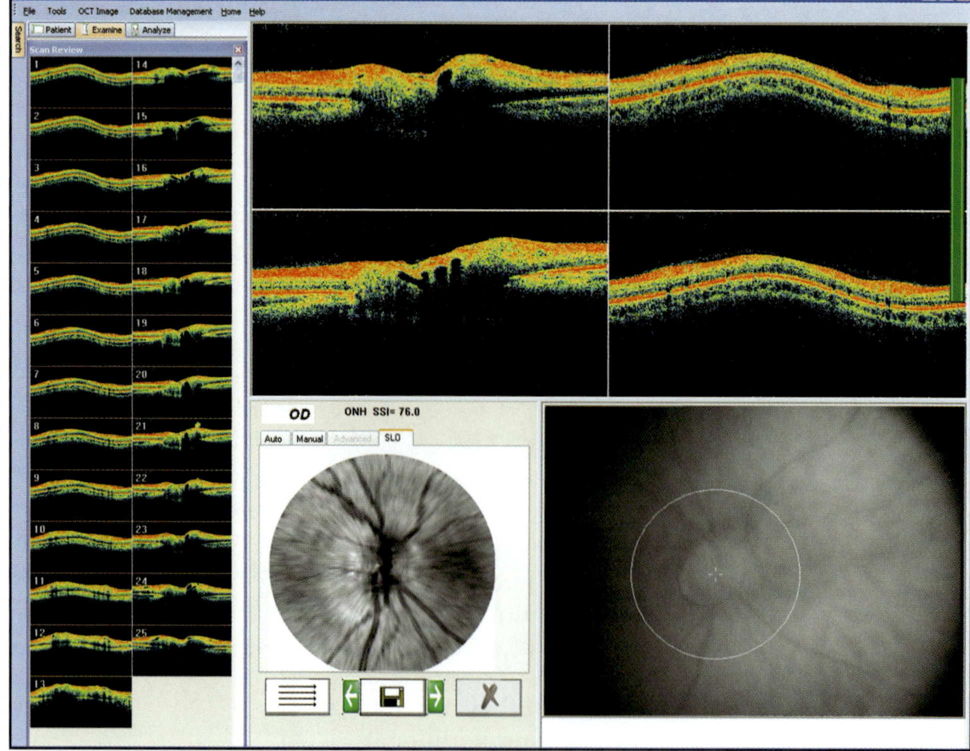

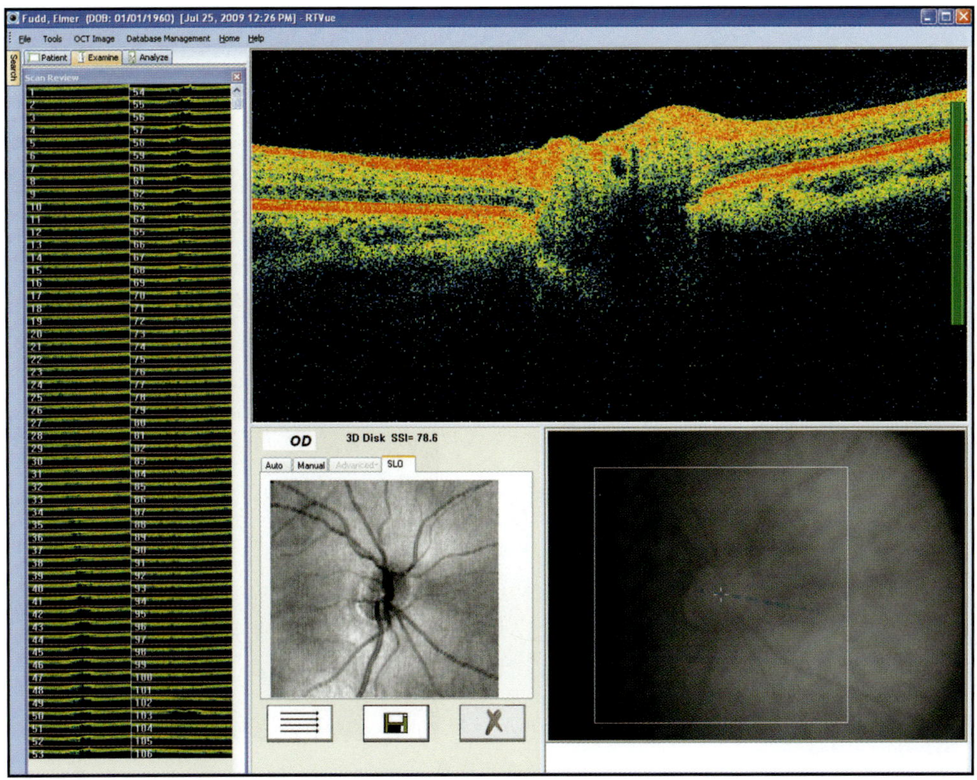

Figure 7-8. Good 3D disk scan example.

- Good SSI (60 or above in patients with clear media)
- No movement shears
- No inverted scan miniatures
- Well-centered optic nerve in infrared and SLO windows

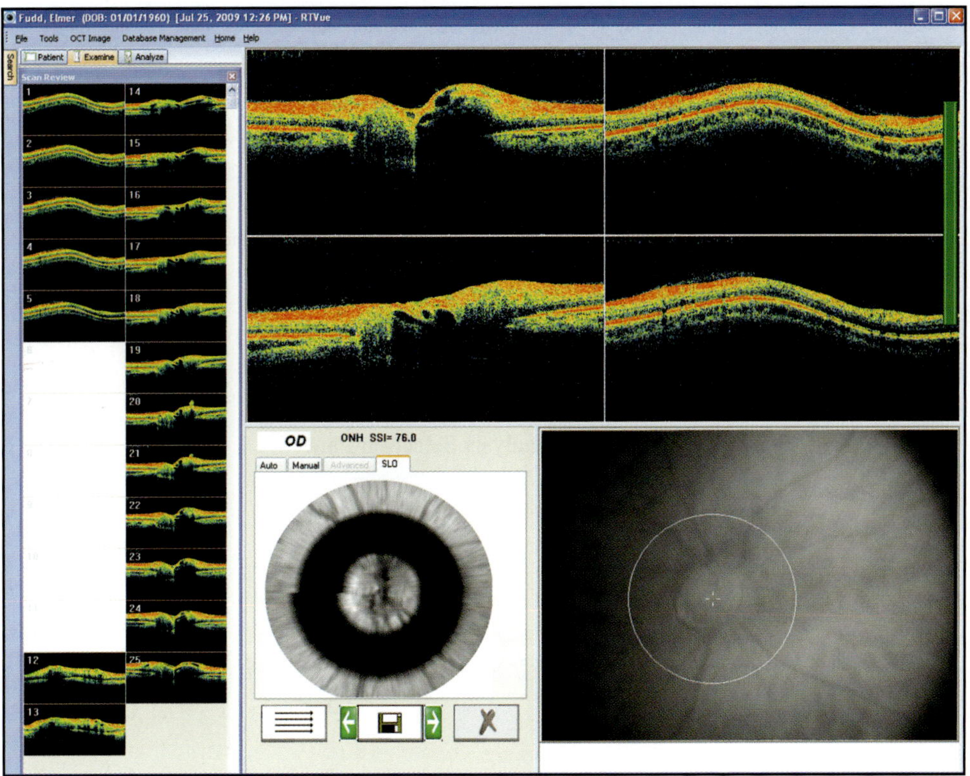

Figure 7-9. Bad ONH scan example. This patient blinked just prior to the scan being captured and several frames are inverted on the left miniatures. The missing data, in this case, show on the SLO as a black ring. The scan is otherwise good, so the user should check to see if there is an alternate set captured that may not have a blink (use the green arrows).

Figure 7-10. Bad 3D disk scan example. In this capture, the signal strength is good, but there are other problems. There are missing frames as shown by the white miniatures and a black band near the top of the SLO image. There are several areas where the SLO image is sheared (blood vessels are not shown as continuous, but rather jump to the side). This indicates patient fixation loss during the 2 seconds prior to ending the capture. Also, the scan was taken too high and the top of the retina was cut off. This scan should be re-acquired.

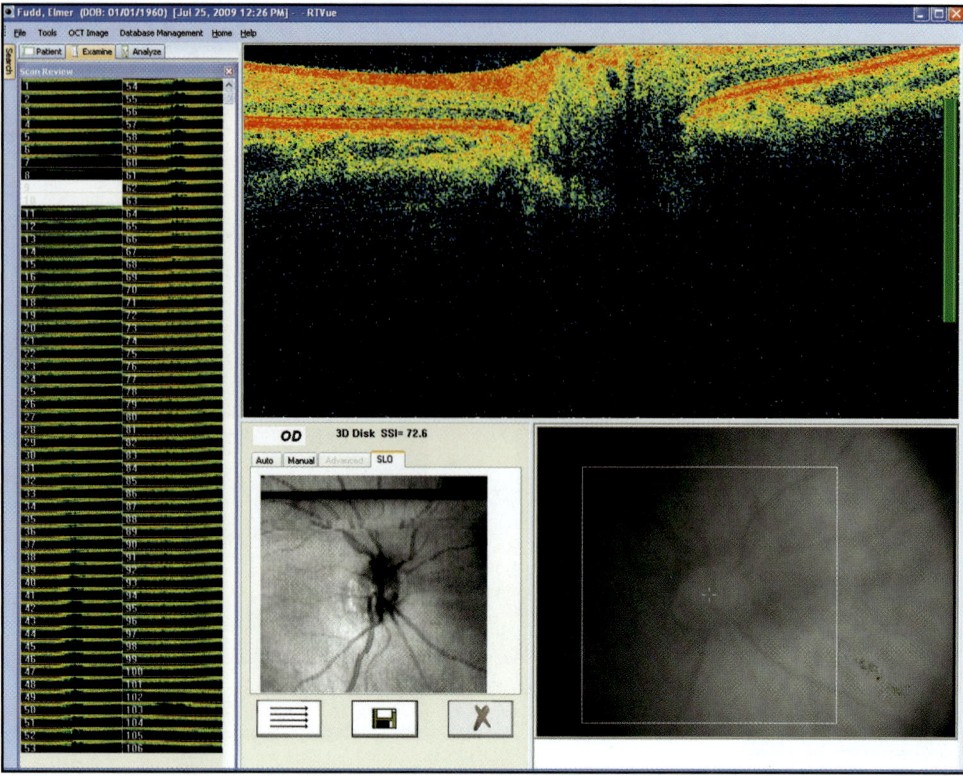

Figure 7-11. Analysis mode screen.

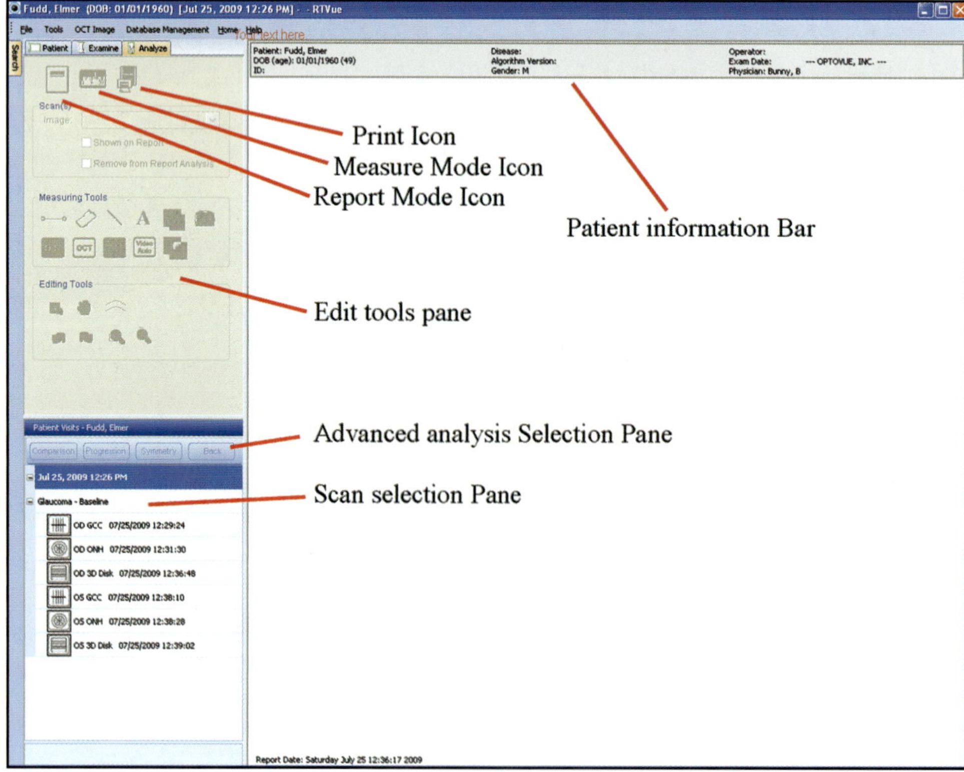

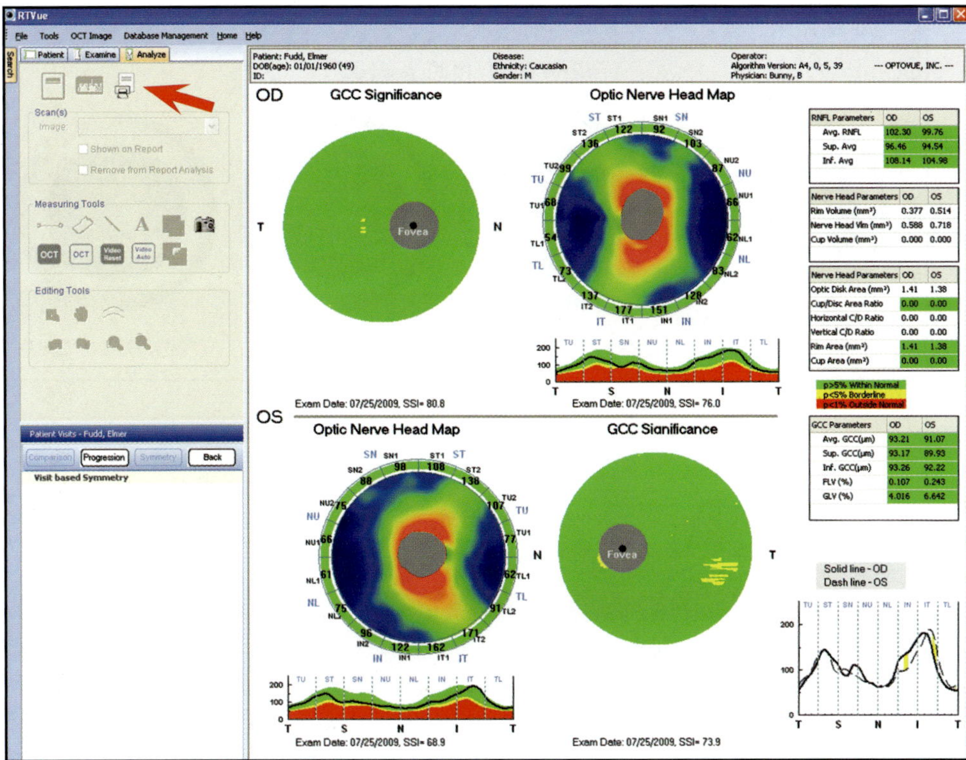

Figure 7-12. Symmetry report.

Symmetry Report—New Patient

On the first visit, the patient is compared to normal patients of similar characteristics. This is best shown using the symmetry report (Figure 7-12). To obtain this report in the Analysis tab, select any GCC or ONH scan. When it is displayed, select the [Symmetry] button above the scan list. The report will process and display.

The report will show both the GCC and the ONH scans for this patient. If more than one of any of these scans has been performed during this visit session, the average of each of these scan types will be displayed. (The chapter on analysis will cover in depth what can be determined from this printout.)

When this report is shown, simply press the [Print] button (red arrow, Figure 7-12). This printout is the primary first visit results for the scan session. To return to the list, just press the [Back] button shown above the scan list.

Progression Report—Optic Nerve Head

On a follow-up visit, the doctor has more information that can be used to assess the damage being done to the retina. In addition to the normative database, he or she also has the previous records of the patient. While the normative database can tell the doctor if the patient is likely to be normal, the progression reports can tell if the patient has gotten any worse (Figure 7-13).

To get a progression report, select either the ONH or GCC scan for the visit. The scan will process and display. Then press the [Progression] button above the scan list. If there are scans from previous visits, the progression report will process.

The report can show the changes in up to 4 visits. By default, the oldest and 3 most recent visits are selected for this report. However, if more than 4 visits are available, the user can select which 4 they would like to see by checking and unchecking the visits in the scan list section of the screen and clicking on the [Progression] button again.

On a follow-up visit, the ONH progression report should be printed for each eye.

Progression Report—Ganglion Cell Complex

Similar to the ONH, the GCC progression report can show up to 4 visits (that can be modified if more than 4 exist) (Figure 7-14). On follow-up visits, the progression report for the GCC should also be printed for each eye.

ADDITIONAL INFORMATION

Due to the limitations of this publication, not all functions or scans have been discussed. As the operators grow in abilities and proficiencies, they will be able to grow their skill set by reading the RTVue-100 manual and the Optovue Web site.

The Web site offers clinical cases, Web-based training schedules, and is a wealth of information. Check out www.optovue.com or contact the office at 510-623-8868.

Figure 7-13. ONH progression report.

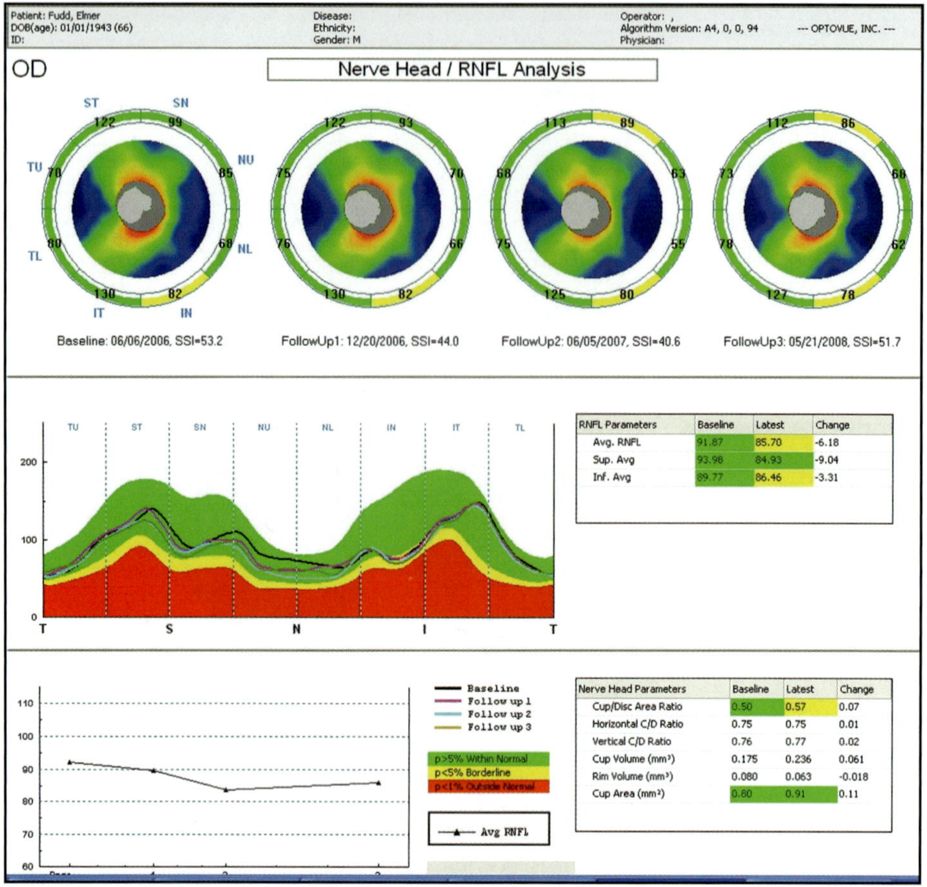

Figure 7-14. GCC progression report.

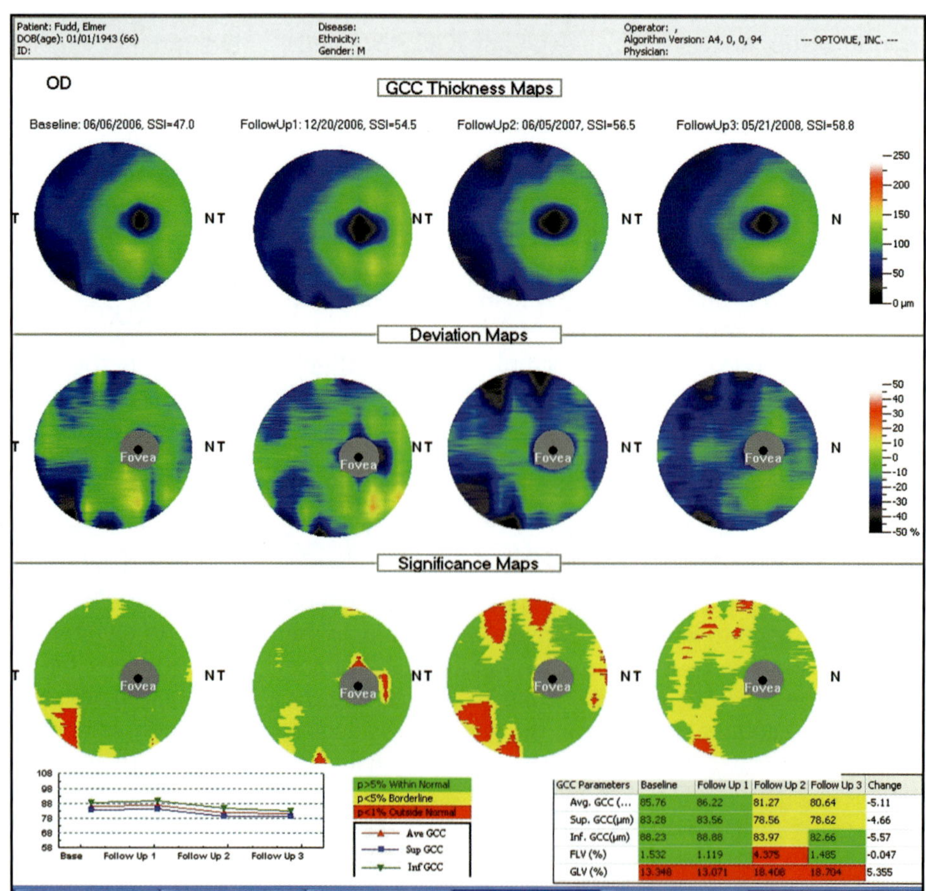

8 | Interpretation of Retinal Images

Andre J. Witkin, MD and Jay S. Duker, MD

OCT allows retinal imaging and calculation of retinal and intraretinal thicknesses with unparalleled precision. Individual high-resolution 2D OCT images analogous to ultrasound B-scans may be acquired, and computer algorithms may be used to extrapolate retinal and intraretinal thickness data from multiple OCT images. Individual high-resolution OCT images allow visualization of intraretinal microstructural details. Computer algorithms extract thickness data from the OCT images based on differences in OCT signal between layers of the vitreous, retina, and choroid in order to provide thickness and volumetric data. Different imaging protocols provide different types of unique information. The interpretation of retinal OCT data involves careful examination of certain aspects of both individual images and thickness calculations. Finally, thickness calculation software may fail due to poor signal quality, disturbances in the normal retinal anatomy, or patient motion and are seen as artifacts or segmentation errors on retinal thickness calculations. It is important for the clinician or researcher to recognize these artifacts and errors in order not to misinterpret OCT findings. This chapter will discuss these various aspects of OCT image interpretation using the RTVue instrument.

INTERPRETATION OF OPTICAL COHERENCE TOMOGRAPHY IMAGES

The relative transparency of the retina allows its imaging via OCT with great detail. Initial studies using postmortem eyes showed good correlation between OCT images and histologic sections. Subsequent *in vivo* and *in vitro* analysis demonstrated the ability of OCT to image the microstructure of the retina.[1-6] The vitreoretinal interface is evident where the minimally reflective vitreous meets the highly reflective internal limiting membrane. General anatomic features of the retinal surface, such as the foveal pit, can then easily be found (Figure 8-1). This highly contrasting interface also lends itself to automatic delineation via computer software.

Individual retinal layers can be visualized using OCT. Histopathologic and clinical correlation to OCT images has helped to interpret which OCT signals represent which corresponding retinal features.[1-7] Increased axial resolution, higher pixel and A-scan density, and image averaging available with the RTVue increases image quality compared to previous time-domain OCT instruments, and allows for clearer separation of individual retinal layers. The nerve fiber layer ([NFL] retinal NFL, inner plexiform layer [IPL], and outer plexiform layer [OPL]) are more highly reflective on OCT, while the cellular layers (ganglion cell layer [GCL], inner nuclear layer [INL], and outer nuclear layer [ONL]) are less highly reflective (see Figure 8-1). The external limiting membrane (ELM) is visible as a moderately reflective line toward the outer part of the outer nuclear layer. The outer aspect of the neurosensory retina is bounded in OCT images by 3 closely spaced reflective layers. The inner of these 3 layers represents part of the photoreceptors, likely the junction between the inner segment (IS) and the outer segment (OS), and the outermost reflective layer represents the retinal pigment epithelium (RPE).[1-3,7] The high contrast between the IS/OS junction and the outer nuclear layer in OCT images provides a

Huang D, Duker JS, Fujimoto JG, Lumbroso B, Schuman JS, Weinreb RN.
Imaging the Eye from Front to Back with RTVue Fourier-Domain Optical Coherence Tomography (pp 75-86).

Figure 8-1. High-definition OCT images from a normal macula (HD line). (A) Note the normal "dip" in retinal contour at the fovea. On the left, the less highly reflective GCL, INL, and ONL are labeled. On the right, the more highly reflective retinal NFL, IPL, and OPL are labeled. (B) Zoomed-in view of the same RTVue image. The individual outer retinal layers are visible: ELM, photoreceptor IS/OS junction, photoreceptor OS/RPE junction, and the RPE. The choroid is also marked (Ch).

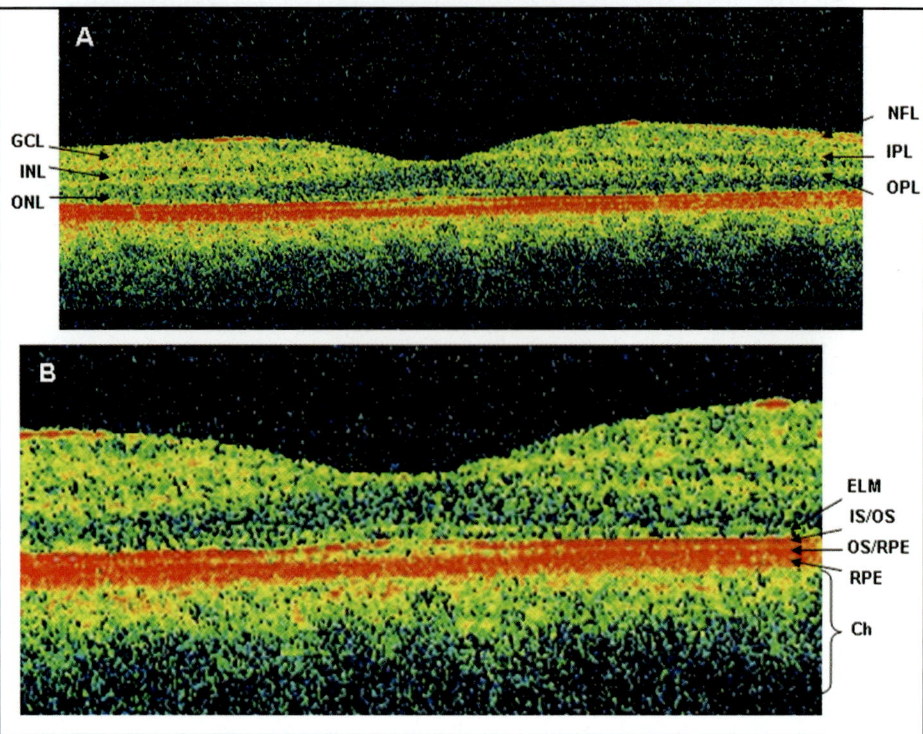

clear boundary in the computation of retinal thicknesses using computer segmentation algorithms. The middle outer retinal layer has been more recently described using higher resolution OCT and spectral-domain OCT and can often be visualized using the RTVue instrument. This layer may represent the interdigitation between the photoreceptor outer segment and the RPE, or the ends of the photoreceptors.[7] Deep to these 3 closely spaced "outer retinal/RPE" layers is the choroid, which is a more highly pigmented and vascular tissue that quickly scatters OCT signal, and appears as a mosaic of moderately and hyporeflective tissue. In some patients, and using specific imaging techniques such as the vitreoretinal/chorioretinal toggle on the RTVue, choroidal details may be more clearly visualized.[8]

Blood and pigment scatter the OCT signal, preventing signal penetration to deeper tissues. Therefore, retinal blood vessels are evident in OCT images by their shadowing of deeper retinal structures. These shadows can be used to register OCT images to the fundus and from visit to visit (discussed later in this chapter). In some cases of posterior vitreous detachment, the posterior hyaloid separates from the retina and can be visualized as a thin line anterior to the retina. Because of its optical reflectivity, the posterior hyaloid may be misinterpreted as the vitreoretinal interface by automated OCT software, and is a common cause of imaging artifacts in OCT thickness calculations. Additionally, each of the retinal layers may be distorted by vitreoretinal pathology. The vitreoretinal interface may be interrupted or distorted in diseases such as macular hole, the individual retinal layers may be thickened due to swelling, or thin or missing due to atrophy or retinal distortion. There may be intraretinal pathologies of differing reflectivity from the surrounding tissue; hyporeflective areas can reflect fluid accumulation, while hyper-reflective lesions may indicate hard exudates or pigment. The outer retina and RPE may be disrupted or elevated by materials of different reflectivities; serous fluid is less reflective, while choroidal neovascular lesions are more highly reflective. Distortion of the outer retina can also disrupt segmentation of the outer retina by computer algorithms, again causing imaging artifacts in OCT thickness calculations.

RTVue Retinal Image Features

Speckle Elimination

Some RTVue scanning protocols, such as the line scan, cross line scan, and grid line, employ a function called speckle elimination or image averaging (Figure 8-2). A number of individual OCT images are taken of the same region, and the images are then overlaid and averaged via automated RTVue software. The contour of the retina is automatically measured by RTVue software before averaging, allowing registration between images.

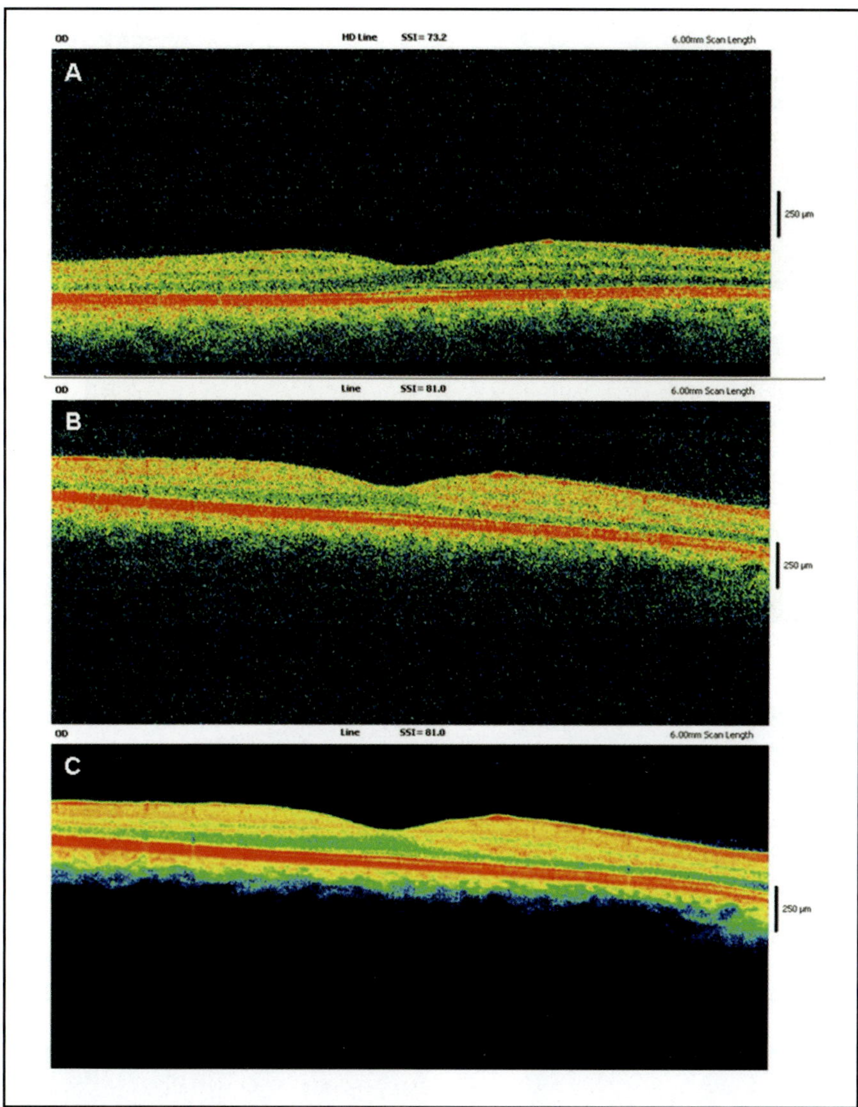

Figure 8-2. (A) A 6-mm 4096-line OCT image from the HD line protocol. Note that the SSI is given at the top of the image. (B) A 6-mm 1024-line single OCT image from the line protocol. (C) A 6-mm OCT image made using image-averaging from 16 1024-line single OCT images from the line protocol.

This technique enhances OCT signal by increasing signal-to-noise ratio, and may be particularly useful in patients with low signal due to media opacity. Increased signal-to-noise ratio offered by image averaging also has been shown to more clearly define intraretinal layers and pathologies.[9-11]

Reflectivity Scale

RTVue software allows the user to toggle between gray and color scale for each OCT image created (Figure 8-3). Images displayed in color scale have been the standard in prior OCT studies and commercial OCT systems. The color scale may allow easier detection of subtle variation in OCT signal to define reflectivity levels, as there are more colors for the viewer's brain to interpret. Conversely, gray scale allows the viewer's brain to visualize contrast more easily, and ultimately allows the viewer to differentiate intraretinal regions on OCT with greater ease.

High Definition

Certain RTVue scanning protocols, such as the HD line and HD cross scan protocols, employ higher A-scan volume per image to create a "high-definition" image by effectively increasing signal-to-noise ratio and enhancing the image (just like high-definition television enhances image quality by increasing pixel density) (see Figure 8-3). Because this is a single image, resolution is not degraded due to patient motion between images, as it may be in some cases where image averaging is employed.

Fundus Reconstruction and Image Registration

The 3D reference protocol can be used to create a reference scanning laser ophthalmoscopy (SLO)-like fundus image using OCT signal averaging; the sum of

Figure 8-3. OCT images from the line protocol, using signal averaging of 16 individual overlaid OCT images. (A) Normal macular image, displayed in color scale. (B) The same image displayed in gray scale. (C) The same image displayed in reverse gray scale.

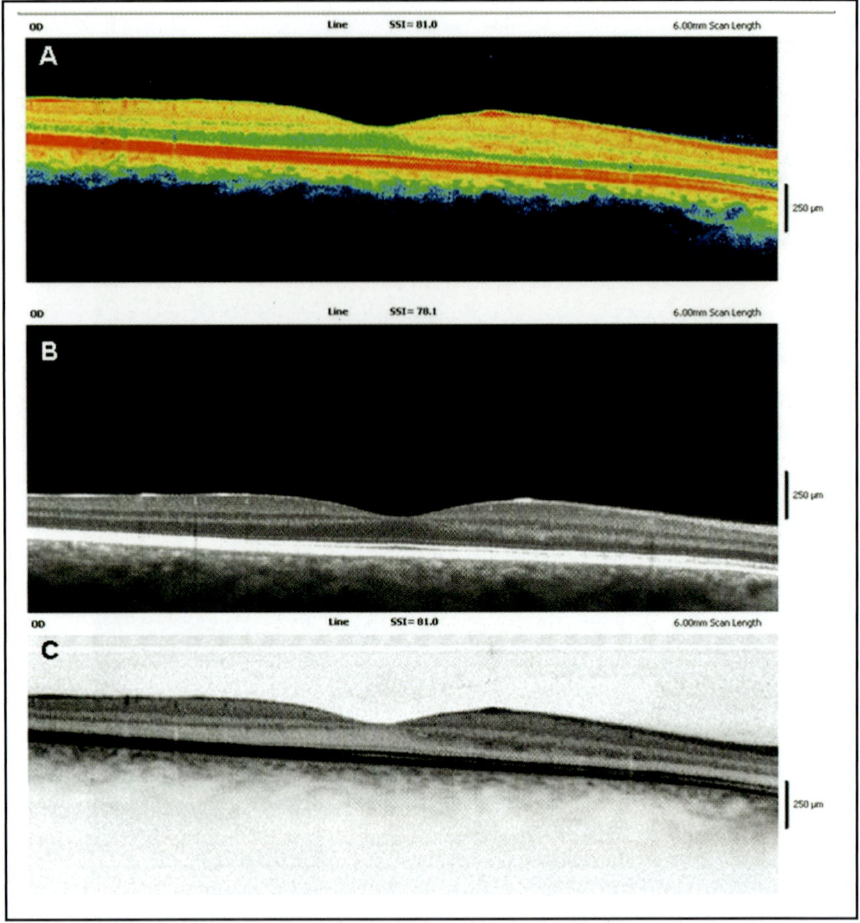

Figure 8-4. Display from the 3D reference protocol. (A) SLO-like fundus view of summated OCT data. The "blood vessels" are traced by RTVue software and used for registration of images from other RTVue imaging protocols. (B) Individual OCT image. RTVue software automatically delineates the vitreoretinal interface (red), IPL (green), IS/OS junction (purple), and outer RPE border (blue). (C) A 3D macular object is created, highlighting retinal topography. The 3D object can be rotated and manipulated on screen.

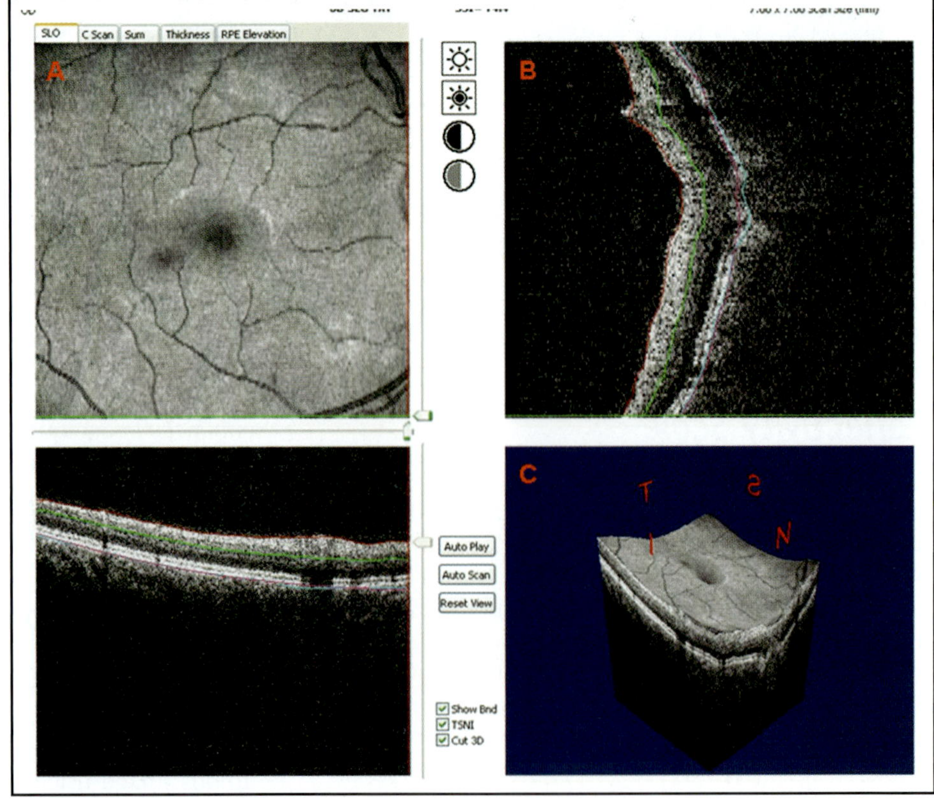

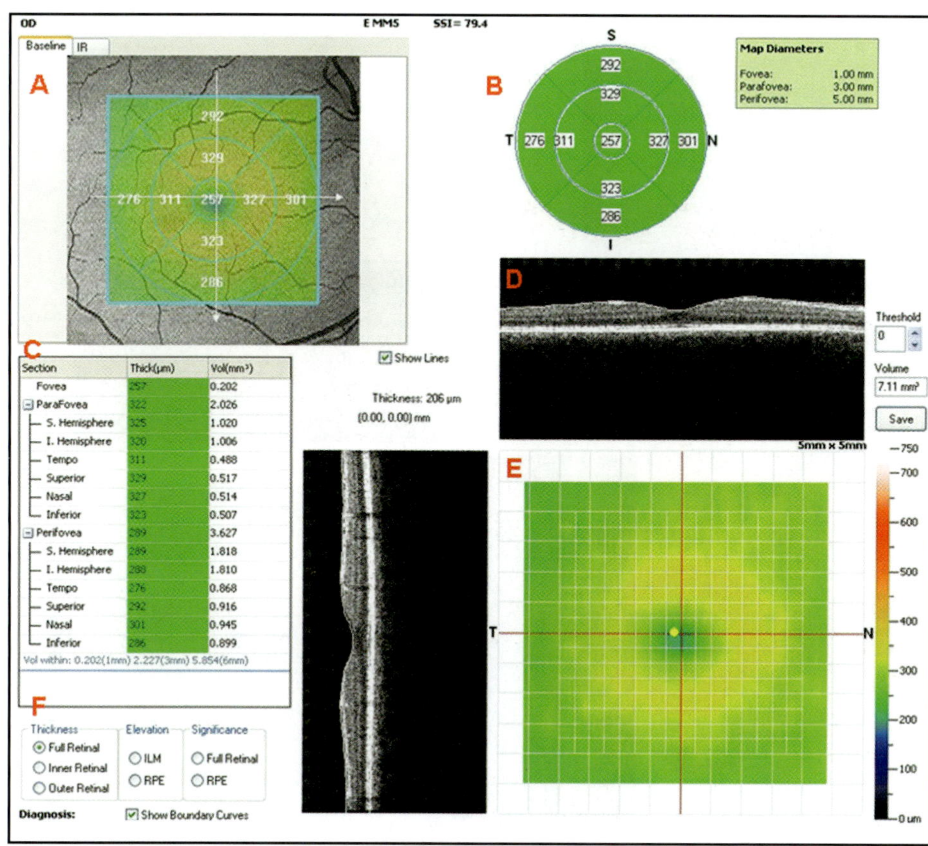

Figure 8-5. Display from the EMM5 protocol. (A) Thickness map along with ETDRS thicknesses is overlayed onto the SLO-type fundus image created from the 3D reference protocol. (B) ETDRS 9-area map of the fundus, with thickness values listed in each area, and significance value compared to the normative database highlighted by color. In this case, the thickness is in the normal range, therefore all ETDRS areas are green. (C) Detailed thickness information from the ETDRS map, and comparison to the normative database highlighted by color. (D) Individual horizontal OCT image used to create the EMM5 map; the image selected is indicated on the thickness map with a red line. (E) More detailed thickness map, highlighting transitions in retinal thickness in high detail. Interpolation between the individual OCT images (lines on the map) is used to calculate transitions in retinal thickness. Differences are highlighted by the color coded legend (at right). (F) Toggle can be used to calculate full retinal, inner retinal, and outer retinal thicknesses, RPE and ILM elevations, and comparison between the normative database and full retinal thickness or RPE elevation.

the OCT signal from each A-scan is used to create the image (Figure 8-4). Because retinal vessels block or scatter OCT signal, they appear hyporeflective on the fundus image, making them easily recognizable. The vessel tracings on this fundus image can then be used to register OCT images from other imaging protocols, such as the EMM5, and can also be used to register OCT images to fundus photographs or angiograms. It is important for the observer to remember that the fundus image is created from OCT data, and is not equivalent to an SLO image or fundus photo.

Thickness and Elevation Calculation

The EMM5 and MM6 protocols create macular maps, and automated OCT software may be used to calculate several retinal layer thicknesses. Full retinal thickness is computed by automatically detecting the vitreoretinal interface as the inner retinal border and the photoreceptor IS/OS junction as the outer retinal border. The area between these 2 lines is then calculated in each OCT image, and the data are summed over the entire image set (Figure 8-5).

In addition, separate inner and outer retinal thicknesses can be measured using software that delineates the boundary between the IPL and INL. The inner retinal thickness is the area between the vitreoretinal interface and the IPL line, and the outer retinal thickness is the area between the IPL line and the photoreceptor IS/OS junction (Figure 8-6). Measurement of the inner retinal thickness may be useful in diseases that affect the NFL and ganglion cells, such as optic nerve disease and retinal vascular disease. In fact, on the newest RTVue software for glaucoma measurements, the inner retinal thickness is referred to as the ganglion cell complex (GCC). Measurement of the outer retinal thickness may be more important in diseases that affect the photoreceptors, such as macular degeneration, retinal dystrophies, and retinal detachment. Clinical utility of these measurements is yet to be shown in a prospective study.

Retinal thicknesses can be displayed in various formats. The Early Treatment Diabetic Retinopathy Study (ETDRS) map displays thicknesses in a 9-zone map, with a 1-mm diameter foveal circle, 4 parafoveal c-shaped areas enclosed in a 3-mm circle, and 4 perifoveal c-shaped areas enclosed in a 5-mm (EMM5) or 6-mm (MM6) circle (Figure 8-7). This map is similar to ETDRS displays provided by prior time-domain OCT instruments and is a format that clinicians are already intimately familiar with. Thickness and volumetric measurements from this ETDRS circle are also displayed to the side of the screen. A color-coded square-shaped thickness map

Figure 8-6. Inner retinal calculations on the EMM5 map. Note that the IPL is delineated by RTVue software to represent the outer boundary of the inner retinal thickness (red arrow), while the vitreoretinal interface is the outer boundary. A normative database is not available for comparison to these thickness values, therefore the ETDRS map and values are no longer color coded to highlight significance.

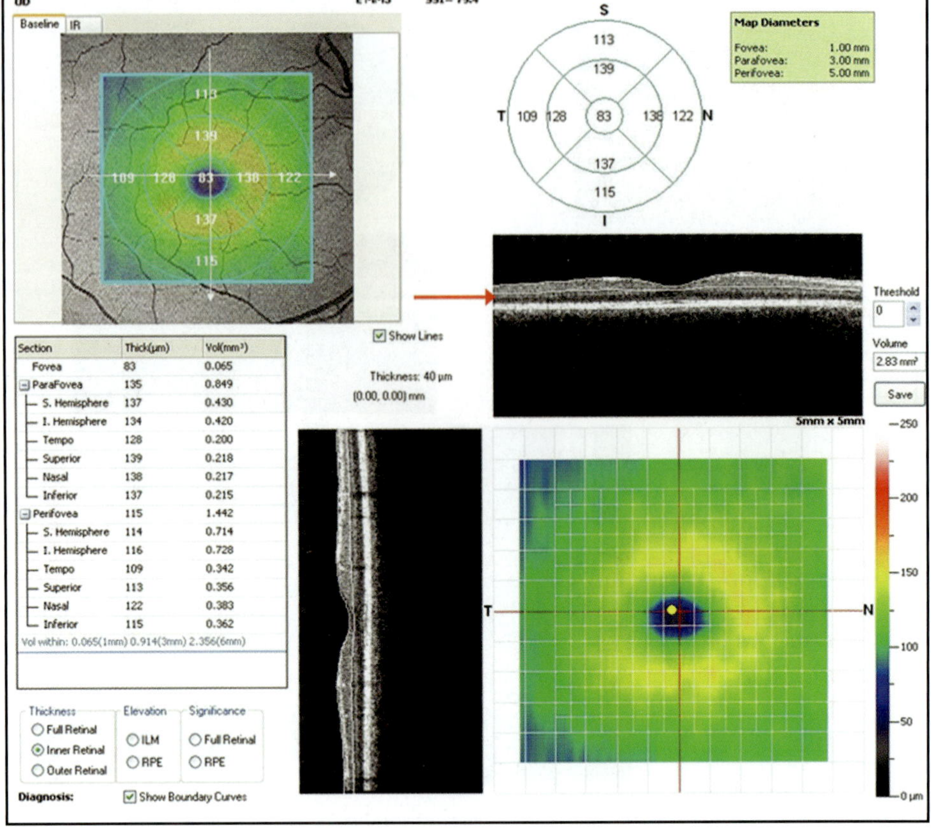

Figure 8-7. Display from the MM6 protocol. The thickness map on the lower right is calculated from 12 radial 1024-line OCT images. Near the center of the map, the radial lines are closely spaced and less interpolation between images is used to calculate retinal thickness; away from the center, the OCT images are spaced further apart and more interpolation is used to calculate the thickness map.

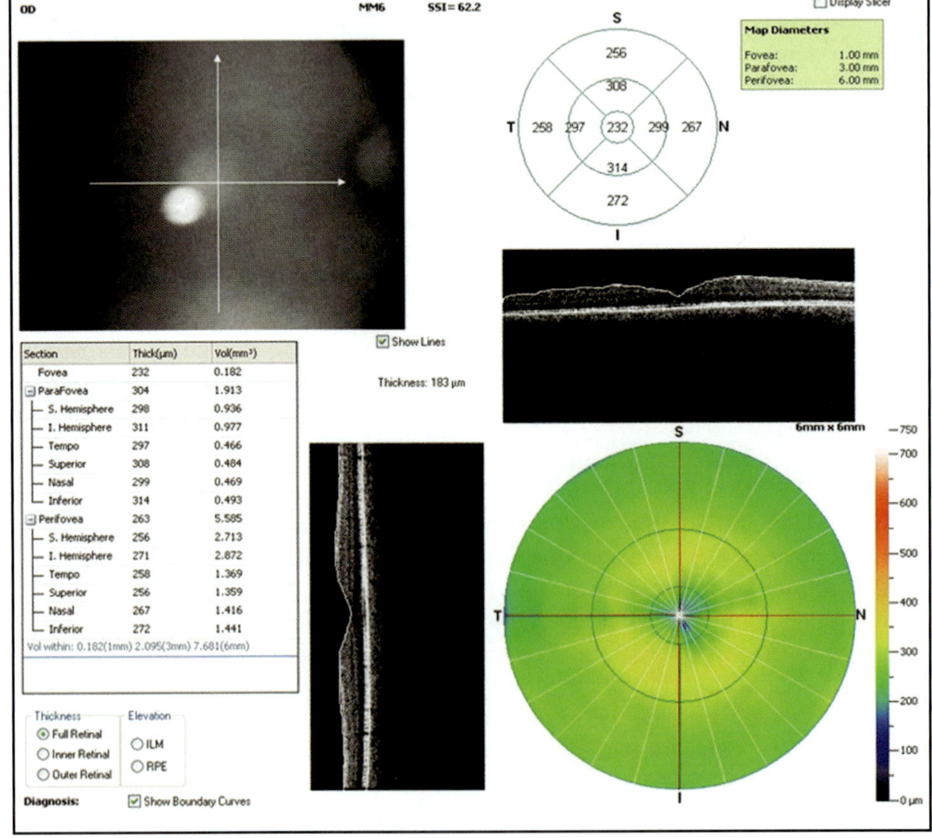

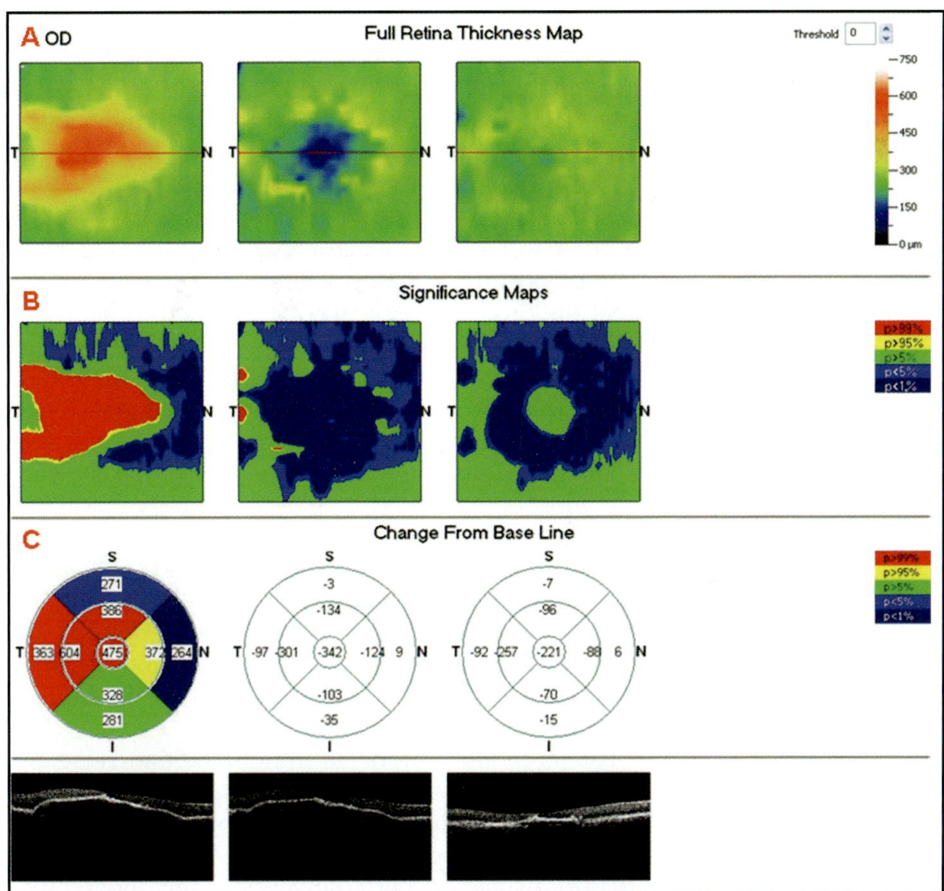

Figure 8-8. Progression analysis report from the EMM5 protocol. This patient had exudative AMD, treated with intravitreal ranibizumab twice and imaged at baseline, at 1 month after first injection, and 1 month after second injection. (A) Display of detailed thickness maps for comparison between visits. (B) Significance maps showing thickness values in comparison to normative database. (C) ETDRS maps showing thickness compared to normative database at first visit and changes from baseline thickness in each of the 9 ETDRS zones at subsequent visits.

is also displayed, which displays greater detail of transitions in retinal thickness and may be used to manually delineate specific areas of retinal elevation or depression. Total volumes above a given macular thickness can also be calculated by defining a thickness "threshold level."

In addition, RTVue software automatically delineates "RPE elevation" and "ILM elevation." These elevation values are calculated as the height of the RPE or ILM compared to a reference place, which is automatically set as the best elliptical fit to the outer RPE boundary. The reference plane can also be manually adjusted. The RPE elevation function, in particular, may be used to calculate sub-RPE thicknesses in diseases where the RPE is elevated from the choroid, such as choroidal neovascularization.

It is important to remember that the MM5 protocol uses more data points (~19,000 A-scans) to calculate a square map, while the MM6/radial slicer protocol uses less data points (~12,000 A-scans) to calculate a circular map. While the center of the MM6 map (where the radial images intersect) uses less interpolation between data points to create the thickness map, more interpolation between data points occurs further from the center of the map.

Normative Databases

Macular thickness data from the EMM5 protocol can also be used for comparison to a normative database (see Figure 8-5). Thickness values from a particular patient are compared to a normative database of normal macular thicknesses, and a color scale is used to indicate which percentile each given thickness value falls into compared to normal thicknesses: red indicates greater than 99% of normals, yellow indicates greater than 95%, green is between 5% to 95%, light blue is less than 5%, and purple is less than 1% of normals. Thickness calculations included in the normative database are full retinal thickness and RPE elevation (total macular and for each individual ETDRS zone).

Progression Analysis

Progression analysis software is available for the EMM5 and MM6 protocols. When patients return for repeat OCT imaging, thickness calculations can be compared from visit to visit using this software. Maps showing absolute thickness values may be displayed. Significance maps detailing thickness values that vary from the normative database are also displayed. ETDRS maps show thickness changes from the baseline OCT data acquired at the patient's first visit to the clinic (Figure 8-8). These values can be used to monitor disease progression or response to treatment.

Three-Dimensional Visualization

The 3D reference protocol can display data in a number of different ways. Similar to the EMM5 and MM6 protocols, the full retinal, outer retinal, and inner retinal thicknesses, as well as RPE elevation, can be extrapolated and viewed in a thickness map. The thickness map can be displayed as a 3D object and examined in 3D, or a movie view can scroll through all the individual OCT images. The thickness map provided by the 3D macular protocol contains the most data of all the thickness maps (51,712 A-scans). The vitreoretinal interface is also viewed in 3D, which may be useful in diseases of the vitreoretinal interface (see Figure 8-4).

En Face *Imaging*

The high A-scan density of the 3D macular imaging protocol allows reconstruction of OCT image data in a plane parallel to the retinal surface, termed *C-scan* or *en face* images (Figures 8-9 and 8-10). The C-scan viewer allows display of a single flat plane of the retina in a top-down view, which can be used to evaluate a retinal lesion in the z-plane in addition to the x- and y-planes viewed in traditional OCT images. The contour of the C-scan can be adjusted to fit different retinal planes, such as the vitreoretinal interface plane or the RPE plane. Using RTVue software, the anterior-posterior height of OCT A-scan data used for summation can be adjusted. This imaging protocol can be used to visualize intraretinal or subretinal pathologies in 3D, which may be useful to evaluate subtle abnormalities in diseases such as age-related macular degeneration (AMD).

Manual Calculations

RTVue software allows manual measurement of intraretinal thicknesses, distances, and areas. This operator-driven software can be used with OCT images from any of the imaging protocols. Distances, areas, and volumes of intraretinal anatomy are delineated by the user, and values are automatically calculated and displayed. Intra- or subretinal lesion volumes for pathologies such as choroidal neovascularization or macular holes may then be more precisely calculated manually.

LIMITATIONS, ARTIFACTS, AND OPTIMIZING DATA

As is the case with all imaging modalities, OCT has some limitations. It is important for the clinician or researcher to keep in mind that although OCT images captured with the RTVue and other spectral-domain OCT systems are of nearly histologic resolution, they do not exactly replicate *in vivo* retinal anatomy, but rather represent changes in optical properties within retinal tissue and tissue effects on backscattering of OCT light signals. The penetration depth of OCT is limited due to the scattering properties of tissue, therefore imaging depths greater than ~1.5 mm are not currently pos-

sible. OCT can be limited by media opacities, such as dense cataracts or vitreous hemorrhage, although the near-infrared property of the OCT light allows better penetration than does visible light through most media opacities.[12]

The quality of OCT images also depends on operator skill. Each scan must be taken in range and in focus. In order to enhance image capture and signal quality, RTVue software includes a number of manual adjustment options. Eye color may be adjusted to help with image capture. Axial length, focus, and polarization can be adjusted automatically or manually to enhance OCT signal quality. The vitreoretinal/chorioretinal toggle may be adjusted depending on the depth of the area of interest within the retina/choroid. Additionally, each scan must be examined for blinks and motion artifacts. If a scan is of poor quality, it should be discarded and retaken by the operator. On the EMM5 and MM6 scans, missing or dropped images due to blink or motion artifact are indicated in gray on the rendered map. The more missing images, the more interpolation of retinal thicknesses between images is used to create the final thickness map (Figure 8-11).

Patient motion can occur in both the transverse and axial directions. Motion is usually negligible in each individual OCT image, given the high speed of spectral-domain OCT imaging utilized in the RTVue instrument (26,000 A-scans/second), however, patient motion may become apparent from image to image. Axial motion correction software is used to align OCT images in the 3D macular protocol, which can distort the true retinal anatomy.

The operator also is responsible for centering the images at the fovea. This can be difficult in patients with pathologies that distort the foveal contour. The EMM5 protocol allows the examiner to recenter the thickness map, but this feature is not available in the other imaging protocols. External fixation with the contralateral eye can be used in some patients with poor fixation.

The use of automated computer algorithms also introduces artifacts in the detection of retinal and intraretinal thicknesses. In patients with macular pathology, retinal boundaries may become unclear. For example, in macular holes, segmentation software often is not able to detect the inner retinal boundary at the area of the retinal defect. In eyes where the posterior hyaloid is visible, the software may mistakenly measure the hyaloid as the inner retinal boundary (Figure 8-12). In patients with choroidal neovascularization, the outer retinal boundary often is blurred, introducing error in the computer detection of this layer (Figure 8-13A).[12-19] These artifacts of segmentation can be observed in individual OCT images, or they can be visualized on the macular map.[14-19] Errors of segmentation also can occur when determining the boundary between the inner and outer retina, as well as the RPE and ILM elevation.

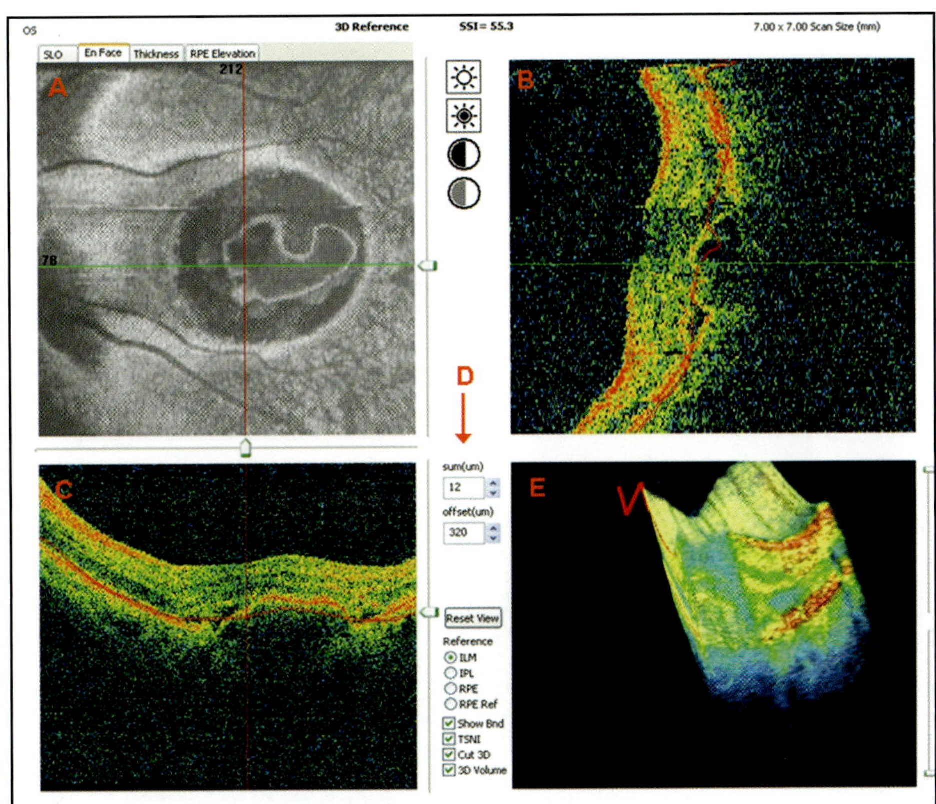

Figure 8-9. Display from the 3D reference protocol, using the *en face* or C-scan view. The patient has exudative AMD with a fibrovascular pigment epithelial detachment as well as subretinal fluid. (A) Reconstructed *en face* or Z-axis OCT image. The C-scan passes through an area of subretinal fluid, visible as a hyporeflective circle on the C-scan, and also through the fibrovascular pigment epithelial detachment, seen as an irregular outlined lesion within the circular area of subretinal fluid. (B) Reconstructed Y-axis OCT image. The image region is indicated with a red line on the *en face* image. The region of the *en face* image is also indicated with a red line on the Y-axis image. (C) X-axis OCT image. The image region is indicated with a green line on the *en face* image. (D) *En face* image parameters can be adjusted. The "thickness" of the C-scan can be adjusted using the [Sum] toggle; in this case, the thickness of the C-scan is 12 µm. The contour of the C-scan can also be adjusted to fit different retinal planes; in this case, the image is fit to the ILM plane. (E) 3D volume reconstruction display.

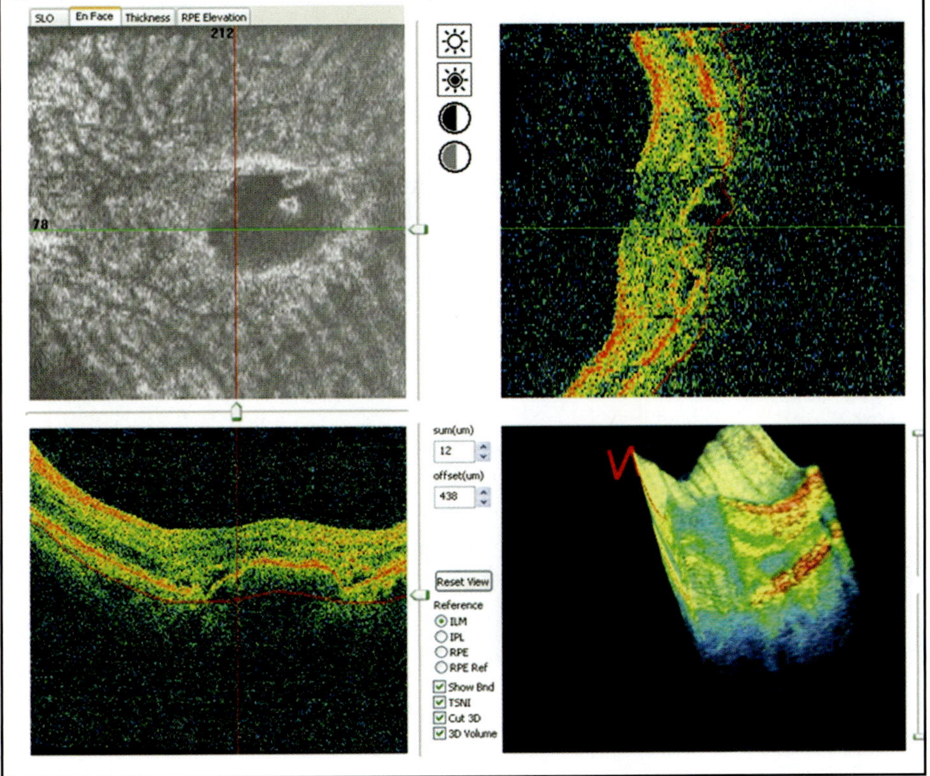

Figure 8-10. *En face* images from the same patient as in Figure 8-9. The C-scan is now moved posteriorly to the choroid, and the outlines of choroidal vessels become visible.

Figure 8-11. Blink artifacts. EMM5 scan of a normal eye taken while blinking. Note that several images were noted as dropped and highlighted with gray lines on the detailed thickness map. Computer software tries to interpolate thickness between dropped images, giving a grid-like appearance of sequential "thin lines" on the detailed thickness map. Inferior and temporal zones on the ETDRS map are measured as "thin" compared to the normative database. The highlighted horizontal OCT image is missing due to blink, while the highlighted vertical OCT image is normal.

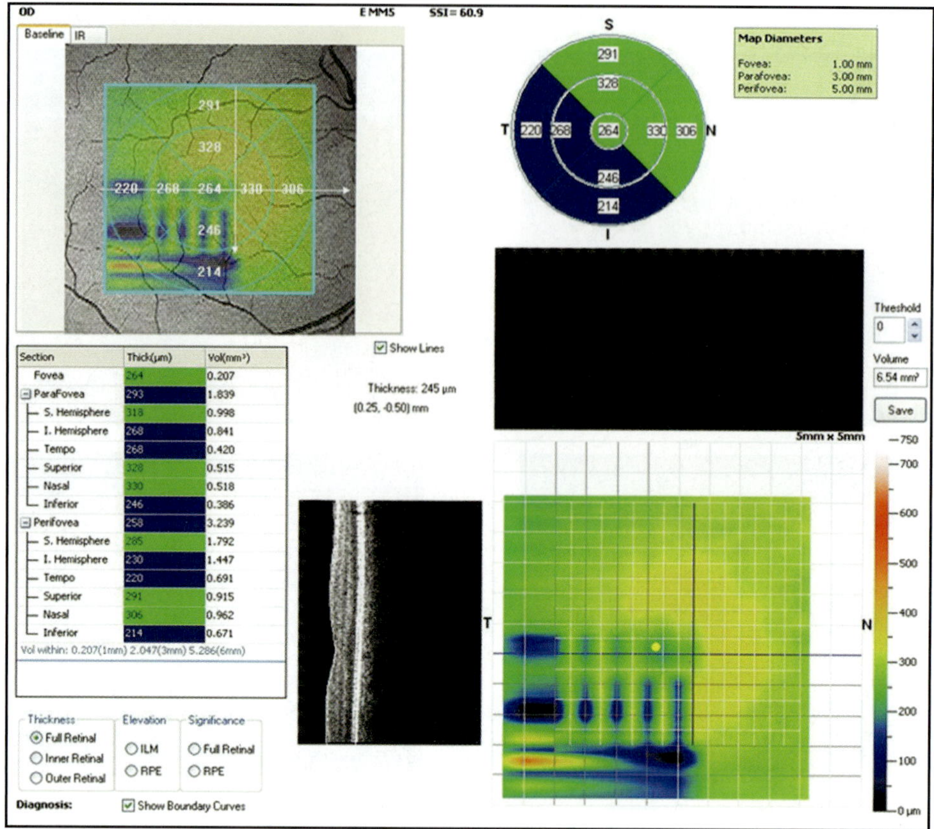

Figure 8-12. Inner retinal segmentation error. EMM5 scan from a patient with vitreomacular traction syndrome. Note there is an area of outer retinal delineation error inferiorly, which is seen on the macular map as a black region where the thickness is measured as 0. In the individual OCT images, it is more apparent that segmentation error is due to the computer having mistaken the detached posterior hyaloid for the vitreoretinal interface.

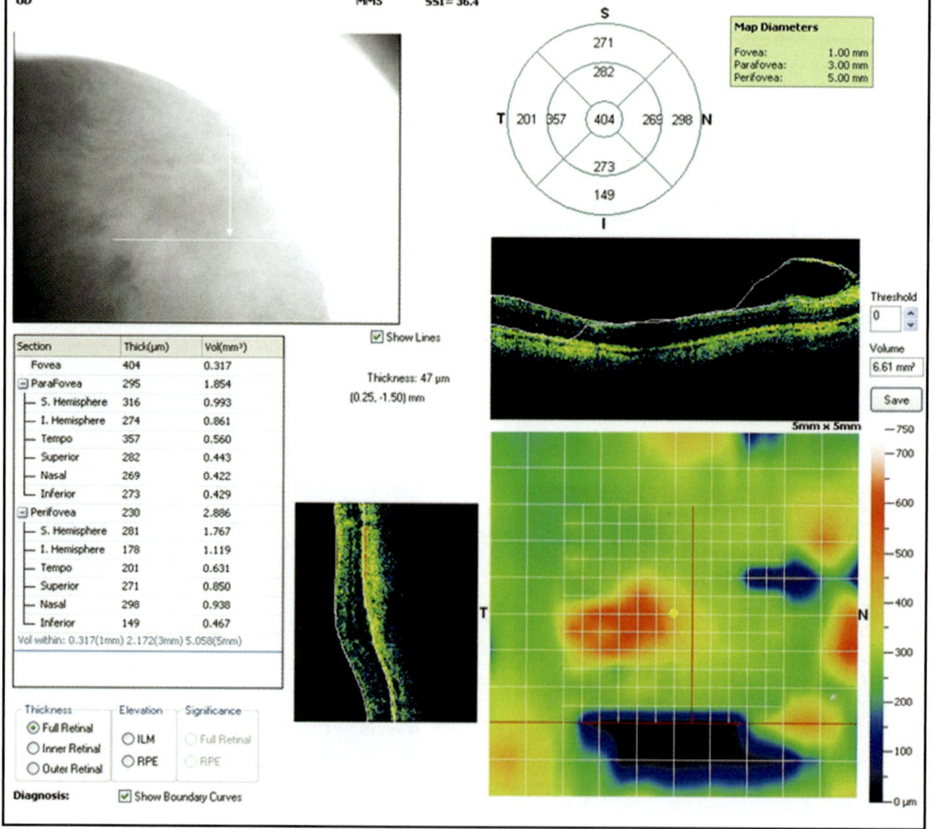

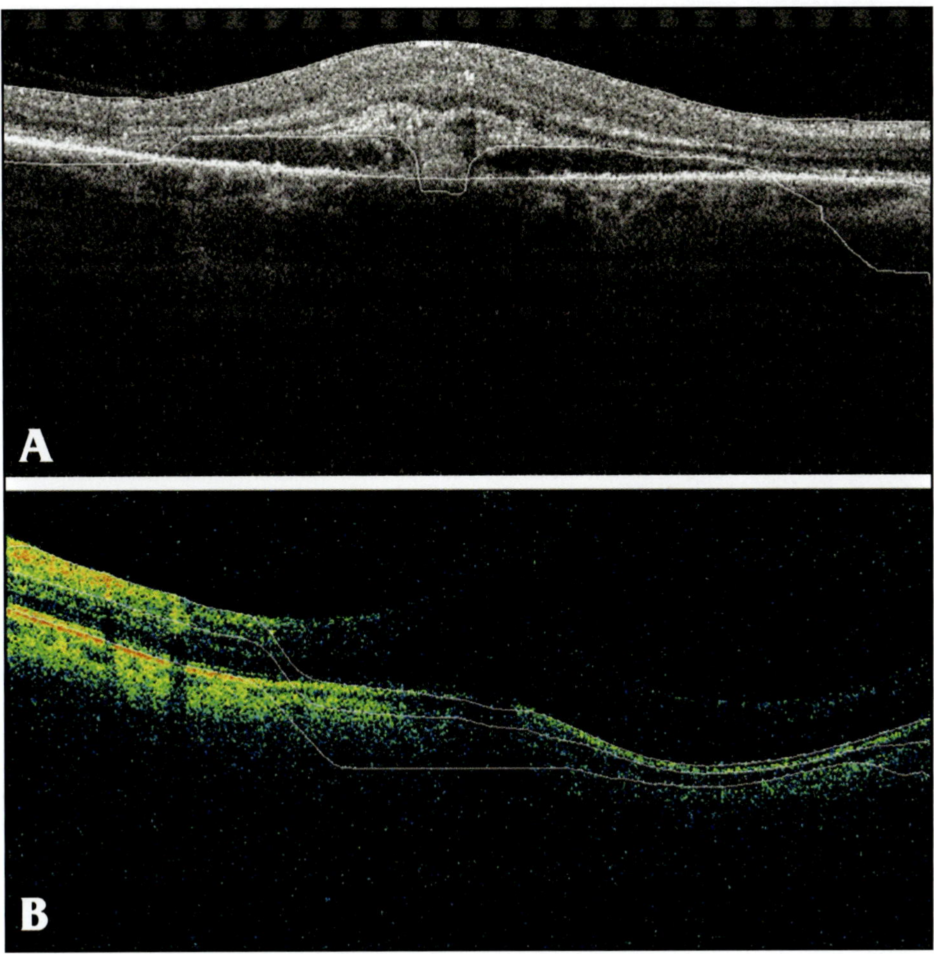

Figure 8-13. Other OCT artifacts. (A) Outer retinal segmentation error. In this image of exudative AMD, the vitreoretinal interface is correctly delineated by computer software, while the computer algorithm incorrectly delineates the outer retinal boundary due to retinal distortion by the subretinal fibrovascular lesion. (B) Poor signal error. In this image, OCT signal is blocked by an opaque preretinal membrane, and the delineation software has broken down.

Signal strength is one of the important aspects to consider when evaluating an individual OCT image or thickness calculation. Decreased signal strength can create difficulty interpreting retinal anatomy in an individual OCT image, and causes an increased rate of breakdown of delineation software (Figure 8-13B). Signal strength is displayed on RTVue images as the Signal Strength Index (SSI), which can range from 1 to 100. A value below 40 should alert the examiner that signal may be too low to obtain reliable thickness data from automated thickness calculation software. Relative signal strength, or signal-to-noise ratio, can be improved by using an RTVue function that utilized image averaging, such as the line, cross scan, or grid line protocol.[9-11] Image averaging, however, can sometimes decrease resolution in patients who move between OCT images.

SUMMARY

The RTVue instrument is capable of providing high resolution OCT images of the retina using a variety of different scanning protocols. Knowledge of the benefits and limitations of each protocol provides the clinician or researcher with the ability to maximize the utility of the RTVue. The line, cross scan, and grid line protocols maximize signal-to-noise ratio per OCT image by employing image averaging, while the HD line and HD cross scan protocols maximize signal-to-noise ratio by providing higher pixel density OCT images. High resolution OCT images allow visualization of intraretinal microstructure with precise detail. The EMM5 and MM6 protocols allow calculation of detailed retinal and intraretinal thickness maps, and both protocols can be used to monitor changes in retinal thickness over time, allowing for monitoring of disease progression or response to treatment. The EMM5 protocol additionally provides a normative database of retinal thickness values to compare with patients. The 3D reference protocol allows maximal 3D visualization of the retina, and allows creation of *en face* OCT images. OCT imaging does have some limitations, as automated delineation software can sometimes fail, introducing imaging artifacts and errors in thickness calculations. Familiarity with all aspects of RTVue capabilities and limitations provides the user with a better understanding of data provided by this instrument.

REFERENCES

1. Drexler W, Sattmann H, Hermann B, et al. Enhanced visualization of macular pathology with the use of ultrahigh-resolution optical coherence tomography. *Arch Ophthalmol.* 2004;121:695-706.

2. Anger EM, Unterhuber A, Hermann B, et al. Ultrahigh resolution optical coherence tomography of the monkey fovea: identification of retinal sublayers by correlation with semithin histology sections. *Exp Eye Res.* 2004;78:1117-1125.

3. Gloesmann M, Hermann B, Schubert C, et al. Histologic correlation of pig retina radial stratification with ultrahigh-resolution optical coherence tomography. *Invest Ophthalmol Vis Sci.* 2003;44:1696-1703.

4. Ko TH, Fujimoto JG, Schuman JS, et al. Comparison of ultrahigh- and standard-resolution optical coherence tomography for imaging macular pathology. *Ophthalmology.* 2005;112(11):1922.e1-15.

5. Srinivasan VJ, Wojtkowski M, Witkin AJ, et al. High-definition and 3-dimensional imaging of macular pathologies with high-speed ultrahigh-resolution optical coherence tomography. *Ophthalmology.* 2006;113(11):2054.e1-14.

6. Chen TC, Cense B, Miller JW, et al. Histologic correlation of in vivo optical coherence tomography images of the human retina. *Am J Ophthalmol.* 2006;141(6):1165.

7. Srinivasan VJ, Monson BK, Wojtkowski M, et al. Characterization of outer retinal morphology with high-speed, ultrahigh-resolution optical coherence tomography. *Invest Ophthalmol Vis Sci.* 2008;49(4):1571-1579.

8 Spaide RF, Koizumi H, Pozonni MC. Enhanced depth imaging spectral-domain optical coherence tomography. *Am J Ophthalmol.* 2008;146(4):496-500.

9. Sander B, Larsen M, Thrane L, et al. Enhanced optical coherence tomography imaging by multiple scan averaging. *Br J Ophthalmol.* 2005;89:207-212.

10. Sakamoto A, Hangai M, Yoshimura N. Spectral-domain optical coherence tomography with multiple B-scan averaging for enhanced imaging of retinal diseases. *Ophthalmology.* 2008;115(6):1071-1078.e7.

11. Christensen UC, Kroyer K, Thomadsen J, et al. Normative data of outer photoreceptor layer thickness obtained by software image enhancing based on Stratus optical coherence tomography images. *Br J Ophthalmol.* 2008;92(6):800-805.

12. Hee MR. Artifacts in optical coherence tomography topographic maps. *Am J Ophthalmol.* 2005;139:154-155.

13. Ray R, Stinnett SS, Jaffe GJ. Evaluation of image artifact produced by optical coherence tomography of retinal pathology. *Am J Ophthalmol.* 2005;139:18-29.

14. Patel PJ, Chen FK, da Cruz L, Tufail A. Segmentation error in Stratus optical coherence tomography for neovascular age-related macular degeneration. *Invest Ophthalmol Vis Sci.* 2009;50(1):399-404.

15. Glassman AR, Beck RW, Browning DJ, et al. Comparison of optical coherence tomography in diabetic macular edema, with and without reading center manual grading from a clinical trials perspective. *Invest Ophthalmol Vis Sci.* 2009;50(2):560-566.

16. Karam EZ, Ramirez E, Arreaza PL, Morales-Stopello J. Optical coherence tomographic artefacts in diseases of the retinal pigment epithelium. *Br J Ophthalmol.* 2007;91(9):1139-1142.

17. Sadda SR, Wu Z, Walsh AC, et al. Errors in retinal thickness measurements obtained by optical coherence tomography. *Ophthalmology.* 2006;113(2):285-293.

18. Jaffe GJ, Caprioli J. Optical coherence tomography to detect and manage retinal disease and glaucoma. *Am J Ophthalmol.* 2004;137:156-169.

19. Ishikawa H, Piette S, Liebmann JM, Ritch R. Detecting the inner and outer borders of the retinal nerve fiber layer using optical coherence tomography. *Graefes Arch Clin Exp Ophthalmol.* 2002;240:362-371.

9. Retinal En Face Scan Analysis

Bruno Lumbroso, MD and Marco Rispoli, MD

RTVue *en face* technology allows one to acquire and analyze *en face* frontal retinal scans adapted to the natural concavity of the eye's posterior pole. These frontal or transverse scans are part of the tridimensional study of the retina, reconstructed from 3D cube. They form a useful complement to the conventional cross-sectional B-scan OCT imaging that is much more intuitive and easier to understand.

Frontal *en face* plane C-scans have been available since 2002 with some time-domain OCT devices. They were of good clinical interest but difficult to read. As the retina is cup-shaped, the flat plane section cuts through all the retinal and choroidal layers—making planar C-scan images difficult to understand.

En face technology is a useful improvement for clinical and research purposes because it brings the possibility to obtain a retinal section that follows the retinal pigment epithelium (RPE) curve to the C-scan.

En face *RTVue scans, frontal scans adapted to pigment epithelium paraboloid shape, are placed at constant depth in the retina or choroid and parallel to RPE.* They bring new elements to the diagnosis and follow-up of retinal diseases and provide a good overview of the retina under study, allowing the assessment of all the scanned area within the retinal cube.

With RTVue technology, *en face* scan procedures are easy to perform and the images are easy to understand after a short learning period. The first times we see *en face* images they seem strange, new, complex, and not as easy to read nor as intuitive to understand as classical cross-section B-scans. In fact, the *only difficulty is in the interpretation* of the scans. This global aspect of the retina was, until recently, unknown to the clinicians.

Not everyone working in OCT yet understands how useful this technique can be and what importance it will have in the near future. Many ophthalmologists and technicians and even retinal specialists say that they do not know how to use *en face* scans or that it is too complicated and time consuming. Frontal *en face* scans are not yet fully appreciated by the clinical ophthalmologists; but, if they give some attention to the problem, most of them will find Optovue *en face* scan technology easy and fast to use and clinically helpful. Even making the technology easier will not (by itself) convince ophthalmologists to use it and to look at *en face* images.

Frontal *en face* scans bring truly a third dimension to the study of retinal diseases in everyday clinical practice. They give clinical information on retinal disturbances and are very important for research. Moreover, the 3D cube allows retinal segmentation, delineating the individual retinal layers.

En face scans (frontal scans adapted to pigment epithelium paraboloid shape) help diagnose retinal diseases, macular holes, lamellar holes, retinoschisis, pigment epithelial detachments (PEDs), retinal serous detachments, macular degenerations, diabetic retinopathies, and many other retinal disturbances.

Cross-section B-scans give a 2D view of the retina. A third dimension is given by frontal *en face* scans or C-scans. C-scans adapt (or not) to the cup-shaped retina and RPE and bring important information to the clinician.

Classical C-scans are perfectly flat and present a clinical interest (Figure 9-1). RTVue *en face* scans are a wonderful improvement on plane C-scans, because they give a retinal section that follows the 3D RPE cup-shaped curve.

Huang D, Duker JS, Fujimoto JG, Lumbroso B, Schuman JS, Weinreb RN.
Imaging the Eye from Front to Back with RTVue Fourier-Domain Optical Coherence Tomography (pp 87-96).
© 2010 SLACK Incorporated.

Figure 9-1. Planar C-scans. Classic C-scans give a frontal, planar, perfectly flat section of the retina and choroid. As the retina is cup-shaped, the flat section cuts diagonally through all the retinal and choroidal layers. The scanning plane intersects with the normally concave retina and choroid layers. The planar C-scans images are difficult to understand. (Illustration by Ms. D. Piccioli.)

RTVue technology can follow the irregularities of a pathologic RPE: obtaining a delineation of the drusen or of the PED. Classical 3D images are scenic and beautiful. Their interest as teaching tools is evident. *En face* scans are clinically more important because they can give quantitative informations.

But it is even more important to have an *en face* frontal scan following the ideal RPE concavity and cutting through pathologic alterations. The frontal scan is parallel to the RPE, at a constant depth in the retina (Figures 9-2 and 9-3).

Transverse OCT imaging has been clinically used since 2002.[1,2] The combined OCT-SLO-System (OPKO-OTI, Miami, FL) allows for confocal *en face* fundus imaging and high resolution OCT scanning at the same time. OCT images were obtained from transversal line scans. Other improvements came from Drexler and his group. Schmidt-Erfurth et al presented images of patients with vitreoretinal diseases and diseases affecting the outer retina/RPE interface with an ultrahigh resolution OCT.[3] Similar to *en face* time-domain OCT systems, the 3D sets (including a 3D reconstruct) provided a better overview of the whole area of interest, allowing identification of small, focal changes and their distribution throughout the scanned area. The devices used produced classic planar C-scans. A few years later some prototype high-speed OCT system was built based on a Fourier-domain OCT technology (a Cirrus prototype). The lack of moving parts and the higher sensitivity of a spectral-domain OCT system compared to time-domain systems allow much higher acquisition speed and improved resolution without loss of image quality. The high acquisition speed made it possible to generate 3D data sets, which led to more precise calculations of retinal thickness, detection of early disease stages, and more accurate follow-up over time.

Spectral-domain or 3D OCT imaging showed improved image resolution, volume rendering of 3D data sets, and mapping of individual retinal layers to create an OCT fundus image and topographic analyses of the macula and optic nerve. The first FDA-approved spectral-domain device was Optovue in 2006. Other spectral-domain OCT devices followed. Only RTVue and Zeiss (Dublin, CA) give *en face* scans adapted to retinal concavity. All the other clinically available devices produce classic planar C-scans.

SCAN PATTERNS FOR RTVUE

Standard and En Face Scan Analysis

Standard macula analysis should include at least one cross scan centered on the fovea, one or more line scans centered on the point we want to study (if it is not located on the fovea), a full retinal map analysis, and an *en face* analysis.

With RTVue OCT, the standard analysis system has to be set with following parameters: line scan, cross scan, grid scan, EMM5 (retinal map), and 3D. During acquisition, the scan has to be centered on the fovea and then on the exact point of study (manually moving on the infrared image).

Macular map analysis has to be accurately assessed, checking that the vitreoretinal, inner/external retina, and RPE markers are correctly traced. Software is able to

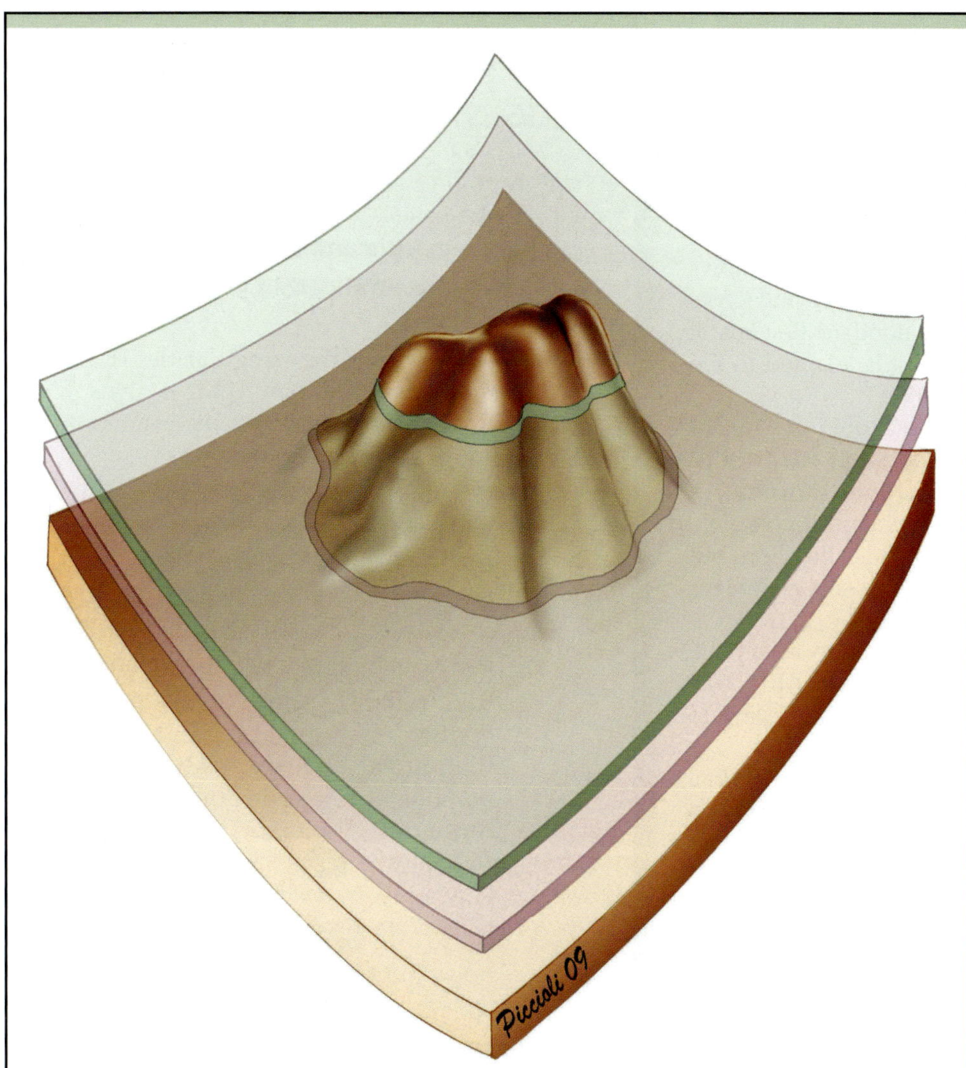

Figure 9-2. RTVue *en face* scans. RTVue *en face* technology is an improvement on planar C-scans because it gives a retinal section that follows the tridimensional RPE curve. *En face* RTVue are frontal scans adapted to the pigment epithelium cup-paraboloid shape. They are parallel to RPE and at a constant depth in the retina. The software allows the user to choose the depth of the scan. It can be shifted forward or back. They are very useful for the diagnosis and follow-up of retinal diseases. *En face* scans are easy to perform. This drawing illustrates *en face* RTVue frontal scans adapted to pigment epithelium cup-paraboloid shape by cutting through a PED in a case of AMD. (Illustration by Ms. D. Piccioli.)

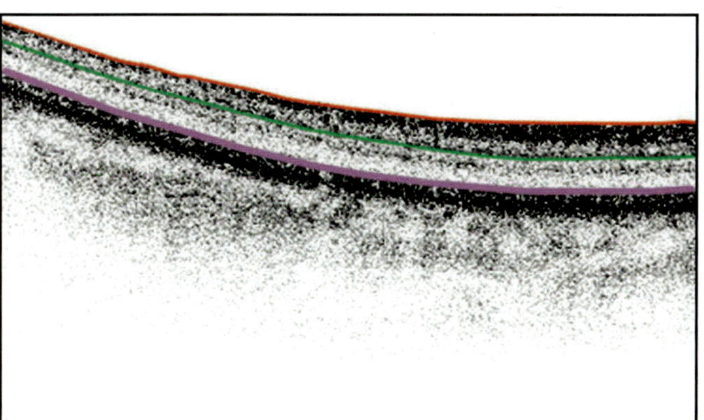

Figure 9-3. RTVue selection of scan profile. Scan profile (ILM, IPL, RPE, or RPE reference): ILM follows the irregularities of normal and pathologic ILM. It gives 3D images to understand the lesion size, shape, and dimensions. IPL follows the irregularities of normal and pathologic IPL. RPE follows the normal RPE profile or the irregularities of a pathologic RPE; it gives 3D images that illustrate the lesion size, shape, and dimensions. RPE reference cancels the pigment epithelium anomalies and gives a theoretical ideal 3D paraboloid reference lamina that cuts through the lesions.

automatically locate the fovea, but in some pathological cases it is necessary to find it manually and reprocess new data. The 3D scans are automatically aligned and the RPE surface is shown to check the process correctly.

En Face *Analysis*

En face analysis is a very important step in OCT study of the retina using RTVue 3D parameters. It allows a frontal, indepth examination of the retina and gives new information not available with standard B-scans. It is superior to flat classic C-scans, as it is adapted to the cup-shaped posterior pole of the retina.

A 3D reference protocol is used to acquire the macular cube. This feature allows the acquisition of 141 horizontal scans of 385 A-scans each, performed over a 7 x 7 mm square (54,285 pixels). The operation is complete in 2.4. It generates a macular cube by adjusting horizontal alignment errors (eye movement). The *en face* images are reconstructed from these data.

En face capability allows the operator to select the *en face* profile following the ILM curvature or RPE real curvature. The profile should be selected according to the retinal or choroidal layers under control.

ILM will be used in case of vitreal or vitreoretinal anomalies; inner plexiform layer (IPL) in case of edema or exudates. RPE reference protocol is used in outer retina or choroid diseases.

En face RTVue technology can follow the irregularities of a pathologic RPE or the ILM (RPE choice). It will give beautiful 3D images that are useful to understand the lesion size, shape, and dimensions, and are perfect as teaching tool.

In case of low signal-to-noise ratio or localized RPE alterations, the device is able to calculate the ideal average RPE shape (RPE reference choice). We normally choose as the reference RPE that gives us the pigment epithelium ideal concavity for the section we want to obtain. *It cancels the abnormalities of the pigment epithelium in the eye under study* and gives a theoretical ideal tridimensional reference lamina.

RPE reference may be tuned by changing the scan thickness (default value 30 μm). Reducing the scan thickness will obtain a thin retinal slice. The image will be a little noisy, but the sensitivity will be very high. If we increase the scan thickness the image will be smoother, but little details may be lost as we will not have a thin lamina but rather a large retinal slab. This procedure should be applied, for example, to visualize better the choroid. Sattler layer should be studied with a thin tissue slice (small vessels), and Haller layer (greater vessels) with a thicker slice.

We can choose the scan depth by shifting the *en face* cup-shaped lamina forward or back to the retinal layer we want to study or to the different layers of the choroid.

En Face Scan Procedure Suggestions and Parameters

High quality *en face* scans can be obtained on a routine basis.

1. Locate macula and lesion.

2. Select scan profile (ILM, IPL, RPE, or RPE reference).

   a. ILM follows the irregularities of normal and pathologic ILM. It gives 3D images to understand the lesion size, shape, and dimensions.

   b. IPL follows the irregularities of normal and pathologic IPL.

   c. RPE follows the normal RPE profile or the irregularities of a pathologic RPE. It gives 3D images that illustrate the lesion size, shape, and dimensions.

   d. RPE reference cancels the pigment epithelium anomalies and gives a theoretical ideal tridimensional paraboloid reference lamina that cuts through the lesions.

3. Select scan thickness (using RPE reference).

   a. At 10 μm or less the image will be noisy but the sensitivity high.

   b. At 16 to 30 μm the image will be smoother but details will be lost.

4. Select scan depth (using RPE reference).

   a. The cup-shaped scan is at a constant depth, parallel to RPE.

   b. The scan is shifted forward to study the retinal layers: 3 to 5 μm to cut through the drusen; 15 to 300 μm to cut through PED, retinal edema, or macular holes.

   c. The scan is shifted back to study the different layers of the choroid. Sattler layer should be studied at a depth of 80 to 90 μm, with a thin 10-μm slice. Haller layer should be studied at a depth of 160 to 190 μm with a 20- to 30-μm slice.

5. Adjust contrast and brightness.

The device automatically gives on the screen the exact depth location inside the retina or the choroid. The retinal or choroidal scan can be shown in black and white or in color. It can be enlarged to full screen and rotated as necessary.

The *en face* frontal scans images we study are new to us, much more complex than classic cross-section B-scans. They are more difficult to read and less intuitive. A special effort has to be made to read them the first times, but the learning curve is easy and fast. C-scans adapted or not to RPE concavity allow appreciation of lesion diameters, visualization of pathologies involving the vitreoretinal interface and the retinal deep layers, and the choroid. They are also useful in quantification of retinal thickness changes.

RTVue *En Face* Scans: Clinical Applications

Clinical studies have confirmed the interest of transverse OCT scans. In central serous chorioretinopathy, frontal scans allow the study of the dimensions of PED and a minute control of the photoreceptors.[4-7]

In polypoidal choroidal vasculopathy, *en face* scans allow the study of PED and of their walls.[8,9]

In diabetic retinopathy, *en face* scans help understand the type and extension of retinal edema (diffuse and cystoid) and exudates, as well as subretinal new vessels.[10,11]

Transverse scans are essential in studying cystoid macular edema, allowing identification of the extension and shape of the edema pseudocysts, their evolution, their pattern, and evolutive stage (early, advanced, regressive). This imaging method illustrates the evolution of cystoid cells volume after intravitreal injections of antiangiogenic drugs (see Chapter 21).[12]

In age-related macular degeneration (AMD), PED *en face* scans allow the study of the shape and dimensions of PED and a precise assessment of their walls. They are essential in studying diffuse and cystoid macular edema, allowing identification of the extension and evolutive stage (early, advanced, regressive) in venous branch occlusion.[12] It is also useful in acute zonal occult outer retinopathy (AZOOR) and in juxtafoveolar retinal telangiectasia.[13,14]

In cases of vitreoretinal interface lesions, epiretinal membranes, and macular holes, 3D imaging enables *in vivo* identification of the extension and dynamics of epiretinal traction. It offers frontal images that facilitate understanding of the abnormalities in the vitreofoveal interface. It provides transverse images that allow much more precise 3D observation of extending intraretinal structural changes associated with a macular hole and vitreoretinal tractions than cross-section conventional OCT imaging does, especially in the photoreceptor inner

segment (IS) and outer segment (OS). Epiretinal membranes are clearly delineated in the *en face* view, and the distribution of traction forces throughout the intraretinal layers is identified down to the level of the RPE. During follow-up, quantification of release in retinal traction is possible.[15-18] In myopic retinoschisis, C-scans allow the study of the patterns of retinal layer delamination.[19,20] This chapter will also present a few cases highlighting the clinical interest of RTVue *en face* scans.

Shifting the *en face* scan forward, we cut through the normal retinal layers and pathological formations: drusen, PEDs, subretinal new vessels, etc. We are able to study the different features of PEDs, their wall shape, thickness dimensions, etc. Lesion shape can be regular or irregular. It is important to measure the diameter and dimensions. The PED walls can be thick or thin, regular or not, smooth or rough, and have spikes. Fluid collection content reflectivity can be assessed (high, medium, low; homogenous or not).

Shifting the *en face* scan back, we cut through choroidal layers to the sclera.

Age-Related Macular Degeneration with Elevated Retinal Pigment Epithelium and Occult Choroidal Neovascularization

En face RTVue technology can follow the irregularities of a pathologic RPE obtaining a delineation of the PED. The 3D image gives a scenic view of the mountain-shaped epithelium detachment.

En face frontal scan follows the RPE cup, ignoring the irregularities and reconstructing the ideal paraboloid concavity. It will cut the PED at different levels, starting immediately in front of RPE and then close to the top. This will clearly show the walls of the PED; these walls are wrinkled and polygonal, irregular in shape with some clusters of irregularities. Some spikes protrude from the walls and a few partitions divide the PED. They are thin, but their thickness is irregular. A transverse section of the irregular detachment of the pigment epithelium, above the pigment epithelium and the frontal scan through the top of the RPE detachment, shows a highly irregular section with a cluster of rounded formations that suggests that the *en face* scans cut through a very irregular polylobed dome (Figure 9-4).

Soft Drusens in Age-Related Macular Degeneration

The RTVue frontal scans are performed by cutting the drusen slightly in front of the RPE (1 to 5 μm). The dissected drusen are seen as rounded formations, slightly irregular with a relatively thick wall (Figure 9-5).

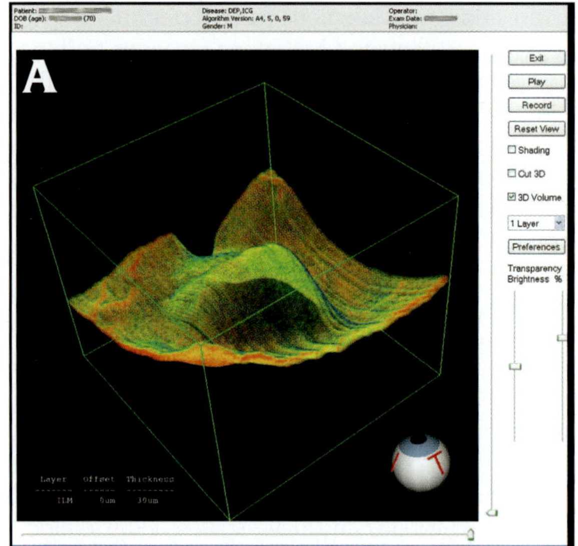

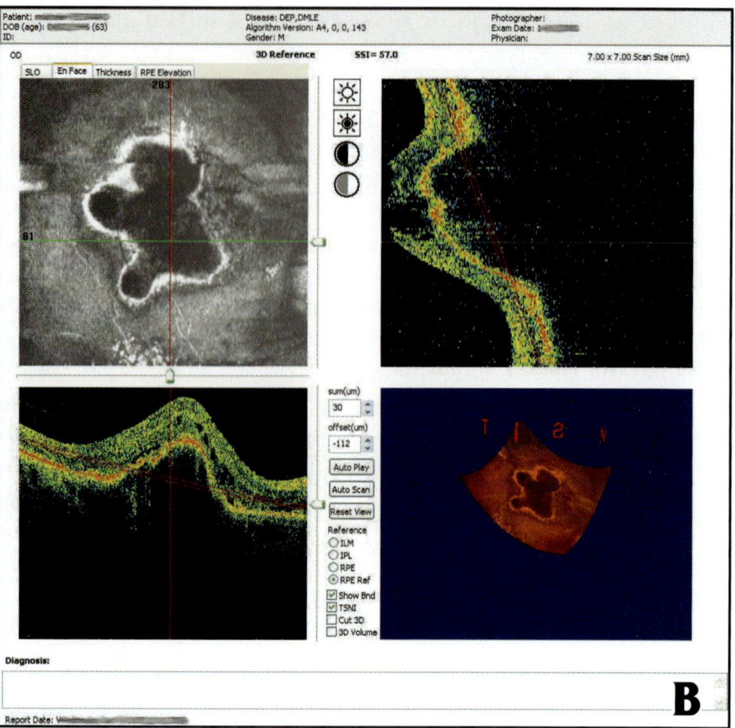

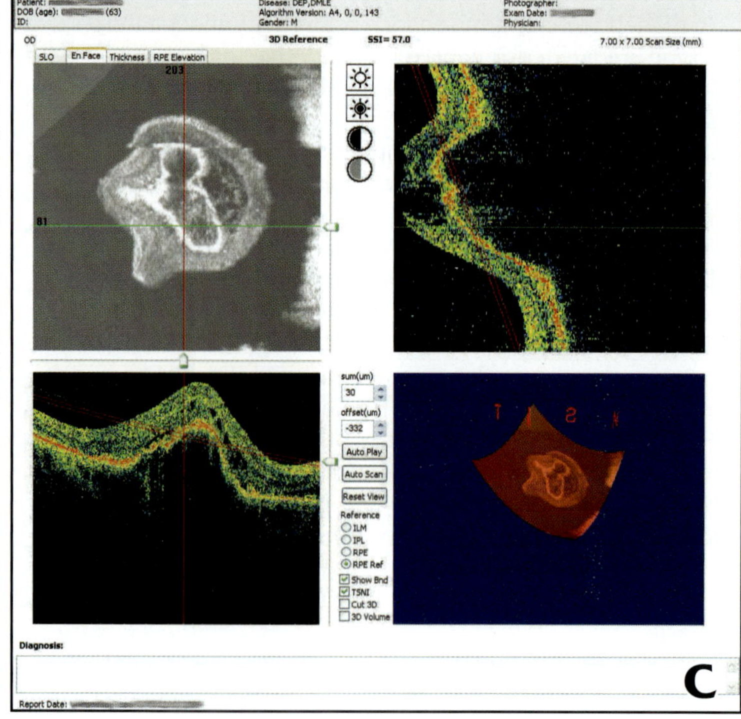

Figure 9-4. AMD with elevated RPE and occult choroidal neovascularization. (A) En face RTVue technology can follow the irregularities of a pathologic RPE: we obtain a delineation of the PED. The 3D image gives a scenic view of the mountain-shaped PED. It is clinically more important to have an en face frontal scan following the ideal RPE cup, ignoring the irregularities and reconstructing the ideal paraboloid concavity. The RTVue frontal scans cut the PED at different levels, starting immediately in front of RPE and then close to the top. (B) Transverse section of the irregular detachment of the pigment epithelium, 112 μm above the pigment epithelium. They show the walls of the PED; these walls are wrinkled and irregular in shape with some clusters of irregularities. Some spikes protrude from the walls and a few partitions divide the PED. The walls are thin and their thickness is irregular. (C) Transverse section of the irregular detachment of the pigment epithelium, 300 μm above the pigment epithelium. The frontal scan through the top of the RPE detachment shows a highly irregular section with a cluster of rounded formations that suggests that the en face scan cuts through a very irregular polylobed dome.

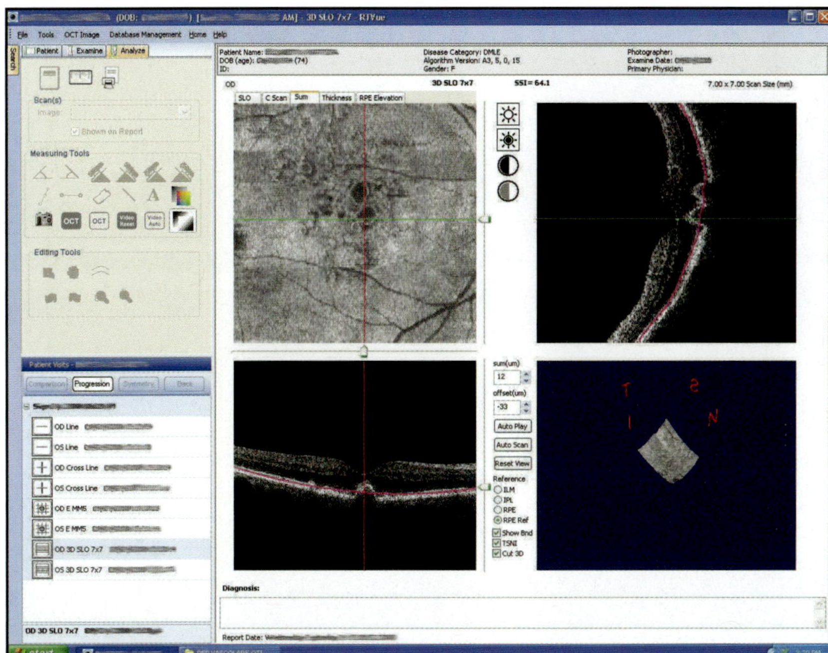

Figure 9-5. Soft drusens in AMD. OCT line scans through the drusen show irregular elevations of RPE. Bruch's membrane can be seen as a horizontal thin line below the elevation of the RPE. The RTVue frontal scans are performed by cutting the drusen slightly in front of the RPE (1 to 5 μm). The dissected drusen are seen as rounded formations, slightly irregular with a relatively thick wall.

Idiopathic Polypoidal Choroidal Vasculopathy

OCT B-scans through the polyps show a big bilobed detachment of the pigment epithelium. Below the macula in the OCT we find multiple detachments of the pigment epithelium with irregular, hyper-reflective areas at the level of choriocapillaris-RPE complex, probably corresponding to polyps. The RTVue frontal scans are performed by cutting the PED at different levels, starting immediately in front of RPE and then close to the top. They highlight the walls of the PED that are smooth, thin, and rounded in shape with some irregularities. The wall thickness is regular (Figure 9-6).

Macular Full-Thickness Hole

The transverse image shows the complexity of the cystoid edema with 2 rows of minute pseudocysts around the hole itself. The 3D imaging and *en face* scans of macular holes give 3D overviews that help to understand the structural pathogenesis of the lesions. This will also provide a stack of consecutive transverse images parallel to RPE that allow a precise observation of intraretinal structural changes in the outer retina and in the photoreceptor IS and OS (Figure 9-7).

Normal and Pathologic Choroid

The frontal scans parallel to RPE can be applied to visualize the choroid layers, and vascular images that can be converted to indocyanine green angiographies are obtained without using invasive techniques. Sattler layer small vessels should be studied with thin scan (10

μm), the Haller layer with a thicker scan (20 to 30 μm). Going deeper to the sclera, short ciliary posterior arteries can sometimes be localized (Figure 9-8).

CONCLUSION

With RTVue technology, en face scans are easy to perform and to read. Their analysis brings new important elements in the study of retinal diseases. Frontal scans adapted to cup-shaped pigment epithelium bring new insights and help diagnose and follow up in a wide range of pathology. They will take a great clinical importance in the near future.

REFERENCES

1. Podoleanu AG, Dobre GM, Webb DJ, Jackson DA. Simultaneous en face imaging of two layers in the human retina by low-coherence reflectometry. *Optic Letters.* 1997;22:1039-1044.

2. Podoleanu AG, Dobre GM, Seeger M, et al. Low coherence interferometry for *en face* imaging of the retina. *Lasers and Light.* 1998;8:187-192.

3. Schmidt-Erfurth U, Leitgeb RA, Michels S, et al. Three-dimensional ultrahigh-resolution optical coherence tomography of macular diseases. *Invest Ophthalmol Vis Sci.* 2005;46:3393-3346.

4. van Velthoven ME. *Combined En-Face Optical Coherence Tomography and Confocal Ophthalmoscopy for Retinal Imaging.* Thesis. University of Amsterdam; 2006.

5. van Velthoven ME, Ongkosuwito JV, Verbraak FD, Schlingemann RO, de Smet MD. Combined *en face* optical coherence tomography and confocal ophthalmoscopy findings in active multifocal and serpiginous chorioretinitis. *Am J Ophthalmol.* 2006;141(5):972-975.

Figure 9-6. Idiopathic polypoidal choroidal vasculopathy. The fundus shows the presence of a PED with hemorrhagic contents and a small subfoveal hemorrhage. (A) Indocyanine green angiography shows a cluster of saccular, hyperfluorescent spots suggestive of a net of idiopathic polypoidal choroidal vasculopathy. Above the cluster, a rounded, densely opaque PED is evident, suggestive of hemorrhagic PED. (B) OCT line scans through these polyps show a big, bilobed detachment of the pigment epithelium. Below the macula in the OCT we find multiple detachments of the pigment epithelium with irregular, hyper-reflective areas at the level of choriocapillaris-RPE complex, probably corresponding to polyps.

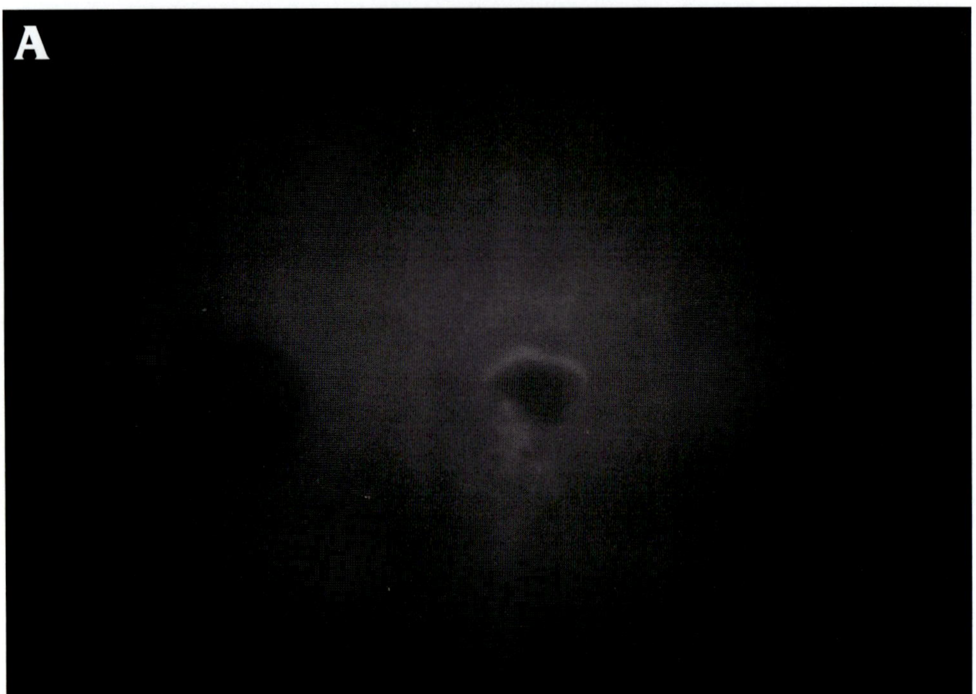

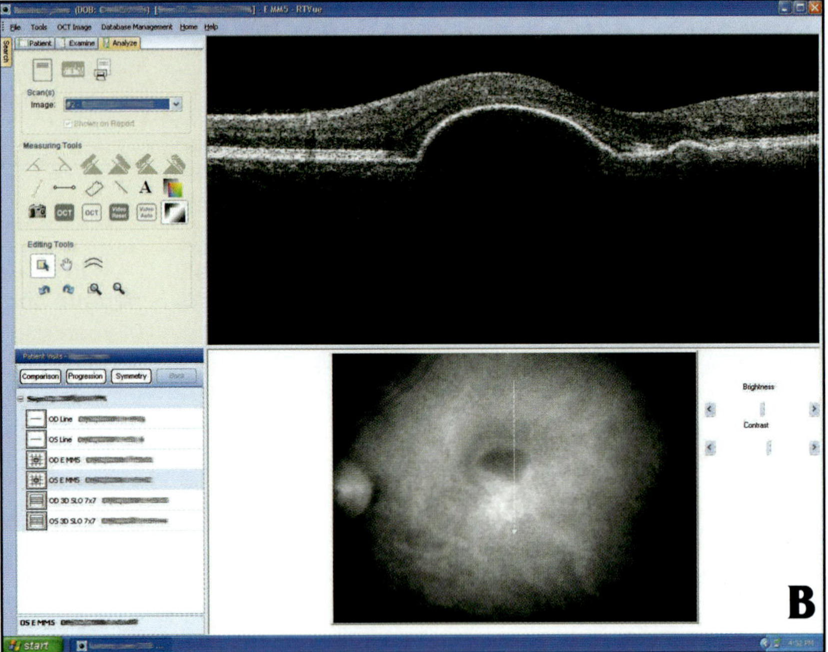

6. van Velthoven ME, Verbraak FD, Yannuzzi LA, et al. Imaging the retina by en face optical coherence tomography. *Retina.* 2006;26(2):129-136.

7. Stock G, Ahlers C, Sayegh R, et al. Three-dimensional imaging in central serous chorioretinopathy. *Ophthalmologe.* 2008;105(12):1127-1134.

8. Saito M, Iida T, Nagayama D. Cross-sectional and en face optical coherence tomographic features of polypoidal choroidal vasculopathy. *Retina.* 2008;28(3):459-464.

9. Kameda T, Tsujikawa A, Otani A, et al. Polypoidal choroidal vasculopathy examined with en face optical coherence tomography. *Clin Experiment Ophthalmol.* 2007;35(7):596-601.

10. Bolz M, Ritter M, Schneider M, et al. A systematic correlation of angiography and high-resolution optical coherence tomography in diabetic macular edema. *Ophthalmology.* 2009;116(1):66-72.

11. de Bruin DM, Burnes DL, Loewenstein J, et al. In vivo three-dimensional imaging of neovascular age-related macular degeneration using optical frequency domain imaging at 1050 nm. *Invest Ophthalmol Vis Sci.* 2008;49(10):4545-4552.

12. Brinkmann CK, Wolf S, Wolf-Schnurrbusch UE. Multimodal imaging in macular diagnostics: Combined OCT-SLO improves therapeutical monitoring. *Graefes Arch Clin Exp Ophthalmol.* 2008;246(1):9-16.

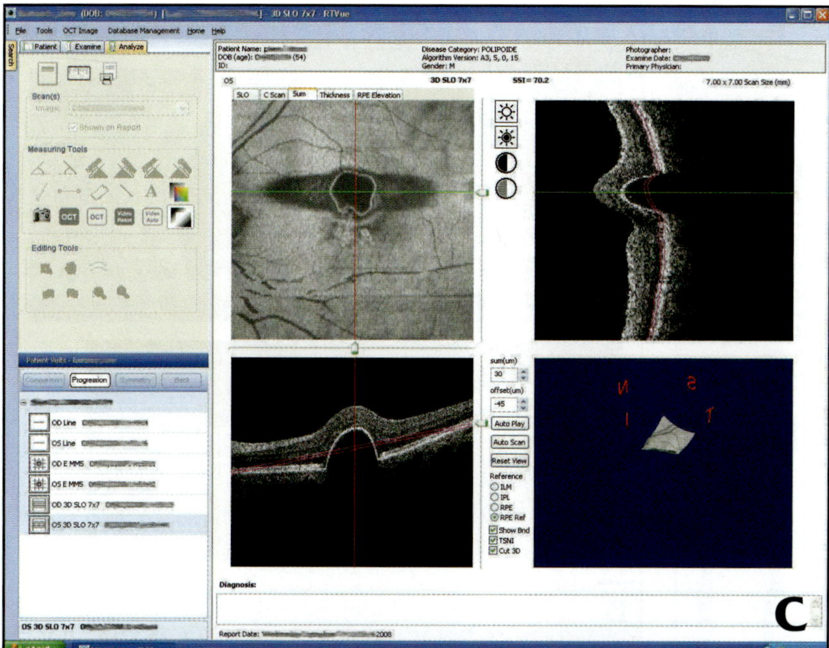

Figure 9-6. (C) The RTVue frontal scan is performed by cutting the PED immediately in front of RPE. It highlights the walls of the PED that are smooth, thin, and rounded in shape with some irregularities. The wall thickness is regular.

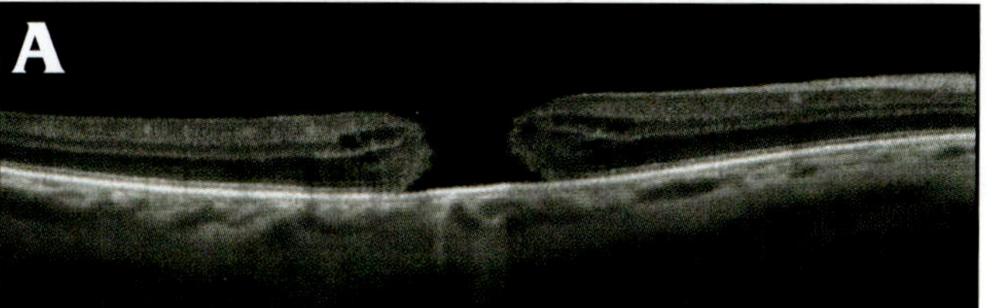

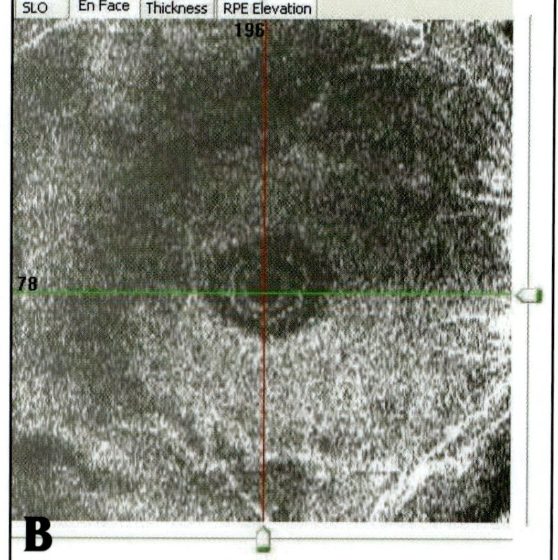

Figure 9-7. Macular hole. OCT line scan (A) and *en face* scan (B) through a full-thickness macular hole. The *en face* image shows 2 rows of minute irregular cystoid macular edema pseudocysts around the hole itself. *En face* scans allow a precise observation of intraretinal structural changes in the outer retina and in the photoreceptor IS and OS.

13. Takai Y, Ishiko S, Kagokawa H, et al. Morphological study of acute zonal occult outer retinopathy (AZOOR) by multiplanar optical coherence tomography. *Acta Ophthalmol.* 2009;87(4):408-418.

14. Koizumi H, Iida T, Maruko I. Morphologic features of group 2A idiopathic juxtafoveolar retinal telangiectasis in three-dimensional optical coherence tomography. *Am J Ophthalmol.* 2006;142(2):340-343.

15. Hangai M, Ojima Y, Gotoh N, et al. Three-dimensional imaging of macular holes with high-speed optical coherence tomography. *Ophthalmology.* 2007;114(4):763-773.

16. Tammewar AM, Bartsch DU, Kozak I, et al. Imaging vitreomacular interface abnormalities in the coronal plane by simultaneous combined scanning laser and optical coherence tomography. *Br J Ophthalmol.* 2009;93(3):366-372.

17. Smith AJ, Telander DG, Zawadzki RJ, et al. High-resolution Fourier-domain optical coherence tomography and microperimetric findings after macula-off retinal detachment repair. *Ophthalmology.* 2008;115(11):1923-1929.

18. Georgopoulos M, Geitzenauer W, Ahlers C, et al. Three dimensional OCT for evaluating the retinal architecture before and after surgery for vitreo retinal traction. *Ophthalmologe.* 2008;8:753-760.

19. Forte R, Pascotto F, Cennamo G, de Crecchio G. Evaluation of peripapillary detachment in pathologic myopia with en face optical coherence tomography. *Eye.* 2008;22(1):158-161.

20. Forte R, Pascotto F, de Crecchio G. Visualization of vitreomacular tractions with en face optical coherence tomography. *Eye.* 2007;21(11):1391-1394.

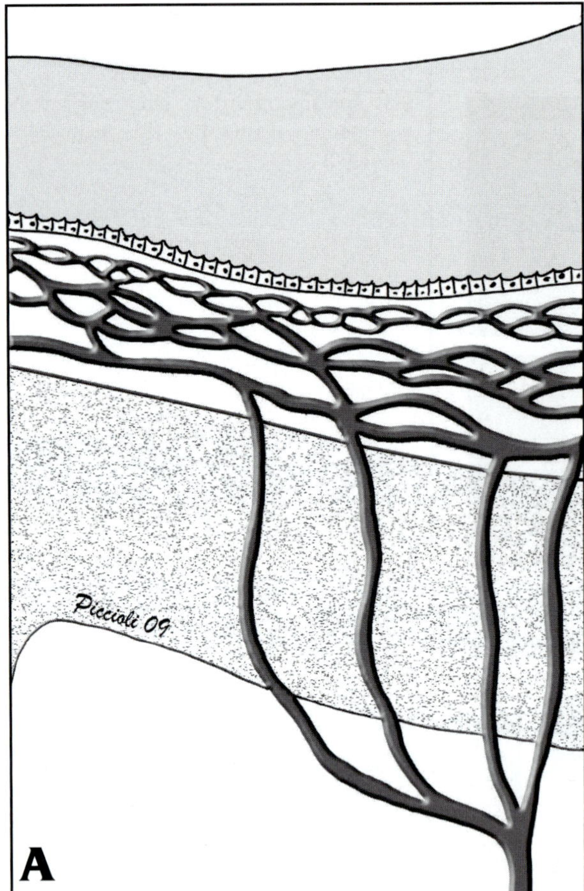

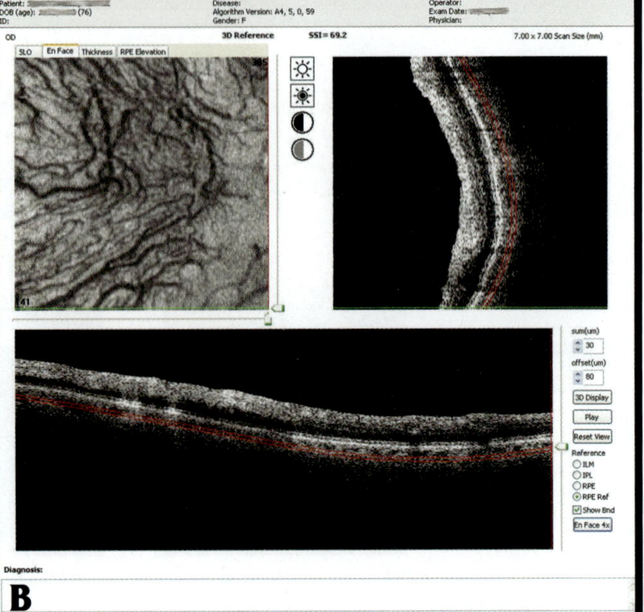

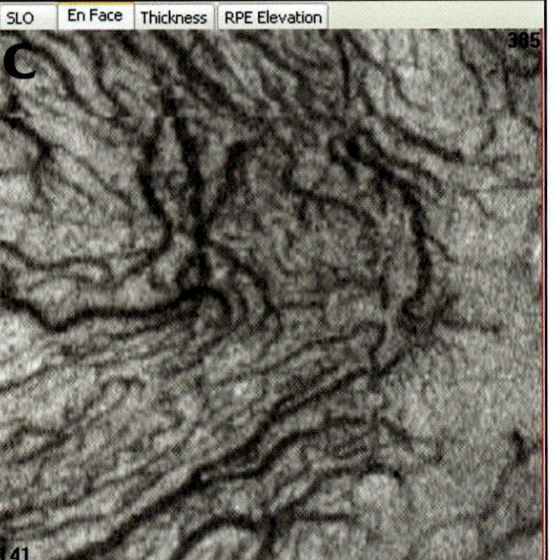

Figure 9-8. Normal choroid. The RTVue frontal scans visualize the choroid layers, acquiring vascular images that can be confronted to indocyanine green angiographies without using invasive techniques. (A) Schematic line drawing of sclera and choroidal layers. Short posterior ciliary arteries penetrate the sclera to reach the choriocapillaris. (B) Choroid Sattler layer. Transverse scan parallel to RPE is taken at a constant depth in the choroid, 83 µm below the pigment epithelium. The small choroidal vessels are seen as dark hyporeflective shadows, smaller and more irregular than the bigger vessels. Sattler layer small vessels should be studied with thin tissue scan (10 µm). (C) Choroid Haller layer. Transverse scan parallel to RPE is taken at a constant depth in the choroid, 169 µm below the pigment epithelium. The big choroidal vessels are seen as dark hyporeflective regular shadows. In the Haller layer, we can use a thicker scan slice (20 to 30 µm). Going deeper to the sclera, short ciliary posterior arteries can sometimes be localized. (Figure 9-8A illustration by Ms. D. Piccioli.)

Imaging of Retinal Blood Flow

Yimin Wang, PhD and David Huang, MD, PhD

Doppler OCT[1,2] is based on the principle that moving particles, such as red blood cells inside a blood vessel, cause a Doppler frequency shift (Δf) (Figure 10-1) to the light scattered according to:

$$\Delta f = \frac{1}{2\pi}(\vec{k_s} - \vec{k_i}) \bullet V \qquad \text{(Equation 1)}$$

where k_i and k_s are wave vectors of incident and scattered light, respectively, and V is the velocity vector of the moving particles. In an OCT system, only backscattered light is detected. Given the angle θ between the scanning beam and the flow direction, the Doppler shift is simplified to:

$$\Delta f = -2Vn\cos\theta / \lambda_0 \qquad \text{(Equation 2)}$$

where λ_0 is the center wavelength of the light source, and n is the refractive index of the medium.

In Fourier-domain OCT, the frequency shift Δf will introduce a phase shift in the spectral interference pattern captured by the line camera. With fast Fourier transform (FFT), the transform result is a complex function characterized by amplitude and phase. The phase difference between sequential axial scans at each pixel is calculated to determine the Doppler shift.[3-5] One limitation of phase-resolved flow measurement is an aliasing phenomenon caused by 2π ambiguity in the arctangent function. This limits the maximum determinable Doppler shift to $\Delta f = 1/(2\tau)$, where τ is the time difference between sequential axial lines. Thus, the maximum detectable speed is $V = \lambda_0/(4n\tau\cos\theta)$. For RTVue OCT, at a line rate of 27,230 Hz, the maximum detectable axial velocity component is 4.21 mm/s.

If flow speed can be measured, the volumetric blood flow can be determined. However, from Equations 1 and 2, it can be seen that in Doppler OCT the frequency shift is caused by flowing particles with a velocity component in the direction of the illuminating beam. Doppler OCT only measures the value of $V\cos\theta$. To determine real flow speed V, the angle θ between probe beam and blood flow must be decoupled from the measured Doppler frequency shift. Therefore, more measurements are needed to determine angle θ.

SCANNING PATTERN

The RTVue OCT system uses a double circular scan pattern (DCSP) to measure retinal blood flow.[6,7] The method was initially developed at the Center for Ophthalmic Optics and Lasers (www.COOLlab.net) at the University of Southern California (USC) by the authors. The Doppler scan pattern is currently available only to research centers participating in a collaborative study. It has not yet been publicly released.

In DCSP pattern, the OCT sampling beam scans a circular pattern on the retina (Figure 10-2A) at radii of $r_1 = 1.7$ mm and $r_2 = 1.875$ mm. A small radius difference $\Delta r_0 = 0.175$ mm is chosen so that the course of the blood vessel segment between the scanning circles is approximately a straight line. The relative positions of a blood vessel on the 2 OCT images are used to determine the Doppler angle θ between the probe beam and blood flow (Figure 10-2B). Then the total flow speed can be determined for volume flow calculation.

Huang D, Duker JS, Fujimoto JG, Lumbroso B, Schuman JS, Weinreb RN.
Imaging the Eye from Front to Back with RTVue Fourier-Domain Optical Coherence Tomography (pp 97-102).

Figure 10-1. In OCT, a probe beam is reflected from a sample. In a moving sample, such as cells in a blood vessel, the frequency of reflected light is Doppler shifted in proportion to the component of the flow velocity in the direction of beam propagation.

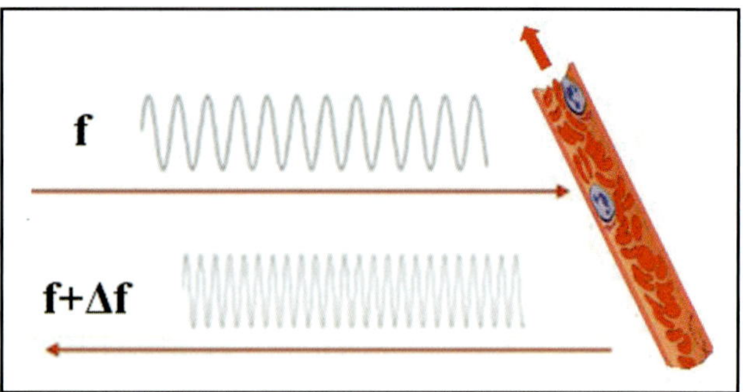

Figure 10-2. Diagram of DSCP. (A) The scan is centered on the optic nerve head. (B) The scans transect the major retinal vessels. The angle θ between the OCT beam and blood flow is determined by the relative positions of the vessel in the 2 cylindrical sections.

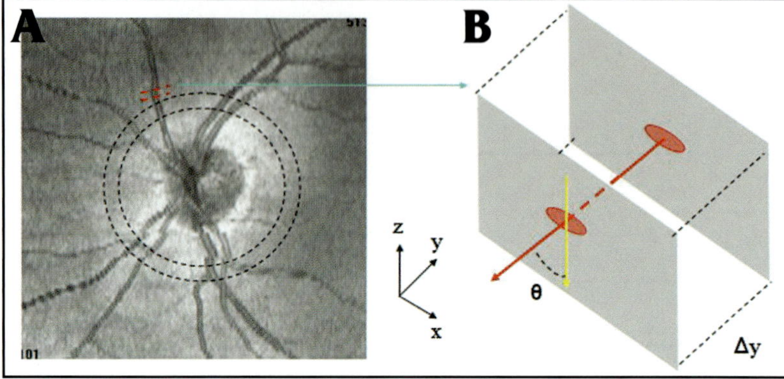

There are 4000 A-lines sampled in each circle. The phase differences for every 4 A-lines are calculated to get the Doppler frequency shift. Thus, each frame consisted of 1000 vertical lines. There are 6 pairs of Doppler Fourier-domain OCT images sampled for each flow measurement. Six are sampled at radius r_1, and 6 are from radius r_2. The total recording time for the 12-image set is approximately 2 seconds.

Figure 10-3 shows examples of Doppler OCT images sampled in 2 circles around the optic disc (dashed circles in Figure 10-3A). The color signal shows flow information, and the gray signal shows retinal structure. In Figures 10-3B and C, the lateral axis is the angular distribution α from 0 to 360 degrees, while the vertical axis shows the depth information in the retina. According to Figure 10-3, the 2 positions of the blood vessel segment on the 2 scanning circles have coordinates $P_1(r_1, \alpha_1, z_1)$, and $P_2(r_2, \alpha_2, z_2)$, respectively. The vector of the blood vessel can be expressed as: r_b ($\Delta x = r_1\cos\alpha_1 - r_2\cos\alpha_2$, $\Delta y = r_1\sin\alpha_1 - r_2\sin\alpha_2$, $\Delta z = z_1 - z_2$).

For OCT scans at radius r, when the probe beam scans to the angle α, the vector of the scanning beam is s ($r\cos\alpha$, $r\sin\alpha$, $-h$), where h is the distance from the nodal point to the retina. With vector s and rb, the angle θ between the OCT probe beam and blood flow can be calculated as[7]:

$$\cos\theta = (\vec{r_b} \bullet \vec{s})/(R_b R_s)$$

$$R_b = \sqrt{\Delta x^2 + \Delta y^2 + \Delta z^2} \qquad \text{(Equation 3)}$$

$$R_s = \sqrt{r^2 + h^2}$$

where R_b is the length of the vector r_b, and R_s is the length of vector s. Because the difference in radii between the 2 scanning circles is small, the radius r in Equation 3 is chosen as r = $(r_1 + r_2)/2$.

FLOW CALCULATION

Assuming the relative flow profile does not change during the short sampling period, the flow velocity V can be expressed as

$$V(x, z, t) = A(x_v, z_v)P(t) \qquad \text{(Equation 4)}$$

where $A(x_v, z_v)$ is the speed distribution in the cross-section of the blood vessel, and P(t) shows the pulsation time course of the flow. The volume flow is calculated through integrating the velocity within the vessel cross-sectional areas and recorded time.[7]

Influence of Sampling Density on Doppler Shift Measurement

In Doppler OCT, the phase difference between sequential axial scans is calculated to determine Doppler frequen-

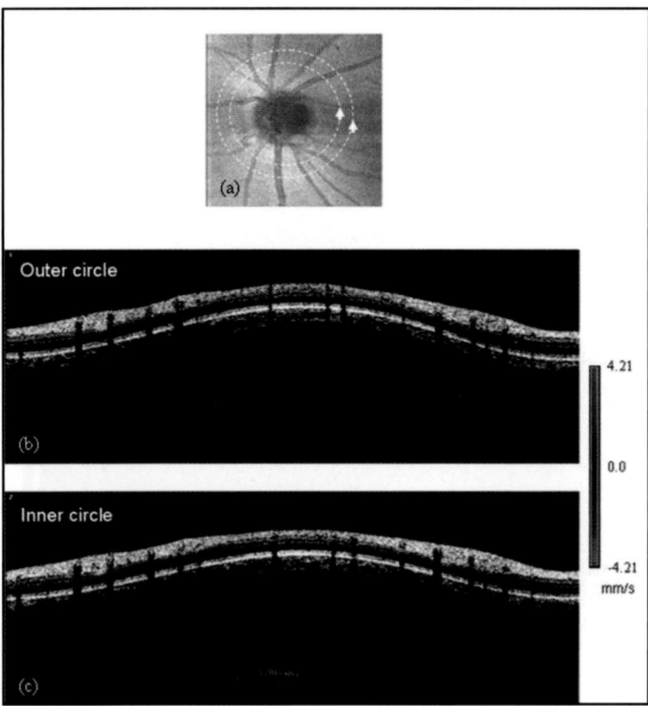

Figure 10-3. Doppler OCT images sampled with DSCP method. (A) Scan pattern. (B and C) Inner and outer cylindrical OCT sections.

cy shift. For retinal OCT systems, the probe beam scans continuously across the retina and there is a small displacement between sequential axial scans. If the sampling locations were not negligibly small relative to the beam diameter, phase decorrelation would decrease the measured Doppler shift. In the RTVue Fourier-domain Doppler OCT system, there are 4000 axial lines for each circle. At a scanning radius of 1.8 mm (circumference 11.30 mm), the sampling step is about 2.8 µm. We use a calibrated factor of 0.8 to correct the measured flow result.

Removal of Background Doppler Signal Due to Eye Motion

The Doppler shift measured by OCT is generated by both the blood flow and the motion of the eye relative to the OCT system. To isolate the blood flow component, the background axial motion of the retinal tissue proximal to each blood vessel is measured and subtracted from the total measured Doppler shift.

After background noise removal, the retinal blood flow F can be calculated through integration within the vessel region in the sampled Doppler Fourier-domain OCT image as[7]:

$$\overline{F} = k|\cos \beta| \sum_S rA_p \Delta\alpha\Delta D \qquad \text{(Equation 5)}$$

$$k = \frac{1}{T}\int_0^T P(t)dt$$

$$\cos \beta = (\vec{r_s} \bullet \vec{r_b}) / R_b$$

where k is the cardiac pulsation factor. We choose to measure flow in the branch veins rather than arteries because the arteries have higher flow velocities that can cause excessive phase wrapping and signal fading. The identification of branch veins among the other vessels distributed around the optic nerve head is based on the recorded Doppler frequency shift and the calculated angle between the probe beam and blood vessel. For each vessel, the inner blood vessel diameters (D) are measured from the Doppler OCT image with the highest Doppler signal. For each vein, the average flow velocity (V_{ave}) is calculated by: $V_{ave}=F/S$, where S is the vessel area size: $S=\pi(D/2)^2$. Blood flows for the individual branch veins in superior and inferior retina are summed separately to obtain the hemispheric flow rate.

CLINICAL STUDIES

Clinical studies have been carried out on normal subjects and glaucoma patients as part of the Advanced Imaging for Glaucoma (AIG) study (www.AIGStudy.net) at USC. The inclusion and exclusion criteria for normal and perimetric glaucoma (PG) subjects are described in the AIG Study Manual of Procedures, which is available at the Web site.

Normal Subjects

Ten healthy human subjects (4 male, 6 female) between the ages of 35 to 69 years (average 57.4 years) participated in the study at the USC Doheny Eye Institute (DEI).[8] The right eyes were measured.

In our study, retinal blood flow could be determined for 8 of the 10 normal subjects. We were unable to obtain retinal blood flow measurements from 2 subjects; one was due to poor fixation, and the other was due to a tilted disc. The total venous blood flow ranged from 40.8 to 52.9 µL/min.[8] Table 10-1 lists the statistics of these measurements. The mean ± SD value (standard deviation) for the flow results for 8 normal subjects was 45.6 ± 3.8 µL/min. The coefficient variation (CV) for total venous flow was 10.5%.

Blood flow in the superior (F_{sup}) and inferior (F_{inf}) retinal hemispheres were examined separately. Venous flow had a mean ± SD of 23.5 ±3.0 µL/min for the superior retina and 22.2 ± 2.6 µL/min for the inferior retina. There was no statistically significant difference (p = 0.4) between the flows for the superior and inferior retinal hemispheres based on a paired t test. The inner diameters (D) of retinal veins were measured from the Doppler OCT images and ranged from 33.3 to 155.4 µm. The average flow velocity in the veins was 19.6 ± 2.9 mm/s for the combined data from all subjects. Linear regression did not show any significant correlation between flow velocity and vein diameter.

Distribution and Reproducibility Statistics of Retinal Vein Measurements				Table 10-1
Parameters	**Distribution** **Mean ± SD [Range]**		**Reproducibility** **Pooled SD**	**CV (%)**
Total Venous Flow Rate (µL/min)	45.6 ± 3.8 [40.8, 52.9]		4.8	10.5

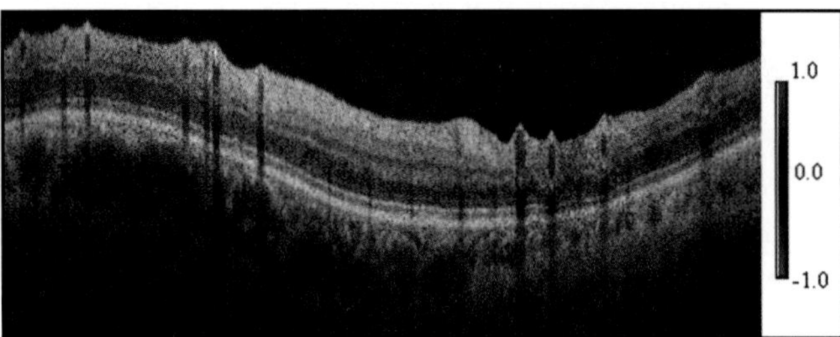

Figure 10-4. Doppler OCT image of 1 PG patient.

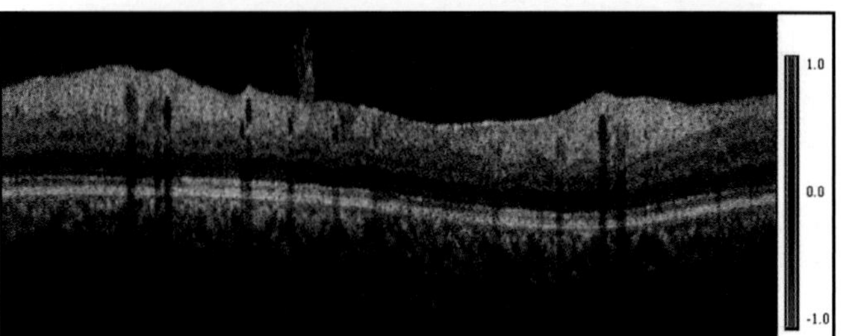

Figure 10-5. Doppler image of the PDR patient.

Glaucomatous Subjects

Fifteen PG patients (8 male, 7 female) between the ages of 49 to 69 years (average 62.4 years) were studied at DEI.[9] The right eyes were measured for each patient. For PG patients, retinal blood flow could be determined from 12 patients among the 15 subjects. Examinations for the other 3 patients were failed due to poor fixation. Figure 10-4 shows the Doppler OCT image from 1 PG subject. His visual field test results were 14.99 (pattern standard deviation or PSD) and -12.25 (mean deviation or MD).

Our preliminary study showed that retinal blood flow in PG patients was significantly lower than that of normal group, and only 1 of 12 PG eyes had retinal blood flow within the normal range. Among the PG eyes, visual field mean deviation was highly correlated with total retinal blood flow—the lower the flow, the greater the visual field loss.[9]

Diabetic Retinopathy

Two diabetic patients with and without retinopathy were recruited to participate in the study at DEI.[10] The first patient was a 26-year-old Hispanic female with Type 2 diabetes of 8 years' duration, with no retinopathy. The second patient was a 37-year-old Hispanic female, with Type 1 diabetes of 34 years' duration, with quiescent proliferative diabetic retinopathy (PDR), status postvitrectomy and postendolaser panretinal photocoagulation. For each subject, 1 eye was measured in this study. Figure 10-5 shows the Doppler OCT image of the PDR patient.

The measured total retinal blood flows of each subject were averaged and the CV was calculated. Table 10-2 shows the result for the 2 patients. The first patient (#1) had diabetes but no retinopathy. Her measured total retinal blood flow had a range from 37.4 to 49.2 µL/min. The mean value was 43.3 µL/min, and the standard deviation was 4.4 µL/min. The CV was 10.2%. For the second patient (#2) with PDR, the measured total retinal blood flow had a range from 30.1 to 39.6 µL/min. The mean value was 33.2 µL/min, the SD was 3.7 µL/min, and the CV was 11.3%.

The averaged superior and inferior hemispheric flow for the first patient was 21.9 and 21.4 µL/min separately. The flow difference ∆F between superior and inferior was 0.5 µL/min. This value showed there was no big difference comparing the blood flow between superior and inferior retina, and her flow distribution was balanced. For the PDR eye, the superior hemispheric flow was 20.4 µL/min while the inferior hemispheric flow was 12.8 µL/min. The flow difference between superior and inferior was 7.6 µL/min. The inferior hemispheric blood flow of the PDR eye was well below the normal range. Averaged flow speed of the PDR eye was also below the normal range. This study shows that Doppler OCT is able to detect decreased blood flow in an eye with treated proliferative diabetic retinopathy.

CONCLUSION

Retinal blood flow measurement with the DCSP is an experimental procedure on the RTVue Fourier-domain OCT system. Preliminary data show that the total retinal venous flow can be measured with a good reproducibility of 10%. Our technique is unique in that

Table 10-2

TOTAL RETINAL BLOOD FLOW OF DIABETIC SUBJECTS

PATIENTS	NUMBER OF VEINS	FLOW RANGE (µL/MIN)	MEAN	SD	CV (%)
#1	6	37.4 ~ 49.2	43.3	4.4	10.2
#2	7	30.1 ~ 39.6	33.2	3.7	11.3

it is able to measure the blood flow in all retinal branch veins several times per second. The measurements are in absolute numbers that can be compared between eyes, between subjects, and between visits. The background retinal motion can be measured from the OCT image and is subtracted from the calculation of blood flow motion. This key difference with laser Doppler flowmetry may allow OCT to provide higher precision and accuracy. This promising new technique is being developed and validated within the Doppler Optical Coherence Tomography Retinal Circulation Research Group (DOCTRC) and will be released for general use on the RTVue platform in the future. By offering a rapid and accurate method of quantifying retinal blood flow, Doppler OCT may help us assess and develop treatment for glaucoma, diabetic retinopathy, and other optic nerve and retinal diseases associated with abnormal blood flow.

REFERENCES

1. Huang D, Swanson EA, Lin CP, et al. Optical coherence tomography. *Science.* 1991;254:1178-1181.

2. Wang XJ, Milner TE, Nelson JS. Characterization of fluid flow velocity by optical Doppler tomography. *Optic Letters.* 1995;20:1337-1339.

3. Zhao Y, Chen Z, Saxer C, et al. Phase resolved optical coherence tomography and optical Doppler tomography for imaging blood flow in human skin with fast scanning speed and high velocity sensitivity. *Optic Letters.* 2000;25:114-116.

4. White BR, Pierce MC, Nassif N, et al. In vivo dynamic human retinal blood flow imaging using ultra-high-speed spectral domain optical Doppler tomography. *Optic Express.* 2003;11:3490-3497.

5. Leitgeb RA, Schmetterer L, Hitzenberger CK, et al. Real-time measurement of in vitro flow by Fourier-domain color Doppler optical coherence tomography. *Optic Letters.* 2004;29:171-173.

6. Wang Y, Bower BA, Izatt JA, Tan O, Huang D. In vivo total retinal blood flow measurement by Fourier domain Doppler optical coherence tomography. *J Biomed Optics.* 2007;12:041215-041222.

7. Wang Y, Bower BA, Izatt JA, Tan O, Huang D. Retinal blood flow measurement by circumpapillary Fourier domain Doppler optical coherence. *J Biomed Optics.* 2008;13:064003-1~9.

8. Wang Y, Lu A, Gil-Flamer J, et al. Measurement of total blood flow in the normal human retina using Doppler Fourier-domain optical coherence tomography. *Br J Ophthalmol.* 2009;93:634-637.

9. Wang Y, Tan O, Huang D. Investigation of retinal blood flow in normal and glaucoma subjects by Doppler Fourier-domain optical coherence tomography. *SPIE Proceeding.* 2009;7168.

10. Wang Y, Fawzi A, Tan O, Gil-Flamer J, Huang D. Retinal blood flow detection in diabetic patients by Fourier domain optical coherence. *Optic Express.* 2009;17:4061-4073.

Keratoconus Screening

Yan Li, PhD; Maolong Tang, PhD; and David Huang, MD, PhD

BACKGROUND

Keratoconus is characterized as a bilateral, noninflammatory, progressive thinning and protrusion of the cornea.[1,2] Moderate to advanced keratoconus is easily recognizable by several distinctive clinical features, but the diagnosis of forme fruste keratoconus (FFK) can be very challenging.[3] This identification is especially important in preoperative screening for laser refractive surgeries. Undetected corneal ectatic disorders can result in accelerated, progressive keratoectasia and unpredictable outcomes after LASIK and photorefractive keratectomy.

Placido disk-based corneal topography is the current standard for detecting corneal ectasia.[4-10] Normal, suspicious, or abnormal topography patterns of these diseases have been classified,[11,12] and quantitative topographic indices have been developed to help diagnose keratoconus.[13-16] However, it is likely that corneal topography does not capture all cases of FFK because eyes with normal topography and no other known risk factors could still develop post-LASIK keratectasia.[11] It is also difficult for topography alone to distinguish keratoconus from contact lens-induced warpage, nonpathological asymmetric astigmatism, subepithelial deposits or scarring, uneven tear film, and other causes of corneal distortion.[17-19] Keratoconus is characterized by focal thinning as well as focal topographic changes. A corneal thickness map can serve as an important adjunct to provide further improvement in diagnostic accuracy. We have developed a pachymetry map-based method to detect abnormal corneal thinning in keratoconic eyes.[20]

OCT has a higher resolution than other imaging modalities and can accurately map the corneal thickness of normal, postoperative, and opacified corneas.[21,22] In this chapter, we will discuss how to use OCT pachymetry map to detect keratoconus.

RTVUE OCT FOR PACHYMETRY MAPPING

The RTVue OCT can map the corneal thickness with its corneal adaptor module (CAM). The mapping scan pattern is called *pachymetry*. It comprises 6-mm radial lines on 8 meridians. We recommend centering the scan pattern on the pupil (see Chapter 3).

The RTVue OCT software automatically processes the OCT image and calculates the corneal pachymetry map (Figure 11-1). The maps are divided into octants pie sectors (superior, superotemporal, temporal, inferotemporal, inferior, inferonasal, nasal, superonasal) and annular rings (2, 5, and 6 mm diameters). The average corneal thickness of each zone is presented in a zonal average map. The zonal corneal thickness analysis results are listed in a table (see Figure 11-1, upper left).

PACHYMETRIC CHARACTERISTICS FOR KERATOCONUS SCREENING

Eccentric focal corneal thinning is a characteristic feature of keratoconus.[1,23,24] Studies of eyes of keratoconus

Huang D, Duker JS, Fujimoto JG, Lumbroso B, Schuman JS, Weinreb RN.
Imaging the Eye from Front to Back with RTVue Fourier-Domain Optical Coherence Tomography (pp 103-108).

Figure 11-1. The pachymetry map report of a normal eye. Statistical parameters relevant to keratoconus diagnosis are displayed in a table (upper right). The position of the thinnest point is marked by an asterick "*" on the color map (lower right). All pachymetric parameters (upper right) were in the normal range: SN – IT = 21 µm; S – I = 35 µm; minimum = 547 µm; minimum – maximum = -51 µm. The thinnest cornea is located inside a central 2-mm diameter zone.

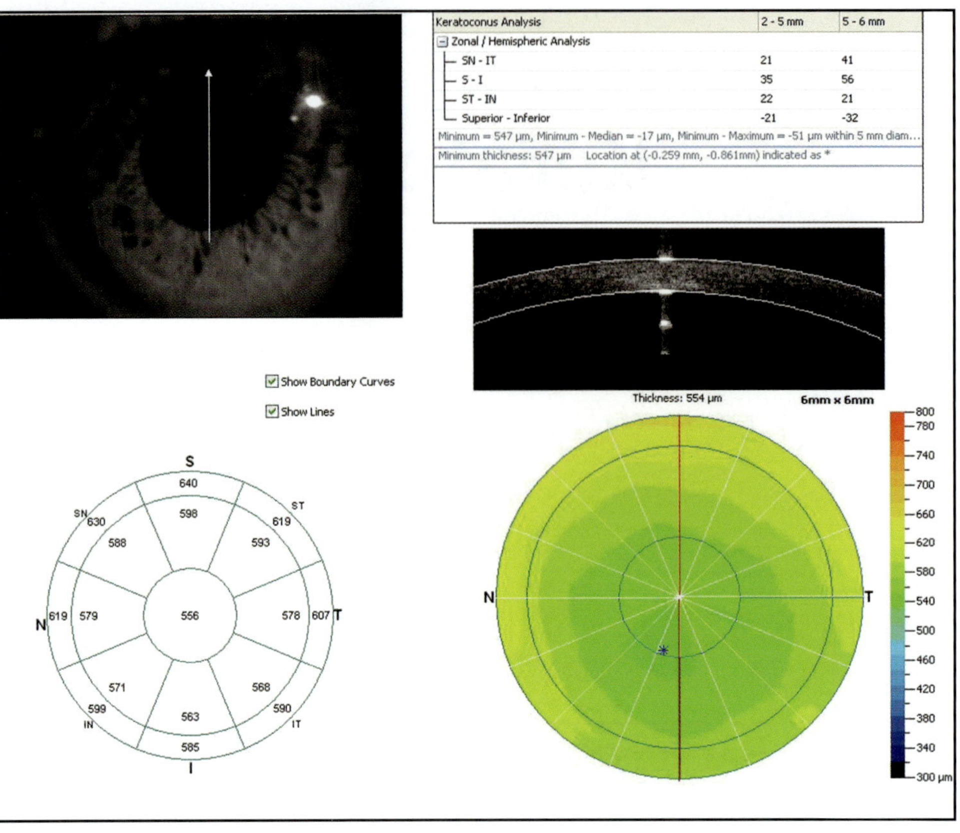

Table 11-1. Cutoff Values for Ketratoconus Screening Based on 1% of Normal Population Distribution (Unit: µm)

Parameter	SN – IT	S – I	Minimum	Minimum – Maximum
Normal Reference (mean ± standard deviation)	28.9 ± 9.6	25.2 ± 11.4	529.7 ± 24.6	-42.3 ± 8.3
Cutoff Value for Abnormality	>51	>52	<472	<-62

and unaffected patients showed that presence of the following pachymetric parameters in the central 5-mm area of an unoperated cornea should raise concerns for keratoconus.[20]

- SN – IT: the average thickness of the superonasal (SN) octant minus the average thickness of the inferotemporal (IT) octant.

- S – I: the average thickness of the superior (S) octant minus the average thickness of the inferior (I) octant.

- Minimum: the thinnest corneal thickness.

- Minimum – maximum: the thinnest corneal thickness minus the thickest corneal thickness.

We conducted a preliminary study to establish a provisional normative standard for detecting keratoconus using RTVue OCT mapping of corneal thickness. The data were collected from 36 eyes in 18 normal subjects. All pachymetry scans were centered on the pupil. The criteria for detecting keratoconic abnormality, based on the first percentile or 99th percentile cutoff points of the normal range, are listed in Table 11-1. In a previous study,[20] we established that if any one of the pachymetric parameters is abnormal, one should suspect keratoconus (sensitivity = 0.97 and specificity = 0.97). If 2 parameters are abnormal, then the diagnosis was highly reliable (specificity = 1.00).[20]

To detect keratoconus, look for:

- SN – IT is greater than 51 or S – I value is bigger than 52 µm. SN – IT and S – I are both asymmetric parameters.

- The minimum corneal thickness less than 472 µm.

- Minimum – maximum (focal thinning parameter) value less than -62 µm.

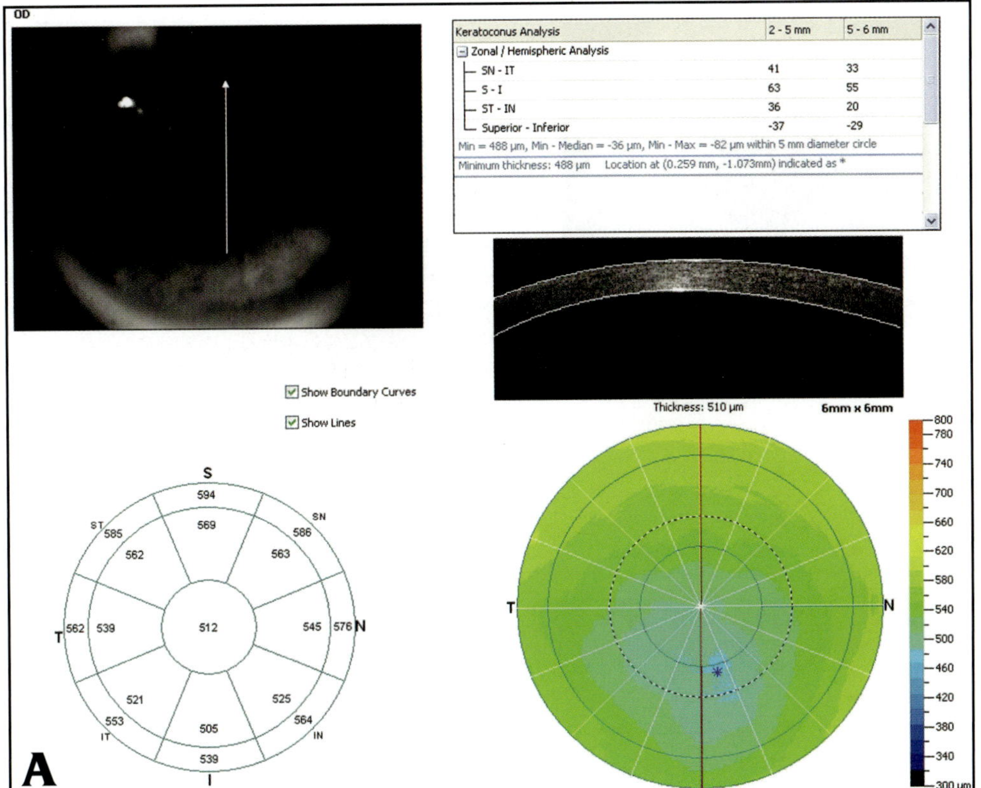

Figure 11-2. An early keratoconus case with BSCVA of 20/25. (A) The pachymetry map report. Two parameters were abnormal (keratoconic): S – I = 63 µm; minimum – maximum = 82 µm. The other parameters were within normal: SN – IT = 41 µm; minimum = 488 µm. (B) Axial power map shows the asymmetric skewed bowtie pattern that is characteristic of keratoconus.

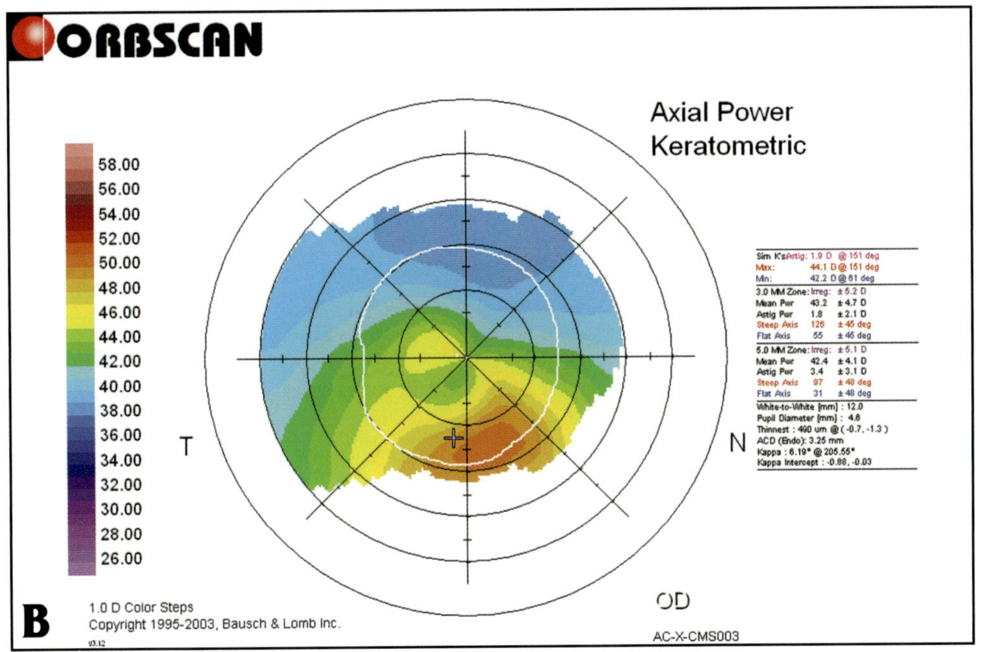

If even 1 parameter is abnormal, the cornea is likely to have keratoconus. If 2 or more are abnormal, then the eye is very likely to have keratoconus or other ectatic conditions such as pellucid marginal degeneration.

Examples for applying these diagnostic criteria are shown in Figure 11-1 (normal), Figure 11-2 (mild keratoconus), and Figure 11-3 (moderate keratoconus). The pachymetric diagnostic system for keratoconus can also be helpful in the diagnosis of keratectasia after LASIK. But one must also take into account the expected corneal thinning from the laser ablation in applying the diagnostic criteria. In keratectasia, the surgically induced corneal thinning should be greater than that the laser ablation. Note that the residual stromal bed thickness may be thinner than the commonly accepted safety limit of 250 µm (see Chapter 1).

Figure 11-3. A moderate keratoconus case with BSCVA of 20/40. (A) The pachymetry map report. All pachymetric parameters were abnormal: SN – IT = 92 µm; S – I = 77 µm; minimum = 362 µm; minimum – maximum = -139 µm. (B) Axial power map shows the asymmetric bow tie pattern.

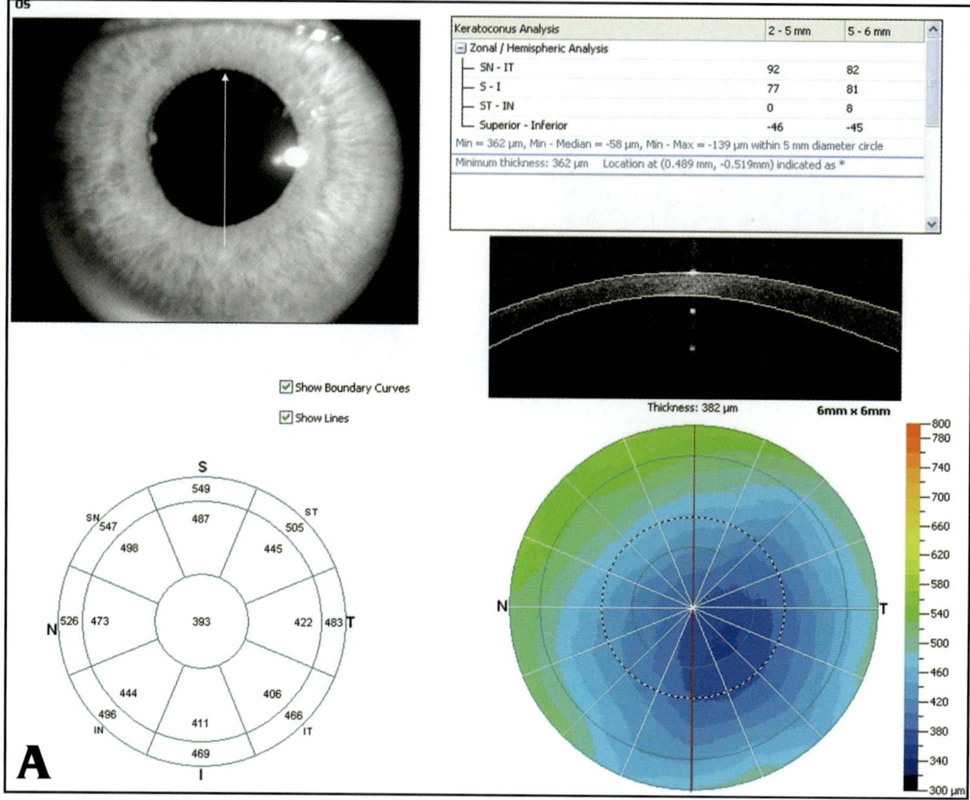

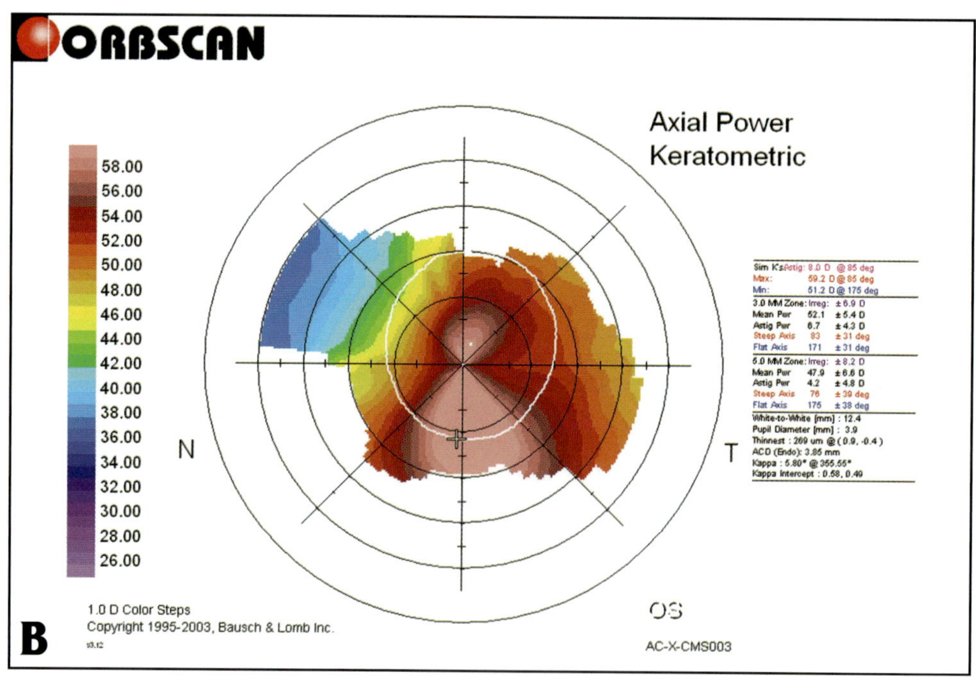

QUALITY CONTROL

The RTVue software may misidentify the corneal boundaries under certain circumstances. For example, the cornea may be shadowed by eyelashes or the OCT signal may be low in the peripheral cornea. The pachymetry map was calculated by measuring the distance between the anterior and posterior corneal surfaces. Therefore, it is recommended to review individual cross-sectional OCT images in cases where the cornea looks normal on topography and slit lamp examination, but the OCT pachymetry map is abnormal. The RTVue OCT software provides an [Edit Surface] tool to manually edit the corneal boundaries and correct the pachymetry map. Rarely, the OCT map may need to be recaptured if the quality of the OCT images is poor and the corneal surface boundaries cannot be recognized even by a human grader. With these corrective measures in place, the physician should find OCT to be a reliable tool for keratoconus screening.

SUMMARY

It is difficult for the surgeon to distinguish FFK from other types of mildly irregular topography maps. The pachymetry map can provide important additional data to confirm or reject the diagnosis and determine whether the patient is a good candidate for refractive surgery. OCT provides precise pachymetry mapping and is useful in the pre-LASIK screening examination. This chapter provides guidelines for detecting abnormal patterns of corneal thinning using the RTVue OCT system. We recommend routine measurement of corneal pachymetry map before LASIK and other laser refractive surgeries.

REFERENCES

1. Krachmer JH, Feder RS, Belin MW. Keratoconus and related noninflammatory corneal thinning disorders. *Surv Ophthalmol.* 1984;28(4):293-322.
2. Rabinowitz YS. Keratoconus. *Surv Ophthalmol.* 1998;42(4):297-319.
3. Ambrosio R Jr., Alonso RS, Luz A, Coca Velarde LG. Corneal-thickness spatial profile and corneal-volume distribution: tomographic indices to detect keratoconus. *J Cataract Refract Surg.* 2006;32(11):1851-1859.
4. Avitabile T, Franco L, Ortisi E, et al. Keratoconus staging: a computer-assisted ultrabiomicroscopic method compared with videokeratographic analysis. *Cornea.* 2004;23(7):655-660.
5. Li X, Rabinowitz YS, Rasheed K, Yang H. Longitudinal study of the normal eyes in unilateral keratoconus patients. *Ophthalmology.* 2004;111(3):440-446.
6. Maguire LJ, Bourne WM. Corneal topography of early keratoconus. *Am J Ophthalmol.* 1989;108(2):107-112.
7. Rabinowitz YS, Garbus J, McDonnell PJ. Computer-assisted corneal topography in family members of patients with keratoconus. *Arch Ophthalmol.* 1990;108(3):365-371.
8. Wilson SE, Lin DT, Klyce SD. Corneal topography of keratoconus. *Cornea.* 1991;10(1):2-8.
9. Maeda N, Klyce SD, Tano Y. Detection and classification of mild irregular astigmatism in patients with good visual acuity. *Surv Ophthalmol.* 1998;43(1):53-58.
10. Ambrosio JR, Klyce SD, Smolek MK, Wilson SE. Pellucid marginal corneal degeneration. *J Refract Surg.* 2002;18(1):86-88.
11. Randleman JB, Woodward M, Lynn MJ, Stulting RD. Risk assessment for ectasia after corneal refractive surgery. *Ophthalmology.* 2008;115(1)37-50.
12. Demirbas NH, Pflugfelder SC. Topographic pattern and apex location of keratoconus on elevation topography maps. *Cornea.* 1998;17(5):476-484.
13. Rabinowitz YS. Videokeratographic indices to aid in screening for keratoconus. *J Refract Surg.* 1995;11(5):371-379.
14. Maeda N, Klyce SD, Smolek MK, Thompson HW. Automated keratoconus screening with corneal topography analysis. *Invest Ophthalmol Vis Sci.* 1994;35(6):2749-2757.
15. Schwiegerling J, Greivenkamp JE. Keratoconus detection based on videokeratoscopic height data. *Optom Vis Sci.* 1996;73(12):721-728.
16. Rabinowitz YS, Rasheed K. KISA% index: a quantitative videokeratography algorithm embodying minimal topographic criteria for diagnosing keratoconus. *J Cataract Refract Surg.* 1999;25(10):1327-1335.
17. Cheng HC, Lin KK, Chen YF, Hsiao CH. Pseudokeratoconus in a patient with soft contact lens-induced keratopathy: assessment with Orbscan I. *J Cataract Refract Surg.* 2004;30(4):925-928.
18. Wilson SE, Lin DT, Klyce SD, et al. Topographic changes in contact lens-induced corneal warpage. *Ophthalmology.* 1990;97(6):734-744.
19. Maeda N, Klyce SD, Smolek MK. Comparison of methods for detecting keratoconus using videokeratography. *Arch Ophthalmol.* 1995;113(7):870-874.
20. Li Y, Meisler DM, Tang M, et al. Keratoconus diagnosis with optical coherence tomography pachymetry mapping. *Ophthalmology.* 2008;115(12):2159-2166.
21. Khurana RN, Li Y, Tang M, et al. High-speed optical coherence tomography of corneal opacities. *Ophthalmology.* 2007;114(7):1278-1285.
22. Li Y, Shekhar R, Huang D. Corneal pachymetry mapping with high-speed optical coherence tomography. *Ophthalmology.* 2006;113(5):792-799, 9 e1-2.
23. Vinciguerra P, Camesasca FI. Prevention of corneal ectasia in laser in situ keratomileusis. *J Refract Surg.* 2001;17(2 Suppl):S187-S189.
24. Binder PS, Lindstrom RL, Stulting RD, et al. Keratoconus and corneal ectasia after LASIK. *J Cataract Refract Surg.* 2005;31(11):2035-2038.

12

Refractive Surgery

Camila Haydée Rosas Salaroli, MD; Stephen L. Trokel, MD;
Perry S. Binder, MD; Yan Li, PhD; and David Huang, MD, PhD

For refractive surgery, the RTVue represents an improvement over previous OCT and slit-scanning systems for corneal tomography. The RTVue's higher speed and resolution leads to greater precision in the mapping of corneal thickness and measurement of LASIK flap thickness. The applications of these capabilities in LASIK and its complications, photorefractive keratectomy (PRK), phototherapeutic keratectomy (PTK), and phakic intraocular lens implantation are illustrated later.

LASIK PLANNING

The RTVue greatly improves the scanning speed by using Fourier-domain OCT technology, further improving the precision of pachymetry mapping.

Many refractive surgeons use ultrasound pachymetry or slit-based topography systems to measure central corneal thickness before LASIK. Although these are time-tested standards, the ultrasound probe only measures 1 point on the cornea at any one time. Any decentration in the manual probe placement could produce an overestimation of the central pachymetry, because the peripheral cornea is thicker, or, alternatively, it could miss peripheral thinning in cases of keratoconus. The RTVue OCT pachymetry map offers coverage over the central 6-mm diameter and automatically calculates the minimum thickness and the central thickness averaged over a central 2-mm diameter circle. Importantly, the measurement is centered over the pupil, which is also the landmark used for the excimer laser ablation. Pachymetric map-

ping by OCT is important in the planning of LASIK. It helps screen out patients with forme fruste keratoconus (FFK) by detecting corneal focal thinning (see Chapter 10). It is also helpful in calculating the expected residual stromal bed thickness, which in standard practice is at least 250 μm, in hopes of preventing post-LASIK ectasia. The reliability of time-domain OCT pachymetric mapping has already been well established.[1] Although slit-scanning technology such as the Orbscan (Bausch & Lomb, Rochester, NY) or the Pentacam (Oculus, Germany) can also produce a pachymetry map, these technologies have a tendency to underestimate corneal thickness in eyes that have undergone PRK[2-4] and LASIK[3-7] and in eyes with a corneal opacity.[8] This problem is especially relevant in retreatment procedures, when underestimation of corneal thickness could lead to an erroneous conclusion that further laser ablation would not be possible. In a representative case, the OCT measurement of minimum corneal thickness after LASIK was 47 μm greater than Orbscan (Figure 12-1), allowing for retreatment if needed.

LASIK FLAP MEASUREMENT

To predict the residual stromal bed thickness, the surgeon needs to know the corneal thickness as well as the mean and standard deviation of the LASIK flap thickness for the microkeratome that will be used. We advise routine OCT scanning 1 week after LASIK to monitor the created flap thickness. At 1 week, surgically induced corneal hydration changes have mostly resolved, and the

Huang D, Duker JS, Fujimoto JG, Lumbroso B, Schuman JS, Weinreb RN.
Imaging the Eye from Front to Back with RTVue Fourier-Domain Optical Coherence Tomography (pp 109-122).
© 2010 SLACK Incorporated.

Figure 12-1. Pachymetry maps taken 3 weeks after LASIK. (A) Orbscan measured a minimum thickness of 353 µm. (B) OCT showed the minimum thickness to be 400 µm.

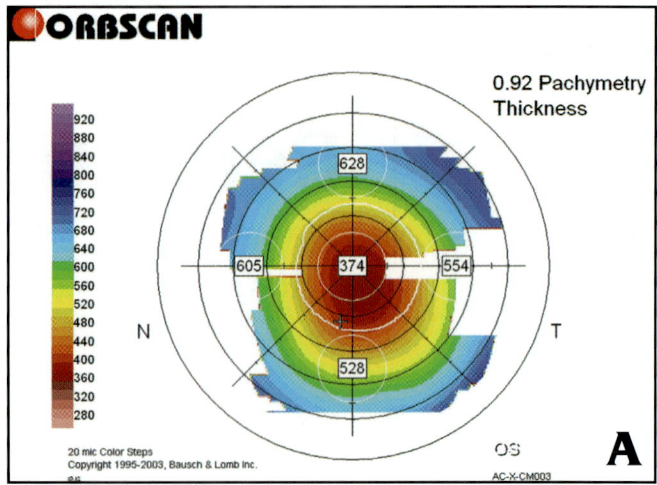

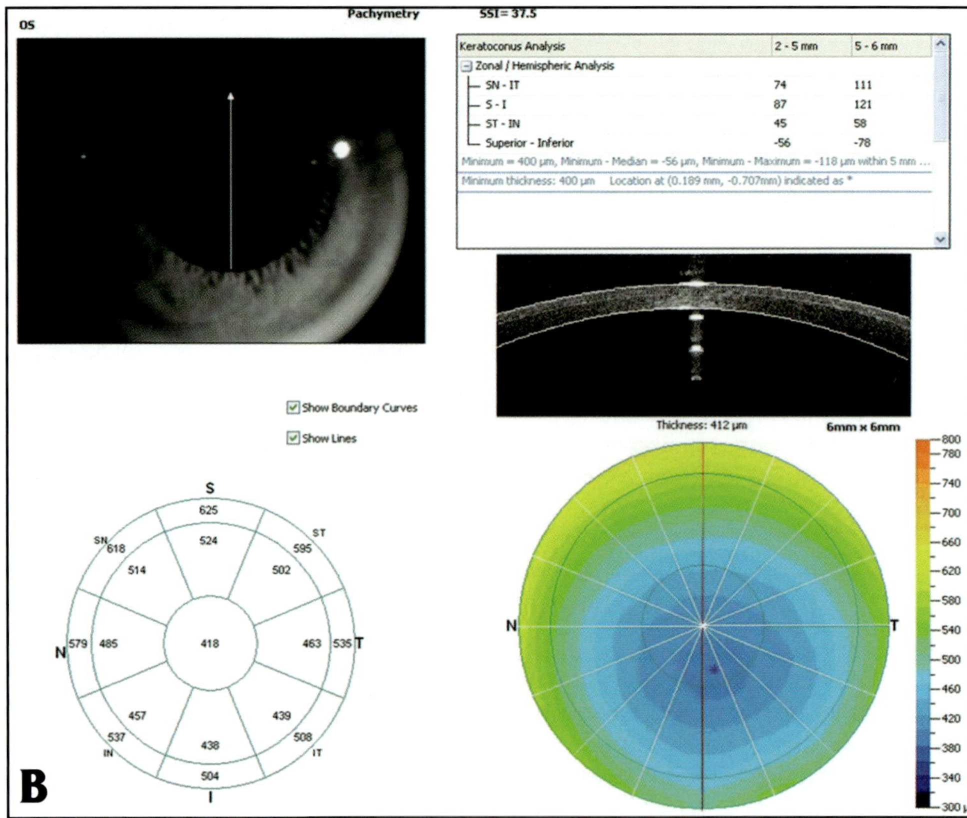

flap interface is still easily visible.[9] It is best to measure the flap thickness at this time before the interface reflectivity peak fades (Figure 12-2).

If the flap thickness is not measured at the 1-week post-LASIK visit, it is still possible to measure it at a later time. This is extremely useful to determine residual bed thickness calculations for any LASIK retreatment. The RTVue is capable of averaging many frames in a line scan. This suppresses the speckle noise and enhances visualization of the flap interface (Figure 12-3). With this technique, it is sometimes possible to visualize the LASIK flap even 10 years later (Figure 12-4). In this case, the flap thickness could be measured only in the periphery. Because the flap is relatively uniform, the surgeon could still estimate the central and minimum bed thickness by subtracting the flap thickness from the central and minimum corneal thicknesses on the pachymetry map.

Another trick in measuring the flap thickness long after LASIK is to utilize the 8 meridians available in the pachymetry scan pattern of the RTVue (Figure 12-5). Often, there are 1 or more meridians in which the interfaces are readily visible.

It is important to routinely monitor LASIK flap thickness, because the actual result is often different from

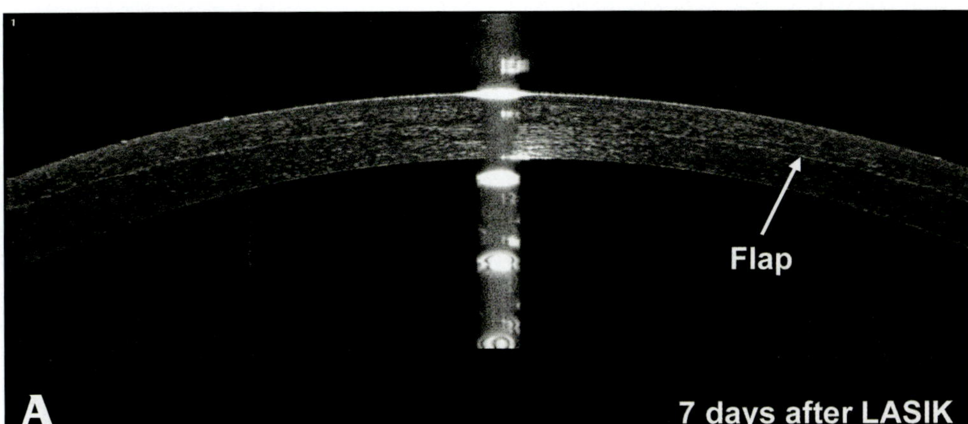

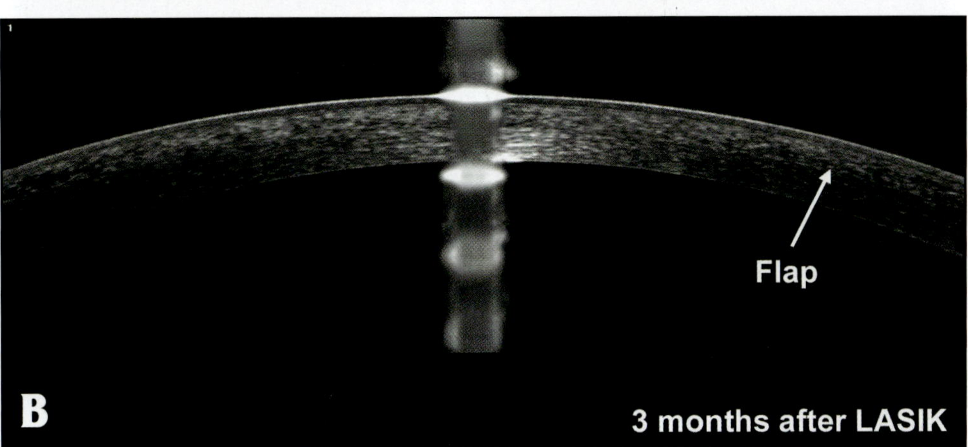

Figure 12-2. Flap interface created with a mechanical microkeratome imaged by the RTVue. (A) The interface has strong reflectivity 7 days after LASIK. (B) The interface is much fainter 3 months after LASIK, but can still be seen and the flap thickness measured.

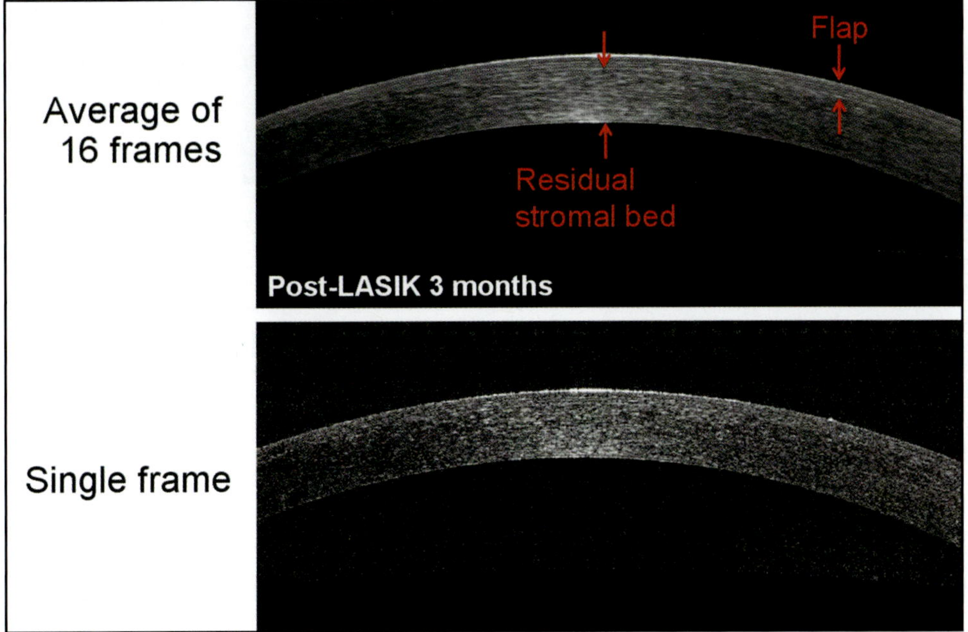

Figure 12-3. LASIK flap imaging after a mechanical microkeratome created the flap. The LASIK flap is more easily visible in the frame-averaged image (top; arrows) compared to the single-frame image (bottom). At 3 months after LASIK, the interface reflectivity has faded, but the flap can still be detected by its higher internal reflectivity relative to the stromal bed and therefore can be measured to be 598 − 210 = 388 µm.

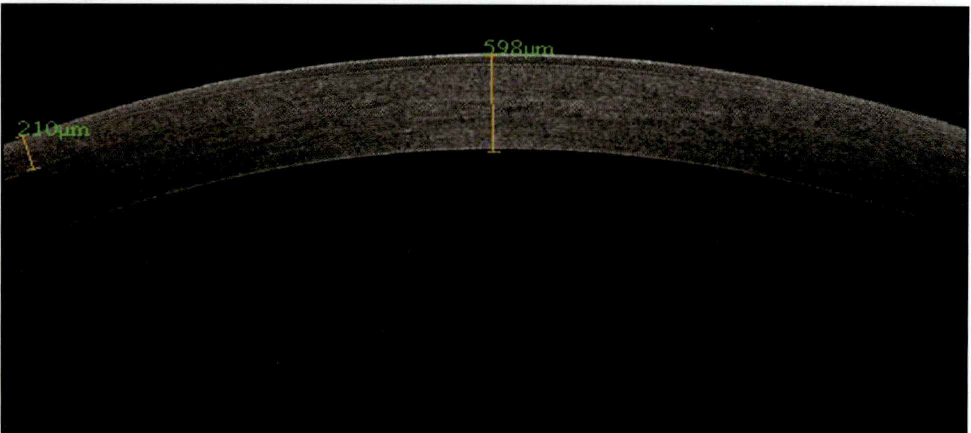

Figure 12-4. RTVue scan 10 years after LASIK. The flap interface is still visible in the periphery allowing measurement. (Courtesy of Dr. Perry S. Binder.)

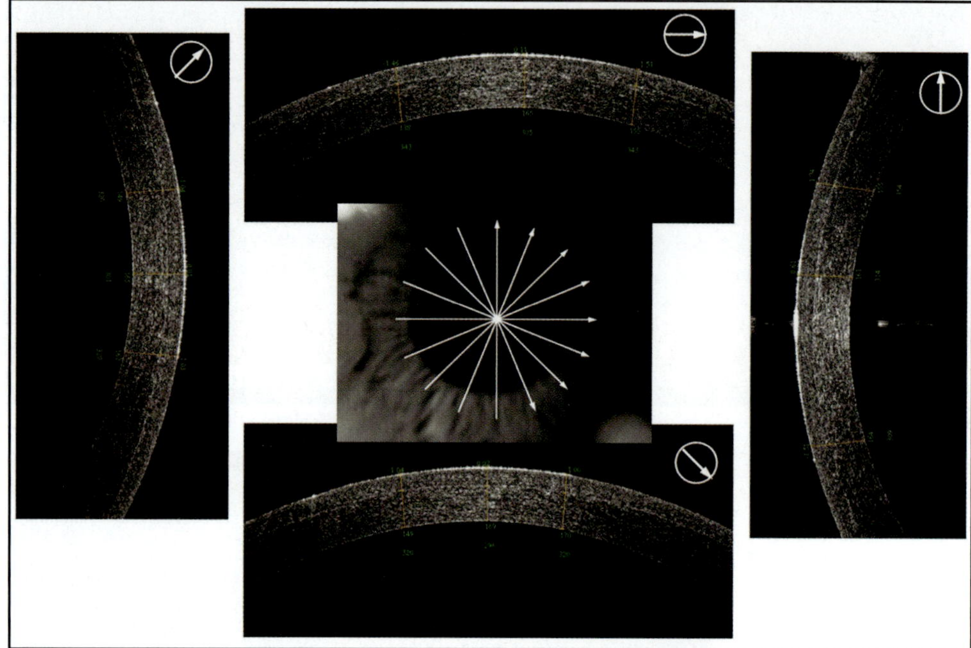

Figure 12-5. Flap thickness measurement can also be obtained from some of the meridional line components of the pachymetry scan when a single line scan is not able to easily show the flap interface.

the microkeratome setting. For example, the Hansatome (Bausch & Lomb, Rochester, NY) typically produces flaps thinner than the nominal setting.[9] Even the femtosecond laser can occasionally create flaps that deviate significantly from the intended thickness, though on the whole the laser is more repeatable. In a consecutive series of 17 eyes, we found that the central flap thickness was 122.5 ± 8.5 μm (mean ± standard deviation), with a range of 113 to 144 μm (Figure 12-6), using the 60 kHz Intralase femtosecond laser (Advanced Medical Optics Inc, Santa Ana, CA) set at 110 μm depth.

LASIK COMPLICATIONS

Post-LASIK Ectasia (Keratectasia)

Keratectasia after LASIK is most often caused by preexisting FFK or an achieved residual posterior stromal bed of less than 250 μm.[10,11] It is similar to keratoconus in that the focal bulging and thinning tend to occur inferiorly or inferotemporally. However, the degree of thinning tends to be more severe in post-LASIK ectasia, as it is likely the result of both ectasia and laser ablation. In our example (Figure 12-7), the mechanically created flap was thick, 204 to 225 μm, and the minimal corneal

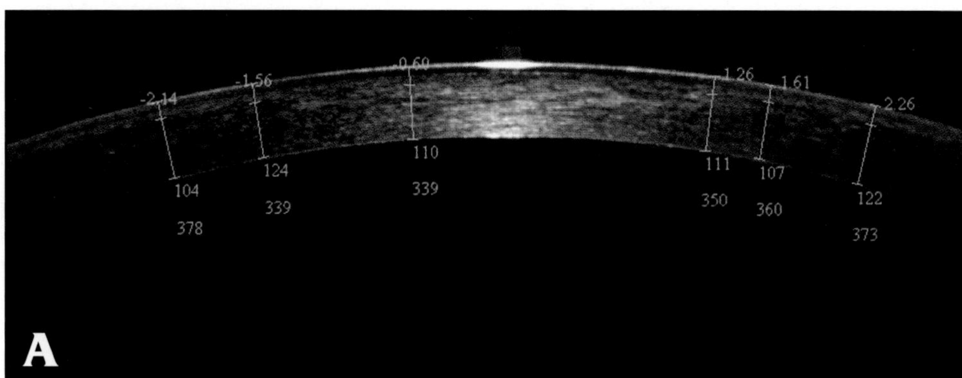

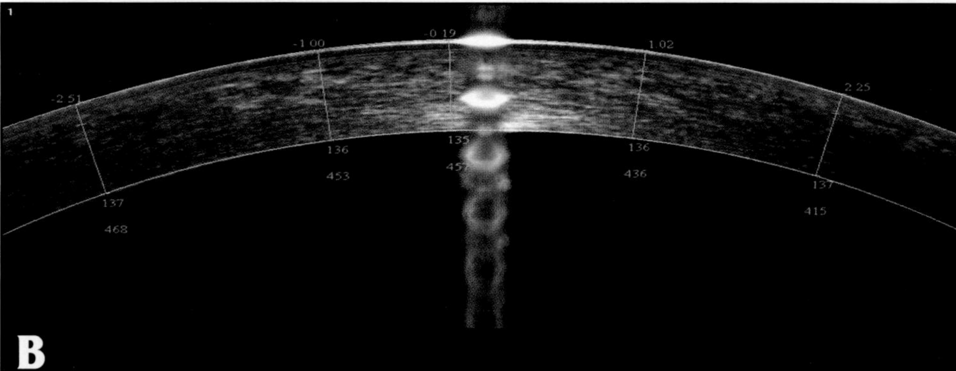

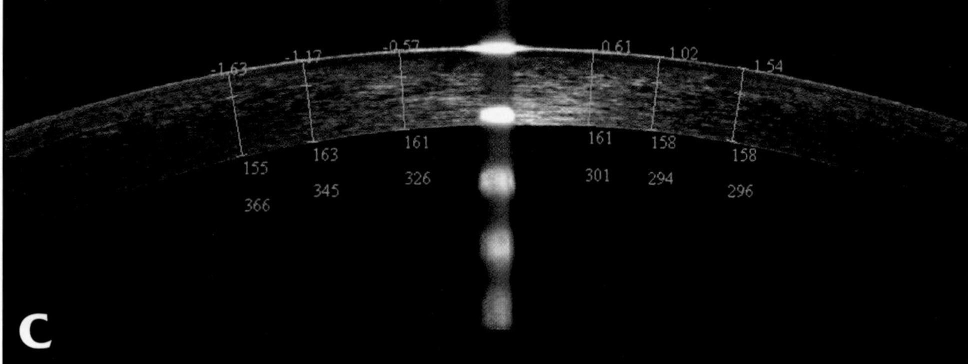

Figure 12-6. RTVue measurements of LASIK flap thickness. The Intralase femtosecond laser was set at 110 µm depth. The flap thickness ranges were (A) 109 to 115 µm, (B) 126 to 139 µm, and (C) 140 to 148 µm. The flap thickness measurements were made with the [Flap] tool in the RTVue CAM software. The user only has to position the center caliper mark at the flap interface. The software automatically identifies the anterior and posterior corneal boundaries and output the measurements in the green labels. The labels are, from top to bottom: transverse location, flap thickness, and bed thickness.

Figure 12-7. A case of keratectasia imaged 2 years after LASIK. (A) Orbscan showed marked inferior steepening with minimum corneal thickness of 320 µm. (B) RTVue pachymetry map showed a minimum central thickness of 398 µm. Inferocentral bulging is visible on the OCT vertical section in the middle-right panel.

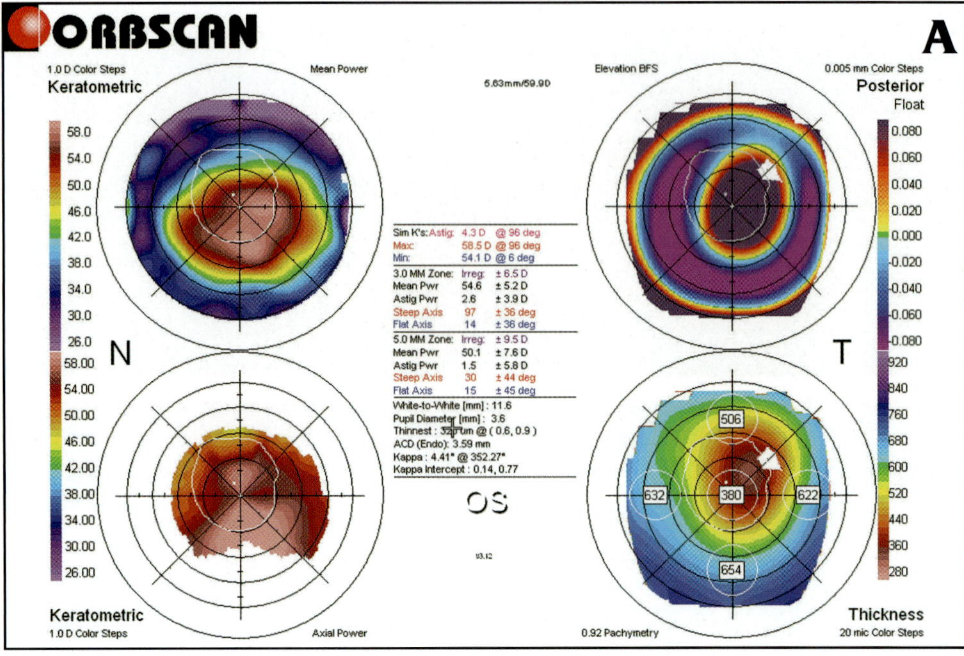

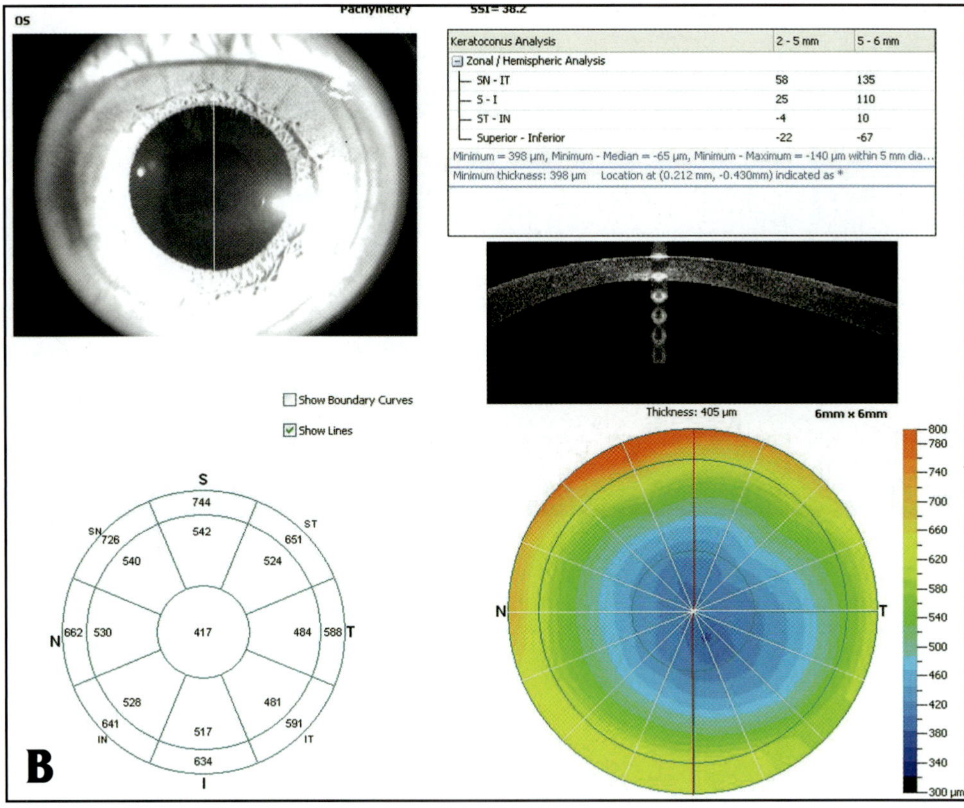

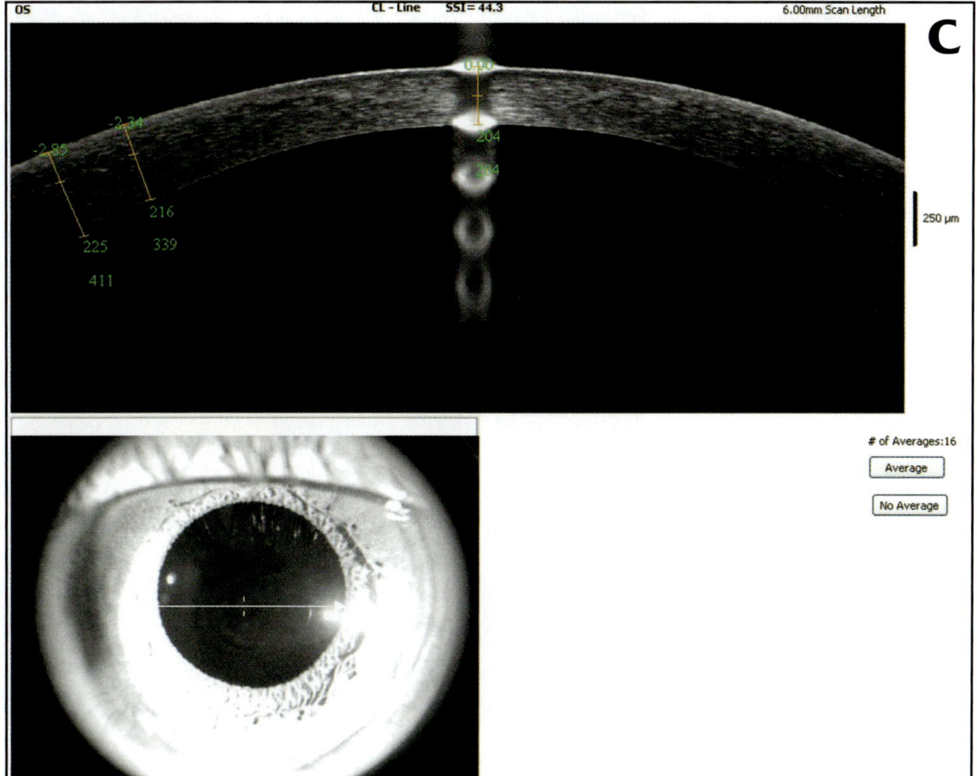

Figure 12-7. A case of keratectasia imaged 2 years after LASIK. (C) RTVue frame-averaged horizontal section showed the central flap thickness was 204 μm and stromal bed thickness was also 204 μm. The estimate of minimal stromal bed thickness was 194 μm (398 to 204).

thickness was 194 μm based on OCT, far below the safe lower limit of 250 μm. This suggests that an excessively thick flap and insufficient residual stromal bed may have caused the keratectasia. It is important to note that the minimum corneal thickness on the Orbscan (Figure 12-7A) was only 320 μm, which usually indicates the ectasia was too advanced to be treated by intracorneal ring (ICR) segment implantation. However, the RTVue pachymetry map showed the minimum thickness to be 398 μm, which indicates a less severe degree of ectasia that might be amenable to ICR treatment rather than corneal transplantation. Of course, a rigid gas-permeable contact lens should be tried before any surgical intervention is considered. This case demonstrates the value of the RTVue in the evaluation and management of keratectasia.

Diffuse Lamellar Keratitis

Diffuse lamellar keratitis (DLK) is a noninfectious inflammation primarily occurring at the interface of the LASIK flap. When the flap is cut with a mechanical microkeratome, the inflammatory cells concentrate in ridges or waves, also known as a "sand of Sahara" pattern. The pattern can be different when the femtosecond laser was used to create the flap. In a case of bilateral DLK after femtosecond laser-assisted LASIK, high resolution RTVue cross-sectional imaging of the more severely affected eye (Stage 3, Figure 12-8) revealed inflammation that involved the center of the cornea and the stroma on both sides of the flap interface. On postoperative day 2, the flap was lifted and the interface irrigated. In the less affected eye (Stage 1, Figure 12-9), the inflammation did not involve the center of the cornea and was less intense, and flap lifting was not needed. Both eyes were treated with intensive topical corticosteroid and a short course of oral corticosteroid. Corneal clarity was restored and both eyes recovered 20/20+ uncorrected vision. The incidence of DLK when a femtosecond laser is used to create the flap has been reported to be rare, with one study reporting an incidence rate of 0.1%.[12]

Epithelial Ingrowth

Epithelial ingrowth under the LASIK flap occurs in a small percentage of cases after enhancements, such as flap relift and retreatment. Rarely, it can occur after primary treatment. OCT shows epithelial ingrowth as a highly reflective area in the interface and quantifies its thickness.[13] In a case of stable ingrowth 2 years after LASIK enhancement (Figure 12-10), the extent of the ingrowth was measured periodically by the referring surgeon. The ingrowth appeared thicker (>100 μm) on OCT compared to the slit lamp impression, and the overlying flap appeared thinned or compressed. However, the ingrowth did not significantly affect vision because of the peripheral location; thus removal was not necessary. Another case exhibited more extensive ingrowth with overlying scarring (Figure 12-11). In both cases, RTVue OCT helped with the quantitative measurement of the extent of ingrowth

Figure 12-8. Diffuse lamellar keratitis, grade 3, 2 days after LASIK with femtosecond laser flap creation. (A) Slit lamp photograph. (B) RTVue OCT section showing a hyper-reflective flap interface. Stromal reflectivity was also increased in both the flap and bed, including the center.

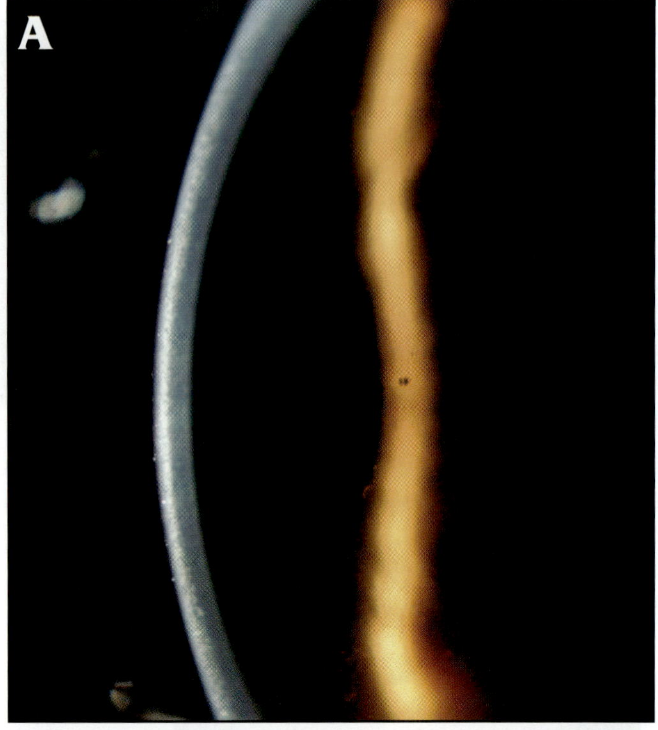

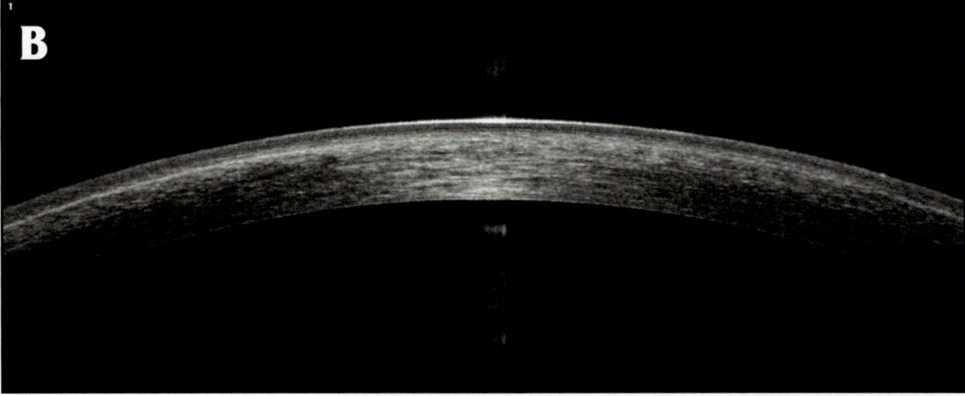

Figure 12-9. Diffuse lamellar keratitis, grade 1, 2 days after LASIK with femtosecond laser flap creation. (A) Slit lamp photograph.

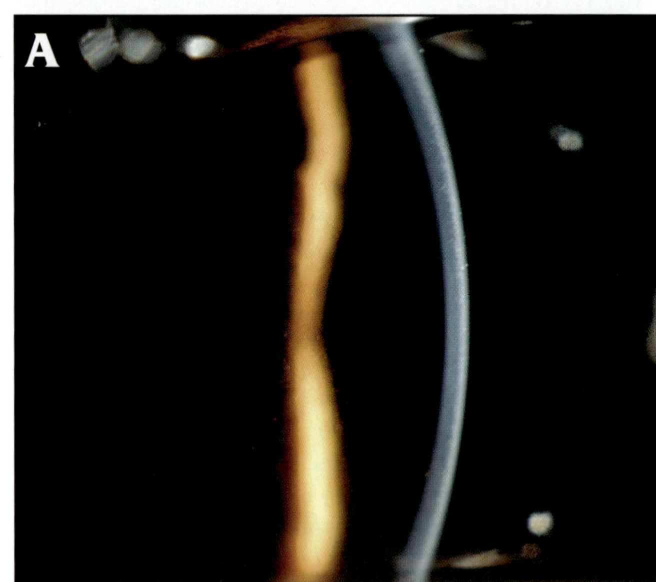

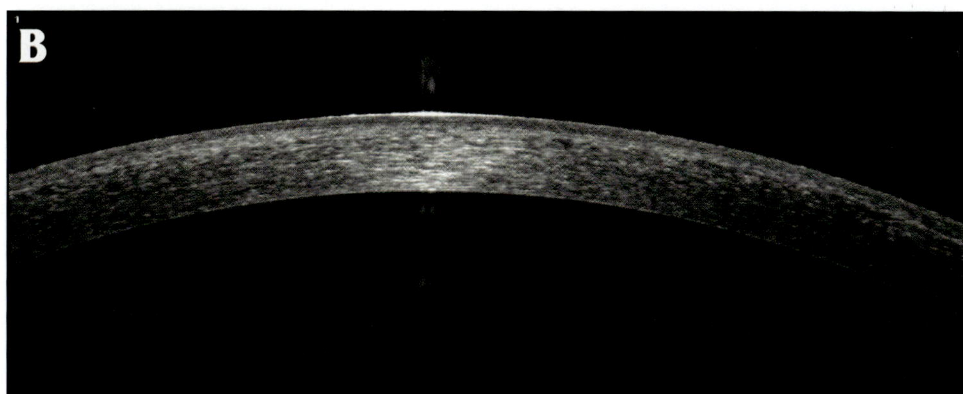

Figure 12-9. Diffuse lamellar keratitis, grade 1, 2 days after LASIK with femtosecond laser flap creation. (B) RTVue OCT section showing a mildly hyper-reflective flap interface not involving the center.

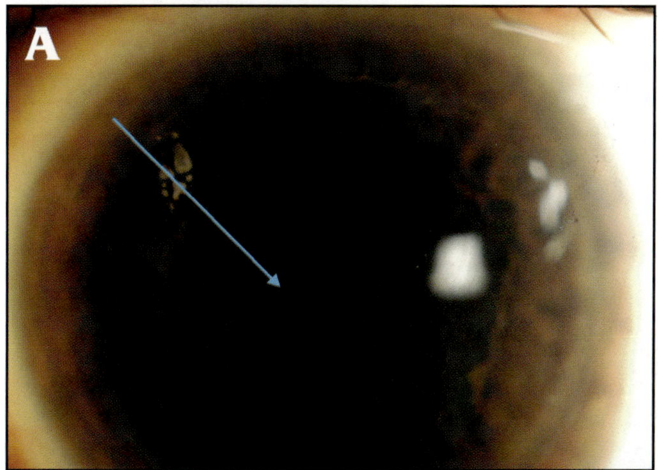

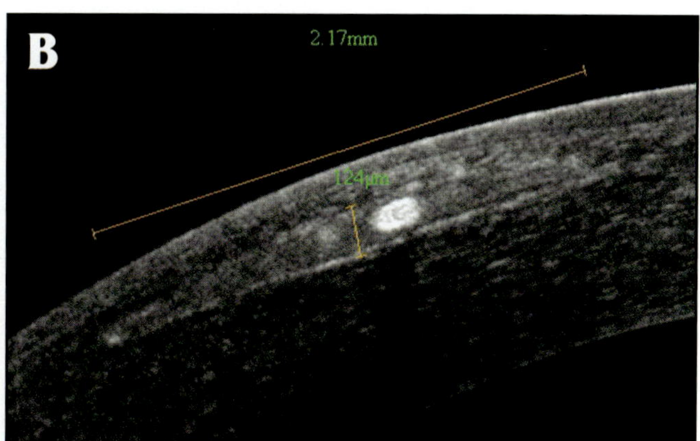

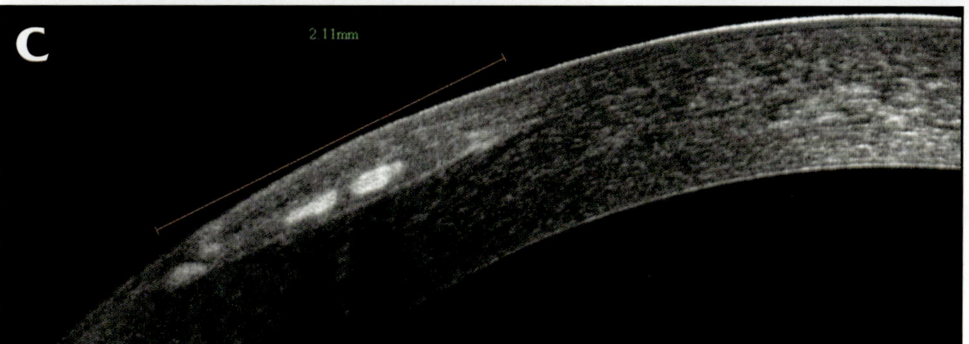

Figure 12-10. Epithelial ingrowth under a nasally hinged LASIK flap. (A) Slit-lamp photograph. (B, C) RTVue OCT sections along radial lines. The computer caliper measured the ingrowth to be 2.1 mm transversely and 124 µm thick. The characteristic opaque pearls within the ingrowth are visible in both images.

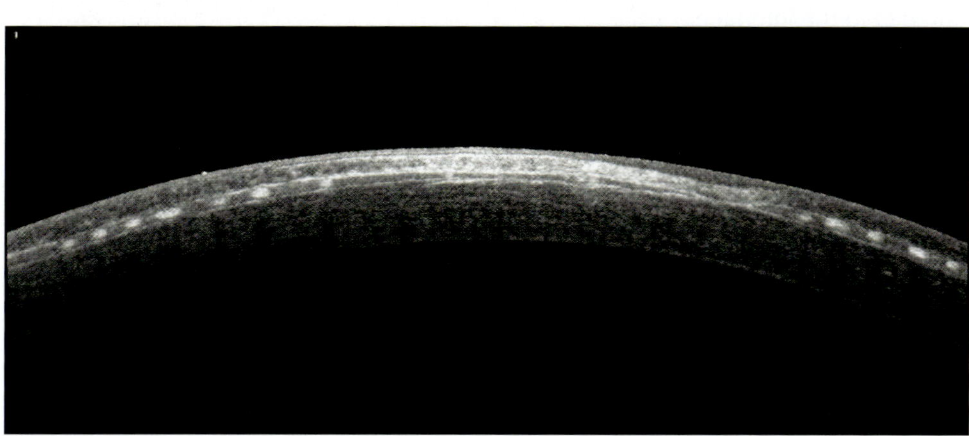

Figure 12-11. Diffuse interface epithelial ingrowth and central scar several months after a mechanical microkeratome LASIK procedure complicated by DLK. (Courtesy of Dr. Perry S. Binder.)

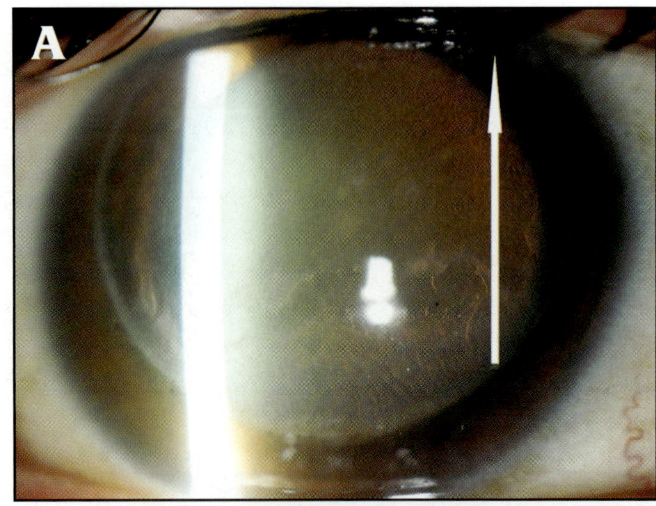

Figure 12-12. Interface fluid syndrome. (A) Slit lamp photography on post-LASIK day 49 shows flap edema, interface fluid, and several zones of opacity. Inferiorly, there was a dense opacity. The arrow shows the OCT scan length and direction. (B) RTVue OCT shows areas of epithelial ingrowth and fluid under the LASIK flap interspersed with streaks and pearls of clear material. The central and superior areas of fluid accumulation are associated with scattered granular haze. (Reprinted with permission from Ramos JL, Zhou S, Yo C, Tang M, Huang D. High resolution imaging of complicated LASIK flap interface fluid syndrome. *Ophthalmic Surg Lasers Imaging.* 2008;39[4 Suppl]:S80-S82.)

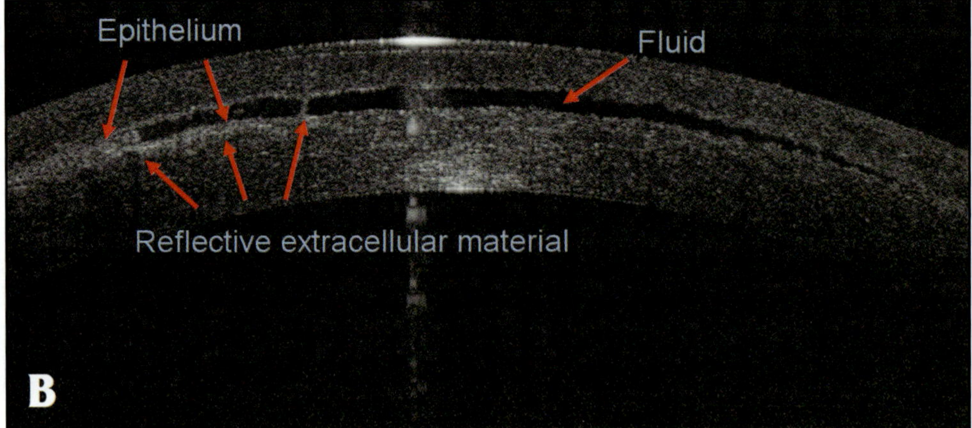

and associated stromal opacification or loss, as well as the health of the overlying flap. This could be helpful in deciding if surgical removal is necessary.

Interface Fluid Syndrome

Interface fluid syndrome (IFS) is an uncommon complication of LASIK that is associated with ocular hypertension or endothelial cell dysfunction. The most commonly reported etiology is steroid-induced ocular hypertension.[14] With its characteristic appearance, including a diffuse haze and a fluid gap in the interface, IFS can be easily mistaken for diffuse lamellar keratitis (DLK, described previously),[15] leading to continued treatment with topical corticosteroids and an increased intraocular pressure.

In a case of post-LASIK IFS[16] that led to epithelial ingrowth, the interface fluid was caused by steroid-induced ocular hypertension (Figure 12-12). On post-LASIK day 49, the interface fluid, epithelial ingrowth, and noncellular reflective deposits were visualized by confocal microscopy and high resolution Fourier-domain OCT using the RTVue. No inflammatory cells, fungal hypheme, or acanthamoeba cysts were seen.

In this case, the higher resolution and definition of the RTVue CAM allowed us to visualize several regions and layers in the interface (see Figure 12-12). The area of epithelial ingrowth had a zone of medium internal reflectivity and a zone of low internal reflectivity. These corresponded to the opaque and clear areas on slit lamp examination (arrows, Figure 12-12). The fluid space had no reflectivity. There were thin layers of medium reflectivity that sandwiched both the epithelial ingrowth and the interface fluid. These layers corresponded to the granular haze observed on the slit lamp examination.

Photorefractive Keratectomy

Subepithelial haze and regression are related problems that occur after excimer laser PRK. The high resolution of the RTVue provides the means for measuring epithelial hyperplasia and deposition of extracellular material that can be associated with clinical "haze." In a representative case, subepithelial scarring after PRK led to an irregular anterior stromal border masked by epithelial hyperplasia (Figure 12-13). The thickness of the epithelium, as measured by the RTVue, increased to 95 µm in the most hyperplastic region. The measurement of epithelial thick-

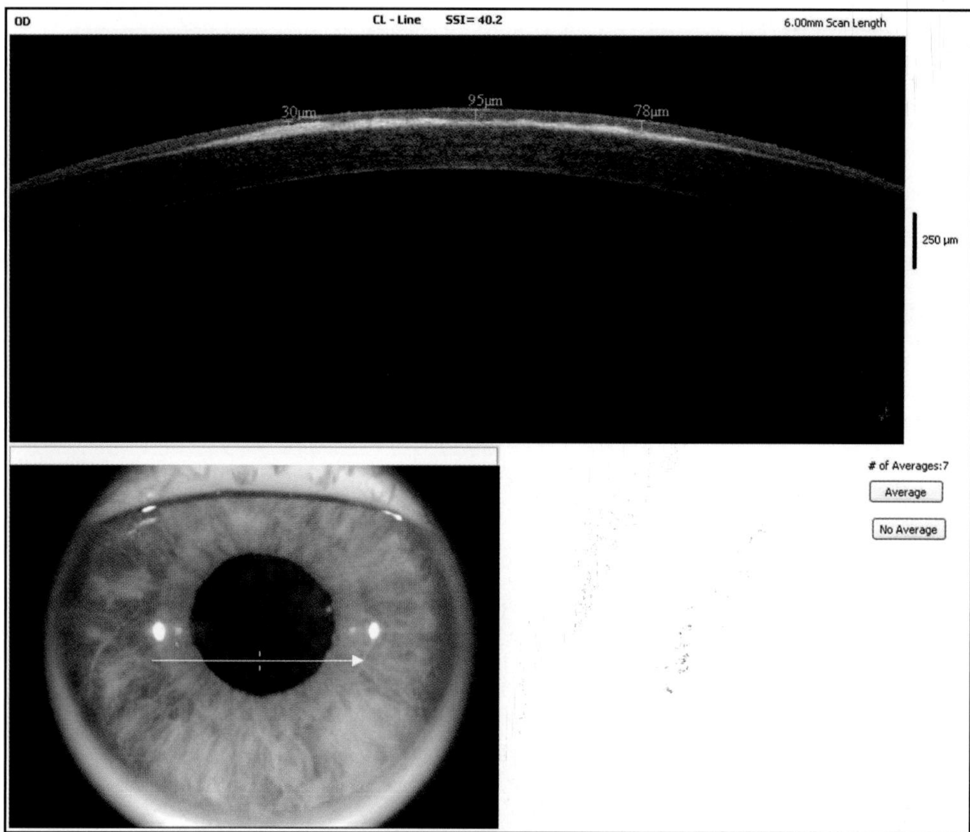

Figure 12-13. RTVue horizontal OCT section (top) after PRK showing severe subepithelial fibrosis under a thickened epithelium. The epithelial thickness varied from 30 to 95 μm. The direction of the scan is shown by the arrow on the video image (bottom). (Courtesy of Harkness Eye Institute, Columbia University.)

ness and the opacity's depth could be helpful in the planning of a transepithelial PRK or PTK to correct irregular astigmatism or a residual refractive error.

Central epithelial hyperplasia is a common response after myopic LASIK and PRK, and may explain some cases of regression. Typically, the OCT image shows central epithelial hyperplasia, as in this case that presented 10 years after PRK for high myopia (Figure 12-14). There was no significant haze.

Phakic Intraocular Lenses

Phakic intraocular lens (IOL) implantation is a surgical option for treating high refractive errors. It differs from laser refractive procedures because it involves implanting an IOL in the eye, not reshaping the cornea.

Besides imaging the cornea and angle, the RTVue can also image the posterior chamber and lens. For example, the RTVue documented the presence of a good vault between an implantable collamer lens such as Visian (STAAR Surgical Co, Monrovia, CA) and the endogenous crystalline lens (Figure 12-15). An adequate distance between the implanted lens and the crystalline lens is needed to avoid cataract formation.

Intracorneal Implants

Intracorneal implants provide a means of changing the refractive property of the cornea without the need for tissue ablation. ICRs have been used to treat myopia, keratoconus, keratectasia, and pellucid marginal degeneration.[17-20] Intrastromal implants have recently been employed to treat presbyopia. Ensuring that the implant is placed at a proper depth is important to avoid complications. Although slit lamp examination can provide a rough sense of implant depth, the RTVue OCT instrument provides high resolution images in which the depth can be precisely measured. Our experience shows that OCT provides a more accurate measurement of depth than slit lamp estimation.[21] This knowledge will help cornea surgeons improve their techniques while addressing depth-related problems earlier.

ICRs, such as Intacs (Addition Technology Inc, Fremont, CA) or Ferrara rings (Mediphacos, Belo Horizonte, Minas Gerais, Brazil), were used initially to treat low myopia, but are now primarily used to treat keratoconus and other corneal diseases. The implant is placed in the midperipheral cornea within surgically created channels at approximately 70% depth. Intacs ICRs consist of a pair of 150-degree polymethylmethacrylate arc segments designed for implantation inside the cornea. The Intacs manufacturer recommends an implant depth of greater than 40% to avoid complications such as epithelial breakdown, anterior stromal melt, and extrusion. The implanted segments serve as a spacer between the corneal collagen fibers, causing these fibers to displace and travel a longer path. This

Figure 12-14. RTVue horizontal OCT section (upper) 10 years after PRK for high myopia showing central epithelial hyperplasia. The direction of the scan is shown by the arrow on the video image (lower). Note the lack of subepithelial scarring compared to the previous case. (Courtesy of Harkness Eye Institute, Columbia University).

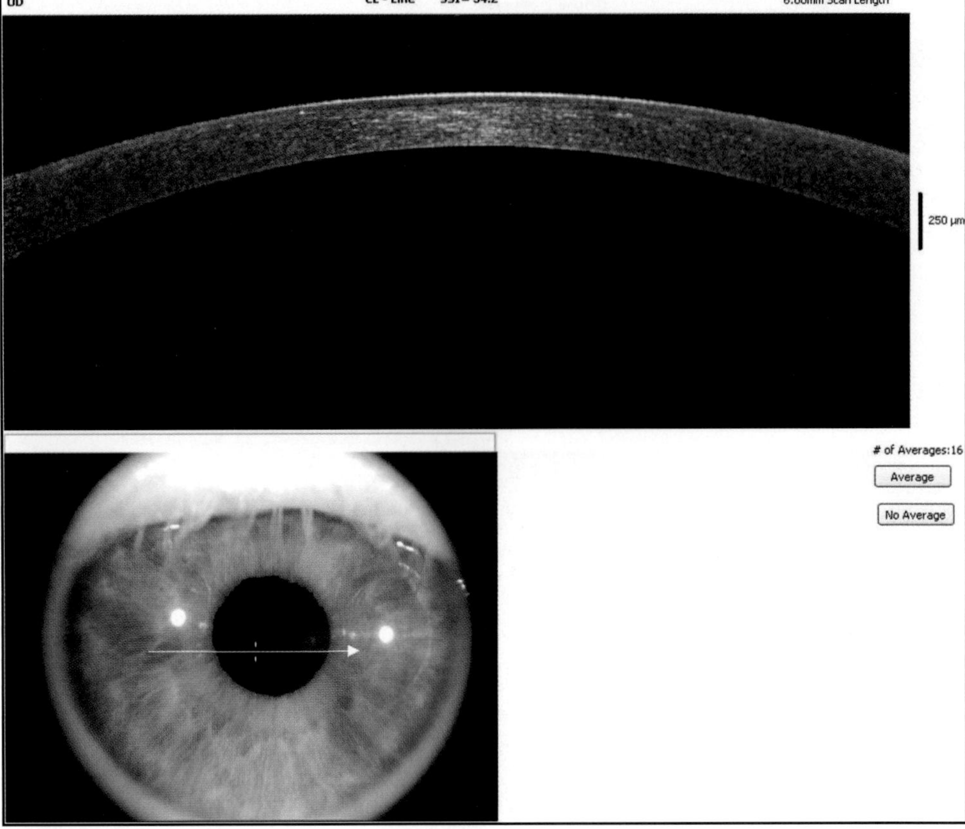

Figure 12-15. RTVue horizontal OCT section showing good clearance of the sulcus-based phakic IOL over the natural crystalline lens. The central IOL thickness was 161 µm and the distance between the IOL and the crystalline lens was 471 µm, as measured by computer calipers (gold lines). (Courtesy of Juan Gabriel Ortiz [Imex].)

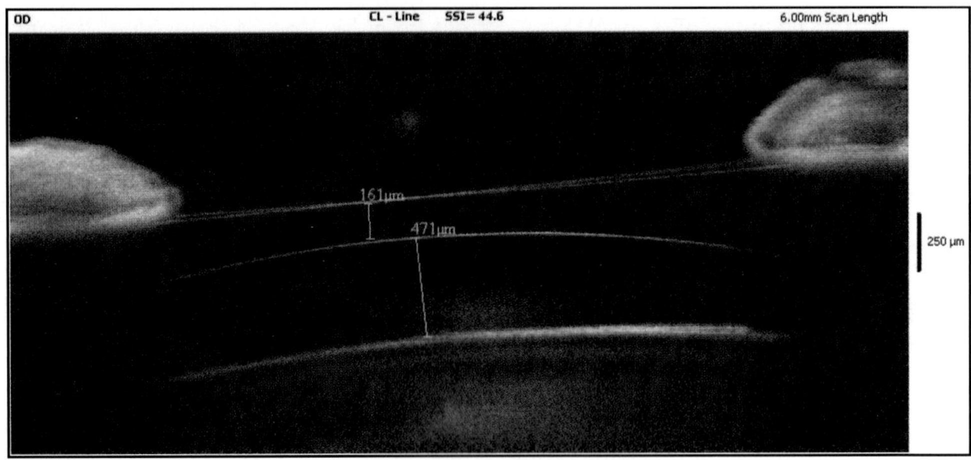

results in a shortened fiber arc length over the visual axis and therefore flattens the central cornea. The flattening effect is proportional to the thickness and depth and length of the implant, resulting in a greater degree of correction.[18] Proper implant depth is critical for ensuring good visual outcome. The use of anterior segment RTVue OCT corneal analysis to help plan the procedure and its follow-up can improve outcomes over the years. Intacs segments appear as dark spaces within the corneal stroma on OCT.

Representative slit lamp and OCT images of an Intacs segment implanted for the treatment of keratoconus show that the absolute implant depth was 370 µm at the center and 513 µm at the inner edge (Figure 12-16). This agrees with the 400 µm setting that was used for the Intralase channel dissection. The fractional depth of the implant was 62% according to the RTVue OCT measurement, close to the 70% recommended depth.

The ACI 7000 corneal implant (AcuFocus Inc, Irvine, CA and Bausch & Lomb, Rochester, NY) is a ring made

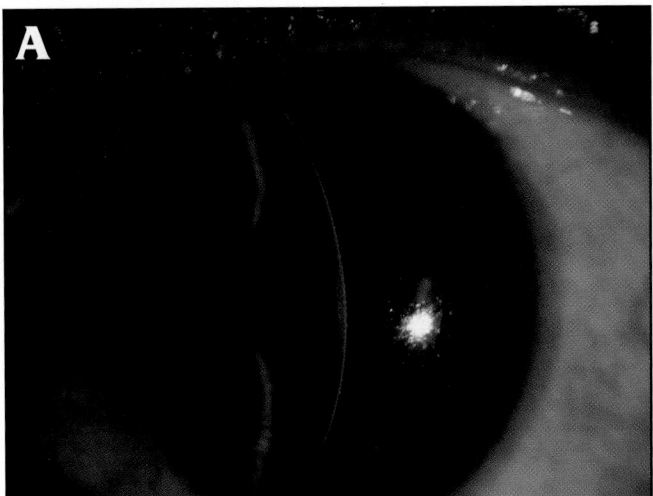

Figure 12-16. A single Intacs segment was implanted inferiorly to treat keratoconus. (A) Slit lamp photograph shows the depth to be approximately 70% as graded by the surgeon. (B) RTVue radial section of the implant. Implant depth is best measured at the inner edge to avoid distortion caused by refraction at the cornea-implant boundary. The depth from the anterior corneal surface to the inner edge of the segment was 513 µm, and the distance from the edge to the endothelium was 313 µm. The fractional depth was 62% (513 / [513 + 313]). Measurements above and below the segment showed a depth of 66% (370 / [370 +192]). The total corneal thickness (total above and below the segment) was 562 µm, while the total was 826 µm next to the inner edge. This indicates stromal compression above and below the segment and stromal expansion lateral to the segment.

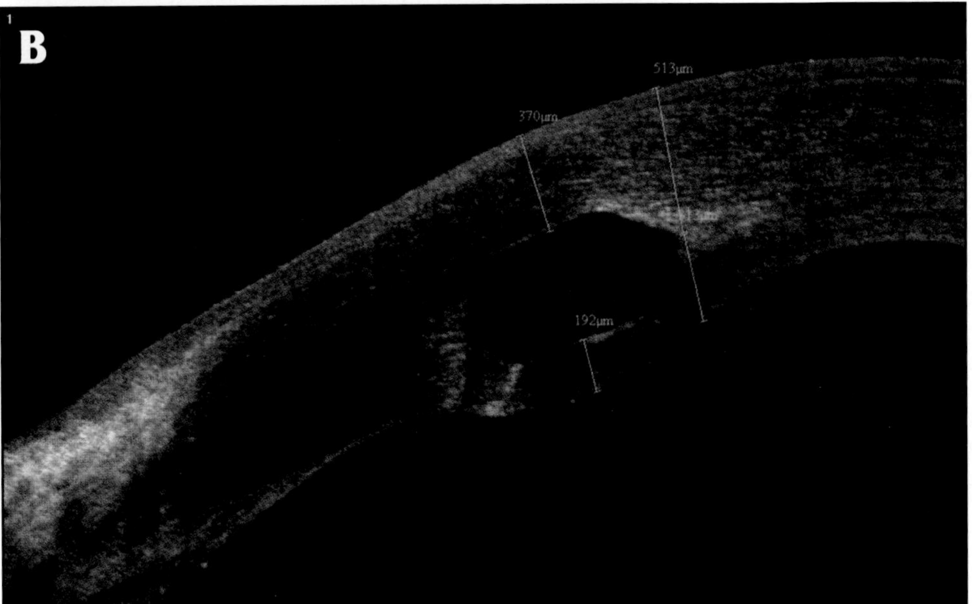

of an opaque biocompatible polymer that increases the depth of focus through pinhole effect. It has shown promising results in an initial trial.[22] The ACI 7000 corneal implant is placed under a corneal flap cut by a microkeratome or femtosecond laser, and it is not designed to permanently alter the cornea. The device appears as a highly reflective band with shadowing of the underlying corneal stroma on OCT (Figure 12-17). OCT can be used to evaluate the implant depth and centration *in situ,* and RTVue provides a way of evaluating the implants *in situ.*

The RTVue OCT images provide a unique tool to visualize and quantify the depth of various corneal implants. This is very helpful in preventing and identifying depth-related complications and minimizing the risks.

REFERENCES

1. Li EY, Mohamed S, Leung CK, et al. Agreement among 3 methods to measure corneal thickness: ultrasound pachymetry, Orbscan II, and Visante anterior segment optical coherence tomography. *Ophthalmology.* 2007;114(10):1842-1847.
2. Boscia F, La Tegola MG, Alessio G, Sborgia C. Accuracy of Orbscan optical pachymetry in corneas with haze. *J Cataract Refract Surg.* 2002;28(2):253-258.
3. Fakhry MA, Artola A, Belda JI, et al. Comparison of corneal pachymetry using ultrasound and Orbscan II. *J Cataract Refract Surg.* 2002;28(2):248-252.
4. Prisant O, Calderon N, Chastang P, et al. Reliability of pachymetric measurements using orbscan after excimer refractive surgery. *Ophthalmology.* 2003;110(3):511-515.
5. Chakrabarti HS, Craig JP, Brahma A, et al. Comparison of corneal thickness measurements using ultrasound and Orbscan slit-scanning topography in normal and post-LASIK eyes. *J Cataract Refract Surg.* 2001;27(11):1823-1828.

Figure 12-17. OCT image showing ACI 7000 corneal implant. The implant appears as a highly reflective band in the flap interface with optical shadowing underneath the implant.

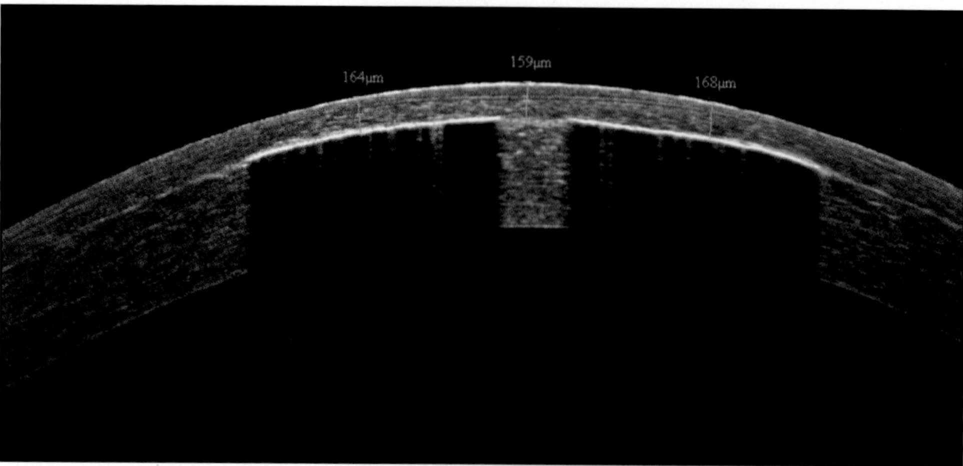

6. Iskander NG, Anderson Penno E, Peters NT, et al. Accuracy of Orbscan pachymetry measurements and DHG ultrasound pachymetry in primary laser in situ keratomileusis and LASIK enhancement procedures. *J Cataract Refract Surg.* 2001;27(5):681-685.

7. Kawana K, Tokunaga T, Miyata K, et al. Comparison of corneal thickness measurements using Orbscan II, non-contact specular microscopy, and ultrasonic pachymetry in eyes after laser in situ keratomileusis. *Br J Ophthalmol.* 2004;88(4):466-468.

8. Khurana RN, Li Y, Tang M, Lai MM, Huang D. High-speed optical coherence tomography of corneal opacities. *Ophthalmology.* 2007;114(7):1278-1285.

9. Li Y, Netto MV, Shekhar R, Krueger RR, Huang D. A longitudinal study of LASIK flap and stromal thickness with high-speed optical coherence tomography. *Ophthalmology.* 2007;114(6):1124-1132.

10. Randleman JB, Russell B, Ward MA, et al. Risk factors and prognosis for corneal ectasia after LASIK. *Ophthalmology.* 2003;110(2):267-275.

11. Randleman JB, Woodward M, Lynn MJ, Stulting RD. Risk assessment for ectasia after corneal refractive surgery. *Ophthalmology.* 2008;115(1):37-50.

12. Chang JSM. Complications of sub-Bowman's keratomileusis with a femtosecond laser in 3009 eyes. *J Refract Surg.* 2008;24:S97-S101.

13. Ustundag C, Bahcecioglu H, Ozdamar A, Aras C, Yildirim R, Ozkan S. Optical coherence tomography for evaluation of anatomical changes in the cornea after laser in situ keratomileusis. *J Cataract Refract Surg.* 2000;26(10):1458-1462.

14. Dawson DG, Schmack I, Holley GP, et al. Interface fluid syndrome in human eye bank corneas after LASIK: causes and pathogenesis. *Ophthalmology.* 2007;114(10):1848-1859.

15. Fogla R, Rao SK, Padmanabhan P. Interface fluid after laser in situ keratomileusis. *J Cataract Refract Surg.* 2001;27(9):1526-1528.

16. Ramos JL, Zhou S, Yo C, Tang M, Huang D. High-resolution imaging of complicated LASIK flap interface fluid syndrome. *Ophthalmic Surg Lasers Imaging.* 2008;39(4 Suppl):S80-S82.

17. Schanzlin DJ, Asbell PA, Burris TE, Durrie DS. The intrastromal corneal ring segments. Phase II results for the correction of myopia. *Ophthalmology.* 1997;104(7):1067-1078.

18. Colin J, Cochener B, Savary G, Malet F. Correcting keratoconus with intracorneal rings. *J Cataract Refract Surg.* 2000;26(8):1117-1122.

19. Lovisolo CF, Fleming JF. Intracorneal ring segments for iatrogenic keratectasia after laser in situ keratomileusis or photorefractive keratectomy. *J Refract Surg.* 2002;18(5):535-541.

20. Rodriguez-Prats J, Galal A, Garcia-Lledo M, et al. Intracorneal rings for the correction of pellucid marginal degeneration. *J Cataract Refract Surg.* 2003;29(7):1421-1424.

21. Lai MM, Tang M, Andrade EM, et al. Optical coherence tomography to assess intrastromal corneal ring depth in keratoconic eyes. *J Cataract Refract Surg.* 2006;32(11):1860-1865.

22. Yilmaz OF, Bayraktar S, Agca A, et al. Intracorneal inlay for the surgical correction of presbyopia. *J Cataract Refract Surg.* 2008;34(11):1921-1927.

13 Corneal Pathologies and Surgeries

Martin Heur, MD, PhD; Maolong Tang, PhD;
Stephen L. Trokel, MD; and David Huang, MD, PhD

INTRODUCTION

RTVue OCT is a very useful imaging modality for evaluating the cornea. OCT is helpful in the evaluation of corneal opacities (such as scars) because, in addition to providing a pachymetric map, the OCT can also be used to measure the depth of the opacities and alterations in overlying epithelial thickness. Although Orbscan can also provide similar information, previous studies have shown that OCT is more reliable because Orbscan underestimates the corneal thickness in the presence of central corneal scars.[1] OCT also has much higher resolution compared to Scheimpflug imaging devices such as the Pentacam (Oculus Inc, Lynnwood, WA). The depth data can help the surgeon decide between performing an ablative procedure for superficial opacities versus lamellar keratoplasty for deeper corneal opacities. OCT imaging is also useful in imaging the donor graft in posterior lamellar keratoplasty and the interlocking incision configurations created in advanced keratoplasty techniques. With the release of the latest software, OCT can also be used to calculate the corneal power by averaging the instantaneous curvature of the central 3-mm diameter containing the entrance pupil of the eye. The new algorithm allows for the estimation of the anterior, posterior, and net corneal powers.

CORNEAL OPACITIES

OCT is a useful adjunct in evaluation of corneal opacities, because in addition to providing a pachymet-ric map of the cornea, it also provides the depth of the opacity. This can be helpful in the preoperative planning by allowing the surgeon to choose between an ablative procedure (such as phototherapeutic keratectomy [PTK]) if the residual stromal bed thickness will be greater than 250 μm, or an procedure therapy (such as anterior lamellar keratoplasty) if the residual stromal bed thickness will be less than 250 μm. OCT imaging can be used to determine the layer of the corneal pathology, such as in a case of long-standing keratoconus in which the opacity was localized to the epithelium (Figure 13-1).

A 41-year-old man was referred for subepithelial haze 11 months after photorefractive keratectomy (PRK) enhancement for myopic regression in the left eye (Figure 13-2). The best spectacle-corrected visual acuity was reduced to 20/100. RTVue OCT images showed that the opacity was relatively deep, 122 to 163 μm, and the central cornea was thin, 365 to 368 μm (Figure 13-3). It was not possible to perform an ablative procedure to remove most of the opacity without maintaining at least 250 μm of residual stromal thickness. The eye was treated with topical corticosteroid and the resulting best-corrected visual acuity (BCVA) was 20/25.

A 70-year-old woman presented with a long-standing corneal scar OS secondary to contact lens-related bacterial keratitis. BCVA was 20/20 OD and 20/70 OS. OCT showed a full-thickness corneal opacity (Figure 13-4). The patient underwent penetrating keratoplasty (PK) OS and had a BCVA OS of 20/60 2 years after the surgery. The visual acuity was limited by age-related macular degeneration.

A 50-year-old woman presented with granular corneal dystrophy and recurrent corneal erosion in both

Huang D, Duker JS, Fujimoto JG, Lumbroso B, Schuman JS, Weinreb RN.
Imaging the Eye from Front to Back with RTVue Fourier-Domain Optical Coherence Tomography (pp 123-126).

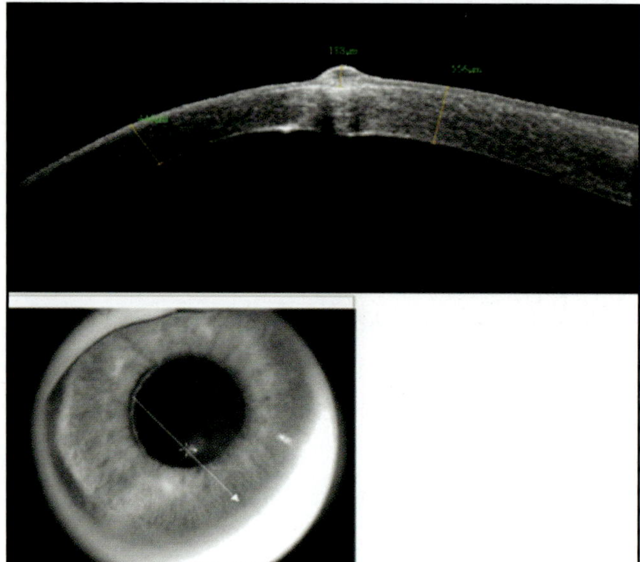

Figure 13-1. Diagonal OCT section of a cornea in long-standing keratoconus with focal epithelial bulla. The scan shows the focal epithelial bulla measuring 188 µm in thickness.

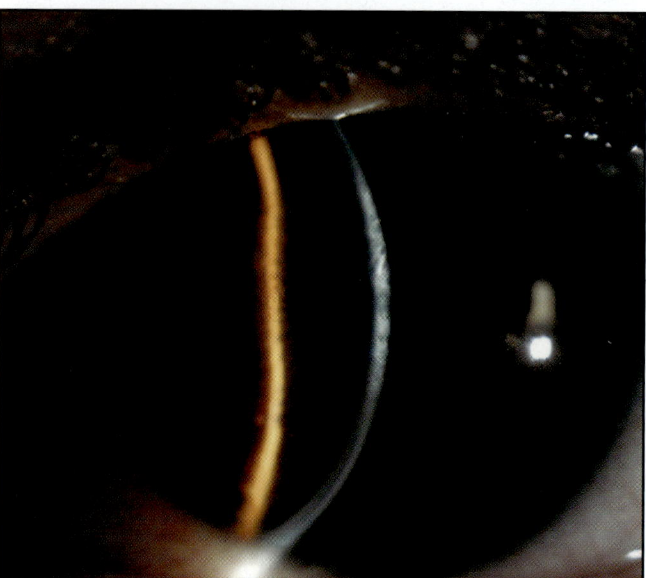

Figure 13-2. Slit lamp photograph of post-PRK haze. The photograph shows the presence of haze in the anterior stroma.

Figure 13-3. OCT of horizontal (top) and vertical (bottom) sections of the cornea from Figure 13-2. Computer calipers were used to measure the opacity depth, epithelial thickness, and corneal thickness. The measured opacity depth ranged from 122 to 163 µm.

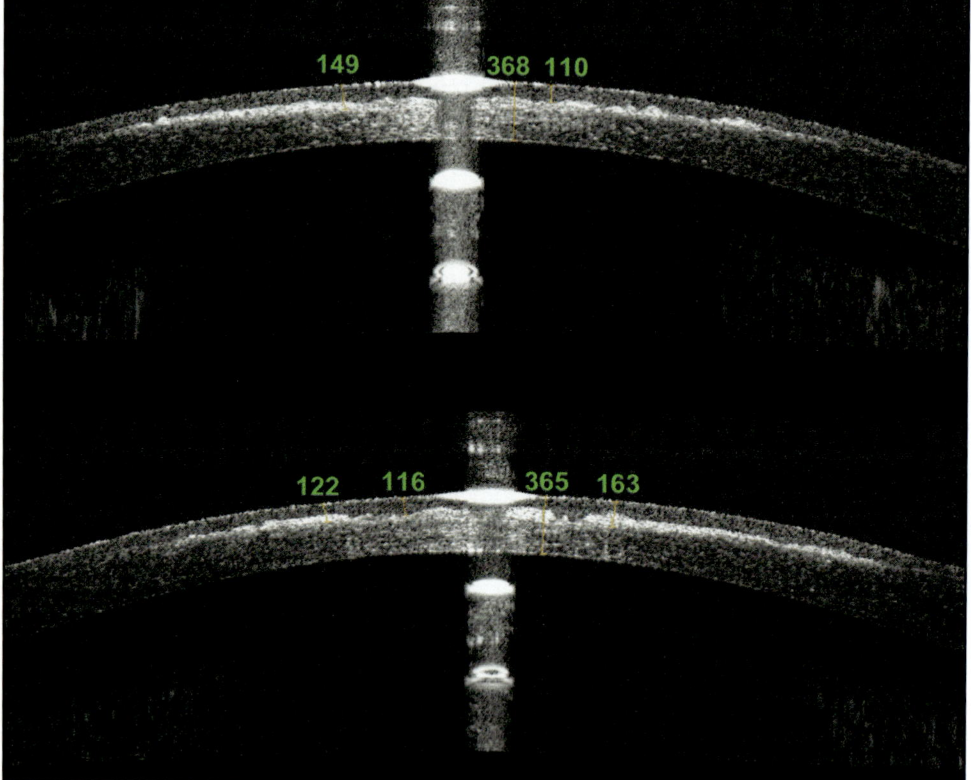

Figure 13-4. Horizontal OCT section of a corneal opacity secondary to contact lens-related bacterial keratitis. The scan showed that the opacity was full thickness.

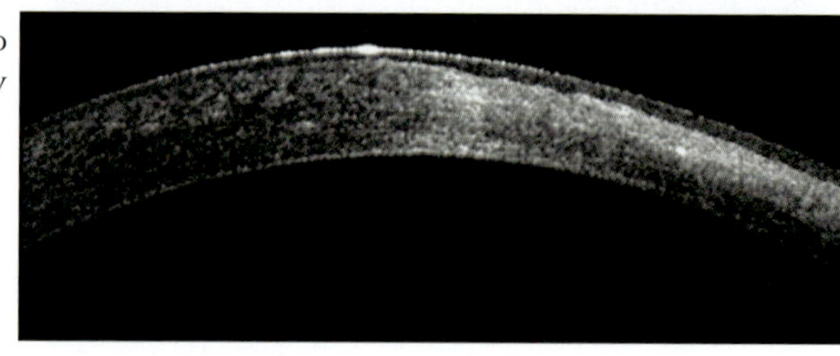

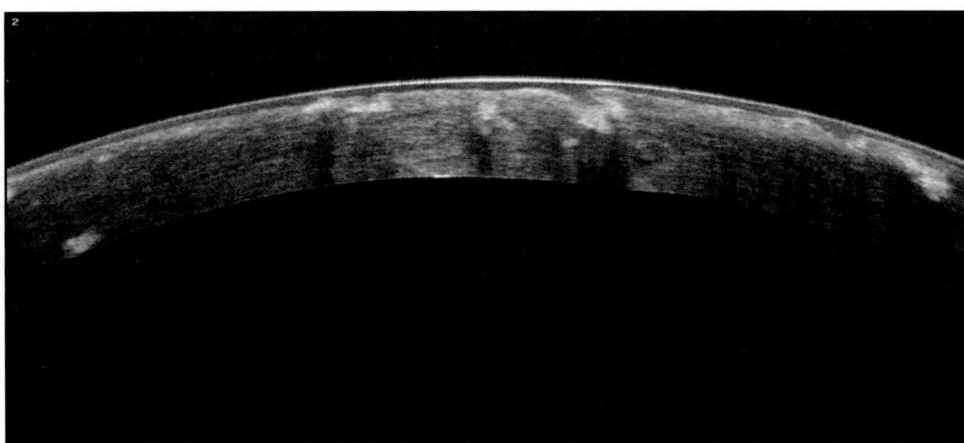

Figure 13-5. Horizontal OCT section from a granular dystrophy patient. The scan shows the presence of deposits in the anterior stroma.

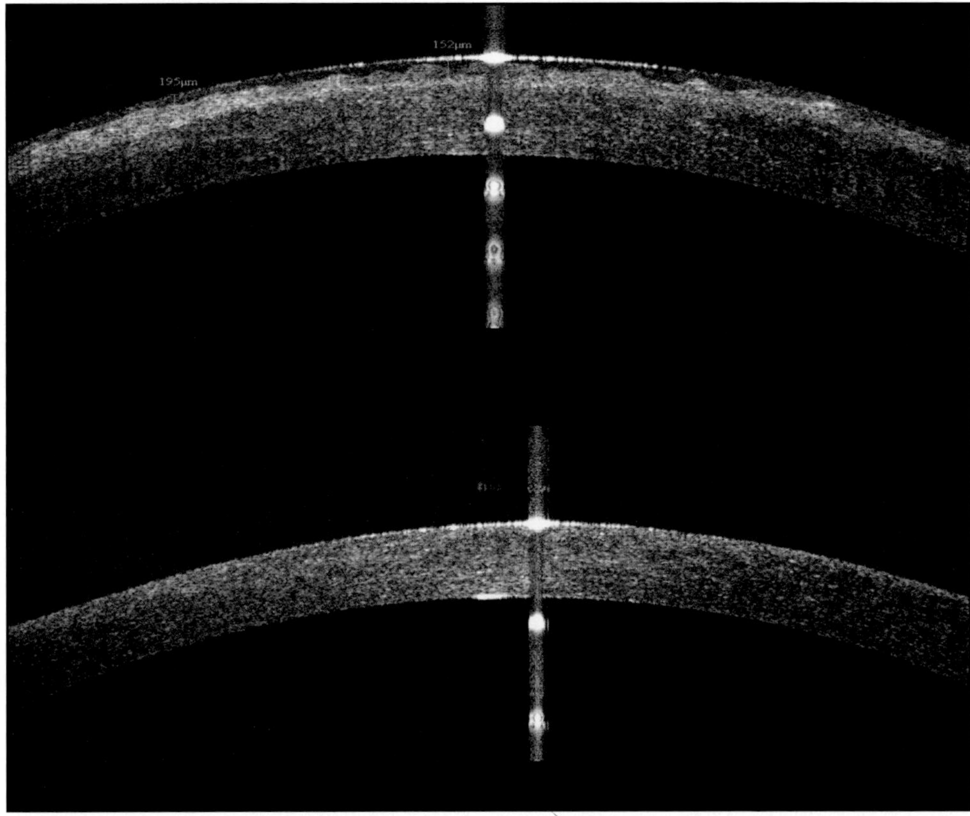

Figure 13-6. Horizontal OCT sections from a Reis-Bucklers dystrophy patient. Top: image of the cornea before PTK shows the presence of deposits in the anterior stroma. Bottom: image of the cornea 2 months after PTK shows that the deposits have not recurred.

eyes. The visual acuity OS was worse, with a BCVA of 20/80. RTVue OCT was performed to plan the PTK treatment. The OCT image showed multiple, hyper-reflective deposits in the anterior stroma under a layer of a subepithelial opacity (Figure 13-5). It also confirmed PTK as a safe option because of adequate residual bed thickness. Transepithelial PTK with adjunctive mitomycin-C was performed to remove the subepithelial opacity and smooth the anterior stromal surface. The postoperative BCVA OS was 20/30.

A 68-year-old woman was referred for treatment of Reis-Bucklers corneal dystrophy OU. BCVA was 20/100 in the OD and 20/80 OS. RTVue OCT was performed to plan the PTK treatment. The OCT image showed the depth of the opacity under the epithelium and in the anterior stroma, indicating that it was safe to perform an ablative procedure to remove the corneal opacities. Two months after the procedure, OCT imaging demonstrated absence of anterior stromal deposits with a BCVA of 20/40 OU (Figure 13-6).

ENDOTHELIAL KERATOPLASTY

Since the inception of modern posterior lamellar keratoplasty in the late 1990s by Melles, the technique has undergone several cycles of refinement to Descemet's

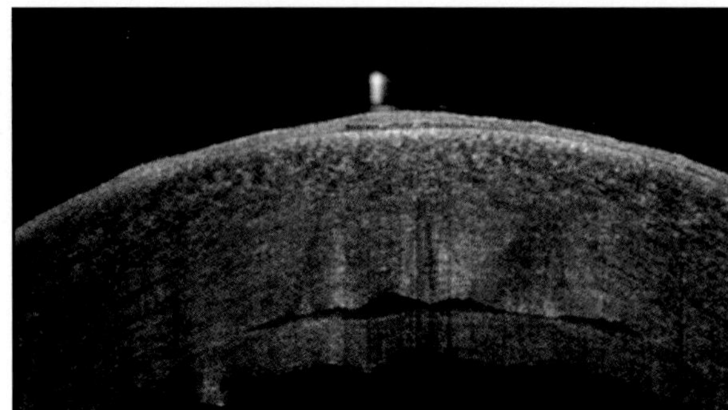

Figure 13-7. Horizontal OCT section showing detachment of donor lenticule 2 weeks following DSAEK. This 72-year-old female continued to have persistent postoperative edema 2 weeks after DSAEK OD. The scan shows a subtle detachment of the donor lenticule with overlying epithelial and stromal edema. The lenticule attached spontaneously, resulting in an improvement in the edema at 1 month postoperative.

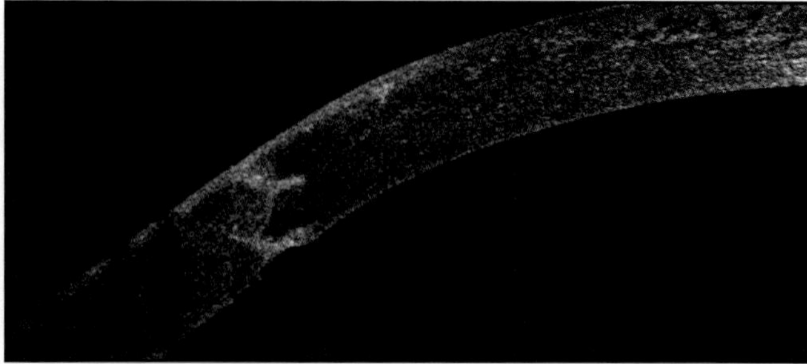

Figure 13-8. Horizontal OCT section of graft host junction 1 month after suture removal following femtosecond laser-assisted keratoplasty. The 24-year-old male presented with advanced keratoconus in the right eye and elected to undergo femtosecond laser-assisted keratoplasty. All sutures were removed 4 months postoperative. The zigzag wound architecture is visible on OCT imaging.

stripping automated endothelial keratoplasty (DSAEK).[2] The primary advantages of DSAEK over PK are a smaller and more secure wound, lower postoperative astigmatism, and faster visual rehabilitation.[3] Postoperative OCT scans can be used to monitor the detachment of the donor lenticule in patients with persistent postoperative corneal edema (Figure 13-7). It is also useful in tracking the postoperative corneal deturgescence, assessing the uniformity of the endothelial graft thickness, and measuring the postoperative posterior corneal curvature.

FEMTOSECOND LASER-ENABLED KERATOPLASTY

PK is undergoing a phase of evolution thanks to the ability to make shaped incisions using a femtosecond laser such as the Intralase. A concern of traditional PK performed using trephine blades is the slow wound healing and the weak host-donor junction. Shaped incisions such as the zigzag, top hat, and mushroom configurations created using the femtosecond laser result in greater resistance to wound leakage when compared to traditional PK.[4] The wound may also be stronger and heal faster due to the interlocking host-donor junction and greater wound contact area. Proper apposition of the interlocking wound can be visualized with OCT (Figure 13-8).

REFERENCES

1. Khurana RN, Li Y, Tang M, et al. High-speed optical coherence tomography of corneal opacities. *Ophthalmology.* 2007;114(7):1278-1285.
2. Melles GR, Wijdh RH, Nieuwendaal CP. A technique to excise the descemet membrane from a recipient cornea (descemetorhexis). *Cornea.* 2004;23(3):286-288.
3. Price MO, Price FW. Descemet's stripping endothelial keratoplasty. *Curr Opin Ophthalmol.* 2007;18(4):290-294.
4. Farid M, Kim M, Steinert RF. Results of penetrating keratoplasty performed with a femtosecond laser zigzag incision initial report. *Ophthalmology.* 2007;114(12):2208-2212.

Non-Exudative Age-Related Macular Degeneration

Leonardo C. Castro, MD

Age-related macular degeneration (AMD) is the leading cause of central visual loss among individuals 65 years of age and older in developed countries.[1-3] The disease adversely affects quality of life and activities of daily living, causing many individuals to lose their independence in their retirement years. It is estimated that more than 8 million people in the United States have some stage of the disease,[4,5] and this prevalence is expected to rise dramatically as a result of the increasing percentage of elderly persons and the improved management of other eye diseases.

The non-exudative (or non-neovascular, or dry) form of AMD accounts for approximately 80% to 90% of all diagnosed cases. Geographic atrophy, the most severe non-exudative manifestation of AMD, is responsible for approximately 21% of the cases of legal blindness in North America.[2]

PATHOGENESIS

Although much progress has been made over the past decades, finding the causes and mechanisms of AMD remains a challenge. Abnormalities in the enzymatic activity of aged retinal pigment epithelium (RPE) cells lead to accumulation of metabolic by-products, interfering with their normal metabolism, and leads to the accumulation of residual bodies that contain lipofuscin. In addition, progressive accumulation of lipids in Bruch's membrane with age[6,7] alters normal photoreceptor function, which is dependent of free diffusion through Bruch's membrane from the choriocapillaris to the RPE.[8,9] Deposits of extracellular material, typically lying between the basement membrane of the retinal pigment cells and the inner collagenous zone of Bruch's membrane, form drusen.

SYMPTOMS

Though symptoms vary, people with the dry form of AMD usually first experience blurred near vision, distortion, and difficulty with reading or distinguishing faces. Early symptoms of vision loss from dry AMD include shadowy areas in the central vision. Usually, the visual loss is slow and progresses gradually, not nearly as severe as wet AMD.

CLASSIFICATION

The following is an AMD classification list.
- No AMD: no drusen or only a few small drusen (<63 µm) in the absence of any other stage of AMD

- Early stage of AMD: few (approximately <20) medium-size drusen (63 to 124 µm) or pigment abnormalities (increased pigmentation or depigmentation) and no other stage of AMD

- Intermediate stage of AMD: usually the presence of at least 1 large drusen (at least 125 µm, approximately the width of a retinal vein as it crosses the optic nerve), but can also be numerous medium-

Huang D, Duker JS, Fujimoto JG, Lumbroso B, Schuman JS, Weinreb RN.
Imaging the Eye from Front to Back with RTVue Fourier-Domain Optical Coherence Tomography (pp 127-136).
© 2010 SLACK Incorporated.

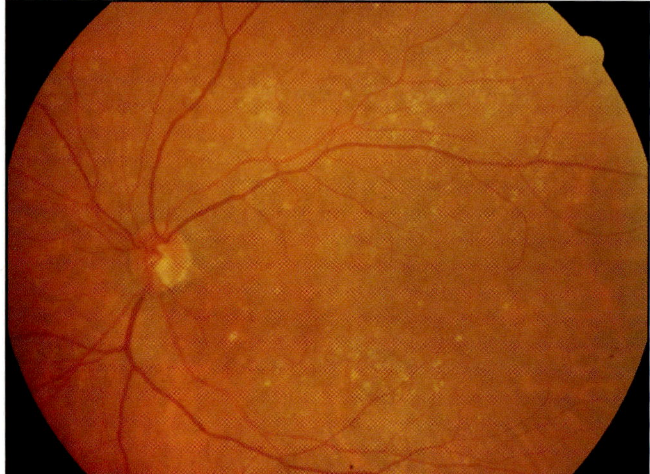

Figure 14-1. Color fundus photo of a dry AMD patient with hard drusen. The drusen are disposed in wide bands outside the vascular arcades and passing on the nasal side of the disc, sparing the inner macula.

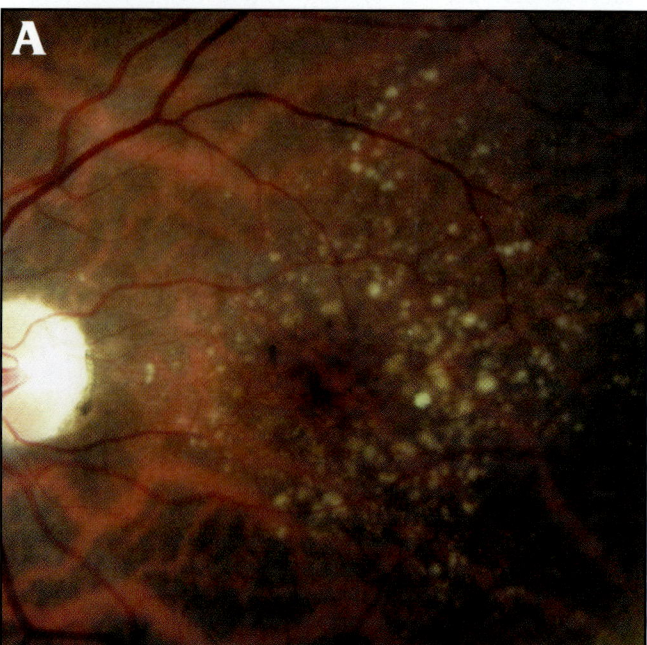

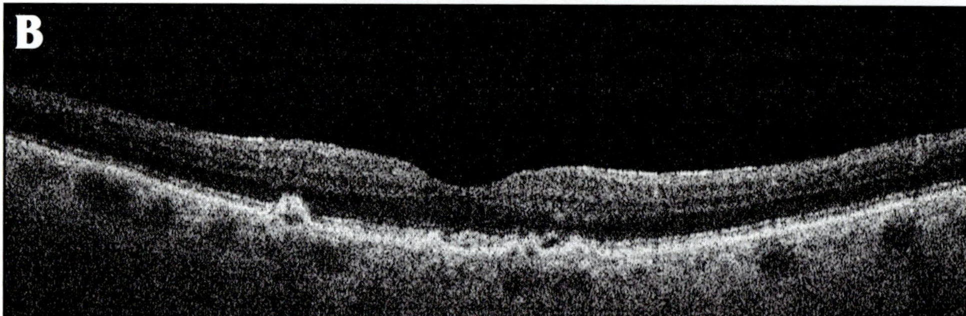

Figure 14-2. Dry AMD patient with soft and hard drusen. (A) Color fundus photo: the soft drusen accumulate close to the center of the macula, while the smaller and the hard drusen more frequently appear in the periphery of the macula. (B) Vertical OCT scan showing 1 hard drusen inferior to the fovea (left in the scan), small drusen and disruption of the photoreceptor IS/OS junction under the fovea. Note that the RPE layer is not clearly defined under the hard drusen.

size drusen (approximately 20 or more when the drusen boundaries are indistinct or amorphous or soft; and approximately 65 or more when the drusen boundaries are distinct or sharp or hard) or the presence of geographic atrophy that does not extend under the center of the macula

- Advanced stage of AMD: geographic atrophy extending under the center of the macula or signs of choroidal neovascularization

OCULAR FINDINGS

Small, hard drusen may first be noted in the central area, 1000 µm in diameter,[10] but when numerous they are most common on the temporal side of the fovea, tending to occur in clusters. Another common pattern consists of a wide band outside the vascular arcades and passing on the nasal side of the disc, with sparing of the inner macula. Toward the equator they assume a linear arrangement in relation to a polygonal pattern of hyperpigmented lines, giving rise to the picture of reticular (honeycomb) degeneration of the pigment epithelium (Figure 14-1).

However, the most important ocular findings of the dry AMD are the soft drusen, for which any change in the pattern is an important factor in the analysis of the risk to develop choroidal neovascularization. Usually they are bigger than 63 µm; have a yellow, solid appearance; and their confluence results in crescentic or sinuous shapes and indistinct margins. Large drusen tend to accumulate close to the center of the macula, while the smaller and the hard drusen more frequently appear in the periphery of the macula (Figure 14-2).

The natural history of the soft drusen is variable. They usually progressively enlarge and merge, occupying a large area in the posterior pole, and eventu-

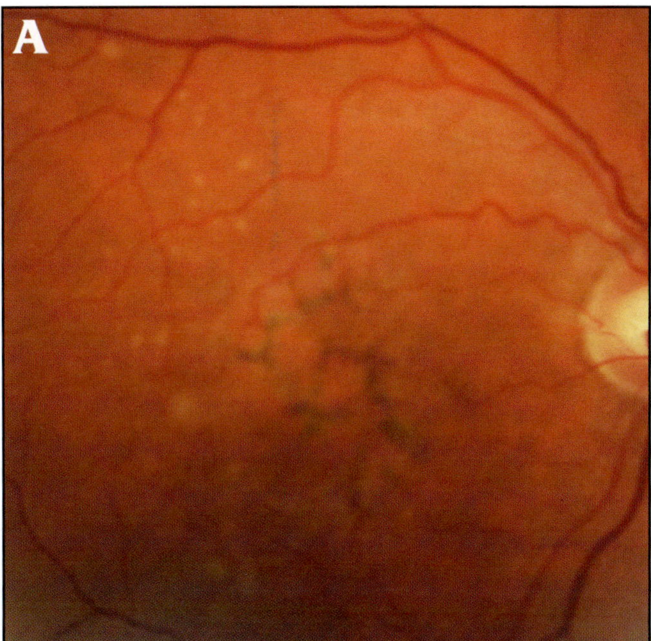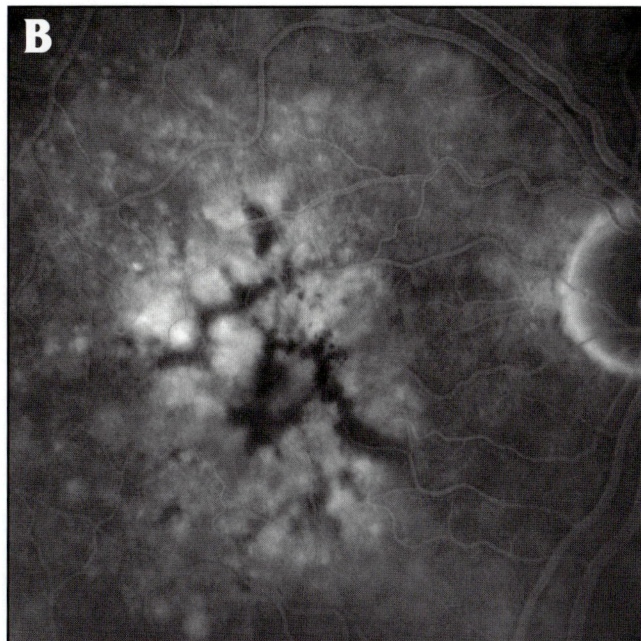

Figure 14-3. Hyperpigmentation. (A) Color fundus photo: large soft drusen, confluent in the center of the fovea, with radial hyperpigmentation. (B) Fluorescein angiography shows late staining of the soft drusen and hypofluorescence due to blockage by the hyperpigmentation.

ally form a drusenoid pigment epithelial detachment. Hyperpigmentation can gradually develop in association with the confluent soft drusen, usually in a radial pattern (Figure 14-3).

The drusenoid pigment epithelial detachments have an irregular shape and multiple small elevations in the surface. In general, drusenoid pigment epithelial detachments have a slow natural history of progression and merging. After a variable period, they start a process of involution and gradually become atrophic.

Immediately preceding geographic atrophy, incipient atrophy is a useful means of predicting its rate and direction of spread. Areas of RPE thinning or depigmentation (incipient atrophy) can be recognized, although not as sharply defined as actual geographic atrophy. The affected fundus appears paler than the normal fundus background due to the loss of RPE (Figure 14-4), and any drusen present appear whiter and harder before fading.

The end result of the atrophic form of AMD is the geographic atrophy, which is currently defined[11] as any sharply delineated round or oval area of hypopigmentation or depigmentation or apparent absence of the RPE, in which choroidal vessels are more visible than in surrounding areas and must be at least 175 μm in diameter (Figure 14-5).

Dystrophic calcification in hard or soft drusen leads to calcified drusen, which have a glistening appearance on the fundoscopy (see Figure 14-10A).

ANCILLARY EXAMS

Fluorescein Angiography

On fluorescein angiography, hard drusen are readily demonstrated, even when small, fluorescing brightly in the mid venous phase and fading soon after the background choroidal fluorescence (Figure 14-6). In contrast, soft drusen, as well as the drusenoid pigment epithelial detachment, are hypofluorescent and almost invisible in the early phases, gradually staining and becoming hyperfluorescent in the late phases (sometimes 10, 20, or 30 minutes) (Figure 14-7).

Incipient atrophy appears as diffuse hyperfluorescence beginning on early phases of the exam, but not as bright as an area of geographic atrophy (see Figures 14-5B and C). The areas of atrophy may be more easily delineated on autofluorescence.

Indocyanine Green Angiography

On indocyanine green angiography, hard drusen become hyperfluorescent 2 to 3 minutes after dye administration, and this persists through the middle and late phases. The soft drusen and the drusenoid pigment epithelial detachments are hypofluorescent during the entire exam. As the choroidal vessels become more hyperfluorescent during the exam, they can easily be seen as hypofluorescent areas blocking the choroidal hyperfluorescence.

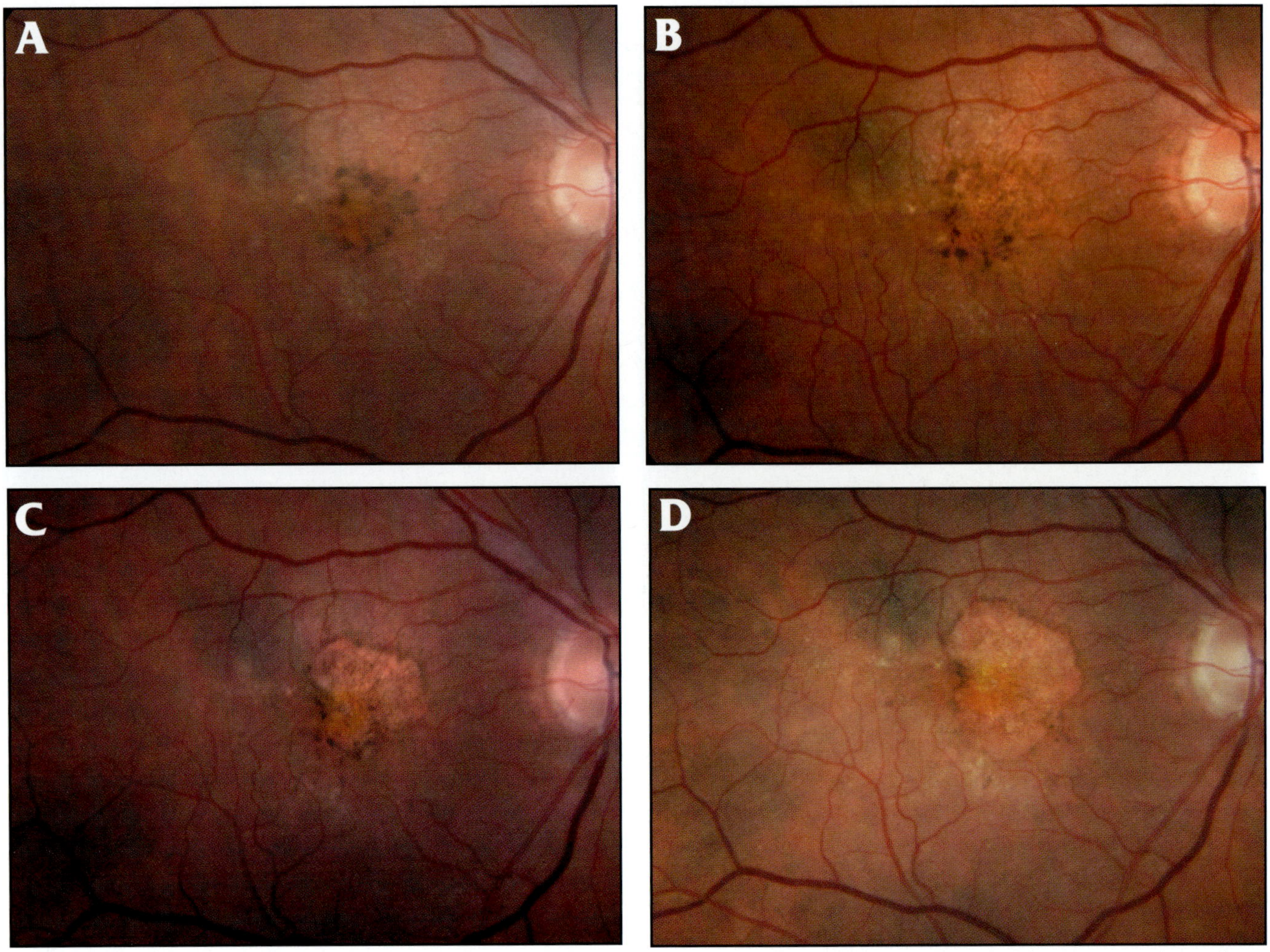

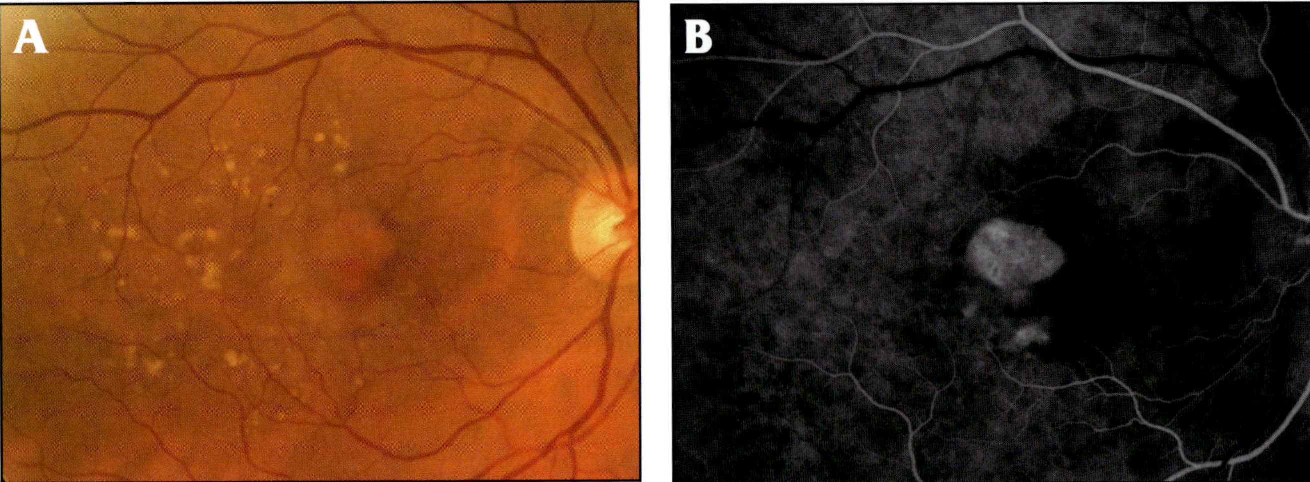

Figure 14-4. Color fundus photos of a patient with dry AMD initially with incipient atrophy (pink area in the superonasal periphery of the hyperpigmentation) (A), which progressively enlarged to form a geographic atrophy (B through D). Each photo is 1 year apart.

Figure 14-5. Geographic atrophy. (A) Color fundus photo: geographic atrophy with large soft drusen around. (B, C) Fluorescein angiography exam: the area of geographic atrophy is highly bright since early phases of the exam (A), staining in the late phase (B). Note that the soft drusen are hypofluorescent in the early phase, fluorescing in the late phase.

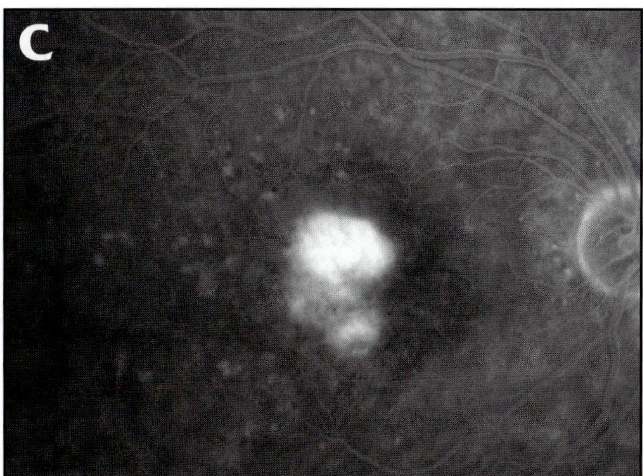

Figure 14-5. Geographic atrophy. (B, C) Fluorescein angiography exam: the area of geographic atrophy is highly bright since early phases of the exam (A), staining in the late phase (B). Note that the soft drusen are hypofluorescent in the early phase, fluorescing in the late phase.

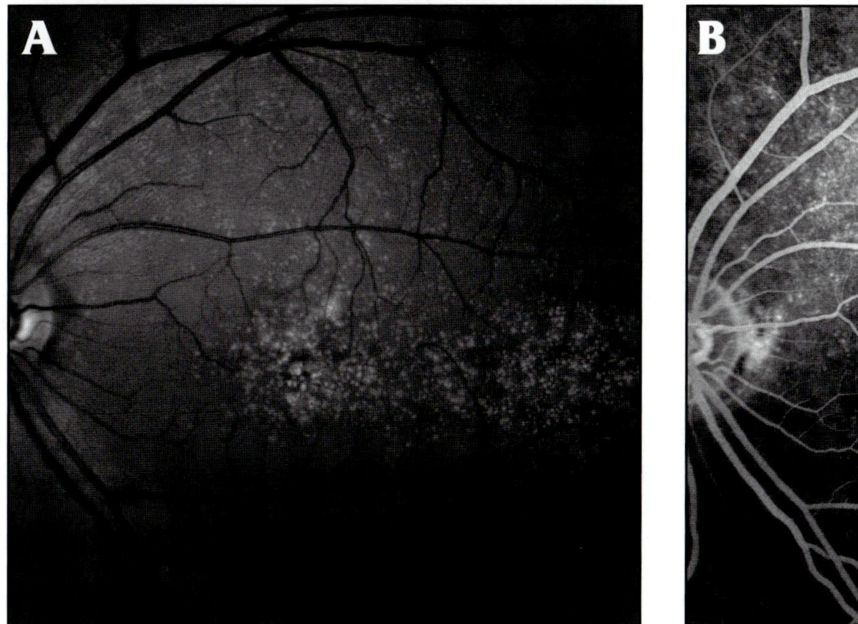

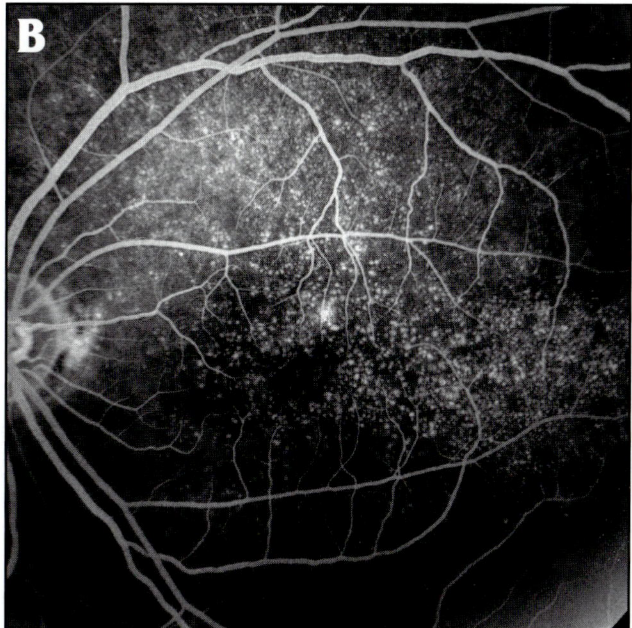

Figure 14-6. Hard drusen. (A) Red free photo showing innumerous small hard drusen in the macula. (B) On fluorescein angiography, the hard drusen fluoresce brightly in the mid venous phase, and several drusen not seen on the fundoscopy highlight during the exam.

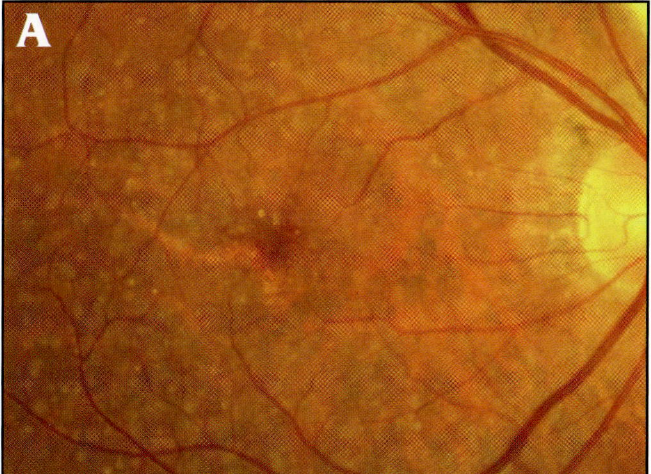

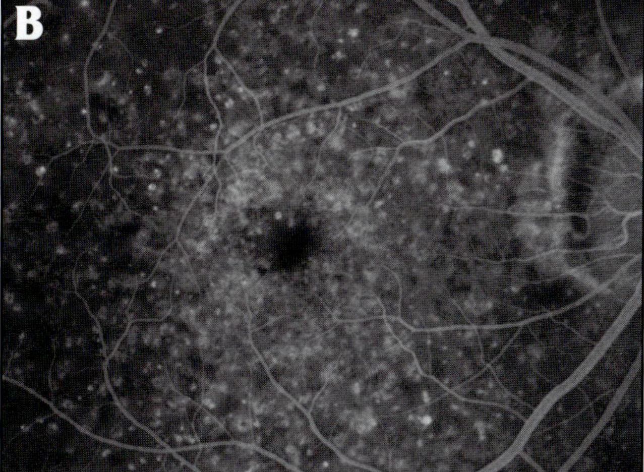

Figure 14-7. Soft drusen. (A) Color fundus photo: innumerous soft drusen in the macula. (B) Fluorescein angiography: late hyper-fluorescence due to staining of the soft drusen.

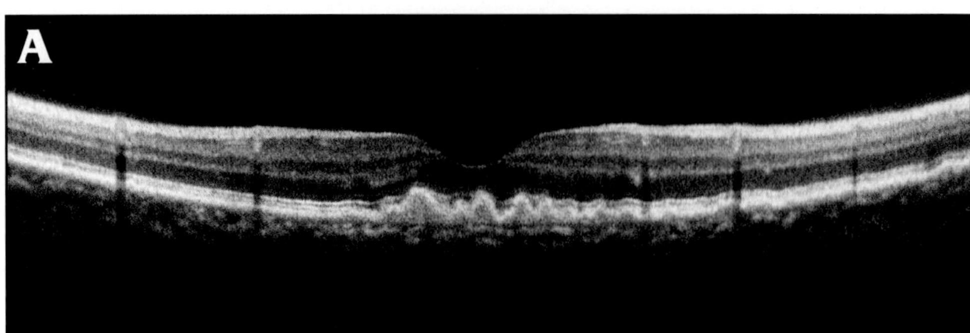

Figure 14-8. Dry AMD with soft drusen. (A) OCT B-scan shows the drusen as elevations of the RPE level, in this case with homogeneous medium internal reflectivity. (B) A 3D macular reconstruction segmented to visualize the RPE level: the multiple elevations in the RPE correspond to the drusen.

Fundus Autofluorescence

Drusen visible on fundus photography are not necessarily correlated with notable fundus autofluorescence changes. Areas of increased fundus autofluorescence may or may not correspond with areas of hyperpigmentation or soft or hard drusen.[12] Fundus autofluorescence signal is usually low over drusen, but not as low as in areas of RPE atrophy, which are typically characterized by a very dark signal. This observation of fundus autofluorescence over drusen would be therefore more consistent with a peripheral displacement of the overlying RPE cytoplasm and/or lipofuscin granules rather than with manifest RPE atrophy.[12] Overall, larger drusen are associated more frequently with more pronounced fundus autofluorescence abnormalities than smaller ones.

Areas of hypopigmentation on fundus photographs tend to be associated with a corresponding decreased fundus autofluorescence signal, suggesting the absence of RPE cells or degenerated RPE cells with reduced content of lipofuscin granules. In contrast, areas of hyperpigmentation can exhibit a higher fundus autofluorescence signal, which may be due to a higher content of autofluorescent melanolipofuscin.[13]

Compared to drusen, atrophic areas typically show an even stronger reduction of fundus autofluorescence.[12] The high-contrast difference between atrophic and non-atrophic retina allows delineating the area of atrophy much better than from conventional fundus photographs. In geographic atrophy areas, it is frequently the visualization of high intensity levels surrounding the atrophic patches in the junctional zone of atrophy.[14]

OPTICAL COHERENCE TOMOGRAPHY

On OCT cross-sectional images, hard drusen appear as discrete nodules of high reflectivity and moderately reflective material in the RPE level. Differently from soft drusen, the RPE layer may not be clearly defined in the area of these nodules, showing discontinuity of the RPE where there is accumulation of material (see Figure 14-2B).[15]

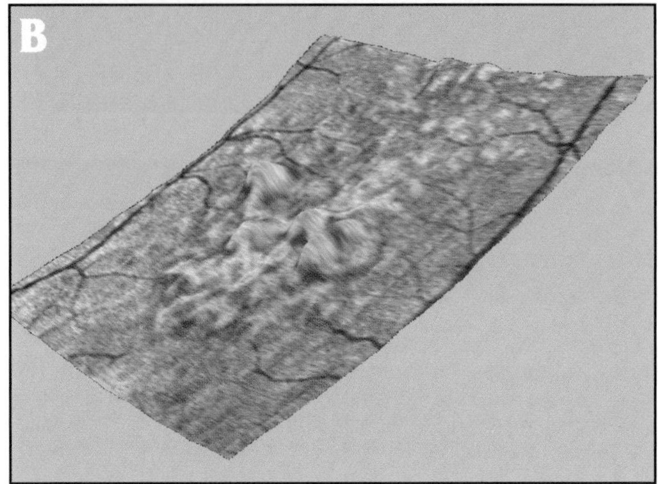

Soft drusen are imaged as elevations of the RPE level, typically with homogeneous medium internal reflectivity (hyper-reflective relative to the photoreceptor layer and hyporeflective relative to the RPE), although low reflectivity (similar to a small pigment epithelial detachment) or high reflectivity (similar to the RPE reflectivity) can be found (Figure 14-8).[16] Confluent soft drusen usually have weakly scattering deposits forming "dome-like" RPE elevations. The RPE between the drusen is detached from Bruch's membrane, indicating confluence of deposits. The photoreceptor IS/OS (inner/outer segment) junction and external limiting membrane can be absent above the large RPE elevations, indicating photoreceptor impairment.[17]

Significant and profound decrease in thickness of the photoreceptor layer (distance between the top of the RPE and the outer plexiform layer [OPL]) focally over the drusen compared with that of age-matched normal eyes is found in most of the eyes, whereas the inner retinal layer remains unaffected.[18]

Pigmentary abnormalities appear in the OCT as a hyper-reflective feature visible in the optic nerve layer (ONL), producing shadowing in deeper layers (Figure 14-9) consistent with the pigmentary changes in the fundoscopy. The finding of pigment in the ONL suggests pigment migration into the retina.

Geographic atrophy OCT images show overlying photoreceptors drop out and/or thinning, leading to the

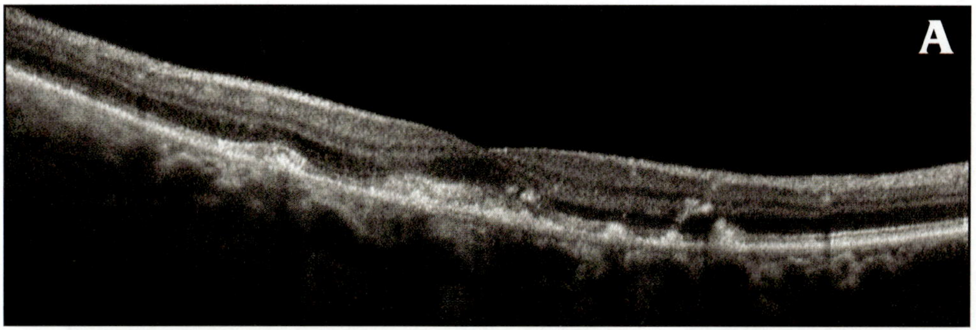

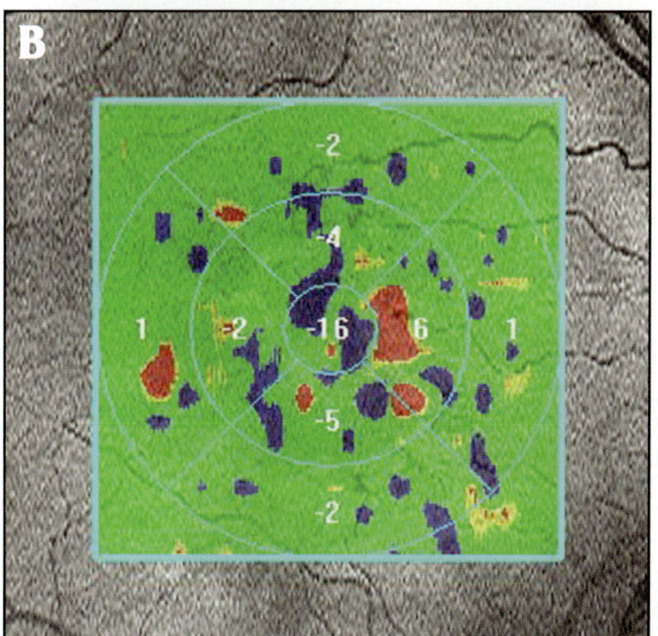

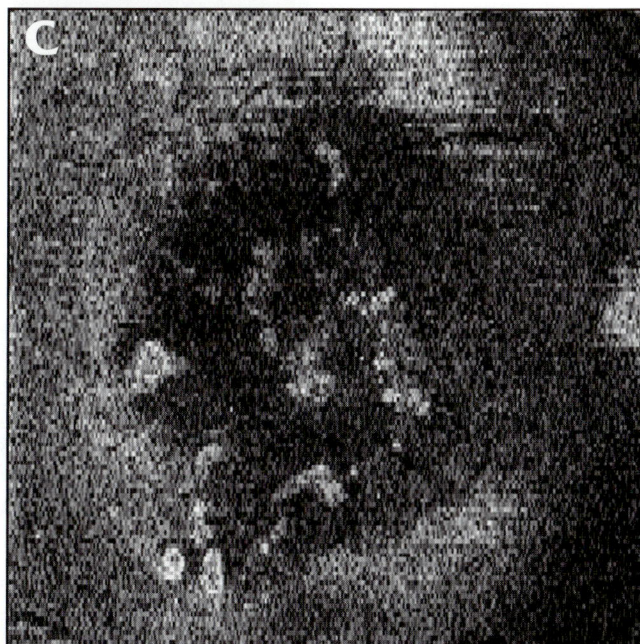

Figure 14-9. OCT imaging from the patient presented in Figure 14-3. (A) Vertical B-scan of the macula: multiple hyper-reflective elevations of the RPE corresponding to the drusen. The areas of hyper-reflectivity in the OPL and ONL superior to the fovea (right), consistent with pigmentary changes in the fundus, suggest pigment migration into the retina. (B) RPE significance map overlay on the scanning laser ophthalmoscopy-like image of the retina: the areas in yellow and red represent elevation of the RPE (namely drusen), and areas in blue represent thinning of the RPE (namely atrophy). (C) C-scan at the IS/OS junction level: the drusen appear as hyper-reflective areas in the macular area (small arrows).

replacement of the normal masking of choroidal reflectivity by the RPE and choriocapillaris by bright areas of increased transmission (bright shadowing) (Figures 14-10B through E). On reconstructed OCT fundus images (*en face* visualization of the summation of the signals from each A-scan acquired with the 3D protocol), the areas of geographic atrophy appear brighter than the relatively darker areas where the RPE is intact (Figure 14-11). The size and shape of these areas correlate well to the areas of geographic atrophy seen on autofluorescence images.[19]

TREATMENT

No proven treatment exists for the visual loss associated with non-neovascular AMD. However, the Age-Related Eye Disease Study (AREDS) gave substantial evidence that the use of high-dose vitamins and antioxidants on a regular basis can decrease the risk of progression of AMD in patients with high-risk characteristics.[20] There is also some evidence that oral supplementation of high doses of macular xanthophylls (lutein and zeaxanthin) and omega-3 long chain polyunsaturated fatty acids (DHA and EPA) may also be of benefit. The results of the AREDS-2 trial will confirm or disprove this hypothesis.

Several treatments have failed to prove their efficacy in prevent or retard the progression of the non-neovascular AMD. Laser application in patients with drusen have shown no beneficial effect on the patients' vision and a higher number of neovascularizations in the treated groups.[21,22] Rheopheresis for extracorporal blood filtration also did not show benefits.[23]

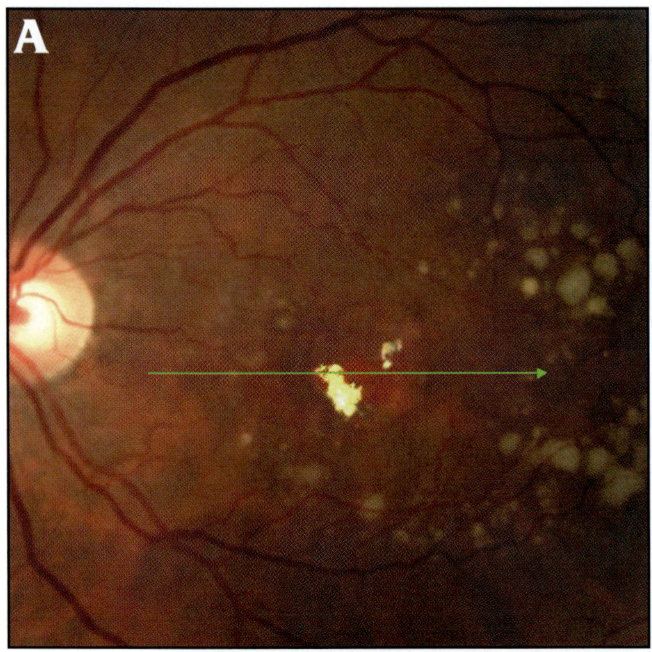

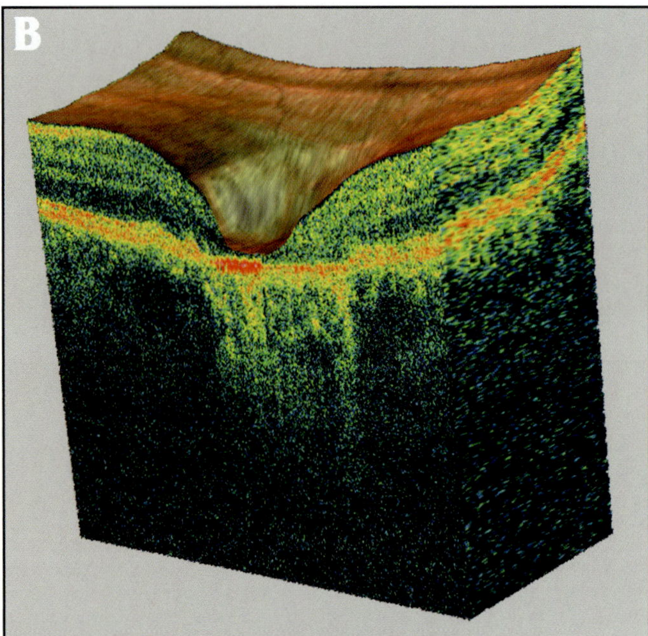

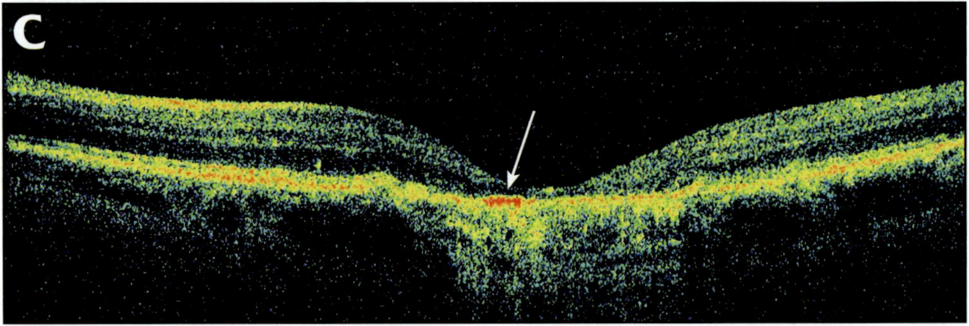

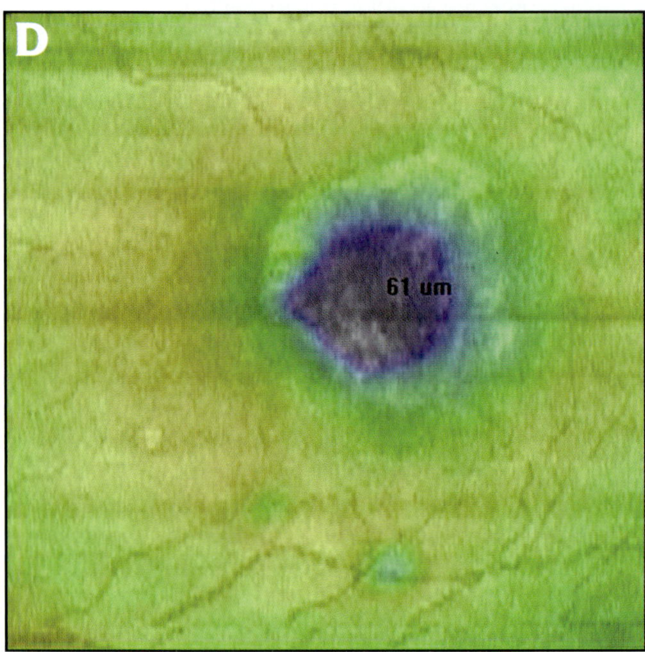

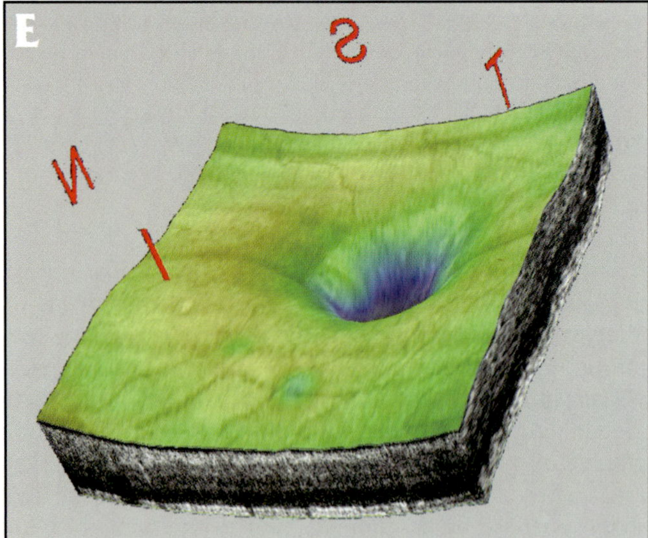

Figure 14-10. Geographic atrophy with calcified drusen. (A) One large and one small calcified drusen can be seen in the geographic atrophy. The arrow shows the locations of the OCT B-scan. (B) A 3D reconstruction of the macula, showing the relation of the geographic atrophy with the macula. (C) Horizontal B-scan of the macula: geographic atrophy is characterized by the dropout and/or thinning of the overlying photoreceptors, leading to bright areas of increased transmission (bright shadowing). The calcified drusen is seen as a spot of high reflectivity in the RPE level (arrow). (D, E) Retina thickness map overlay on the scanning laser ophthalmoscopy-like image (D) pointing the thickness of the retina in the geographic atrophy area, and on the 3D reconstruction of the macula (E).

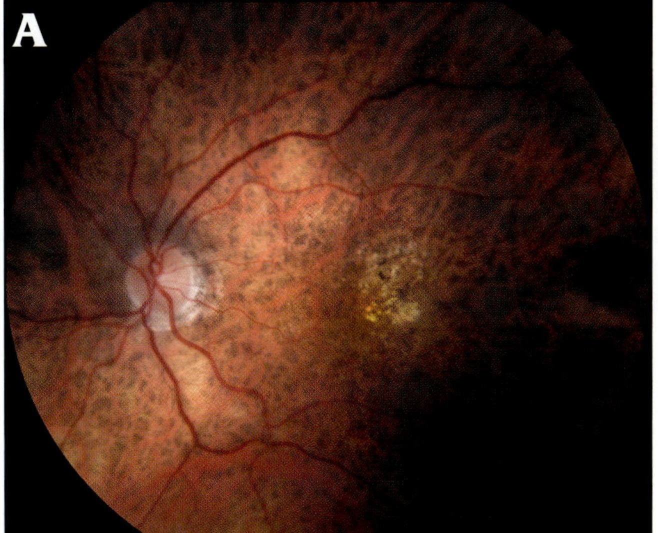
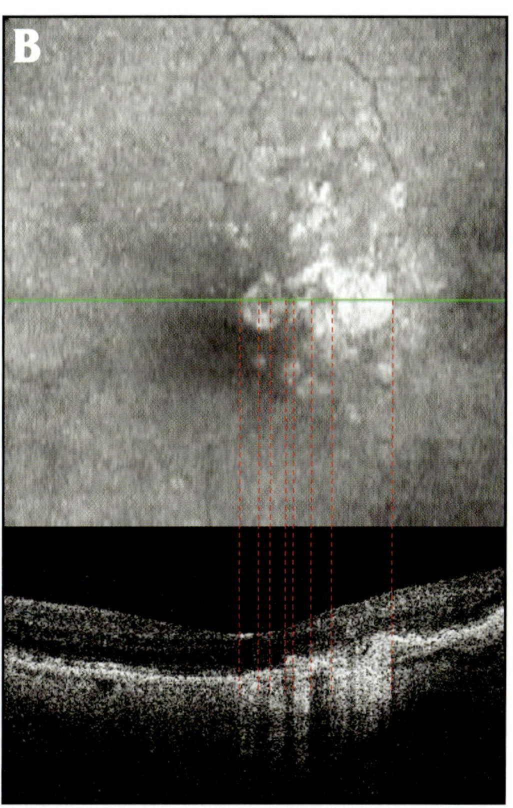

Figure 14-11. Geographic atrophy. (A) Color fundus photo shows the geographic atrophy. (B) At the reconstructed OCT fundus image of the macular area (superior), the area of geographic atrophy appears brighter than the relatively darker areas where the RPE is intact. The B-scan (inferior from the position) shown by a green line demonstrates the increased signal in areas of RPE atrophy due to choroidal reflectance. The areas of RPE loss line up exactly with areas that appear bright in the reconstructed OCT fundus image.

Future Perspectives

Future treatments for non-neovascular AMD still under investigation include:

- Weekly injections of Copaxone (glatiramer acetate), which showed in a preliminary study the ability to decrease the total drusen area[24]

- Intraocular implant of encapsulated genetically engineered cells capable of producing human ciliary neurotrophic factor (CNTF) is undergoing a Phase II clinical trial in patients with geographic atrophy associated with dry AMD

- Intravitreal injection of sirolimus (rapamycin) to preserve vision in patients with geographic atrophy associated with non-neovascular AMD (Phase I/II clinical trial)

- Oral administration of fenretinide in the treatment of geographic atrophy (Phase II clinical trial)

- Topical use of OT-551 (an anti-oxidant, anti-inflammatory, and antiangiogenic drug that protects against sunlight oxidative damage; Othera Pharmaceuticals, Conshohocken, PA) in the treatment of geographic atrophy (the OMEGA study, a Phase II clinical trial)

REFERENCES

1. Klein R, Klein BE, Jensen SC, Meuer SM. The five-year incidence and progression of age-related maculopathy: the Beaver Dam Eye Study. *Ophthalmology.* 1997;104:7-21.

2. Leibowitz HM, Krueger DE, Maunder LR, et al. The Framingham Eye Study monograph: an ophthalmological and epidemiological study of cataract, glaucoma, diabetic retinopathy, macular degeneration, and visual acuity in a general population of 2631 adults, 1973–1975. *Surv Ophthalmol.* 1980;24:335-610.

3. Rahmani B, Tielsch JM, Katz J, et al. The cause-specific prevalence of visual impairment in an urban population. The Baltimore Eye Survey. *Ophthalmology.* 1996;103:1721-1726.

4. The Age-Related Eye Disease Study Research Group. Potential public health impact of AREDS results: AREDS report no. 11. *Arch Ophthalmol.* 2003;121:1621-1624.

5. The Eye Diseases Prevalence Research Group. Prevalence of age-related macular degeneration in the United States. *Arch Ophthalmol.* 2004;122:567-572.

6. Pauleikhoff D, Harper CA, Marshall J, et al. Aging changes in Bruch's membrane: A histochemical and morphological study. *Ophthalmology.* 1990;97:171-178.

7. Ruberti J, Curcio CA, Milican L, et al. Quick-freeze/deep etch visualization of age-related lipid accumulation in Bruch's membrane. *Invest Ophthalmol Vis Sci.* 2003;44:1753-1759.

8. Bok D. Retinal photoreceptor-pigment epithelium interactions. *Invest Ophthalmol Vis Sci.* 1985;26:1659-1694.

9. Bok D, Heller J. Transport of retinol from the blood to the retina: An autoradiographic study of the pigment epithelial cell surface receptor for plasma retinol-binding protein. *Exp Eye Res.* 1976;22:395-402.

10. Wang Q, Chappell RJ, Klein R, et al. Patterns of age-related maculopathy in the macular area, the Beaver Dam Eye Study. *Invest Ophthalmol Vis Sci.* 1996;37:2234-2242.

11. Bird AC, Bressler NB, Bressler SB et al. An international classification and grading system for age-related maculopathy and age-related macular degeneration. *Surv Ophthalmology.* 1995;39:367-374.

12. Delori FC, Fleckner MR, Goger DG, et al. Autofluorescence distribution associated with drusen in age-related macular degeneration. *Invest Ophthalmol Vis Sci.* 2000;41(2):496-504.

13. Bindewald A, Bird AC, Dandekar SS, et al. Classification of fundus autofluorescence patterns in early age-related macular disease. *Invest Ophthalmol Vis Sci.* 2005;46(9):3309-3314.

14. Holz FG, Bellman C, Staudt S, et al. Fundus autofluorescence and development of geographic atrophy in age-related macular degeneration. *Invest Ophthalmol Vis Sci.* 2001;42(5):1051-1056.

15. Pieroni CG, Witkin AJ, Ko TH, et al. Ultrahigh resolution optical coherence tomography in non-exudative age related macular degeneration. *Br J Ophthalmol.* 2006;90(2):191-197.

16. Khanifar A, Koreishi A, Izatt J, Toth C. Drusen ultrastructure imaging with spectral domain optical coherence tomography in age-related macular degeneration. *Ophthalmology.* 2008;115(11):1883-1890.

17. Gorczynska I, Srinivasan VJ, Vuong LN, et al. Projection OCT fundus imaging for visualizing outer retinal pathology in non-exudative age related macular degeneration. *Br J Ophthalmol.* 2008;Epub Jul 28.

18. Schuman SG, Koreishi AF, Farsiu S, et al. Photoreceptor layer thinning over drusen in eyes with age-related macular degeneration imaged in vivo with spectral-domain optical coherence tomography. *Ophthalmology.* 2009;Epub Jan 29.

19. Lujan BJ, Rosenfeld PJ, Gregori G, et al. Spectral domain optical coherence tomographic imaging of geographic atrophy. *Ophthalmic Surg Laser Imaging.* 2009;40(2):96-101.

20. The Age-Related Eye Disease Study Research Group. A randomized, placebo-controlled, clinical trial of high-dose supplementation with vitamins C and E, beta carotene, and zinc for age-related macular degeneration and vision loss: AREDS report no. 8. *Arch Ophthalmol.* 2001;119:1417-1436.

21. Friberg TR. Laser photocoagulation of eyes with drusen: will it help? *Semin Ophthalmol.* 1999;14(1):45-50.

22. Friberg TR, Musch DC, Lim JI, et al. Prophylactic treatment of age-related macular degeneration report number 1: 810-nm laser to eyes with drusen. Unilaterally eligible patients. *Ophthalmology.* 2006;113(4):622.e1.

23. Pulido JS, Winters JL, Boyer D. Preliminary analysis of the final multicenter investigation of rheopheresis for age related macular degeneration (AMD) trial (MIRA-1) results. *Trans Am Ophthalmol Soc.* 2006;104:221-231.

24. Landa G, Butovsky O, Shoshani J, et al. Weekly vaccination with Copaxone (glatiramer acetate) as a potential therapy for dry age-related macular degeneration. *Curr Eye Res.* 2008;33(11):1011-1013.

15 Exudative Age-Related Macular Degeneration

Leonardo C. Castro, MD

Age-related macular degeneration (AMD) is the major cause of severe visual loss in older adults.[1] The visual loss usually results from photoreceptor dysfunction due to underlying atrophy or choroidal neovascularization (CNV).[2] Approximately only 10% to 15% of the AMD patients present the neovascular form of the disease, yet more than 80% of the people with severe visual loss (20/200 or worse) due to AMD have the neovascular form.[3]

Exudative AMD (also called neovascular AMD or wet AMD) is characterized by the presence of neovascularization within the macula, which may be CNV under the macula, with associated manifestations such as retinal pigment epithelium detachment (PED), retinal pigment epithelial tears, fibrovascular disciform scarring, and vitreous hemorrhage,[4] or may be predominantly within the retina, known as retinal angiomatous proliferation (RAP). In the advanced forms of neovascular AMD, it is not unusual to see retinal-choroidal anastomoses, a communication between the retinal circulation and the choroidal circulation.

PATHOGENESIS

The stimulus for vascular ingrowth of choroidal vessels remains unknown, but several theories have been advanced. Soft drusen have been associated histopathologically with CNV. Some investigators believe that soft drusen represent extracellular matrix material produced by the retinal pigment epithelium (RPE).[5,6] Deposition of this material may suggest a widespread RPE abnormality,[7] and the drusen may act as an indirect angiogenic factor by attracting macrophages from the choroid.[8] Also, breaks in Bruch's membrane permit ingrowth of new vessels from the choriocapillaris. Following penetration of the inner aspect of Bruch's membrane, the new vessels proliferate laterally between the RPE and Bruch's membrane.[9]

Another theory postulates that hemodynamic alterations in the choroidal circulation is an important pathophysiological mechanism.[10] The postcapillary resistance is increased due to atherosclerotic changes in the ocular vasculature, causing elevated hydrostatic pressure, with exudation of extracellular proteins and lipids; these manifest as basal deposits and drusen. Calcification and fragmentation of the Bruch's membrane are caused by the corresponding degeneration of the elastin and collagen. The relative choroidal ischemia induces an angiogenic stimulus, which results in increased levels of various growth factors, such as vascular endothelial growth factor (VEGF) and platelet-derived growth factor (PDGF). These, in turn, incite neovascular ingrowth from the choriocapillaris through the fractured Bruch's membrane into the sub-RPE space.

SYMPTOMS

AMD may go undetected until abrupt visual loss is noted from the conversion of dry AMD to wet AMD. Wet AMD typically shows a more rapid progression of visual acuity loss relative to its non-neovascular counterpart. Blurred vision and distortion, especially distorted near

Huang D, Duker JS, Fujimoto JG, Lumbroso B, Schuman JS, Weinreb RN.
Imaging the Eye from Front to Back with RTVue Fourier-Domain Optical Coherence Tomography (pp 137-154).

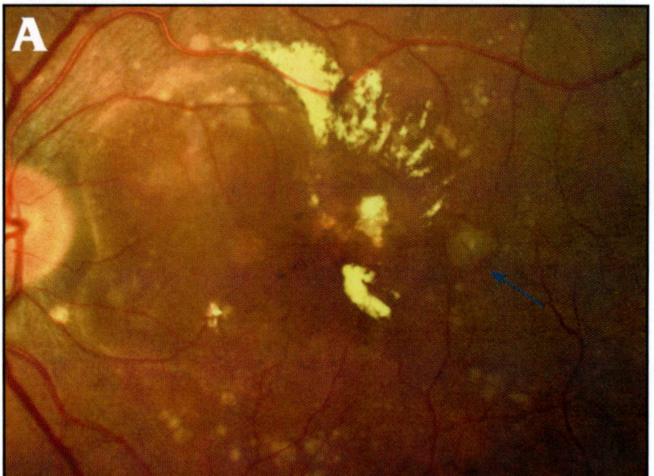

Figure 15-1. Exudative AMD ocular findings. (A) Massive subretinal fluid associated with lipid exudates and a small serous PED (green arrow). Large soft drusen can be seen around the macula. (B) Extensive subretinal (lighter) and sub-RPE (darker) hemorrhage around the choroidal neovascular membrane. (C) Subretinal fibrosis (or disciform scar) in a patient treated with photodynamic therapy with verteporfin 5 years prior.

vision, are the symptoms most patients with CNV notice first. They generally report decreased ability to perform functional tasks (eg, face recognition, reading newspaper) with the affected eye. Patients may also complain of micropsia or a scotoma. Visual acuity, although frequently decreased, may not always be affected.

OCULAR FINDINGS

The clinical manifestations of neovascular AMD can include: subretinal fluid (Figure 15-1A); macular edema; retinal or subretinal lipid exudates (see Figure 15-1A); retinal, subretinal, or sub-RPE hemorrhage (Figure 15-1B); plaque-like membrane or gray or yellow-green discrete discoloration (see Figure 15-1B); RPE detachment (serous, fibrous, drusenoid, or hemorrhagic); RPE tear; subretinal fibrosis; or disciform scar (Figure 15-1C). Associated features of non-neovascular AMD such as drusen, focal pigmentary changes, and geographic atrophy are typically present. However, neovascularization secondary to AMD may occur without any of these precursor lesions.

CLASSIFICATION OF THE CHOROIDAL NEOVASULARIZATION

By Boundaries

Using fluorescein angiography, the CNV can be classified as having well-defined or poorly defined boundaries.

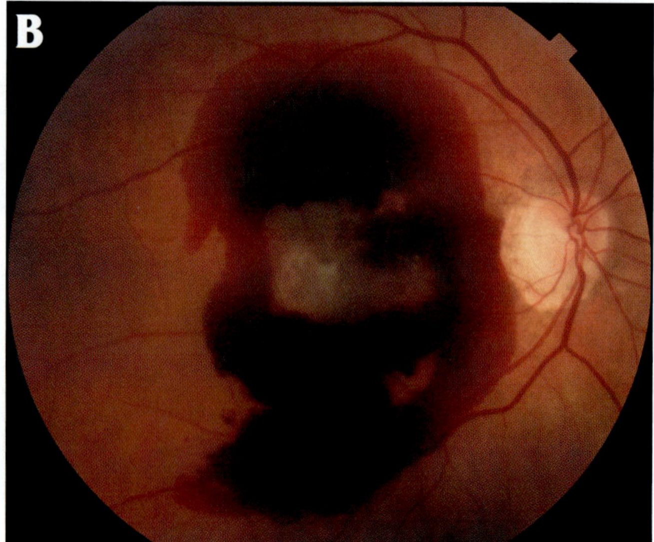

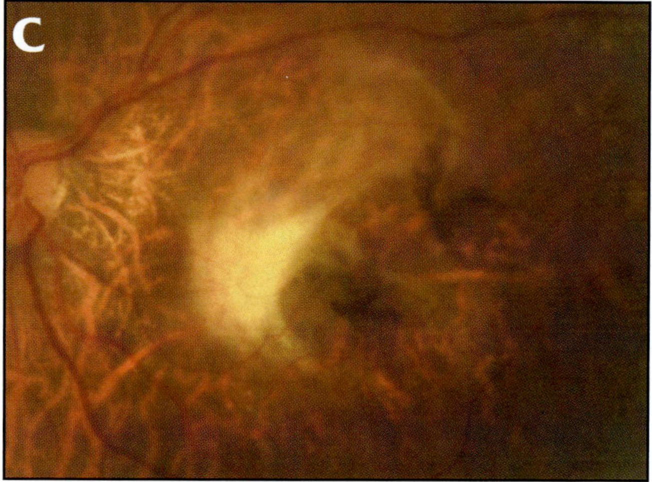

By Location

If the CNV is well demarcated, its location can be determined angiographically by its closest point to the center of the foveal avascular zone (FAZ):

- Extrafoveal (≥200 µm and <2500 µm from the center of the FAZ)
- Juxtafoveal (1 to 199 µm from the center of the FAZ)
- Subfoveal (under the center of the FAZ)

By Angiographic Patterns of Fluorescence

- Classic CNV: bright, uniform, early hyperfluorescence exhibiting leakage in the late phase and obscuring the boundaries (Figure 15-2)
- Occult CNV: one of 2 patterns: fibrovascular PED (area of irregular elevation of the RPE, often with stippled hyperfluorescence present in the midphase of the angiogram and leakage or staining by the late phase) (Figure 15-3) or late leakage from an undetermined source (speckled hyperfluorescence with

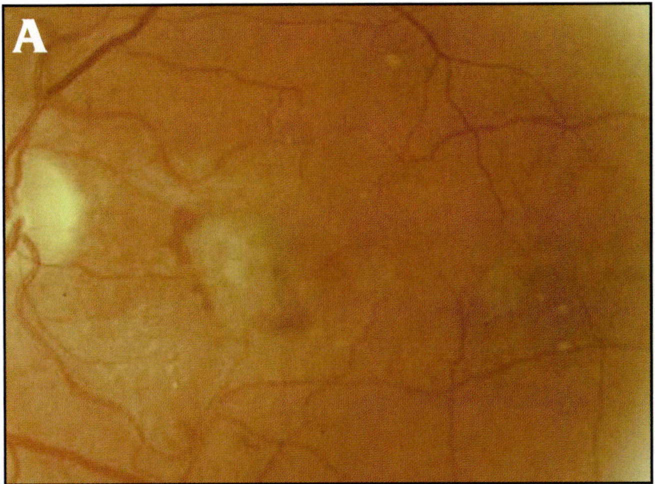

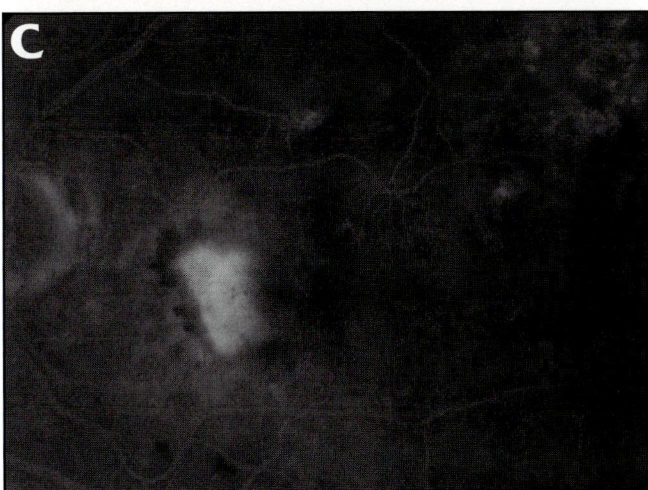

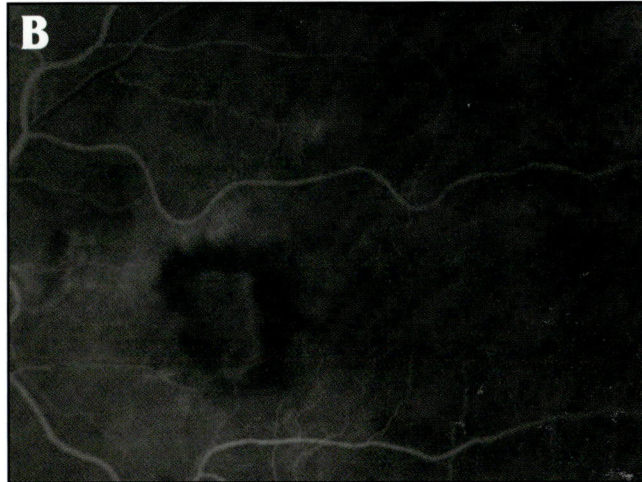

Figure 15-2. Classic CNV membrane. (A) Color fundus photo from a patient with classic CNV membrane. Bright, uniform, early hyperfluorescence of the CNV is seen nasal to the fovea (B), with late leakage of dye obscuring the boundaries (C) during the fluorescein angiogram.

ANCILLARY EXAMS

Fluorescein Angiography

Fluorescein angiography allows the determination of the pattern, composition, boundaries, and location of the CNV (described previously). Beyond the CNV features, 4 other angiographic features should be noted in cases of exudative AMD:

1. Blood that is visible on color fundus photographs and thick enough to obscure the normal choroidal fluorescence (Figure 15-5)

2. Hypofluorescence due to hyperplastic pigment or fibrous tissue, or blood not visible on color fundus photographs (see Figure 15-3)

3. Serous detachment of the RPE

4. Scar from CNV which either stains or blocks fluorescence (depending on the extent of RPE within the scar)

The first 2 of these 4 features block the angiographic view of the choroids, making it impossible to determine whether CNV is located under them, and its size. Serous PED exhibits bright hyperfluorescence with progressive pooling of fluorescence within a fixed sub-RPE cavity. Neovascularization within the area of the serous detachment may not be identifiable due to intense hyperfluorescence.

Fluorescein angiography can also disclose intraretinal vascular structures that anastomose with the CNV, in which case they are called RAP. The vessels can be seen dividing at right angles from the surface of the retina to the neovascular lesion (Figure 15-6).

dye pooled in the subretinal space in the late phase; the source of leakage does not correspond to classic CNV or fibrovascular PED in the early or midphase of the angiogram) (Figure 15-4)

By Composition of the Lesion

- Predominantly CNV
- Predominantly classic CNV: at least 50% of the lesion is composed of classic CNV
- Minimally classic CNV: less than 50% of the lesion is composed of classic CNV
- Occult with no classic CNV
- Predominantly hemorrhagic

By Histopathology

- Type 1: the CNV is under the RPE
- Type 2: the CNV in above the RPE, under the neurosensorial retina

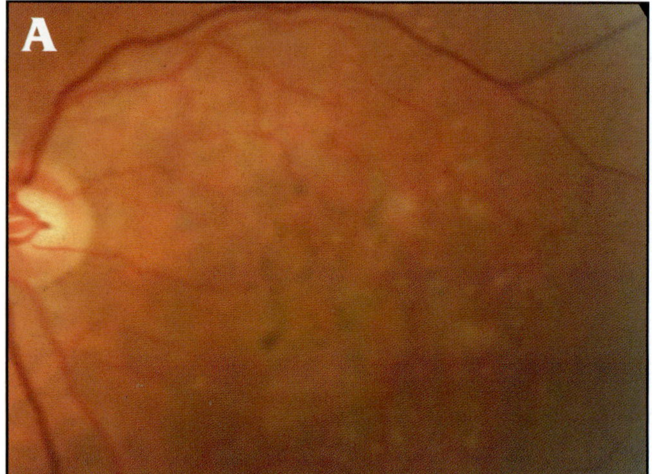

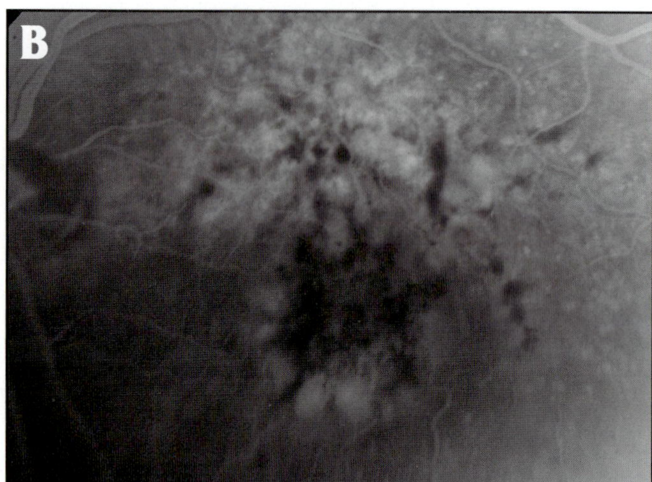

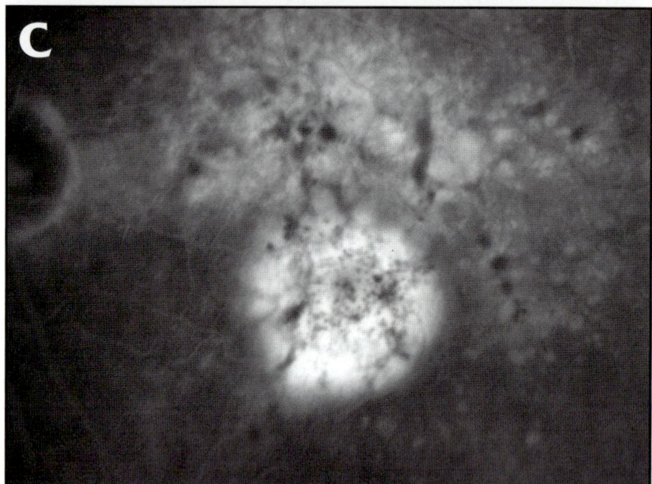

Figure 15-3. Occult CNV membrane: fibrovascular PED. The color fundus photo (A) shows the elevation of the RPE with a solid appearance and radial hyperpigmentation, compromising the foveal area, which on fluorescein angiography presents a stippled hyper-fluorescence in the midphase of the exam (B) with staining in the late phase (C).

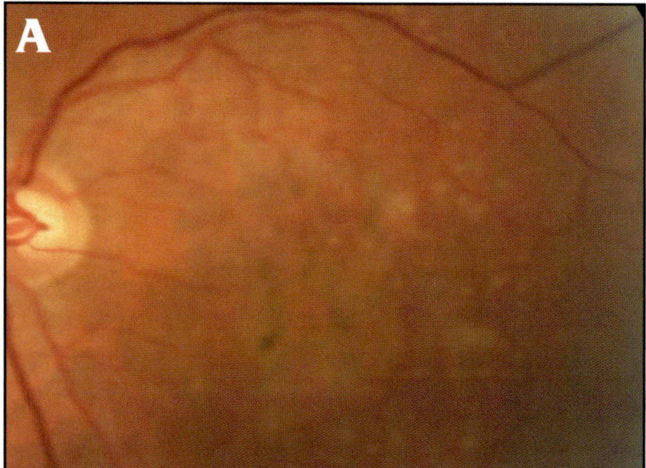

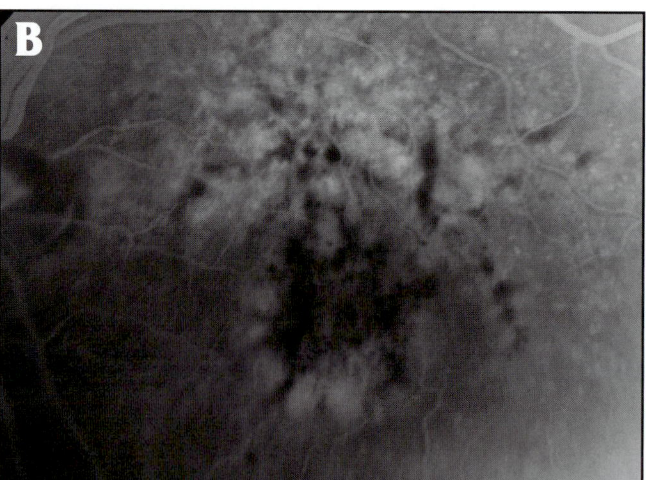

Figure 15-4. Occult CNV membrane: undetermined source of leakage. The color fundus photo (A) shows innumerous soft drusen in the posterior pole and loss of the normal foveal bright. The fluorescein angiogram showed a speckled hyperfluorescence with dye pooled in the subretinal space in the foveal area on the late phase (B) of the exam, without a visible source of leakage visible in the early phase.

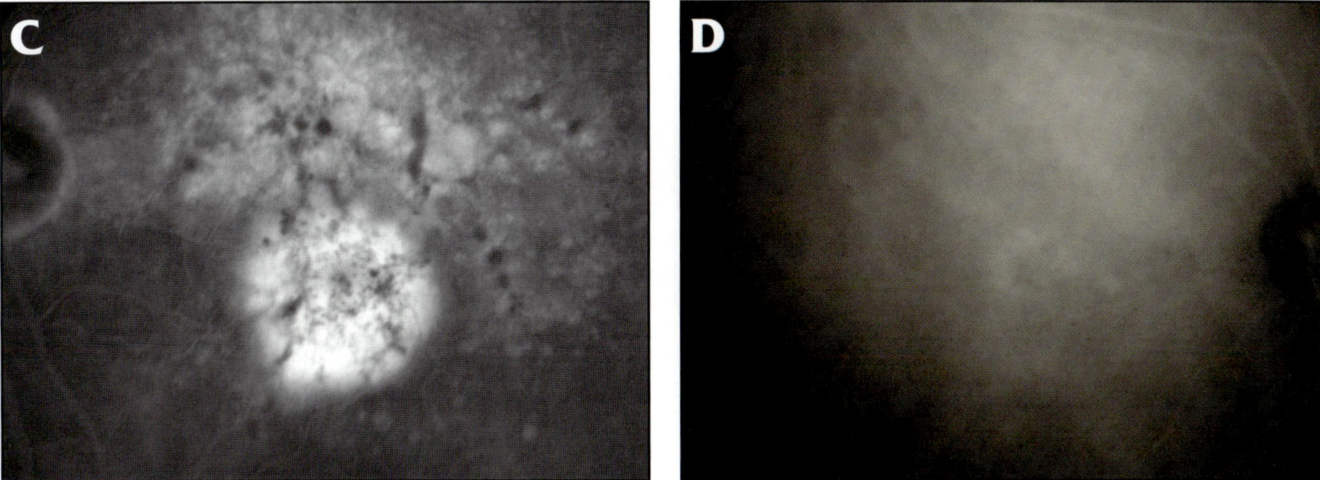

Figure 15-4. Occult CNV membrane: undetermined source of leakage. The fluorescein angiogram showed a speckled hyperfluorescence with dye pooled in the subretinal space in the foveal area on the late phase of the exam, without a visible source of leakage visible in the early phase (C). On the indocyanine green angiography, small spots of bright fluorescence (green arrow) can be seen in the area of speckled hyperfluorescence seen on fluorescein angiography (D).

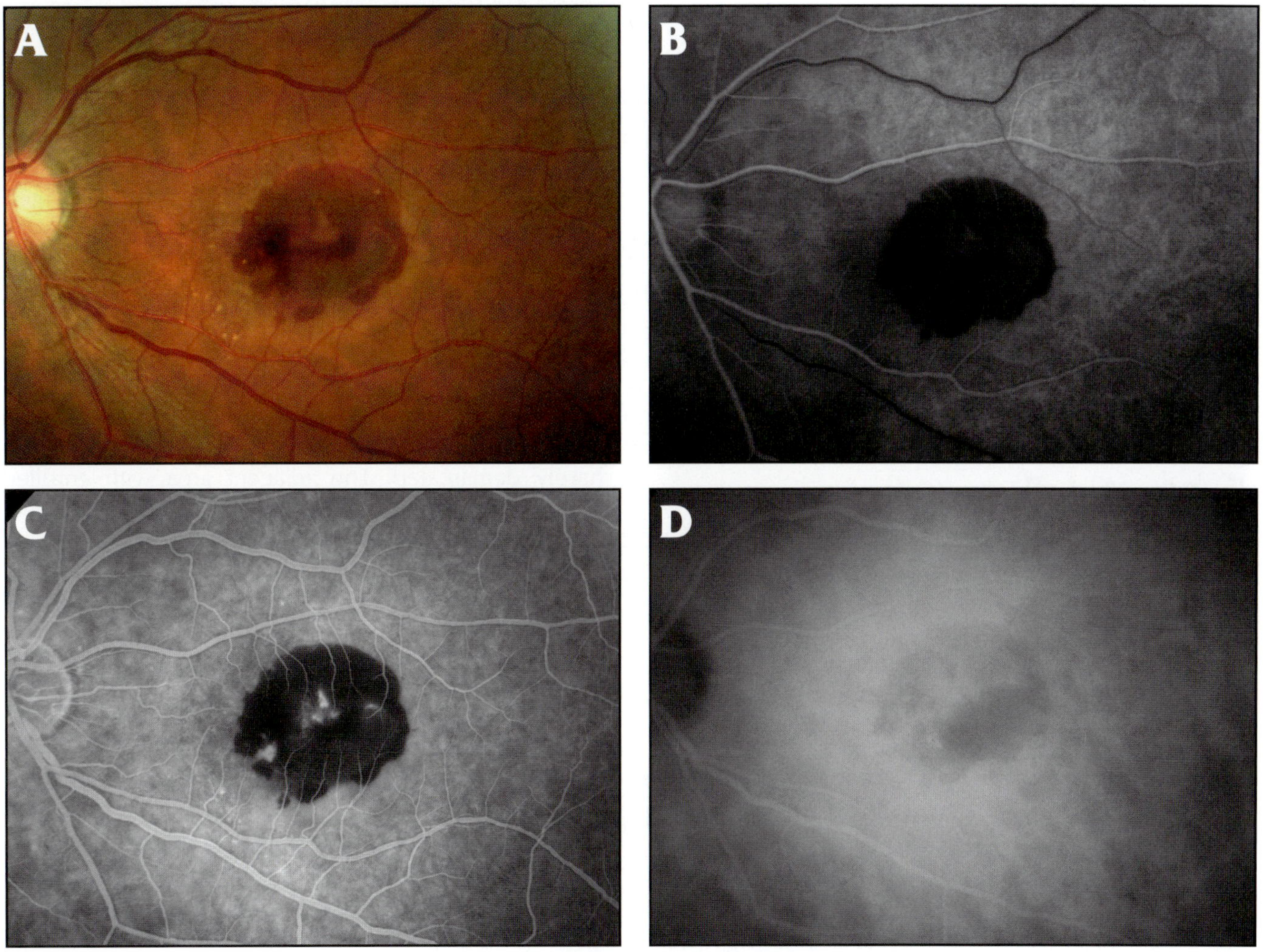

Figure 15-5. Subretinal and sub-RPE hemorrhage (A through D). The thick subretinal hemorrhage obscures the normal choroidal fluorescence during the fluorescein angiogram, making it impossible to determine the extension of the CNV membrane (B, C). On the indocyanine green angiography, the membrane can be better visualized under the hemorrhage (D).

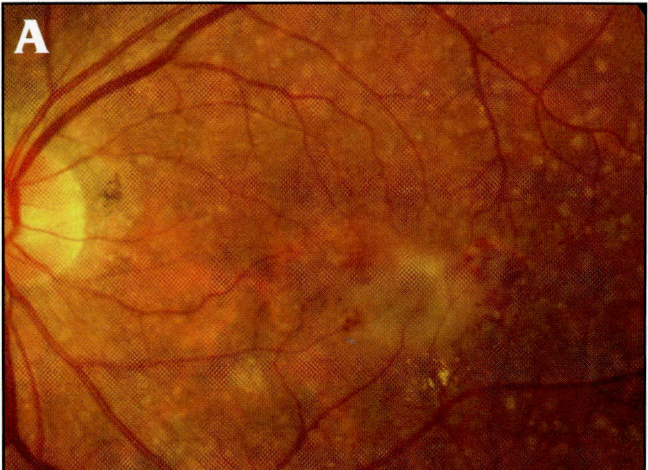

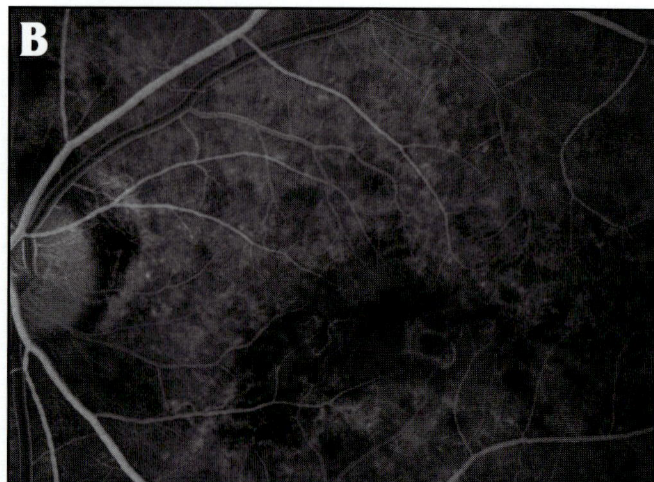

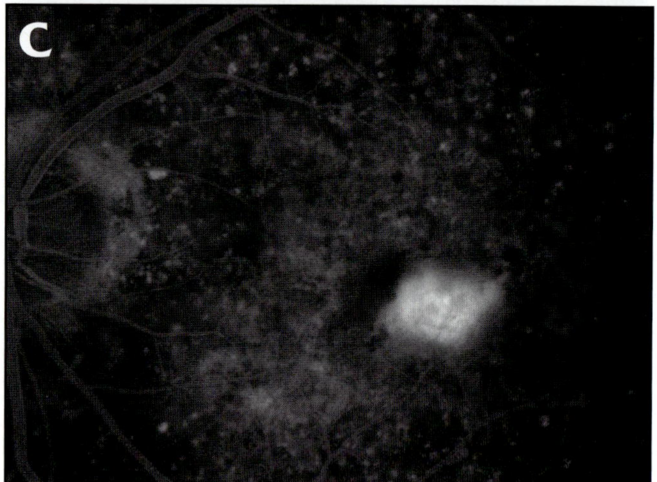

Figure 15-6. Retinal angiomatous proliferation. The color fundus (A) shows several soft drusen in the posterior pole, intraretinal hemorrhages and a yellowish intraretinal lesion in the foveal area. On the fluorescein angiogram, it is possible to see the anastomosis between the retina and the choroids in the yellowish lesion (B), with late leakage of the dye (C).

Indocyanine Green Angiography

Three basic patterns of fluorescence have been reported in indocyanine green angiography of CNV judged to be occult on fluorescein angiography:

- Small, focal "hot spot": a bright area of fluorescence less than 1 disc area that usually shows by the midphase of the angiogram (see Figure 15-4D)
- Plaque: a well-demarcated area of fluorescence more than 1 disc area in size that emerges relatively late in the angiogram (see Figure 15-5D)
- Combination of the two

Fundus Autofluorescence

Although the areas of underlying CNV do not correspond to any specific findings in fundus autofluorescence imaging,[11] areas with high autofluorescence are often found extending beyond the edge of the angiographically defined lesion. As in other exudative retinal diseases, such as central serous chorioretinopathy, areas with increased autofluorescence adjacent to CNV are commonly found inferior to the leakage on fluorescein angiography.[12]

Pigment epithelial detachments found on clinical examination are associated with variable fundus autofluorescence phenomena. The area topographically corresponding to the lesion site seen on funduscopy and fluorescein angiography may have decreased, normal, or increased autofluorescence signal. These variations may reflect different stages of evolution in the development of PEDs.[12]

RPE tears are easily identified with fundus autofluorescence imaging. The areas of RPE loss are represented by loss of signal (due to the loss of lipofuscin) and are sharply demarcated. The adjacent area with enrolled RPE is characterized by heterogeneous signal of distinct increased autofluorescence. Disciform scars present with decreased autofluorescence.

Hemorrhages and intraretinal exudates typically show a decreased fundus autofluorescence signal because of absorption phenomena and may obscure the underlying retinal details. Usually other imaging studies, such as fundus photography, are required to differentiate between hemorrhages, exudates, and RPE atrophy.

OPTICAL COHERENCE TOMOGRAPHY

The interpretation of the OCT imaging of exudative AMD is very difficult. In fact, the neovascular membrane, RPE, and choriocapillaris are structures with similar reflec-

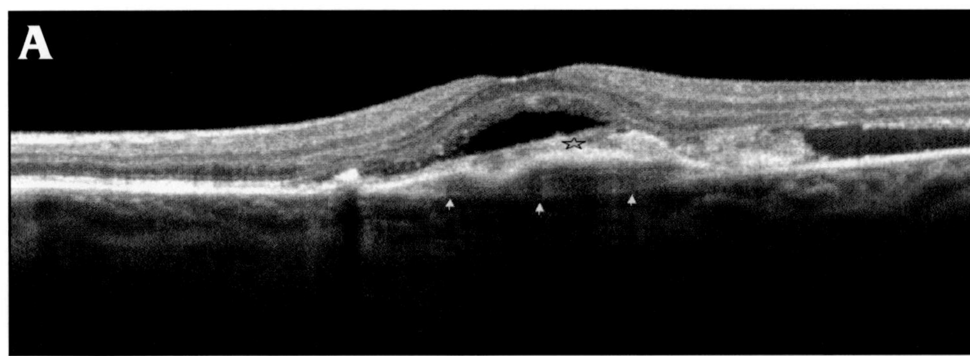

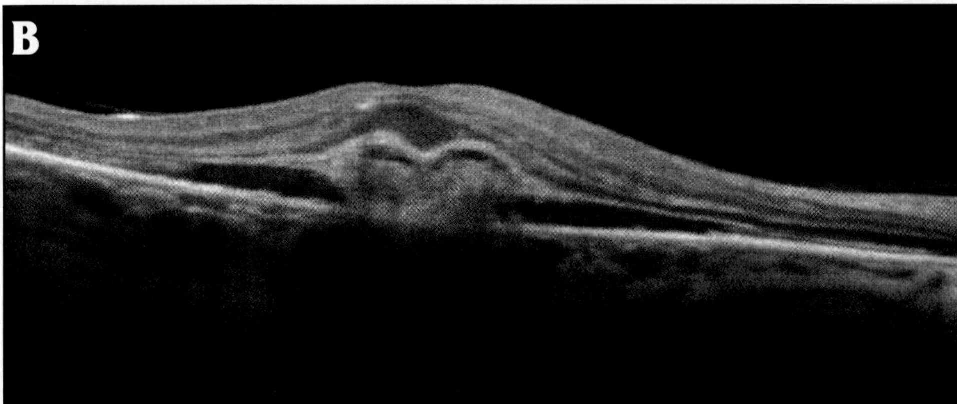

Figure 15-7. Classic CNV membranes seen on OCT. (A) The highly reflective band (RPE/Bruch's/choriocapillaris band) is raised by a non-uniform slightly hyper-reflective formation, and also duplicated with high back-scattering material between the 2 bands (green star). Subretinal fluid appears as an optically empty space above and temporal (right) to the CNV. The Bruch's membrane can be seen as an almost straight hyporeflective line under the elevation of the RPE (arrow heads). (B) The classic CNV can be delineated as a moderate hyper-reflective formation above the RPE (Type 2 CNV).

tivities[13] and the identification of a single structure is often not easy. While OCT cannot predict with certainly the angiographic appearance of neovascular lesions, there are characteristic features observed on OCT that complement angiography, and may even serve as better predictors of disease status. Small changes in the structure of the retinal layers and subretinal space can readily be noted, allowing the detection of structural changes that may signal progression or regression of the lesions.

Classic CNVs appear on OCT imaging as a fusiform enlargement of the RPE/Bruch's membrane/choriocapillaris reflective band with defined borders. The highly reflective band may appear irregular and duplicated, with high backscattering material between the 2 bands (Figure 15-7A). The membrane may be imaged above the RPE (Type 2 CNV) occasionally (Figure 15-7B).[14]

Occult CNV is considered ill-defined on fluorescein angiography and has a nonspecific appearance on OCT.[15] However, its presence is indicated by RPE elevation that is separated from the underlying choroidal plane. The RPE elevation forms a cavity that can be fibrovascular (a mild backscattering structure is seen under the RPE band) or, less frequently, serous (optically empty) with varying degrees of expansion according to the stage of the lesion. OCT imaging of occult lesions of the fibrovascular PED type (Figure 15-8) often show various characteristics within one CNV complex, with thickening of the hyper-reflective RPE/Bruch's/chorio-

capillaris band consistent with the neovascular component of the lesion. Duplications and interruptions of the RPE/Bruch's/choriocapillaris band can also be seen.[16] The choroidal reflection may be enhanced due to increased light penetration.

Serous PEDs are precisely imaged by OCT. They usually appear as a localized prominence with steep borders and a flat, plateau-shaped top. Intraretinal edema and subretinal fluid can be seen at the margins and at the top of the detachments. Duplications of RPE/Bruch's/choriocapillaris band are not present.[16] In fibrovascular PED, the prominence has flat borders that become confluent with the surrounding choriocapillary layer. The surface of the lesion is irregular.

OCT imaging of RAP lesions (Figure 15-9) may reveal increased foveal thickness, cystoid macular edema consisting of large central cysts and smaller cystoid spaces located mainly in the outer retinal layers, serous retinal detachment, and a highly reflective intraretinal mass overlying a highly or moderately elevated RPE. This mass corresponded to the hot spot observed on indocyanine green angiography.

Others features associated with exudative AMD imaged with OCT include:

- Subretinal fluid appears on OCT imaging as an optically empty space adjacent to the presumed CNV (see Figure 15-7A). This collection of fluid can be quantitatively evaluated with OCT.

Figure 15-8. Fibrovascular PED (same patient shown in Figure 15-3) seen on OCT. (A) OCT imaging of the fibrovascular PED depicts a mild backscattering structure under the RPE band causing shadowing of the choroids underneath it. The RTVue-100 OCT device allows the quantitative evaluation of the retinal pigment elevation (B, C).

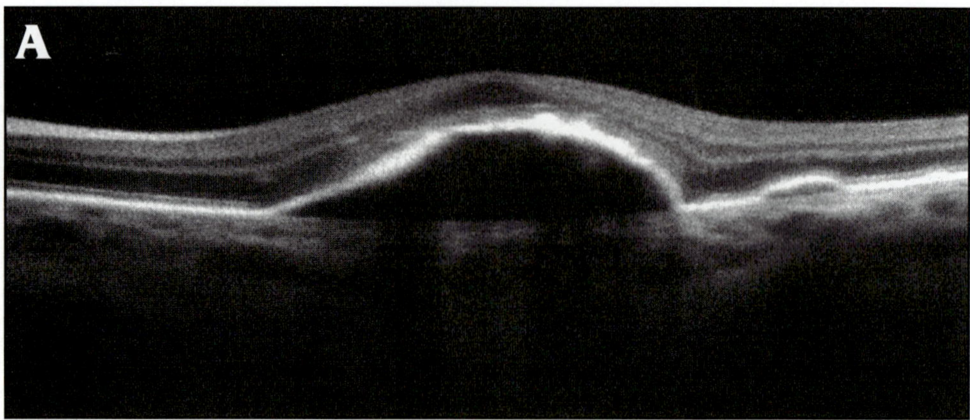

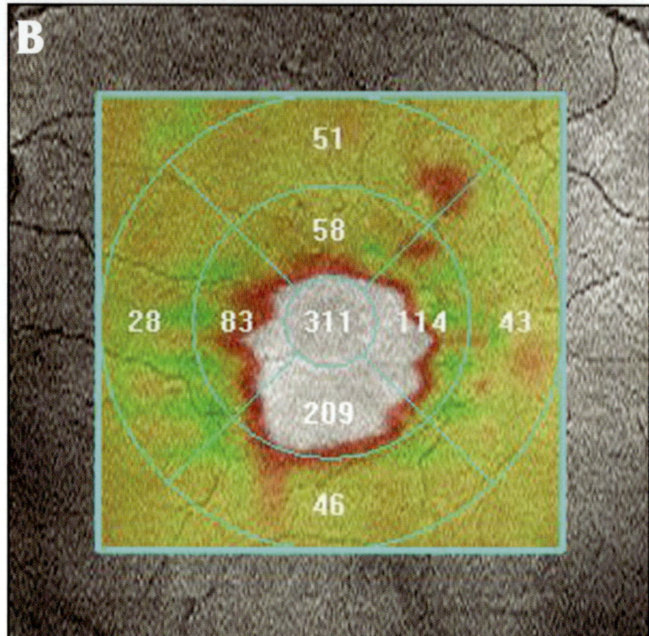

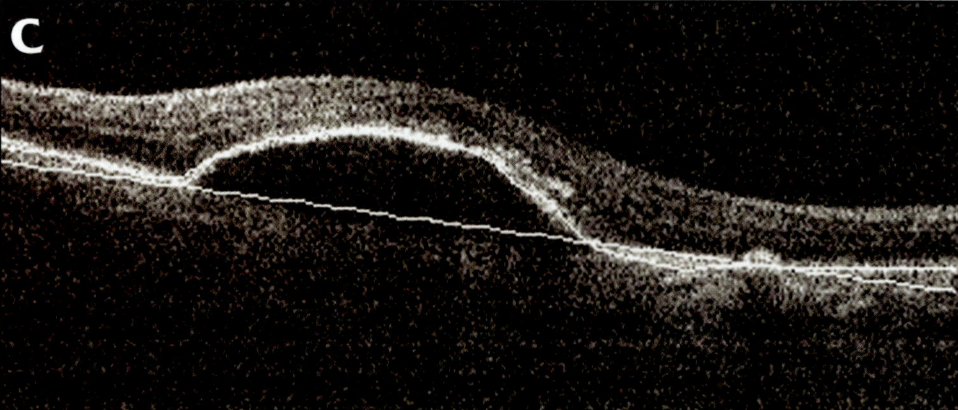

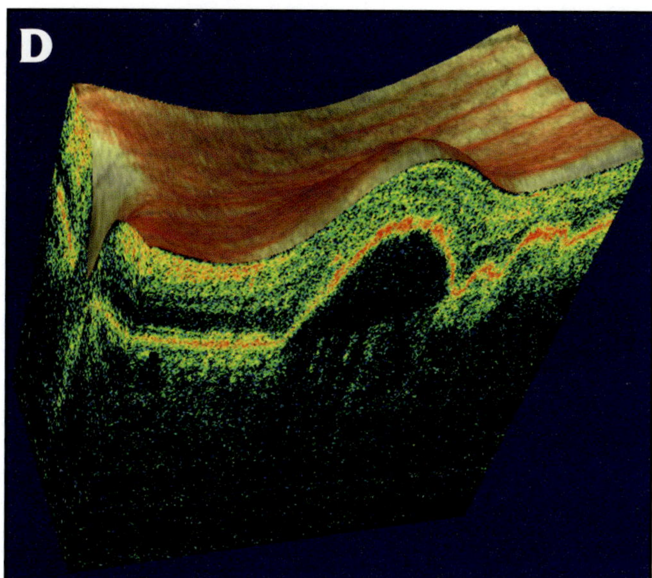

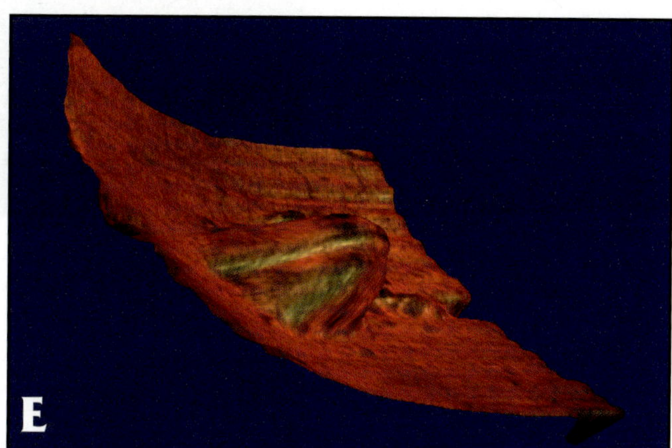

Figure 15-8. Fibrovascular PED (same patient shown in Figure 15-3) seen on OCT. It also allows the visualization of its relationship with the macula (D) or individually (E) (qualitative evaluations).

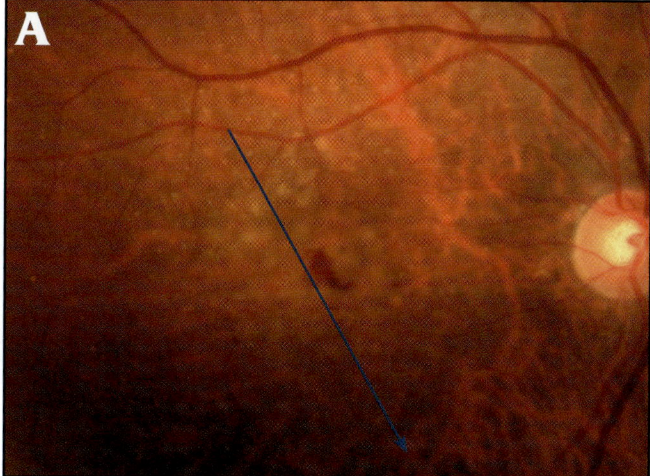

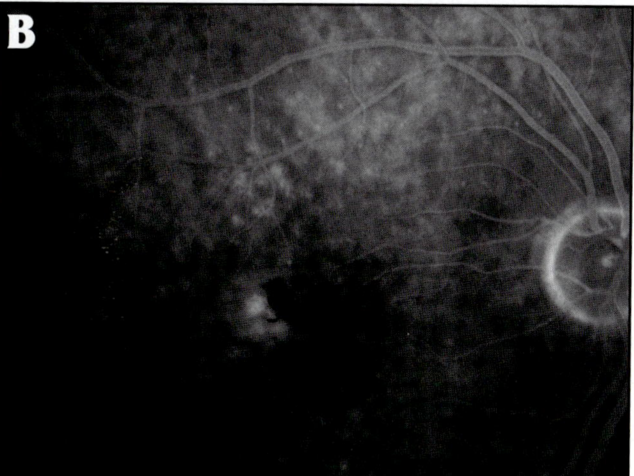

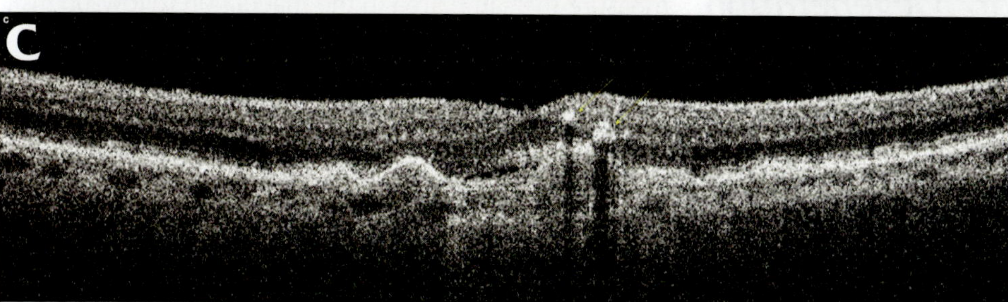

Figure 15-9. Retinal angiomatous proliferation. (A, B) Color fundus photo and late phase fluorescein angiography of a patient with RAP lesion. Several large and confluent soft drusen and intraretinal hemorrhage are seen in the foveal area, with late dye leakage from the RAP lesions. The green arrow represents the location of the OCT scan. (C) The OCT scan depicted the thickening of the RPE/Bruch's/choriocapillaris band (CNV membrane), a small spot of subretinal fluid, and 2 RAP lesions represented by highly reflective intraretinal lesions at the outer plexiform layer, causing shadowing of the others structures (yellow arrows).

- Intraretinal fluid can be seen as focal cysts (such as cystoid macular edema) or diffuse thickening of the neurosensory retina (diffuse macular edema) and can also be quantitatively evaluated with OCT (Figure 15-10).

- RPE tears are shown on OCT as a discontinuation along the highly reflective external band at the edge of the RPE tear. Often, OCT is able to identify the site of separation and the rolling edge of the tear.

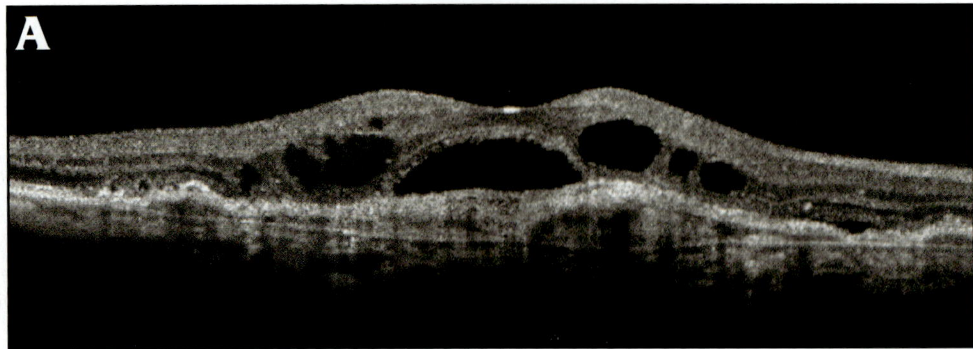

Figure 15-10. CNV membrane with intraretinal and subretinal fluid. (A) The cross-sectional OCT image depicted the thickened RPE raised by a non-uniform slightly hyper-reflective formation (CNV), the Bruch's membrane, and the presence of subretinal fluid and intraretinal cysts (optically empty spaces). (B) The 3D reconstruction of the macula enables the isolation of different layers of the retina. In this case, the summation of 40 µm of thickness at the inner plexiform layer shows the intraretinal cysts (dark areas) present at the macula. The green line represents the location of the cross-sectional image.

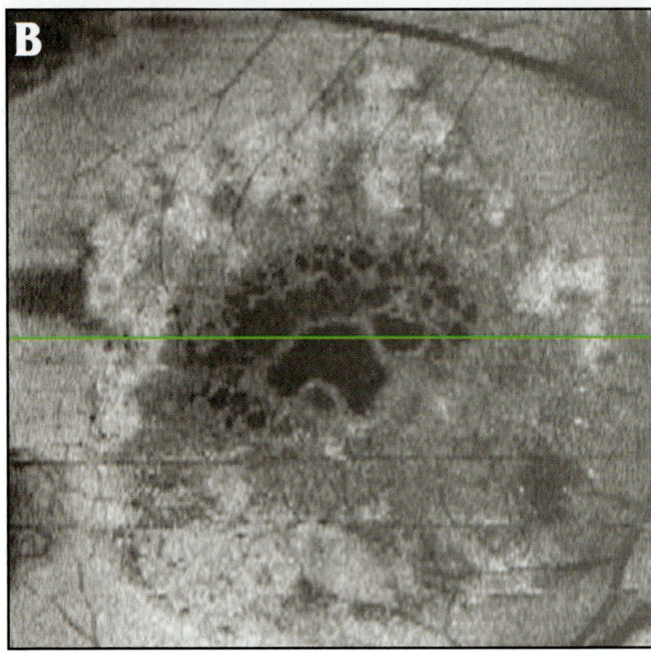

- Subretinal and sub-RPE hemorrhage can be distinguished using the OCT. The subretinal hemorrhage appears as an elevated and often edematous retina with a highly reflective underlying layer (the blood), which shadows the RPE/Bruch's membrane/choriocapillaris band (Figure 15-11). Sub-RPE hemorrhage shows an elevated and intact external band with an underlying area of high reflectivity and shadowing (the blood).

- Subretinal fibrosis or disciform scarring are highly reflective on OCT and are frequently associated with atrophy of the overlying retina (Figure 15-12).

Clinical Applications of Optical Coherence Tomography in Exudative Age-Related Macular Degeneration Patients

One of the most important clinical uses of the OCT is its capability to quantify changes in central retinal thickness and volume, as well as subretinal fluid; which are the parameters used, in combination with visual acuity, to evaluate the response to therapy in several diseases, including exudative AMD.

Differently from time-domain OCT, in which the reproducibility of the images between follow-up visits requires an experienced OCT technician, spectral-domain OCT takes advantage of its high acquisition speed to improve the reproducibility between visits. On the RTVue-100 OCT device, 141 B-scans centered on the macula are used to produce a scanning laser ophthalmoscopy-like reference image of 7 x 7 mm. The "sum of C-scans" produces a high contrast gray scale image of the retina surface detail, which will be used as reference and registration image based on vessel tracing. This allows a more accurate evaluation of the progression of the disease, as the retinal thickness maps will measure the same area in the macula on follow-up visits (Figures 15-13 through 15-15).

The CNV complex and its associated features can be characterized in detail with the different scan protocols of the RTVue-100 OCT device. The improvement in the axial resolution, associated with a higher number of A-scans per B-scan, a denser scan pattern, and the averaging of up to 16 images (to reduce the contrast-to-noise ratio) allow the visualization of small alterations of the retinal layers not seen with the time-domain OCT (Figure 15-16). Small areas of intra- and subretinal fluid are more frequently found, although the clinical relevance of these small amounts of fluid as a sign of activity of the disease to indicate that the treatments still need to be proved.

TREATMENT

Several treatments have been proposed and extensively studied to prevent the severe visual loss in exuda-

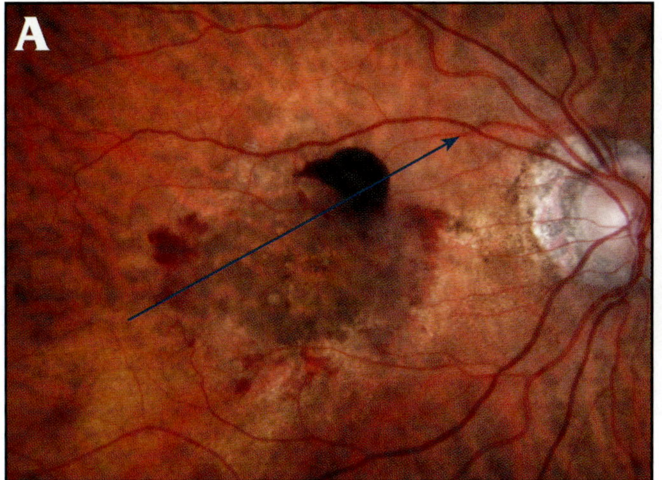

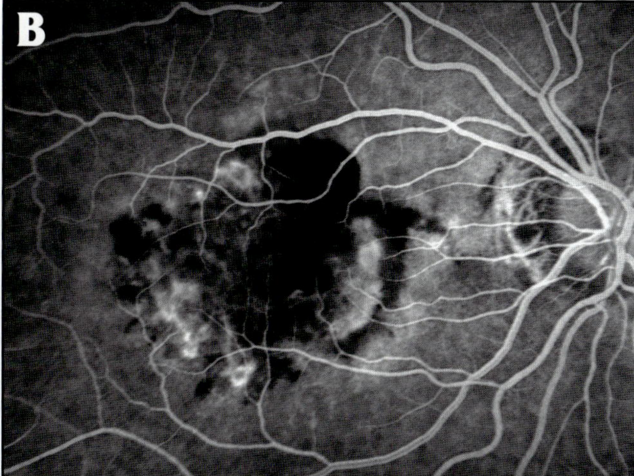

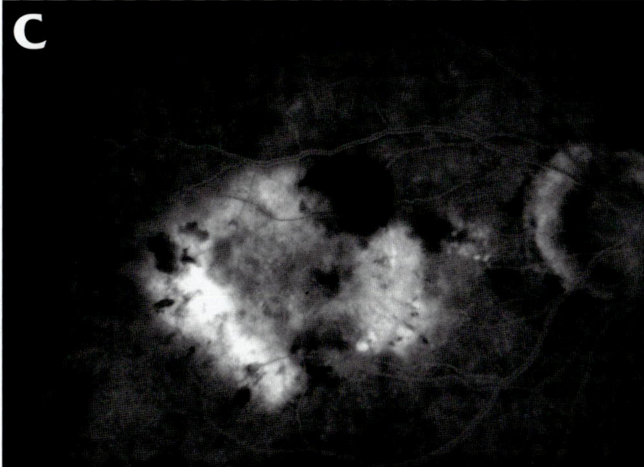

Figure 15-11. Classic CNV membrane with subretinal hemorrhage. Color fundus photo (A), early (B) and late phase (C) of the fluorescein angiography. The green arrow represents the location of the cross-sectional image. (D) The cross-sectional image disclosed the CNV above the RPE (green asterisk), intraretinal cysts, and the subretinal hemorrhage (yellow triangle). The subretinal hemorrhage appears as an elevated retina with a highly reflective underlying layer (the blood), which shadows the RPE/Bruch's membrane/choriocapillaris band.

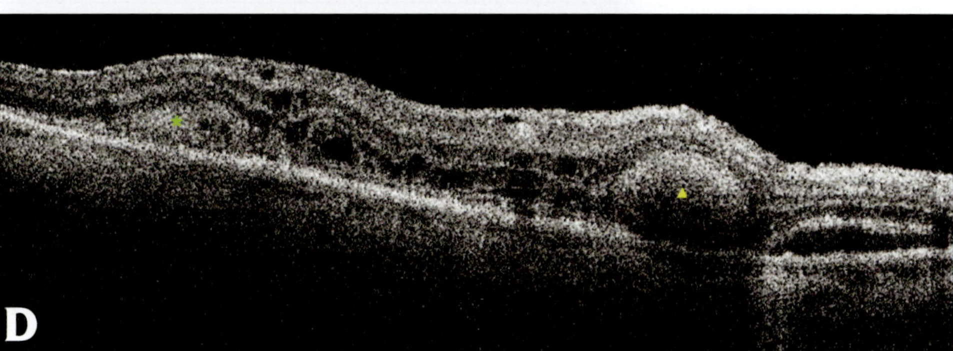

Figure 15-12. Horizontal cross-sectional image of the patient presented in Figure 15-1C. The disciform scar is represented as a highly reflective formation at the level of the RPE, with thinning of the overlying retina and shadowing of the structures underneath.

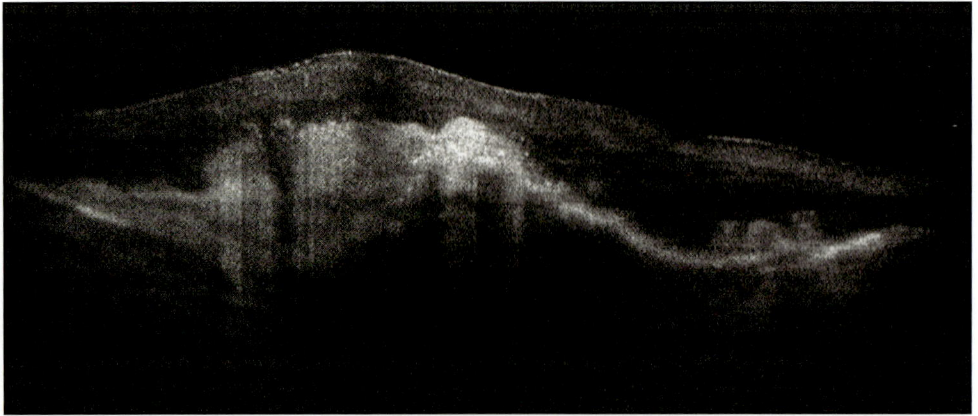

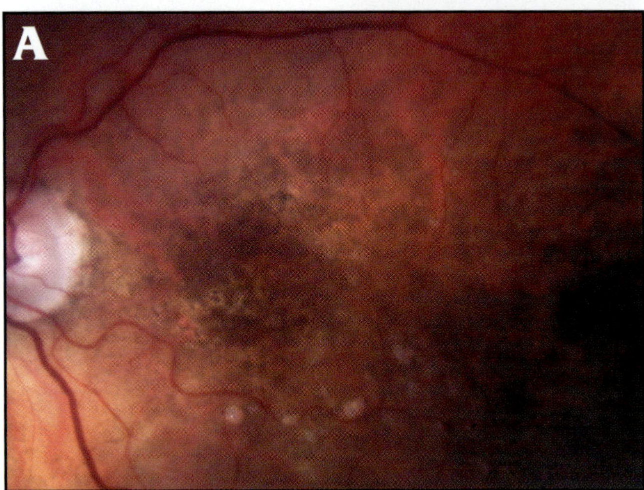

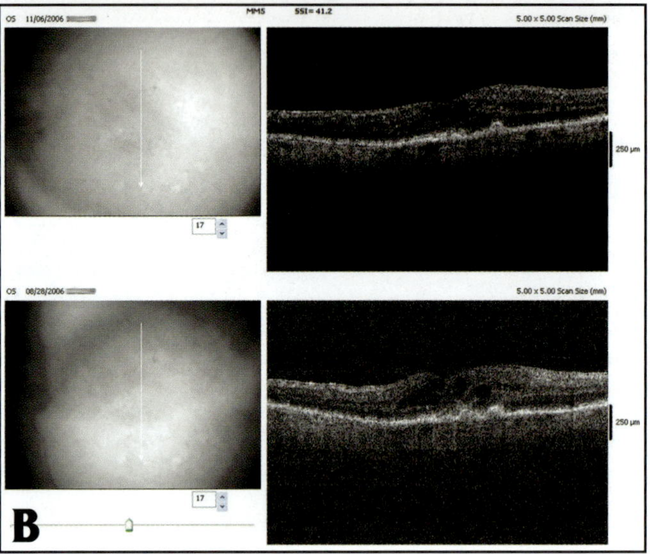

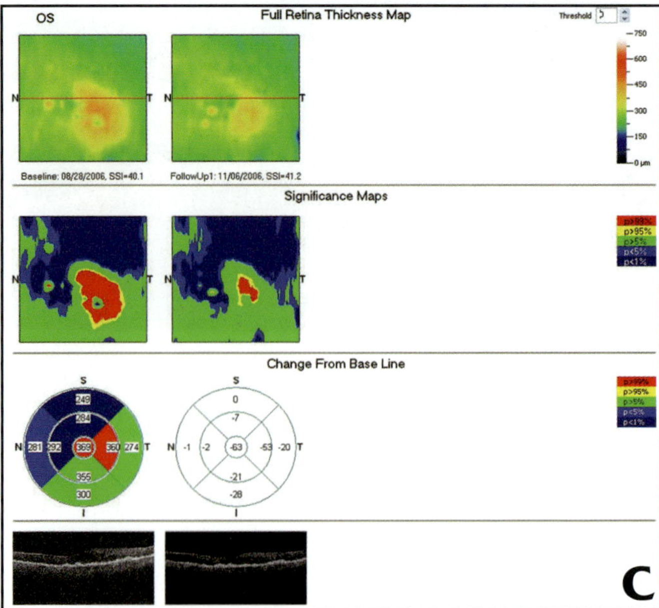

Figure 15-13. The evaluation of the progression of the disease using the EMM5 scan protocol allows the precise registration of the same area scanned among the follow-up visits. The comparison between the cross-sectional images is more accurate (B), as well as the evaluation of the macular thickness (C).

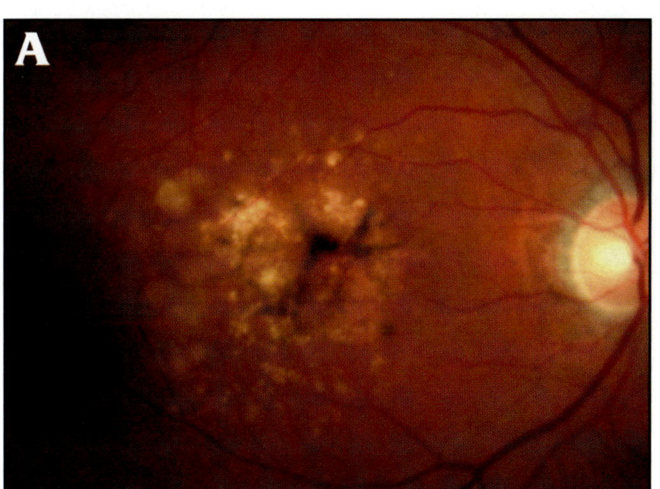

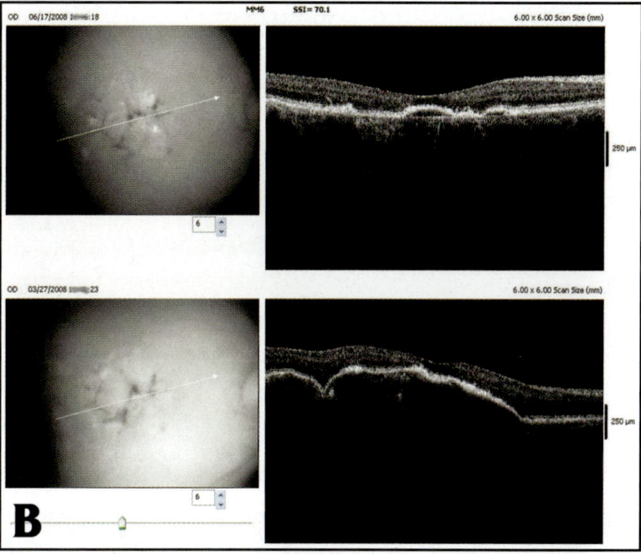

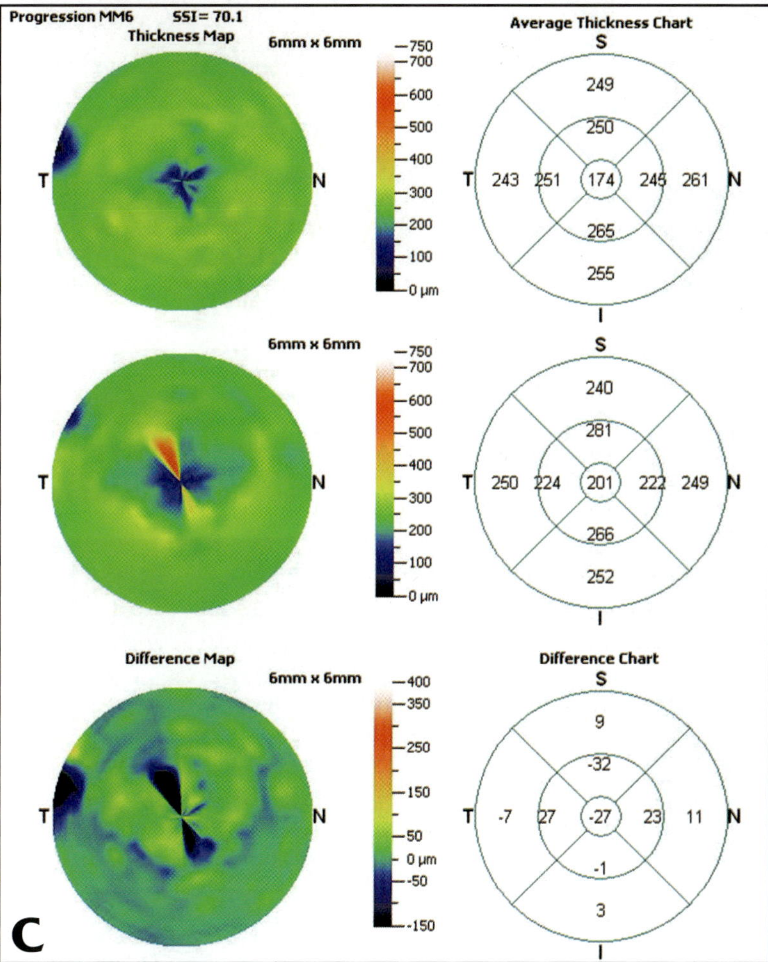

Figure 15-14. Using the MM6 scan protocol, the macula is scanned with 12 radial 6-mm long cross-sectional images, each one 15 degrees apart. The comparison between the same scan between visits is precise: (B) pretreatment on the bottom, post-treatment on the top, allowing monitoring the efficacy of the treatment between visits (C).

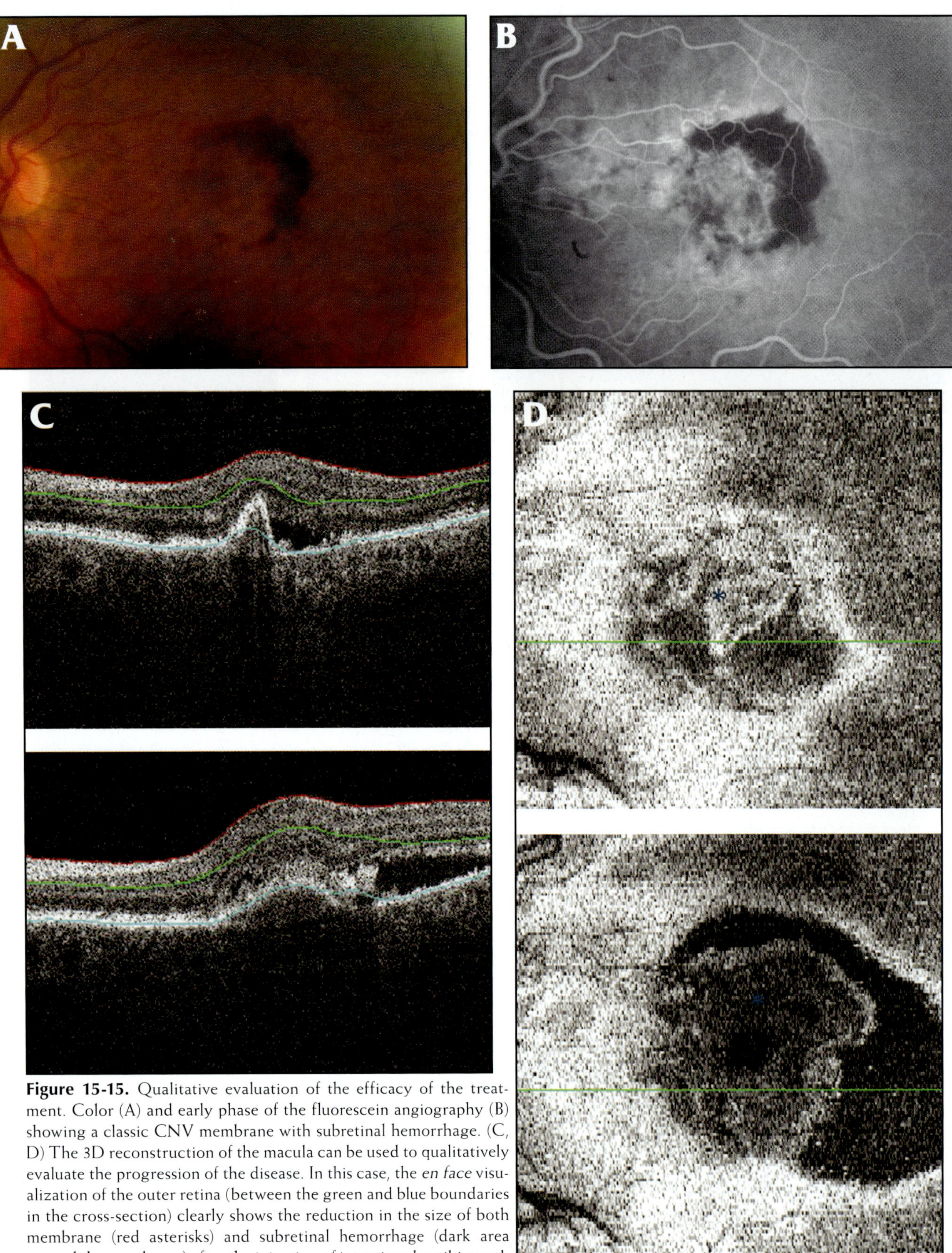

Figure 15-15. Qualitative evaluation of the efficacy of the treatment. Color (A) and early phase of the fluorescein angiography (B) showing a classic CNV membrane with subretinal hemorrhage. (C, D) The 3D reconstruction of the macula can be used to qualitatively evaluate the progression of the disease. In this case, the *en face* visualization of the outer retina (between the green and blue boundaries in the cross-section) clearly shows the reduction in the size of both membrane (red asterisks) and subretinal hemorrhage (dark area around the membrane) after the injection of intravitreal ranibizumab (before treatment on the bottom, after treatment on the top).

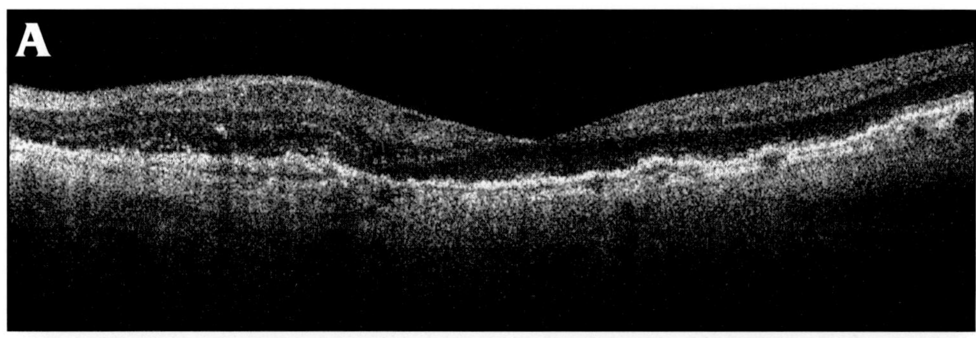

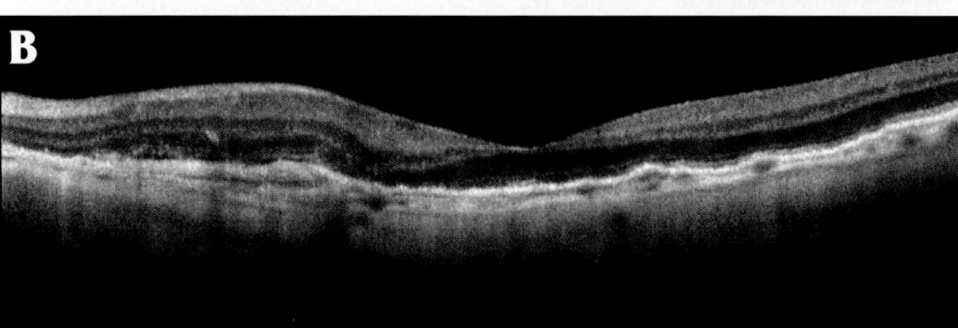

Figure 15-16. Averaging of 16 B-scans in a patient with exudative AMD. (A) B-scan without averaging. (B) Averaging of 16 B-scans from the same patient. Visualization of all retinal layers is improved after averaging of the B-scans. Some retinal layers (external limiting membrane and photoreceptor inner/outer segment junction) and subretinal fluid (arrowheads) are only identified with averaging of the 16 B-scans.

tive AMD patients, including photocoagulation of the CNV, transpupilary thermotherapy (TTT), ocular photodynamic therapy with verteporfin (PDT), indocyanine-mediated photothrombosis (i-MP), submacular surgery, macular translocation, and more recently antivascular endothelial growth factor (VEGF) therapy.

The efficacy of the photocoagulation of CNV was assessed by the Macular Photocoagulation Study (MPS), which demonstrated the best efficacy for extrafoveal lesions with some benefits for juxtafoveal and subfoveal CNV. However, laser photocoagulation is rarely used today in the management of neovascular AMD. Restrictive eligibility criteria, immediate visual loss due to laser-induced scotoma, and high recurrence rates are limiting factors for its use.

Ocular PDT was developed to specifically target CNV tissue while sparing the surrounding and overlying retinal structures. A low-energy light is used to activate an intravenously injected photosensitizing agent (Visudyne [verteporfin]) and induce closure of the neovascular complex. The TAP (Treatment of Age-Related Macular Degeneration with Photodynamic Therapy) study,[17,18] one of the 2 large prospective clinical trials that evaluated the PDT for the treatment of subfoveal CNV due to AMD, showed that it was effective compared to placebo in preventing moderate and severe visual loss; preserving contrast sensitivity; and limiting final lesion size, lesion leakage, and progression in individuals with predominantly classic CNV. The second large prospective clinical trial, the VIP (Verteporfin in Photodynamic Therapy) study,[19] showed that PDT was able to reduce the rate of moderate visual loss after 24 months in

individuals with occult CNV and no classic component compared to placebo treatment. However, 4% of verteporfin-treated eyes experienced an acute severe visual acuity decrease defined as a loss of 20 or more letters within 7 days of treatment.

The combination of PDT with intravitreal injection of steroids have been shown to improve visual acuity outcomes and decrease the treatment burden.[20-25] However, these results have been obscured by the visual acuity and anatomic benefits observed with inhibitors of vascular endothelial growth factor-A (VEGF-A).

The first VEGF-A inhibitor approved by the Food and Drug Administration (FDA) for the treatment of exudative AMD, the pegaptanib sodium (Macugen), was shown to reduce the risk of moderate and severe visual loss in patients who received intravitreal injections every 6 weeks during 2 years, as compared with placebo.[26] However, only 6% of the pegaptanib-treated patients gained 15 letters or more of visual acuity after 1 year.[27]

Ranibizumab (Lucentis) is a monoclonal antibody antigen-binding fragment that binds and inhibits all the biologically active forms of VEGF-A. It received FDA approval in 2006 on the basis of 2 pivotal Phase III trials known as the MARINA (Minimally Classic/Occult Trial of the Anti-VEGF Antibody Ranibizumab in the Treatment of Neovascular AMD)[28] and ANCHOR (Anti-VEGF Antibody Ranibizumab for the Treatment of Predominantly Classic Choroidal Neovascularization in AMD)[29] trials. In these trials, monthly intravitreal injections showed improvement in visual acuity outcomes in most patients and prevented the leakage and growth of neovascularization in all the major angiographic forms of CNV. Alternative dosing strategies have been pursued trying to reduce the burden of monthly injections. In the PIER study, patients were treated monthly for 3 consecutive months, followed by a treatment every 3 months thereafter. Although there was visual acuity improvement after the first 3 consecutive monthly injections, this visual acuity improvement was lost when the injection regimen changed to every 3 months, and the outcomes were not as good as the visual acuity outcomes observed following monthly dosing studies.[30] The PrONTO Study (Prospective OCT Imaging of Patients with Neovascular AMD Treated with Intra-Ocular Ranibizumab) explored whether dosing on the basis of fluid in the macula as detected by OCT could result in fewer injections but achieve visual acuity outcomes similar to the results obtained with monthly dosing.[31] Patients were treated monthly for 3 consecutive months and then examined monthly with OCT. They were retreated only if certain OCT-based criteria were fulfilled. The visual acuity outcomes were comparable to the outcomes in the MARINA and ANCHOR trials after 1-year and 2-year follow-up, however with far fewer injections.

Bevacizumab (Avastin) is a full-length humanized monoclonal antibody that binds and inhibits all the biologically active forms of VEGF-A. It is FDA-approved for the systemic (intravenous) treatment of metastatic colorectal cancer. The first patient treated with off-label intravitreal injection of bevacizumab (in May 2005) had a response identical to the responses previously observed with systemic bevacizumab and intravitreal ranibizumab.[32] Bevacizumab has gained popularity as an alternative to ranibizumab due to its apparent efficacy and safety, its worldwide availability, and its low cost. The Comparison of Age-Related Macular Degeneration Treatment Trial (CATT) will compare the safety and efficacy of bevacizumab and ranibizumab in wet AMD patients with all major lesions subtypes, and the first results are expected in February 2010.

Several others treatments for exudative AMD are under study, and new options for treatment will be available in the future. Among them, the VEGF Trap, a fusion protein of VEGF receptor ligand-binding domains and the Fc region of IgG1, has been studied both as an intravenous and intravitreal drug, and showed good outcomes. A Phase III trial using intravitreal injections is underway with primary outcome measurements expected in July 2011.

Others treatments under study:

- Combination of epiretinal radiation (brachytherapy) of the neovascular membrane associated with intravitreal injection of ranibizumab, seeking to reduce the number of injections needed to control the disease (CABERNET Study—Phase III trial)
- Mecamylamine eyedrops (ATG003) are being studied in a Phase II trial in patients receiving maintenance injections of either ranibizumab or bevacizumab
- Subconjuntival injection of rapamycin (sirolimus), associated or not with intravitreal injection of ranibizumab, in the treatment of exudative AMD (EMERALD Study—Phase II trial)

REFERENCES

1. Eye Diseases Prevalence Research Group. Prevalence of age-related macular degeneration in the United States. *Arch Ophthalmol.* 2004;122:564-572.
2. Young RW. Pathophysiology of age-related macular degeneration. *Surv Ophthalmol.* 1987;31:291-306.
3. Ferris FL III, Fine SL, Hyman L. Age-related macular degeneration and blindness due to neovascular maculopathy. *Arch Ophthalmol.* 1984;102:1640-1642.
4. Bressler NM, Bressler SB, Fine SL. Age-related macular degeneration. *Surv Ophthalmol.* 1988;32:375-413.
5. Bressler NM, Silva JC, Bressler SB, et al. Clinicopathologic correlation of drusen and retinal pigment epithelial abnormalities in age-related macular degeneration. *Retina.* 1994;14:130-142.
6. Kenyon KR, Maumenee AE, Ryan SJ, et al. Diffuse drusen and associated complications. *Am J Ophthalmol.* 1985;100:119-128.
7. Submacular Surgery Trials (SST) Research Group. Surgery for hemorrhagic choroidal lesions of age-related macular degeneration: ophthalmic findings. SST report no. 13. *Ophthalmology.* 2004;111:1993-2006.
8. Killingsworth MC, Sarks JP, Sarks SH. Macrophages related to Bruch's membrane in age-related macular degeneration. *Eye.* 1990;4:613-621.
9. Green WR, Enger C. Age-related macular degeneration histopathologic studies. The 1992 Lorenz E. Zimmerman lecture. *Ophthalmology.* 1993;100:1519-1535.
10. Friedman E. The role of the atherosclerotic process in the pathogenesis of age-related macular degeneration. *Am J Ophthalmol.* 2000;130:658-663.
11. Karadimas P, Bouzas EA. Fundus autofluorescence imaging in serous and drusenoid pigment epithelial detachments associated with age-related macular degeneration. *Am J Ophthalmol.* 2005;140(6):1163-1165.
12. Schmitz-Valckenberg S, Fleckenstein M, Scholl HP, Holz FG. Fundus autofluorescence and progression of age-related macular degeneration. *Surv Ophthalmol.* 2009;54(1):96-117.
13. Hee MR, Baumal CR, Puliafito CA, et al. Optical coherence tomography of age-related macular degeneration and choroidal neovascularization. *Ophthalmology.* 1996;103:1260–1270.
14. Giovannini A, Amato G, Mariotti C, Scassellati-Sforzolini B. OCT imaging of choroidal neovascularisation and its role in the determination of patients' eligibility for surgery. *Br J Ophthalmol.* 1999;83(4):438-442.
15. Coscas F, Coscas G, Souied E, et al. Optical coherence tomography identification of occult choroidal neovascularization in age-related macular degeneration. *Am J Opthalmol.* 2007;144(4):592-599.
16. Ahlers C, Michels S, Beckendorf A, et al. Three-dimensional imaging of pigment epithelial detachment in age-related macular degeneration using optical coherence tomography, retinal thickness analysis and topographic angiography. *Graefes Arch Clin Exp Ophthalmol.* 2006;244(10):1233-1239.
17. Treatment of Age-Related Macular Degeneration with Photodynamic Therapy (TAP) Study Group. Photodynamic therapy of subfoveal choroidal neovascularization in age-related macular degeneration with verteporfin: one-year results of 2 randomized clinical trials: TAP report 1. *Arch Ophthalmol.* 1999;117:1329-1345.
18. Bressler NM. Photodynamic therapy of subfoveal choroidal neovascularization in age-related macular degeneration with vertepor-

fin: two-year results of 2 randomized clinical trials: TAP report 2. *Arch Ophthalmol.* 2001;119:198-207.

19. VIP Study Group. Verteporfin therapy of subfoveal choroidal neovascularization in age-related macular degeneration: two-year results of a randomized clinical trial including lesions with occult with no classic choroidal neovascularization: Verteporfin in Photodynamic Therapy (VIP) report 2. *Am J Ophthalmol.* 2001;131:541-560.

20. Spaide RF, Sorenson J, Maranan L. Combined photodynamic therapy with verteporfin and intravitreal triamcinolone acetonide for choroidal neovascularization. *Ophthalmology.* 2003;110:1517-1525.

21. Arias L, Garcia-Arumi J, Ramon JM, et al. Photodynamic therapy with intravitreal triamcinolone in predominantly classic choroidal neovascularization: one-year results of a randomized study. *Ophthalmology.* 2006;113:2243-2250.

22. Augustin AJ, Schmidt-Erfurth U. Verteporfin therapy combined with intravitreal triamcinolone in all types of choroidal neovascularization due to age-related macular degeneration. *Ophthalmology.* 2006;113:14-22.

23. Augustin AJ, Schmidt-Erfurth U. Verteporfin and intravitreal triamcinolone acetonide combination therapy for occult choroidal neovascularization in age-related macular degeneration. *Am J Ophthalmol.* 2006;141:638-645.

24. Chan WM, Lai TY, Wong AL, et al. Combined photodynamic therapy and intravitreal triamcinolone injection for the treatment of subfoveal choroidal neovascularization in age related macular degeneration: a comparative study. *Br J Ophthalmol.* 2006;90:337-341.

25. Ruiz-Moreno JM, Montero JA, Zarbin MA. Photodynamic therapy and high-dose intravitreal triamcinolone to treat exudative age-related macular degeneration: 2-year outcome. *Retina.* 2007;27:458-461.

26. Chakravarthy U, Adamis AP, Cunningham Jr ET, et al. Year 2 efficacy results of 2 randomized controlled clinical trials of pegaptanib for neovascular age-related macular degeneration. *Ophthalmology.* 2006;113:1508.e1-1508.e25.

27. Gragoudas ES, Adamis AP, Cunningham Jr. ET, et al. Pegaptanib for neovascular age-related macular degeneration. *N Engl J Med.* 2004;351:2805-2816.

28. Rosenfeld PJ, Brown DM, Heier JS, et al. Ranibizumab for neovascular age-related macular degeneration. *N Engl J Med.* 2006;355:1419-1431.

29. Brown DM, Kaiser PK, Michels M, et al. ANCHOR Study Group. Ranibizumab versus verteporfin for neovascular age-related macular degeneration. *N Engl J Med.* 2006;355:1432-1444.

30. Regillo CD, Brown DM, Abraham P, et al. Randomized, double-masked, sham-controlled trial of ranibizumab for neovascular age-related macular degeneration: PIER study year 1. *Am J Ophthalmol.* 2008;145:239-248.

31. Fung AE, Lalwani GA, Rosenfeld PJ, et al. An optical coherence tomography-guided, variable dosing regimen with intravitreal ranibizumab (Lucentis) for neovascular age-related macular degeneration. *Am J Ophthalmol.* 2007;143(4):566-583.

32. Rosenfeld PJ, Moshfeghi AA, Puliafito CA. Optical coherence tomography findings after an intravitreal injection of bevacizumab (Avastin) for neovascular age-related macular degeneration. *Ophthalmic Surg Lasers Imaging.* 2005;36:331-335.

16 Pathologic Myopia

Kaori Sayanagi, MD

High myopia is defined as the refractive error of -6 D or more, or the axial length of 25 mm or more. Several retinal disorders including chorioretinal atrophy, choroidal neovascularization (CNV), macular hole and retinal detachment, myopic foveoschisis (MF), and paravascular abnormarities are usually observed.

Due to the chorioretinal atrophy, it is difficult to observe the morphology of the posterior pole retina using the ophthalmoscope. Recently, OCT has been an essential tool for diagnosing the high-myopia specific diseases and is also useful for the observation during follow-up.

MYOPIC FOVEOSCHISIS

MF was first described by Phillips in 1958 as a shallow retinal detachment at the area of the posterior pole without a macular hole.[1] Takano and Kishi first investigated MF using time-domain OCT and found the splitting of the retina and the column-like structure bridging both the inner and outer layers.[2] On OCT, the inner and outer layer is visualized as a hyper-reflective layer and a thin faintly hyper-reflective layer, respectively, and the column-like structure is visualized as a hyper-reflective line across the hyporeflective spaces.[3] Time-domain OCT also visualized the preretinal structures, such as the epiretinal membrane, vitreous cortex, and internal limiting membrane (ILM), as a hyper-reflective line. These subretinal structures often caused tangential and/or inward traction, resulting in retinal thickening, retinal

detachment, and a macular hole.[3-6] MF is divided into 3 subtypes by OCT depending on the status of photoreceptor and retinal pigment epithelium (RPE) layer, and the presence or absence of a macular hole, foveoschisis type, foveal detachment type, and macular hole type (Figures 16-1 through 16-3).[6,7] In the foveoschisis type, the photoreceptor layer is still attached to the RPE. In the foveal detachment type, the photoreceptor layer is detached from the RPE. In the macular hole type, a macular hole is observed. The foveal status (subtypes) is thought to be a key factor in determining the surgical outcome. The foveal detachment type achieved the most favorable outcome, though others also preserve the vision.[7]

RTVue, spectral-domain OCT, visualized the intraretinal morphology, especially the photoreceptor inner and outer segment (IS/OS) junction more clearly as compared to the conventional time-domain OCT. In MF, the defect of IS/OS is observed and may predict the postoperative visual recovery.[8]

MACULAR HOLE AND RETINAL DETACHMENT

Macular hole and retinal detachment are vision-threatening complications of highly myopic eyes. Preoperatively, OCT can visualize the retinal morphology only when the area of retinal detachment is limited in the staphyloma. OCT reveals the retinal detachment as an absence of reflectivity, the disruption of the neu-

Huang D, Duker JS, Fujimoto JG, Lumbroso B, Schuman JS, Weinreb RN.
Imaging the Eye from Front to Back with RTVue Fourier-Domain Optical Coherence Tomography (pp 155-160).
© 2010 SLACK Incorporated.

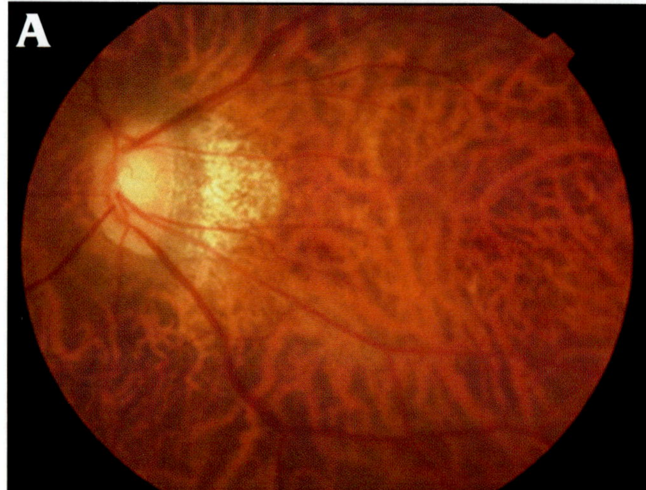

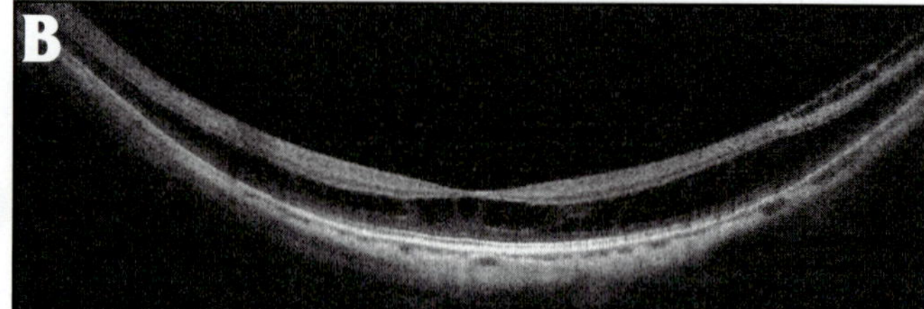

Figure 16-1. Case 1. A 43-year-old woman with MF (foveoschisis type). The column-like structure was clearly visualized between the inner and outer layers. The IS/OS was preserved and BCVA was 20/20.

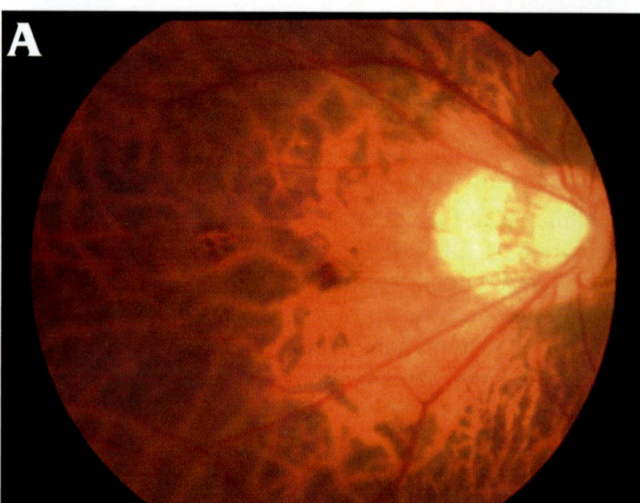

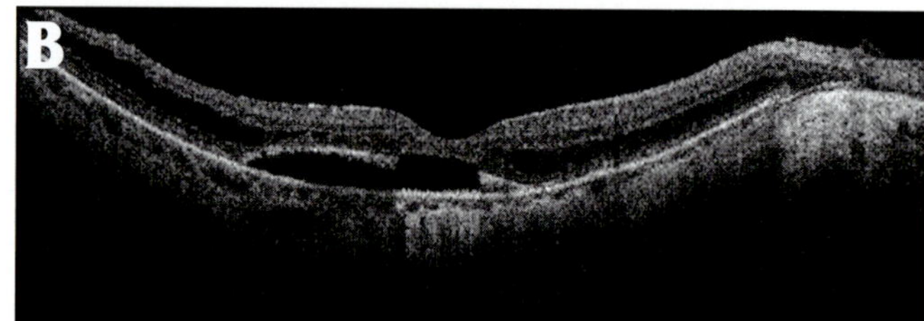

Figure 16-2. Case 2. A 43-year-old man with MF (foveal detachment type). The column-like structure was observed between the inner and outer layer. The defect of IS/OS was detected at the fovea, and BCVA was 20/70.

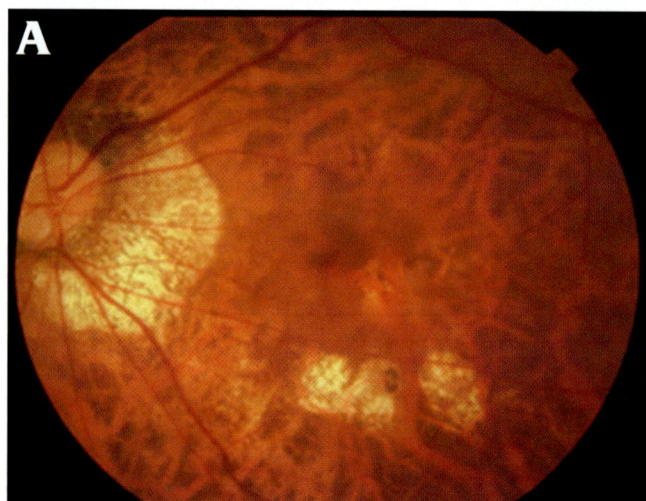

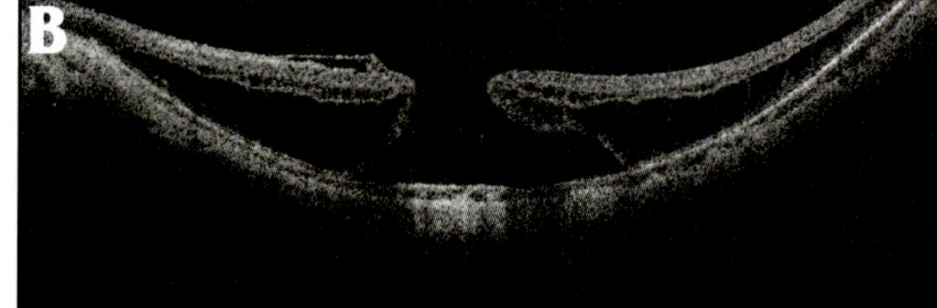

Figure 16-3. Case 3. A 57-year-old man with MF (macular hole type). The defect of IS/OS was observed at the base of the macular hole. The epiretinal membrane was also detected.

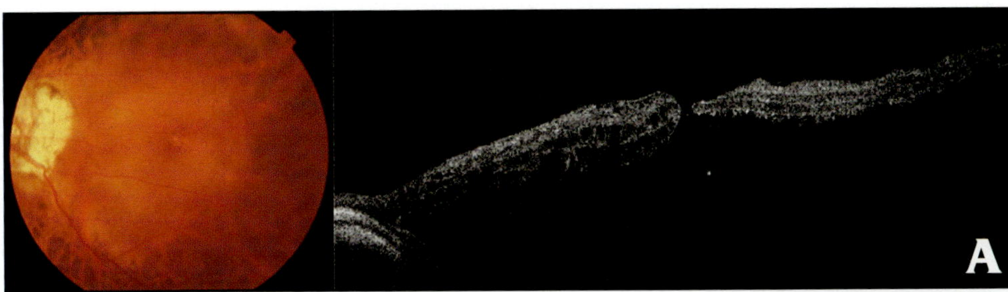

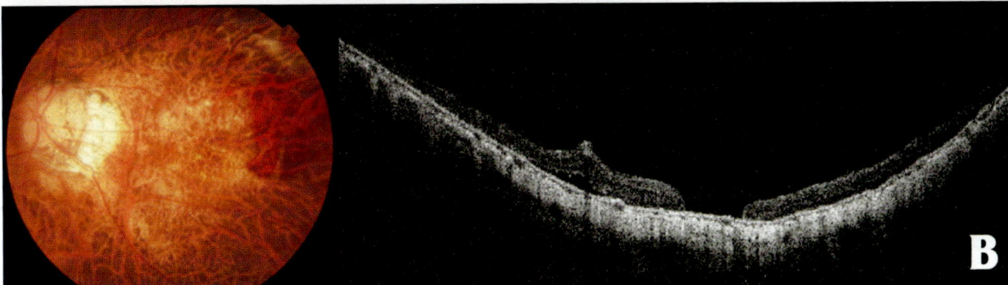

Figure 16-4. Case 4. A 67-year-old man with a macular hole and retinal detachment. Preoperatively (A), the detached retina was visualized on both the fundus photograph and RTVue image; however, macular hole was visualized more clearly on RTVue image. Postoperatively (B), the macular hole closure was not obtained, although the retina reattached. The vertical section detected the retinal microfold, which seemed to cause the inward traction of the retina.

rosensory retina as a hyporeflective intraretinal space, and the vitreous cortex adherent to the inner retinal surface as a hyper-reflective line.[9,10] In spectral-domain OCT, the outer fold of the IS/OS is observed and a 3D image reveals the area of intraretinal cysts, which usually expand outside of the macular hole (Figure 16-4).[11]

Postoperatively, OCT is essential to observe the foveal status. In macular hole and retinal detachment, the macular hole closure is an important factor in achieving favorable surgical outcome. Several investigators reported using time-domain OCT, the macular hole closure rate was 30% to 43% after the vitrectomy.[12,13] Spectral-domain OCT, especially a 3D image, should offer a more precise evaluation of the foveal status, as a line scan might miss the fovea if the patient has poor fixation when 3D scan should not.[14]

MYOPIC CHOROIDAL NEOVASCULARIZATION

Myopic choroidal neovascularization (mCNV) is one of the most severe vision-threatening complications of high myopia. In natural course, almost all patients dropped to 20/200 or less within 5 to 10 years after the onset due to the development of chorioretinal atrophy around CNV.[15] Although the efficacy for the chorioretinal atrophy, which is the main factor decreasing the long-term vision, has not been clarified, photodynamic therapy and intravitreal bevacizumab injection are currently proven to be effective for CNV.[16-20] Since it is difficult to evaluate the activity of mCNV by fluorescence angiography, OCT is essential tool for the determination of the primary and/or retreatment.

Baba observed the morphology of mCNV using time-domain OCT[21] and reported that time-domain OCT demonstrated the stage and the activity of CNV. In the active stage, CNV was visualized as a hyper-reflective dome-like elevation above RPE; in the scar stage, only the surface of CNV showed hyper-reflectivity. In the atrophic stage, CNV had become totally flat and the chorioretinal atrophy around CNV showed hyper-reflectivity. However, they also reported that time-domain OCT did not display the subretinal fluid and intraretinal edema.[21] Spectral-domain OCT displays both subretinal fluid and intraretinal edema clearly. Also, a 3D scan can detect extrafoveal subretinal fluid and intraretinal edema (Figure 16-5). These findings are useful for the precise diagnosis, evaluation, and/or treatment of mCNV.

PARAVASCULAR ABNORMALITIES

Paravascular abnormalities, including the retinal microfold and the paravascular retinal hole and cysts, are not uncommon in high myopia and OCT is an essential tool to detect them.

The retinal microfold was firstly described by Ikuno in the eyes after the vitrectomy for the MF.[22] The microfold is thought to cause the inward traction of the retina resulting in the various diseases specific to high myopia (Figure 16-6). The incidence of this retinal microfold in high myopia without the vitrectomy varies from 2.9% to 44.6%.[23,24] The pathogenesis of the retinal microfold has been unknown; however, the axial length elongation and the inflexibility of the retinal vessel are thought to be related to the development.

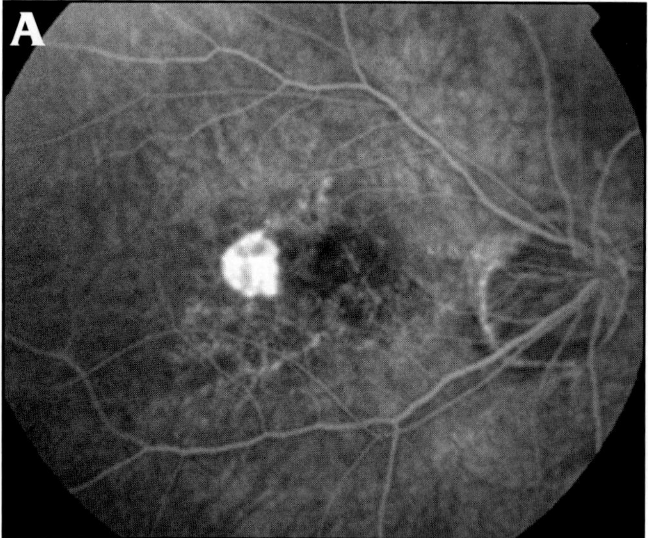

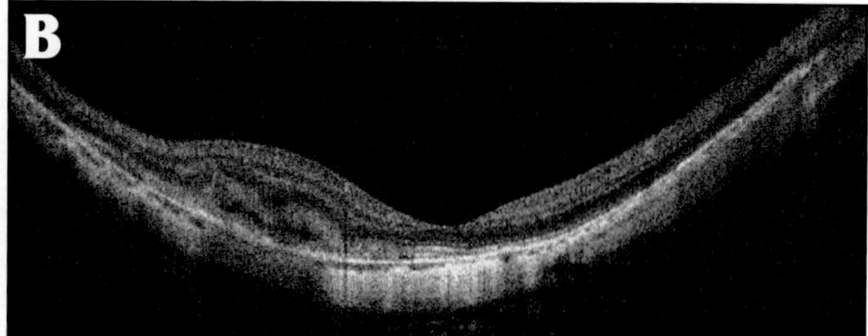

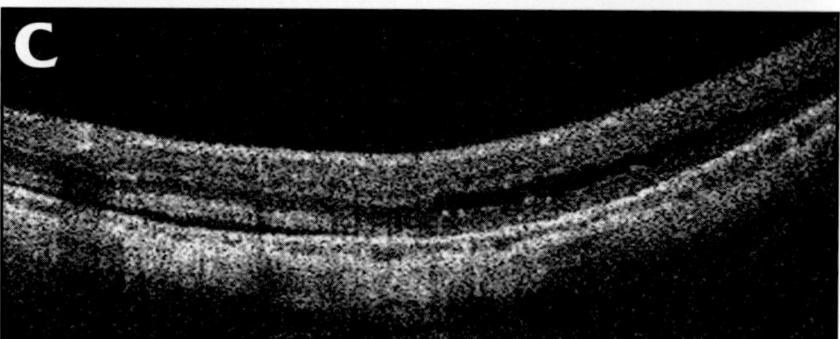

Figure 16-5. Case 5. (A) A 35-year-old man with mCNV. The foveal section (B) revealed CNV at the jaxta-fovea, and IS/OS seemed to be preserved at the fovea. BCVA was 20/20. The section obtained from the 3D image (C) revealed subretinal and intraretinal fluid, which the foveal section missed.

Shimada et al reported the prevalence and the characteristics of the paravascular retinal hole and cysts using time-domain OCT.[24] They indicated the prevalence of paravascular retinal hole and cysts are 26.8% and 49.5% in the eyes with high myopia, respectively. OCT visualized the vitreous traction on the inner walls of retinal cysts in sections adjacent to the retinal holes. Respectively, 44.2% and 19.5% of the eyes with paravascular lamellar holes are accompanied with the internal limiting membrane detachment and MF.[24]

REFERENCES

1. Phillips CI. Retinal detachment at the posterior pole. *Br J Ophthalmol.* 1958;42:749-753.

2. Takano M, Kishi S. Foveal retinoschisis and retinal detachment in severely myopic eyes with posterior staphyloma. *Am J Ophthalmol.* 1999;128:472-476.

3. Gaucher D, Haouchine B, Tadayoni R, et al. Long-term follow-up of high myopic foveoschisis: natural course and surgical outcome. *Am J Ophthalmol.* 2007;143:455-462.

4. Shimada N, Ohno-Matsui K, Baba T, et al. Natural course of macular retinoschisis in highly myopic eyes without macular hole or retinal detachment. *Am J Ophthalmol.* 2006;142:497-500.

5. Sayanagi K, Ikuno Y, Tano Y. Tractional internal limiting membrane detachment in highly myopic eyes. *Am J Ophthalmol.* 2006;142:850-852.

6. Benhamou N, Massin P, Haouchine B, Erginay A, Gaudric A. Macular retinoschisis in highly myopic eyes. *Am J Ophthalmol.* 2002;133:794-800.

7. Ikuno Y, Sayanagi K, Soga K, et al. Foveal anatomical status and surgical results in vitrectomy for myopic foveoschisis. *Jpn J Ophthalmol.* 2008;52:269-276.

8. Sayanagi K, Ikuno Y, Soga K, Tano Y. Photoreceptor inner and outer segment defects in myopic foveoschisis. *Am J Ophthalmol.* 2008;145:902-908.

9. Baba T, Hirose A, Kawazoe Y, Mochizuki M. Optical coherence tomography for retinal detachment with a macular hole in a highly myopic eye. *Ophthalmic Surg.* 2003;34:483-484.

10. Ripandelli G, Parisi V, Friberg TR, et al. Retinal detachment associated with macular hole in high myopia. *Ophthalmology.* 2004;111:726-731.

11. Sayanaagi K, Ikuno Y, Soga K, Wakabayashi T, Tano Y. Outer retinal folds in highly myopic macular hole and retinal detachment. *Retina.* In press.

12. Ikuno Y, Sayanagi K, Oshima T, et al. Optical coherence tomography findings of macular hole and retinal detachment after vitrectomy in highly myopic eyes. *Am J Ophthalmol.* 2003;136:477-481.

13. Kobayashi H, Takahashi K, Ishihara K, Kishi S. Tomographic features of macular hole in high myopic eyes following solution of retinal detachment by vitreous surgery. *Jpn J Clin Ophthalmol (Rionsho Ganka).* 2001;55:799-803.

14. Sayanagi K, Oshima Y, Ikuno Y, Tano Y. Presumed vascular traction-associated recurrence of retinal detachment in patients with myopic macular hole. *Ophthalmic Surg Lasers Imaging.* 2009;40:60-64.

15. Yoshida T, Ohno-Matsui K, Yasuzumi K, et al. Myopic choroidal neovscularization: a 10-year follow-up. *Ophthalmology.* 2003;110:1297-1305.

16. Hayashi K, Ohno-Matsui K, Teramukai S, et al. Comparison of visual outcome and refression pattern of myopic choroidal neovascularization after intravitreal bevacizumab or after photodynamic therapy. *Am J Ophthalmol.* 2009;148:396-408.

17. Blinder KJ, Blumenkranz MS, Bressler NM, et al. Verteporfintherapy of subfoveal choroidal neovascularization in pathologic myopia: two-year results of a randomized clinical trial—VIP Report No. 3. *Ophthalmology.* 2003;110:667-673.

18. Ikuno Y, Sayanagi K, Soga K, et al. Intravitreal bevacizumab for choroidal neovascularization attributable to pathological myopia: one-year results. *Am J Ophthalmol.* 2009;147:94-100.

19. Gharbiya M, Allievi F, Mazzeo L, Gabrieli CB. Intravitreal bevacizumab treatment for choroidal neovascularization in pathologic myopia: 12-month results. *Am J Ophthalmol.* 2009;147:84-93.

20. Chan WM, Lai TYY, Liu DTL, Lam DSC. Intravitreal bevacizumab (Avastin) for myopic choroidal neovascularization: one-year results of a prospective study. *Br J Ophthalmol.* 2009;93:150-154.

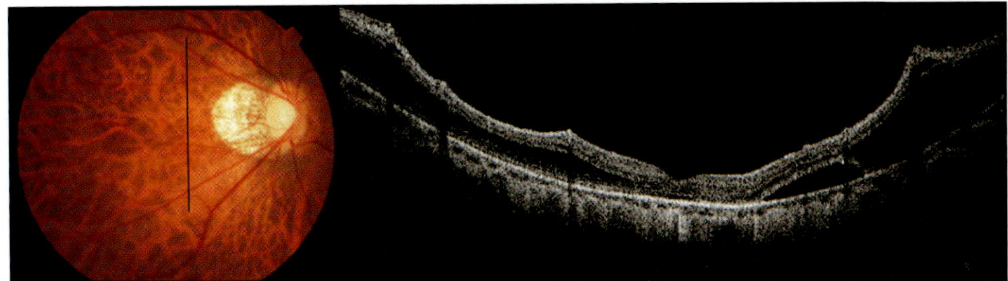

Figure 16-6. Case 6. One year after the vitrectomy for MF of Case 2. The 3 microfolds were detected and all of them corresponded with the retinal artery.

21. Baba T, Ohno-Matsui K, Yoshida T, et al. Optical coherence tomography of choriodal neovascularization in high myopia. *Acta Ophthalmol Scand.* 2002;80:82-87.

22. Ikuno Y, Gomi F, Tano Y. Potent retinal arteriolar traction as a possible cause of myopic foveoschisis. *Am J Ophthalmol.* 2005;139:462-467.

23. Sayanagi K, Ikuno Y, Gomi F, Tano Y. Retinal vascular microfolds in highly myopic eyes. *Am J Ophthalmol.* 2005;139:658-663.

24. Shimada N, Ohno-Matsui K, Nishimura A, et al. Detection of paravascular lamellar holes and other paravascular abnormalities by optical coherence tomography in eyes with high myopia. *Ophthalmology.* 2008;115:708-717.

17 Central Serous Chorioretinopathy

Fumi Gomi, MD, PhD

INTRODUCTION

Central serous chorioretinopathy (CSC) is characterized by serous detachment of the neurosensory retina in the macula, which often causes metamorphopsia and microsomia. Both eyes are sometimes affected. CSC is a self-limiting disease with a majority of patients recovering good vision. A chronic variant of CSC, however, may relate the consequent severe visual decline. Type A personality, systemic use of corticosteroids, and other factors are said to be risk factors for development of CSC, but the details are still unknown.[1,2]

Fluorescein angiography shows focal leakage at the level of the retinal pigment epithelium (RPE). Evaluations using indocyanine green angiography in eyes with CSC have shown multifocal islands of inner choroidal staining at late phase. From those findings, the pathogenesis of CSC is suggested as follows: focally induced permeability of the choroidal vessels is the primary event of CSC and subsequent changes at the RPE, probable defect of the RPE layer, allow the fluid to enter the subretinal space.[2]

OCT is used to diagnose and follow the course of the disease.[3-13] In addition, detailed visualization using progressed OCT apparatus can provide the precise morphologic changes in the intraretina, the subretinal space, and the RPE and prognosticate the vision.[14-16] In this chapter, characteristic OCT images found in CSC eyes are displayed.

RETINAL DETACHMENT IN ACUTE CENTRAL SEROUS CHORIORETINOPATHY

OCT can detect even shallow serous retinal detachment, which is hard to detect by slit lamp biomicroscopy. It is useful to evaluate the change of the amount of subretinal fluid during the clinical course noninvasively (Figure 17-1). Spontaneously, or after successful treatment, the detached retina may decrease its height and subretinal fluid may shift inferiorly because of gravity and finally resolve (Figures 17-2 through 17-4). The vertical OCT line scan easily detects such an inferior shift of subretinal fluid (see Figures 17-2C and 17-3C). Macula map or 3D retinal scan by the RTVue may help to detect extramacular retinal detachment or multiple retinal detachments (Figure 17-5).

DETAILED FINDINGS OF DETACHED RETINA

The posterior surface of detached retina is usually smooth. The line corresponding to the external limiting membrane is usually visible, although the line corresponding to the junction of the photoreceptor

Huang D, Duker JS, Fujimoto JG, Lumbroso B, Schuman JS, Weinreb RN.
Imaging the Eye from Front to Back with RTVue Fourier-Domain Optical Coherence Tomography (pp 161-168).

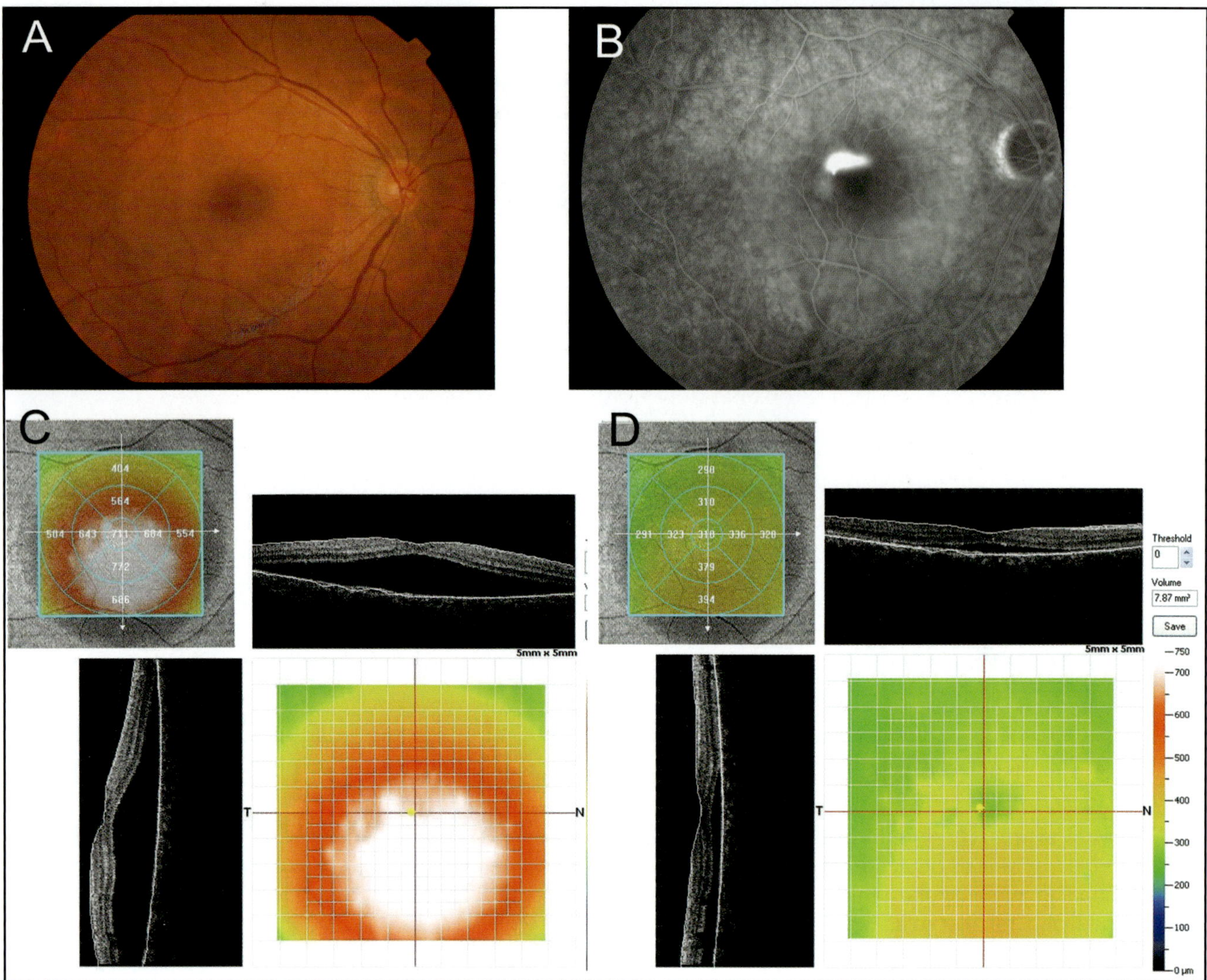

Figure 17-1. Acute CSC. Baseline fundus photograph (A), fluorescein angiograph (B), and EMM5 OCT image from RTVue (C) of a 58-year-old man. Due to subretinal fluid, the retinal surface is elevated. After 3 weeks of observation, the subretinal fluid is reducing spontaneously and the elevation of retina also decreases (D).

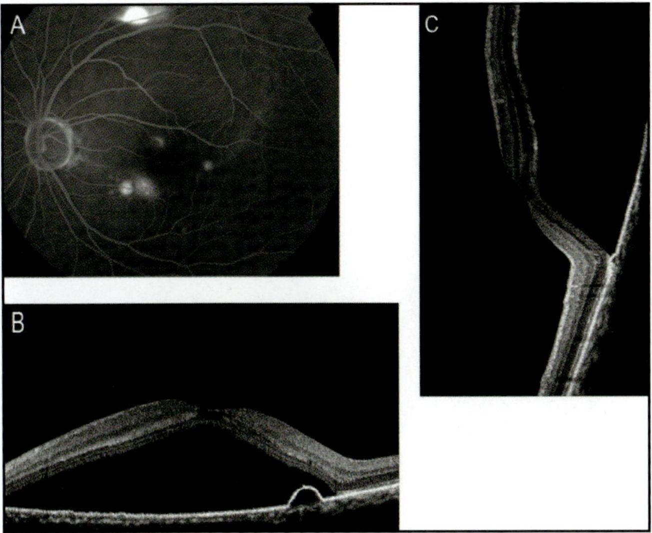

Figure 17-2. Acute CSC. Multiple fluorescein leakages and pigment epithelial detachments are seen in eye of a 60-year-old male (A). Horizontal (B) and vertical (C) OCT scans through the fovea show a highly elevated neurosensory retina and small pigment epithelial detachments.

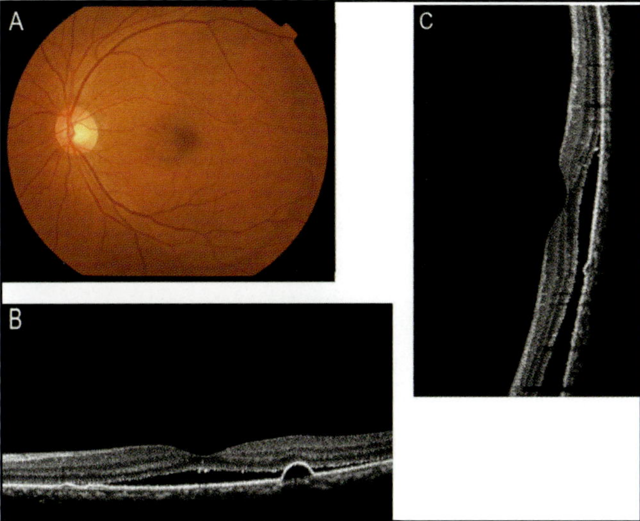

Figure 17-3. One month after the photocoagulation of the same eye as shown in Figure 17-2. Fundus photo (A) shows small white deposits around the fovea. Horizontal OCT (B) shows the decreased subretinal fluid and small granules at the posterior surface of detached retina. Vertical OCT (C) shows the inferior shift of the subretinal fluid.

Figure 17-4. The horizontal OCT images at 3 months (A) and 6 months (B) after the treatment of the same eye in Figures 17-2 and 17-3. The thickness of the photoreceptor outer segment increases (A) and after the reattachment of the retina, the IS/OS line becomes visible (B).

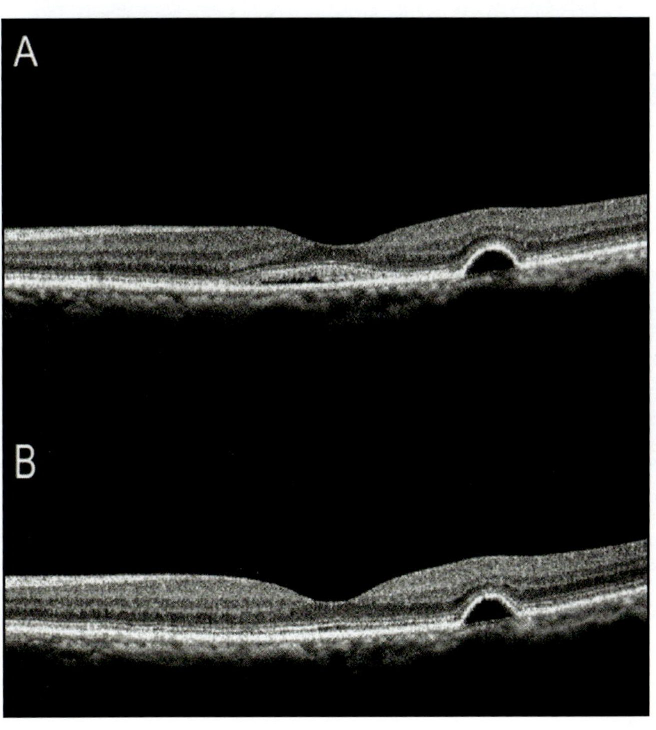

Figure 17-5. A 45-year-old male showing multiple fluorescein leakages (A). The labels B, C, and D in this image refer to Figures 17-5B, 17-5C, and 17-5D. Consecutive 3D OCT scans (B, C, D) corresponding to the lines in A show the isolated retinal detachments around the leakage points.

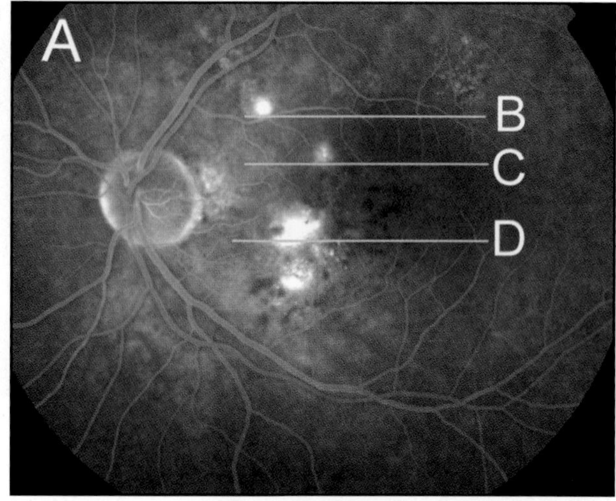

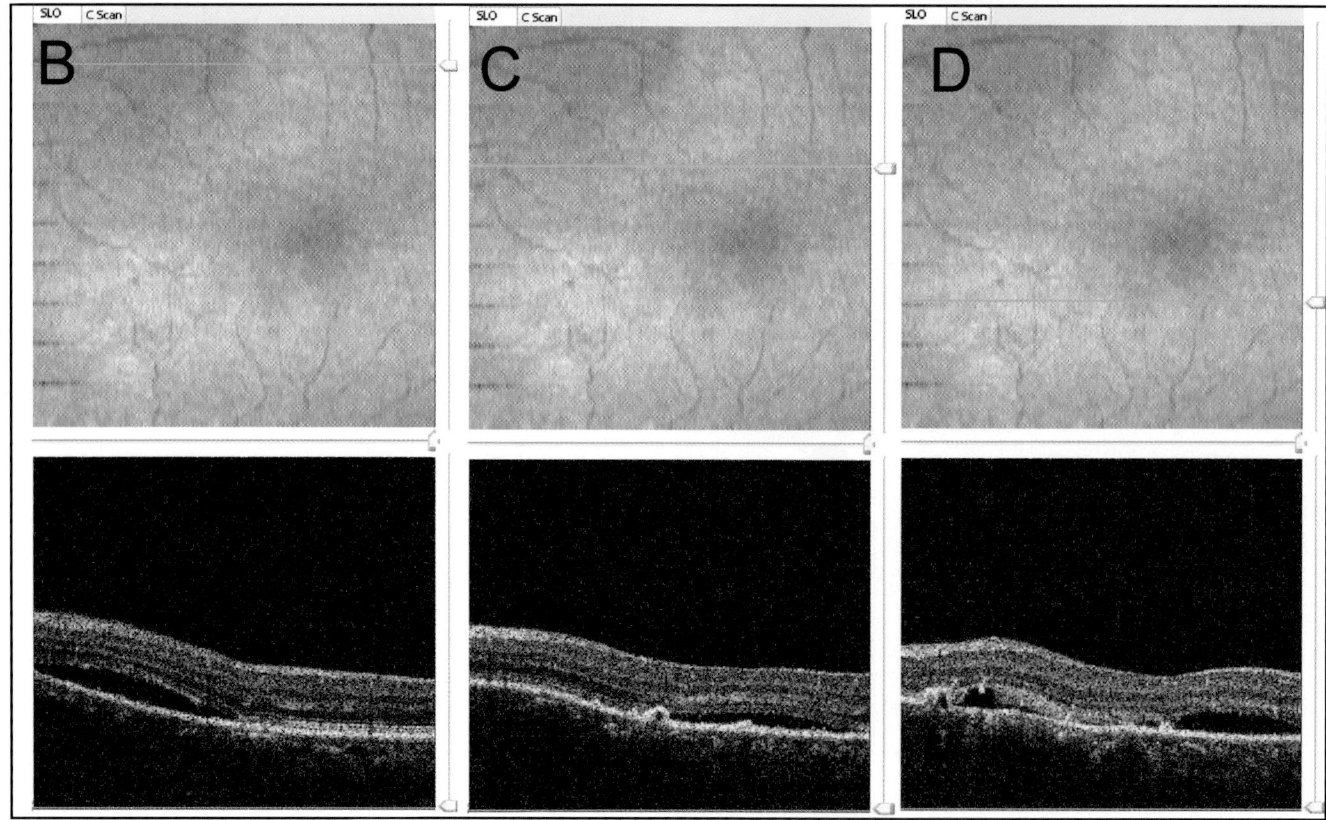

inner/outer segments (IS/OS) is invisible in the detached retina.[14-16] Just above the leakage site, sagging or dipping of the posterior layer of the neurosensory retina, which seems to arise from the swelling of the outer nuclear layer (ONL), may be observed (Figures 17-6B and C).[9,16]

The smooth posterior surface of the detached retina may turn to a granular appearance and increase the thickness of photoreceptor OS with time (see Figures 17-3B, 17-4A).[15] The granulated appearance is more prominent in eyes with long-term subretinal fluid and white punctuate deposits of the fundus.[16] Small granules may also be seen on the anterior surface of the RPE or intra-

retinally. When the neurosensory retina attaches, the IS/OS line is gradually distinct (see Figure 17-4B).[14,16] Subretinal granules may remain just after the resolution of subretinal fluid but become undetectable.

Fibrinous Exudation in the Subretinal Space

In eyes showing yellow-white fibrinous exudation around the leakage point, OCT shows a corresponding

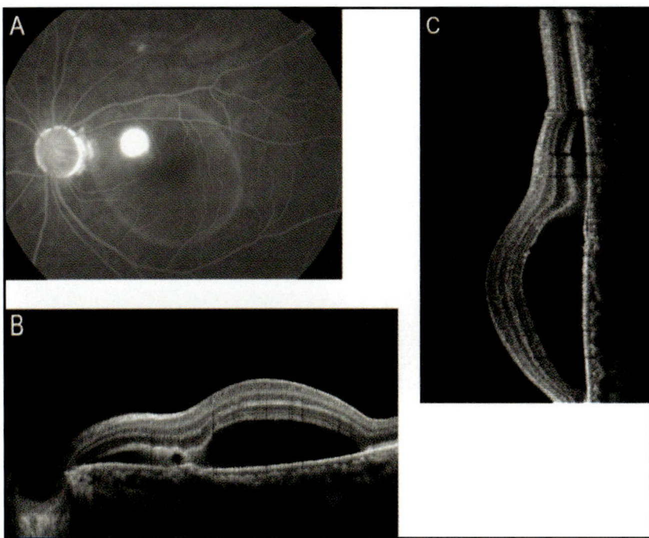

Figure 17-6. OCT images at the leakage point. Through the area of leak on fluorescein angiography (A), the horizontal OCT (B) shows the subretinal fluid and subretinal fibrinous exudation with pellucid core. The RPE shows slight elevation. The vertical OCT (C) shows the focal dipping of the detached retina due to the swelling of outer nuclear layer.

hyper-reflective shadow in the subretinal space (Figure 17-6B).[9,16] Even in eyes without apparent exudation on fundus examination, the slight haze in the subretinal space around the leakage site may be observed. Some eyes show a translucent core within the hyper-reflective exudation corresponding to the orange-red region within the fibrinous exudation on fundus.[7] Because consecutive B-scans confirmed the pellucid core led to the area just above the RPE showing leakage, it is considered to be the subretinal route of the exudative fluid from the defect of RPE leakage. When subretinal fluid reduces, fibrinous exudation decreases and is finally absorbed, but it usually remains as fibrotic scar in the subretinal space.

MORPHOLOGIC CHANGES OF RETINAL PIGMENT EPITHELIUM AT THE LEAKAGE SITE

The retinal pigment epithelial detachments are often seen in eyes with CSC. The RPE elevation map or *en face* OCT image constructed from 3D scan on the RTVue can show the location and the height of each pigment epithelial detachment clearly (Figure 17-7). At a fluorescein leakage site, pigment epithelial detachments exist frequently.[8-10] Using the RTVue, we observed pigment epithelial detachments within or at the edge of the retinal detachment at 61% of leakage sites and an irregular RPE layer including a small RPE protrusion at 35% of leakage sites in eyes with acute CSC.[16] The size or height of the

RPE elevation was unrelated to the disease intensity, the speed or extent of dye leakage, or the amount of subretinal fluid. High quality images of Fourier-domain OCT have enabled the user to depict the small disruption in the RPE layer within the pigment epithelial detachment exactly corresponding to the fluorescein leakage points (Figure 17-8). The repeated OCT examination at the leakage sites showed interesting findings that a defect in the RPE layer might resolve or remain when the subretinal fluid was decreasing. Therefore, the self-limiting but occasionally recurring nature of acute CSC might be explained based on both the RPE integrity and the hydrostatic pressure of the choroid. Some eyes show the residual pigment epithelial detachments even after the resolution of subretinal fluid.

RETINAL CHANGES IN CHRONIC CENTRAL SEROUS CHORIORETINOPATHY

CSC usually resolves spontaneously, but may recur and become chronic and result in persistent visual deterioration. The poor visual acuity has been reported to be associated with atrophy of the outer retinal layers and cystoid macular degeneration in the central macula detected by OCT.[11-14] The reduced thickness of fovea is frequently seen in eyes with longer duration of symptoms and persistent vision reduction, despite resolution of the subretinal fluid. Detailed OCT findings confirm the thinning or loss of photoreceptor layer and the reduced ONL thickness may cause the foveal attenuation (Figure 17-9).

Cystoid macular degeneration in CSC is defined as cystoid spaces shown by OCT with no intraretinal fluorescein leakage and is considered as affecting the foveal area and associated with severe reduction of visual acuity as a result of long-term disease and the persistence of subretinal fluid in the macular region. Besides central macular area, cystoid spaces in CSC may be seen in wider area throughout the posterior pole (Figure 17-10). Thus, OCT can predict the visual recovery after macular reattachment or provide an explanation for poor vision in chronic CSC.

REFERENCES

1. Gass JD. Pathogenesis of disciform detachment of the neuroepithelium. II. Idiopathic central serous choroidopathy. *Am J Ophthalmol.* 1967;63:587-615.

2. Gass JDM. Idiopathic central serous chorioretinopathy. *Stereoscopic Atlas of Macular Diseases: Diagnosis and Treatment.* 4th ed. St. Louis, MO: CV Mosby; 1997:52-58.

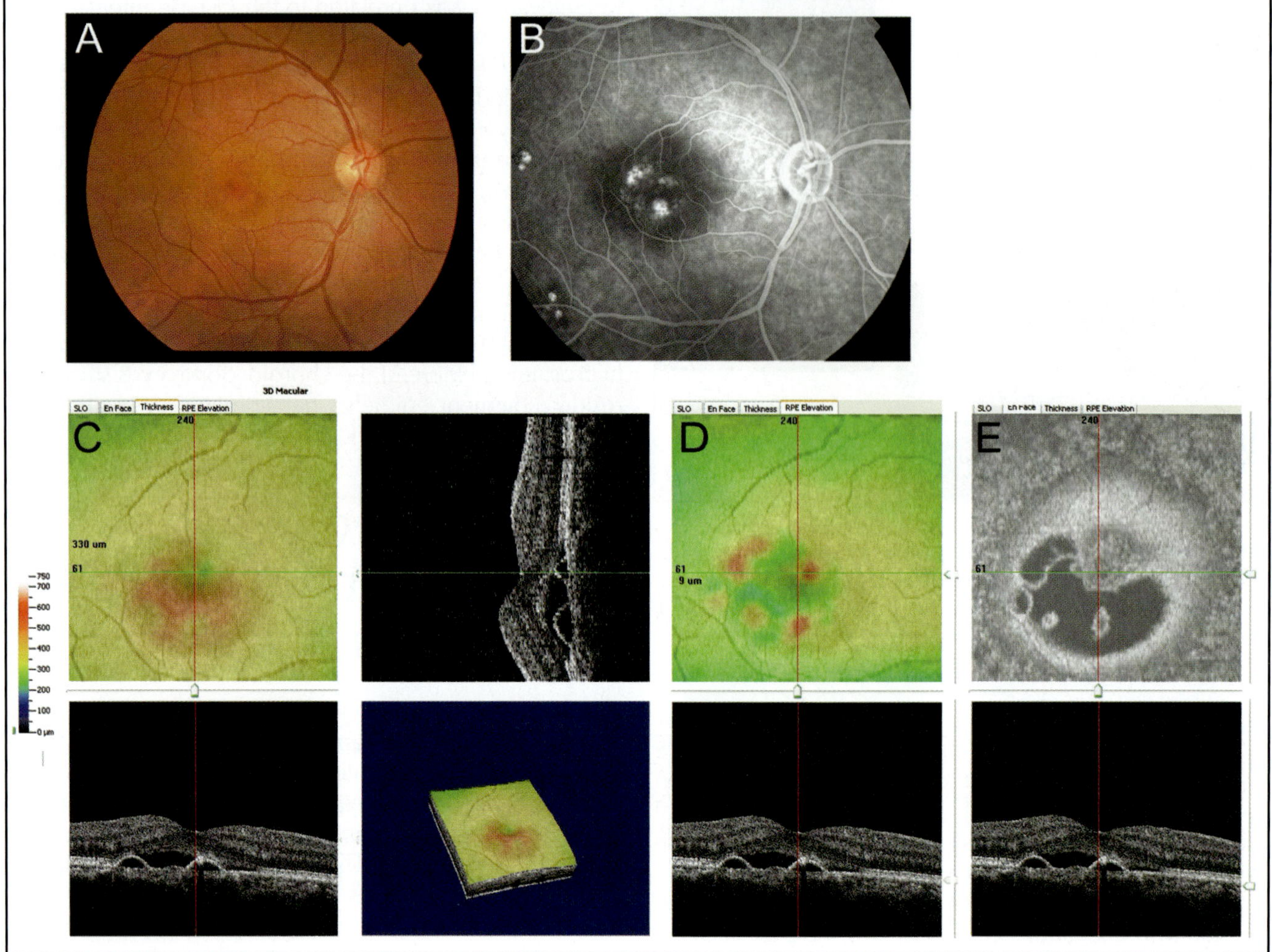

Figure 17-7. Pigment epithelial detachments associated with CSC. Fundus (A) and fluorescein angiograph (B) show subretinal fluid and multiple pigment epithelial detachments. The retinal thickness map (C) shows the elevation of the whole retina, but the RPE elevation map (D) and *en face*/C-scan (E) can image each pigment epithelial detachment.

3. Hee MR, Izatt JA, Swanson EA, et al. Optical coherence tomography of central serous chorioretinopathy. *Am J Ophthalmol.* 1995;120:65-74.

4. Iida T, Hagimura N, Kishi S, et al. Evaluation of central serous chorioretinopathy with optical coherence tomography. *Am J Ophthalmol.* 2000;129:16-20.

5. Kramppeter B, Jonas JB. Central serous chorioretinopathy imaged by optical coherence tomography. *Arch Ophthalmol.* 2003;121:742-743.

6. Montero JA, Ruiz-Moreno JM. Optical coherence tomography characterization of idiopathic central serous chorioretinopathy. *Br J Ophthalmol.* 2005;89:562-564.

7. Saito M, Iida T, Kishi S. Ring-shaped subretinal fibrinous exudate in central serous chorioretinopathy. *Jpn J Ophthalmol.* 2005;49:516-519.

8. Mitarai K, Gomi F, Tano Y. Three-dimensional optical coherence tomographic findings in central serous chorioretinopathy. *Graefes Arch Clin Exp Ophthalmol.* 2006;244:1415-1420.

9. Hussain N, Baskar A, Ram LM, Das T. Optical coherence tomographic pattern of fluorescein angiographic leakage site in acute central serous chorioretinopathy. *Clin Experiment Ophthalmol.* 2006;34:137-140.

10. Hirami Y, Tsujikawa A, Sasahara M, et al. Alterations of retinal pigment epithelium in central serous chorioretinopathy. *Clin Experiment Ophthalmol.* 2007;35:225-230.

11. Wang MSM, Sander B, Larsen M. Retinal atrophy in idiopathic central serous chorioretinopathy. *Am J Ophthalmol.* 2002;133:787-793.

12. Iida T, Yannuzzi LA, Spaide RF, et al. Cystoid macular degeneration in chronic central serous chorioretinopathy. *Retina.* 2003;23:1-7.

13. Piccolino FC, de la Longrais RR, Ravera G, et al. The foveal photoreceptor layer and visual acuity loss in central serous chorioretinopathy. *Am J Ophthalmol.* 2005;139:87-99.

14. Ojima Y, Hangai M, Sasahara M, et al. Three-dimensional imaging of the foveal photoreceptor layer in central serous chorioretinopathy using high-speed optical coherence tomography. *Ophthalmology.* 2007;114:2197-2207.

15. Matsumoto H, Kishi S, Otani T, et al. Elongation of photoreceptor outer segment in central serous chorioretinopathy. *Am J Ophthalmol.* 2008;145:162-168.

16. Fujimoto H, Gomi F, Wakabayashi T, et al. Morphologic changes in acute central serous chorioretinopathy evaluated by Fourier-domain optical coherence tomography. *Ophthalmology.* 2008;115:1494-1500.

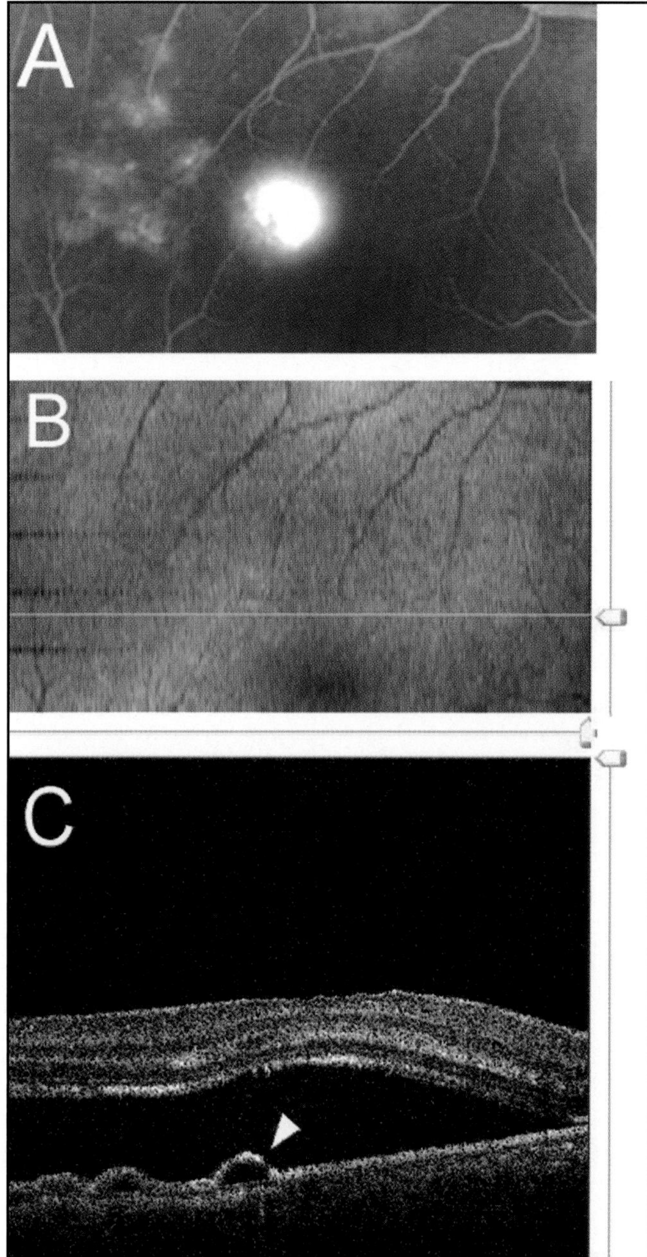

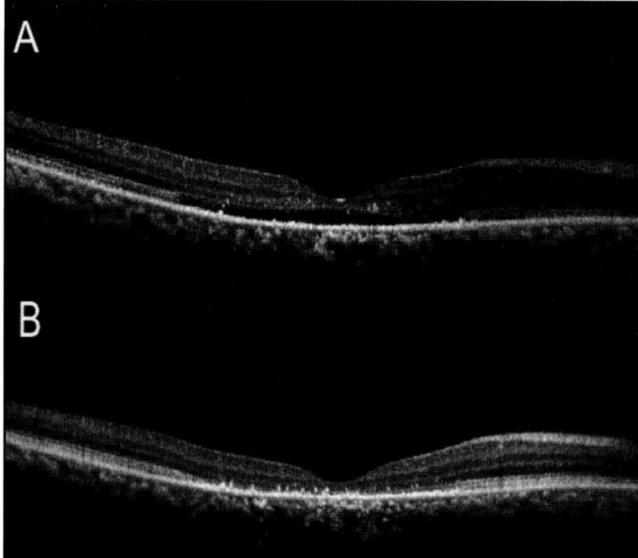

Figure 17-9. Chronic CSC in a 47-year-old male with subjective symptoms more than 5 years. OCT shows the thinning of the detached retina due to the loss of outer retinal layer (A) and after the reattachment (B), the outer retinal layer does not recover.

Figure 17-8. A rip of RPE at the leakage point. Fluorescein angiograph (A) shows one pigment epithelial detachment accompanying leakage among multiple pigment epithelial detachments. Projected fundus image from 3D OCT scans (B) is comparable to the fluorescein angiograph and the OCT scan (C) corresponding to the line in B shows a defect of RPE (arrowhead) within the pigment epithelial detachments, showing leakage.

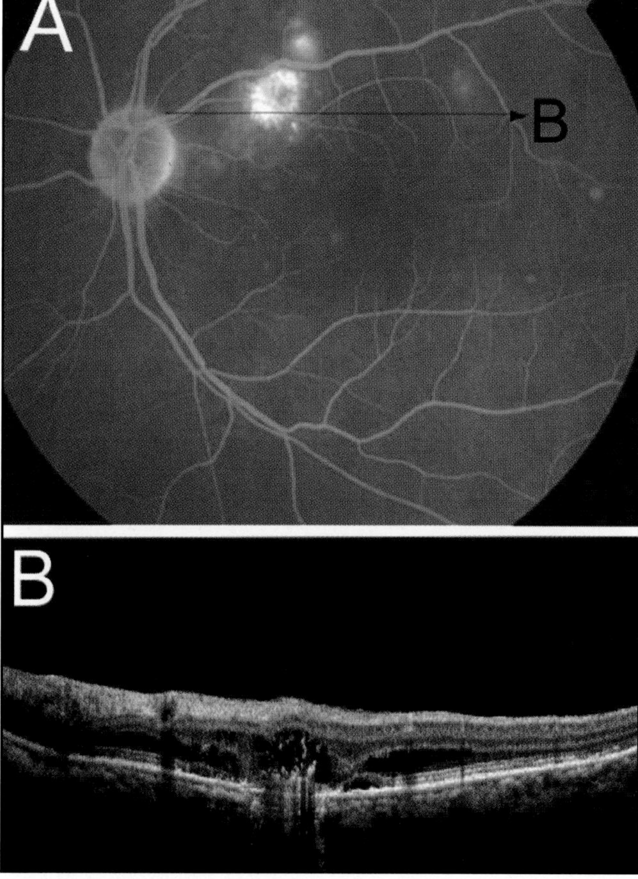

Figure 17-10. Cystoid changes of the retina around the fluorescein leakage site that was treated by photocoagulation. OCT image (B) corresponding to the arrow in fluorescein angiograph (A) shows the subretinal fluid and small cystoid spaces above the area treated by photocoagulation. Diffuse cystic edema is also seen within the ONL.

18 Macular Hole

Yutaka Imamura, MD

INTRODUCTION

OCT is a powerful imaging technology because it performs the real time, *in situ* visualization of the deep structure of sensory retina, retinal pigment epithelium (RPE), and choroid. A 3D OCT data sets can be generated by acquiring sequential cross-sectional images, similar to magnetic resonance imaging or computed tomography scans. OCT enables us to image macular diseases and has been revealing the pathogenesis of macular diseases.[1]

The intraretinal morphologic features such as the inner and outer plexiform layers, external limiting membrane, and inner/outer segments (IS/OS) of the photoreceptors can be clearly differentiated in OCT images. The recent improvements in the resolution of OCT with eye tracking and image averaging have enabled ophthalmologists to observe retinal structures far more precisely.

OCT is the main imaging technique for retinal specialists to evaluate macular holes. Using OCT, retinal specialists can show the shapes of macular hole to their patients, explain their symptoms to them, and predict the prognosis of their visual acuity. OCT can be used for the following purposes for the evaluation of macular hole:

- To make a clinically useful classification of macular hole
- To determine if there is a concurrent epiretinal membrane (ERM)
- To examine the disruption of the junction between the IS and OS of the photoreceptors (this information is useful to predict postoperative visual acuity)
- To evaluate the postoperative morphology of fovea

- To differentiate the diseases similar to macular hole (eg, pseudoholes, inner lamellar holes, cystoid macular edema, geographic atrophy, and ERM)

Clinical Symptoms of Macular Hole

Macular hole is a condition affecting mainly women in the fifth to seventh decade of life. The disease is bilateral in approximately 10% of patients.[2] They often discover blurred vision and metamorphopsia when they cover the fellow normal eyes.[3] Amsler grid is useful to detect for the detection of distorsion, metamorphopsia, and scotoma. Most patients report that both eyes were asymptomatic during their last examination 1 or 2 years previously.

PATHOGENESIS AND OPTICAL COHERENCE TOMOGRAPHY-BASED STAGING OF MACULAR HOLE

Progressive liquefaction of the posterior vitreous occurs with aging, giving rise to a large, optically empty cavity of liquefied vitreous in the premacular space, known as posterior premacular vitreous pocket.[4] Changes in the vitreous may cause traction on the retinal surface and the distortion of macula. The traction by the vitreous leads to the formation of a pseudocyst in the fovea.[5] However, trauma can lead to the formation of macular hole without the formation of a pseudocyst.

Originally, Donald Gass made a classification scheme for staging macular hole based on biomicroscopic

Huang D, Duker JS, Fujimoto JG, Lumbroso B, Schuman JS, Weinreb RN.
Imaging the Eye from Front to Back with RTVue Fourier-Domain Optical Coherence Tomography (pp 169-174).
© 2010 SLACK Incorporated.

observations.[6,7] His theory that tangential vitreous traction causes various manifestations of macular hole has been well accepted. In brief, Stage 1 includes localized shrinkage of prefoveal cortical vitreous causing tractional shallow detachment of the foveola, creating an intraretinal yellow spot (Stage 1A) and small yellow ring (Stage 1B). A Stage 2 hole is typically diagnosed by evidence of a central or eccentric retinal defect. Stage 2 holes generally progress to Stage 3 holes with partial vitreomacular separation around the macula, and then to Stage 4 with complete separation of the vitreous from the entire macular surface and optic disc.[3]

However, one concern for the classification of Gass is that the posterior hyaloid itself is typically invisible on biomicroscopy. Gass thought the process of posterior vitreous detachment does not contribute much to the formation of a macular hole.[3] However, OCT studies showed that the process of forming a posterior vitreous detachment (PVD) was very important in the formation of a macular hole. OCT has enabled clinicians to see the vitreofoveal adhesion clearly and to make a useful classification system of macula hole. A staging system is useful not only to understand the pathophysiology of macular hole, but to obtain prognostic information.

OCT-based staging of macular hole, which was proposed by Spaide et al in 2006, appears simple and convenient for retinal specialists.[8]

- *Stage 0:* Normal foveal contour, normal retinal thickness, and an oblique thin line attached to at least one side of fovea. This was found in 27 of 94 fellow eyes of patients with a unilateral Stage 2, 3, or 4. They are 6 times more likely to develop a macular hole than the eyes without Stage 0 macular hole.[5]
- *Stage 1:* Formation of a foveal pseudocyst. Foveolar or foveal detachment with an underlying yellow spot or ring deposition. OCT shows vitreofoveal traction to pseudocysts. Stage 1 holes frequently resolve spontaneously. Some researchers recommended vitrectomy for the symptomatic eyes with Stage 1 macular hole that fail to spontaneously resolve with observation.[9]
- *Stage 2:* Early lateral separation of the photoreceptor layer. This may occur with or without dehiscence of the internal limiting membrane and may have an ophthalmoscopic appearance of a yellow ring.
- *Stage 3:* A full-thickness macular hole less than 400 µm.
- *Stage 4:* A full-thickness macular greater than 400 µm.[10]

This classification system is unique because it is based on the findings of OCT and on the theory that the vision-defining events are attributable to the separation of the photoreceptor layer, and not the dehiscence of the ILM.

Preoperative Evaluation of Macular Hole Using Optical Coherence Tomography

Vitrectomy and gas tamponade is the first choice of the treatment for Stage 2, 3, and 4 macular hole. An anatomic closure is achieved in most cases, except for long-standing chronic cases. Removal of the ILM promotes the rate of anatomic closure and probably decreases the reopening rate after surgery.[11] Preoperative visual acuity correlates with postoperative visual acuity; however, OCT gives us important information that may be useful to predict macular hole surgery outcome.[8] The information available from OCT, which may be useful for preoperative evaluation, are 1) size of hole and 2) presence of ERM. Representative images of macular hole are shown in Figures 18-1 through 18-7.

It is now generally accepted that the preoperative size of the macular hole is inversely correlated with the anatomic success rate.[10,12] Some eyes with macular holes show the intense signal reflectance of the vitreous surface, probably representing glial proliferation along the surface of posterior vitreous, although seeing nothing in the vitreous space does not mean there is not a sheet of adherent vitreous.[8] ERMs are present in 5 of 20 eyes with Stage 2 macular hole and 53 of 68 eyes with Stage 3 and 4 hole.[13] The recurrence of ERM can reopen the macular hole after vitrectomy, suggesting that removal of internal limiting membrane as well as ERM would be recommended to prevent late reopening. The identification of ERM using OCT may enable surgeons to know preoperatively the extent of the membrane, however during surgery, ERM is often larger than what it appears on OCT. *En face* image of RTVue can clearly detect the area of ERM (see Figure 18-7).

Postoperative Evaluation of Macular Hole Using Optical Coherence Tomography

OCT is expected to be used to predict the prognosis of visual acuity in eyes with macular holes undergoing vitrectomy. Using OCT, the hole is closed in 25 out

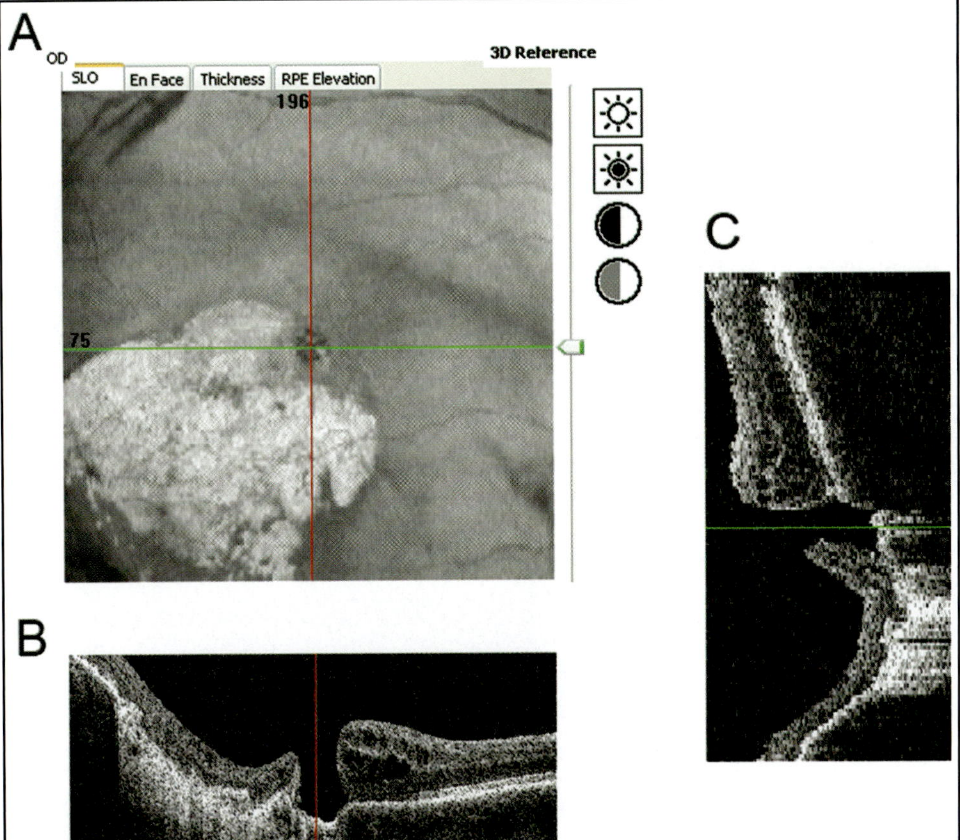

Figure 18-1. The *en face* image of the right eye with a macular hole in a 62-year-old female (Case 1), scanned by RTVue. (A) Image from scanning laser ophthalmoscope. Note the large chorioretinal atrophy at the inferior-temporal area. (B) OCT scan corresponding to the red line in A. (C) OCT scan corresponding to the green line in A.

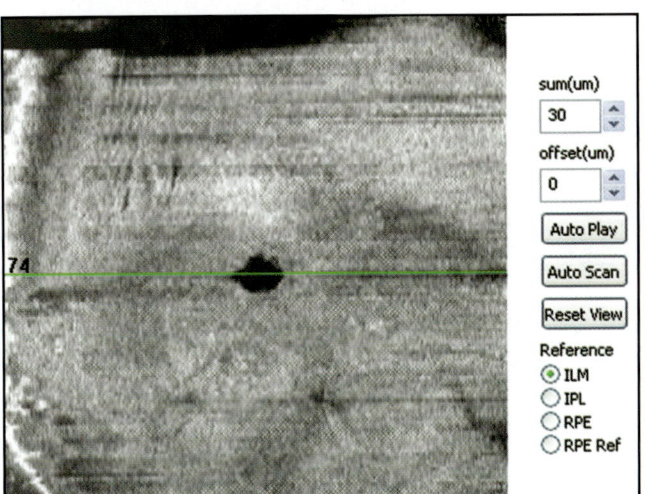

Figure 18-2. The *en face* image of the right eye of Case 1. An image at the level of ILM scanned by RTVue.

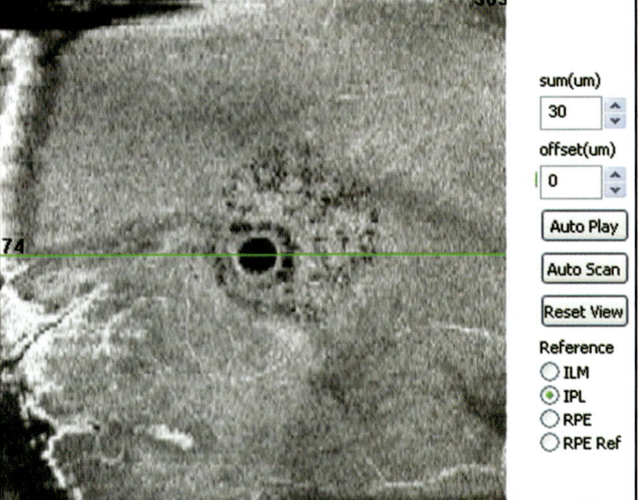

Figure 18-3. An image at the level of the inner plexiform layer scanned by RTVue (Case 1). Cystic changes surrounding the hole are visible.

Figure 18-4. Case 1. An en face image (left) at the level as selected by the red area in the OCT scan (right). Intraretinal vascular structures are visible.

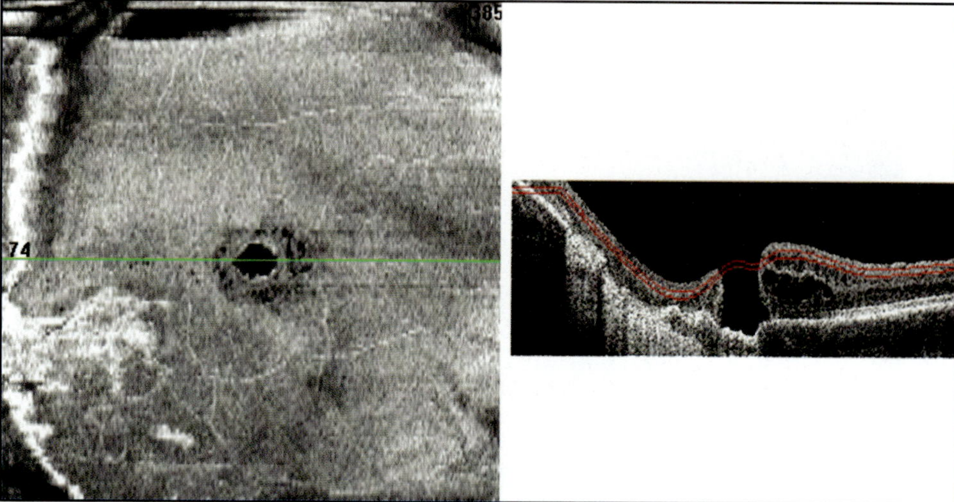

Figure 18-5. Cystic changes are well visible (left), when the appropriate area is selected by OCT scan (right).

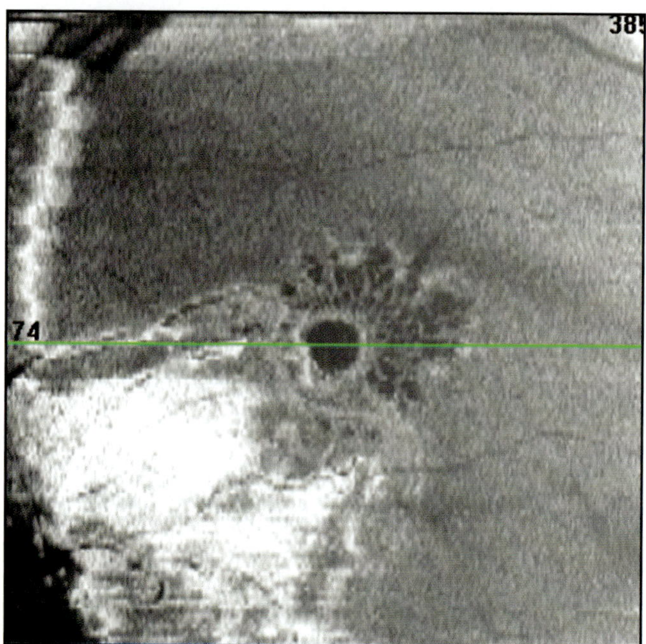

Figure 18-6. Retinal thickness mapping of Case 1 (left). Thickness is measured in the area between red and blue lines in the right panel.

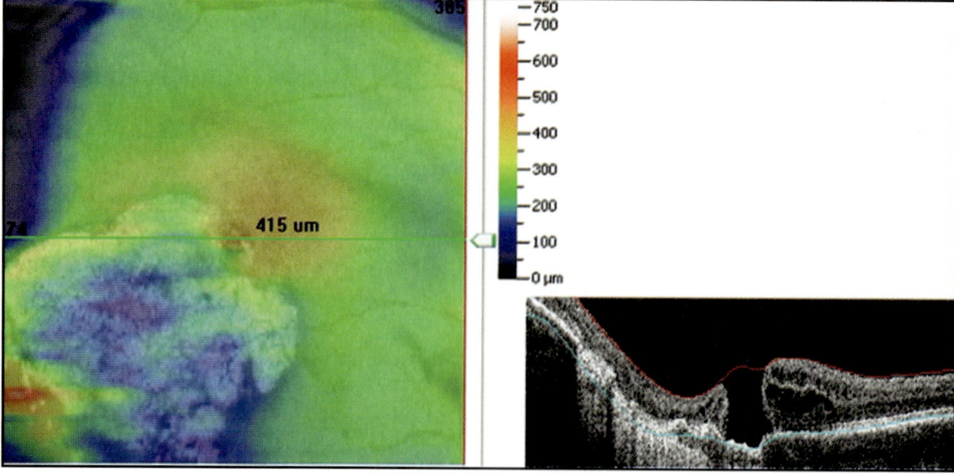

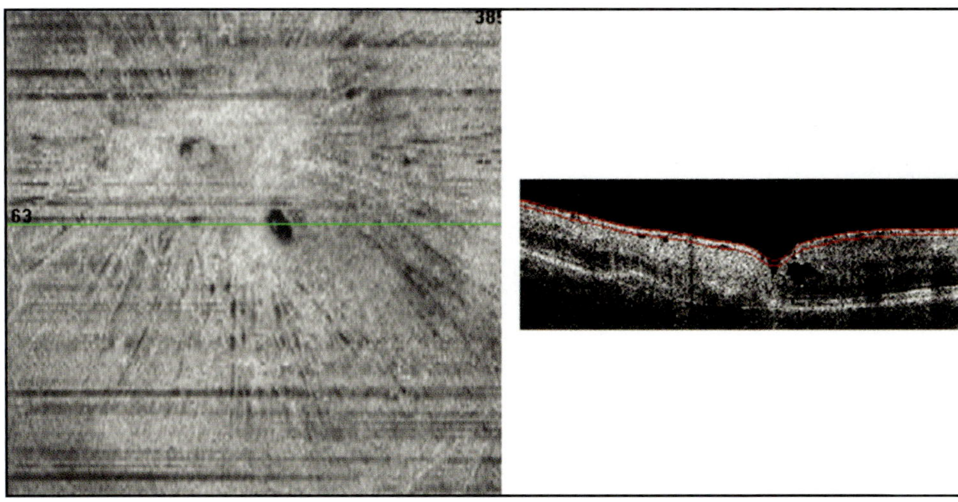

Figure 18-7. The *en face* image of the right eye with a pseudo-hole in a 50-year-old female (left). Shrinkage of epiretinal membrane is clearly visible. The left image is created from the area surrounded by the red lines in the right panel.

of 33 eyes 48 hours postoperatively.[14] The presence of the IS/OS junction indicates a normal alignment of the photoreceptors, and this alignment is essential for normal visual function.[15] It will be interesting to know if there are correlations between the visual function and the area of disturbed IS/OS junctions. There are several reports studying the correlation between disruption of the IS/OS junction and postoperative visual acuity. Baba et al reported that the visual acuity was significantly better in eyes with normal IS/OS junction than in those not detected at 3 and 6 months postoperatively.[16] Inoue et al studied 57 eyes of macular hole undergoing vitrectomy, and found that the mean total area and maximum length of the IS/OS junction defect at 12 months after surgery were significantly negatively correlated with the postoperative visual acuity.[17]

A recent paper by Chang et al demonstrated that the photoreceptor damage induced by macular hole formation was at least partially reversible.[18] They reported 6 of 7 eyes with macular hole closure had a decrease in the size of the IS/OS defect after the hole was closed.

OCT can classify the healing pattern of postoperative macular hole. The closed macular holes were categorized into 2 patterns based on OCT: Type 1 closure (closed without foveal neurosensory retinal defect) and Type 2 closure (closed with foveal neurosensory retinal defect). Interestingly, the visual acuity in eyes with Type 1 closure is higher than that of Type 2.[19] Another study classified the closed hole in U- (normal foveal contour), V- (steep foveal countour), and W-type (foveal defect of neurosensory retina) and was well correlated with postoperative visual acuity.[20]

Among the 448 closed holes after vitrectomy, 49 eyes showed a reopened hole, and factors associated with hole reopening were macular pucker with cystoid macular edema (CME) in 17 eyes (35%), retinal detachment in 6 eyes (12%), and CME following cataract surgery in 2 eyes (4%).[21] Evaluation of the postoperative morphology of the closed hole and the presence of ERM and CME using OCT will be useful to predict the risk of reopening.

Further studies are needed to investigate the time course for IS/OS junction recovery, and preoperative factors influencing the final recovery of IS/OS junction.

CONCLUSION

The application of OCT in the evaluation of macular hole has been revolutionizing our understanding of macular hole. With OCT, we can now see the adhesion of vitreous and fovea clearly, measure the height and length of a macular hole, and measure the area with disrupted IS/OS junction. This information is valuable in predicting the prognosis of visual acuity following the vitreous surgery, since the anatomical closure of the macular hole does not always correlate with a good functional result. Further improvement of resolution of OCT is expected to reveal the real dynamics of photoreceptor cells and Müller cells during the process of macular hole formation and its closure. RTVue, which can visualize both transverse and *en face* images at the desired depth at the same time, is expected to solve the fundamental issues regarding the pathophysiology of macular hole.

REFERENCES

1. Drexler W, Fujimoto JG. *Optical Coherence Tomography Technology and Applications*. Berlin: Springer-Verlag; 2008.
2. Ansari H, Rodriguez-Coleman H, Langton K, Chang S. Spontaneous resolution of bilateral stage 1 macular holes documented by optical coherence tomography. *Am J Ophthalmol*. 2002;134:447-449.
3. Gass JDM. *Stereoscopic Atlas of Macular Diseases*. 4th ed. St. Louis, MO: Mosby; 1997.

4. Kishi S, Shimizu K. Posterior precortical vitreous pocket. *Arch Ophthalmol.* 1990;108:979-982.

5. Chan A, Duker JS, Schuman JS, Fujimoto JG. Stage 0 macular holes: observations by optical coherence tomography. *Ophthalmology.* 2004;111:2027-2032.

6. Gass JDM. Idiopathic senile macular hole: its early stages and pathogenesis. *Arch Ophthalmol.* 1988;106:629-639.

7. Gass JD. Reappraisal of biomicroscopic classification of stages of development of a macular hole. *Am J Ophthalmol.* 1995;119:752-759.

8. Spaide RF, Iranmanesh R. Macular holes. In: Huang D, Kaiser PK, Lowder CY, Traboulsi EL, eds. *Retinal Imaging.* Philadelphia, PA: Elsevier; 2006.

9. Subramanian ML, Truong SN, Rogers AH, et al. Vitrectomy for stage 1 macular holes identified by optical coherence tomography. *Ophthalmic Surg Lasers Imaging.* 2006;37:42-46.

10. Ip MS, Baker BJ, Duker JS, et al. Anatomical outcomes of surgery for idiopathic macular hole as determined by optical coherence tomography. *Arch Ophthalmol.* 2002;120:29-35.

11. Yoshida M, Kishi S. Pathogenesis of macular hole recurrence and its prevention by internal limiting membrane peeling. *Retina.* 2007;27:169-173.

12. Ruiz-Moreno JM, Staicu C, Piñero DP, et al. Optical coherence tomography predictive factors for macular hole surgery outcome. *Br J Ophthalmol.* 2008;92:640-644.

13. Cheng L, Azen SP, El-Bradey MH, et al; Vitrectomy for Macular Hole Study Group. Effects of preoperative and postoperative epiretinal membranes on macular hole closure and visual restoration. *Ophthalmology.* 2002;109:1514-1520.

14. Eckardt C, Eckert T, Eckardt U, Porkert U, Gesser C. Macular hole surgery with air tamponade and optical coherence tomography-based duration of face-down positioning. *Retina.* 2008;28:1087-1096.

15. Ko TH, Fujimoto JG, Duker JS, et al. Comparison of ultra-high- and standard-resolution optical coherence tomography for imaging macular hole pathology and repair. *Ophthalmology.* 2004;111:2033-2043.

16. Baba T, Yamamoto S, Arai M, et al. Correlation of visual recovery and presence of photoreceptor inner/outer segment junction in optical coherence images after successful macular hole repair. *Retina.* 2008;28:453-458.

17. Inoue M, Watanabe Y, Arakawa A, et al. Spectral-domain optical coherence tomography images of inner/outer segment junctions and macular hole surgery outcomes. *Graefes Arch Clin Exp Ophthalmol.* 2009;247:325-330.

18. Chang LK, Koizumi H, Spaide RF. Disruption of the photoreceptor inner segment-outer segment junction in eyes with macular holes. *Retina.* 2008;28:969-975.

19. Kang SW, Ahn K, Ham DI. Type of macular hole closure and their clinical implications. *Br J Ophthalmol.* 2003;87:1015-1019.

20. Imai M, Iijima H, Gotoh T, Tsukahara S. Optical coherence tomography of successfully repaired idiopathic macular holes. *Am J Ophthalmol.* 1999;128:621-627.

21. Sarrafizadeh R, Noffke AS, Williams GA, et al. Factors associated with maculr hole re-opening after successful closure. *Invest Ophthalmol Vis Sci.* 2002;43:E-abstract 2492.

19

Epiretinal Membranes

Giovanni Staurenghi, MD; Alessandro Invernizzi, MD; and Marco Pellegrini, MD

The introduction of OCT technique has revolutionized the assessment of the vitreoretinal interface. The opportunity to analyze, in a noninvasive way, the group of structures that compose this anatomical district, has lead to new insights on the pathogenesis, classification, development, and follow-up of many vitreoretinal diseases.

Until few years ago in fact, most of this pathological conditions were grouped under generic definitions (ie, vitreoretinal interface syndrome) depending on the clinician's judgment.

Nowadays on the contrary, especially with the introduction of the spectral-domain OCT, we are able to visualize and to describe through an objective analysis, any minimal change of each single vitreoretinal structure giving it an appropriate definition and staging.

In this chapter we are focusing our attention on the group of diseases generically defined epiretinal membranes (ERMs).

These structures usually derive from a proliferation upon the retinal surface of different cellular elements, mainly glia and retinal pigment epithelium (RPE) cells, surrounded by a neoformed extracellular matrix accompanied by a posterior vitreous detachment ([PVD] partial or complete).

Except for the idiopathic forms, ERMs usually result from a previous pathological condition affecting the vitreous or inner retina.

The most frequent causes are: vitreomacular traction (Figure 19-1), persistent posterior inflammation (ie, uveitis), trauma, medical/surgical treatments, high myopia, diabetic retinopathy, retinal detachment, proliferative vitreoretinopathy due to a giant retinal tear or proliferative diabetic retinopathy (Figure 19-2), and scars

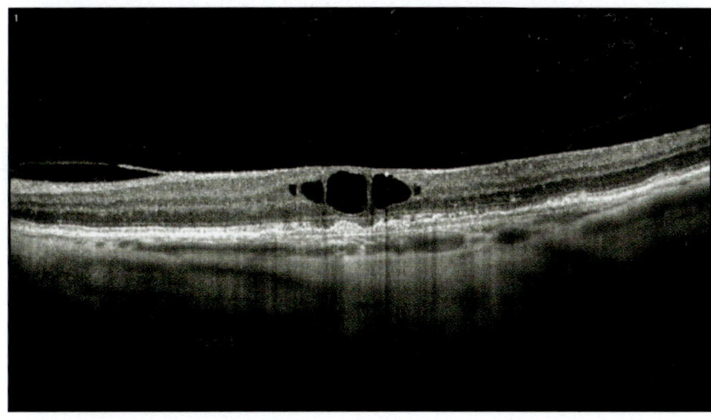

Figure 19-1. An ERM is visible along the whole retinal profile and an incomplete PVD is clearly detectable. On the right side of the image the ERM is easily differentiated from the other structures while, on the left side, the membrane merges with the posterior hyaloid which looks thickened. The foveal region appears altered due to the presence of big cystic spaces, probably generated for the most part by the vitreoretinal traction.

(eg, fibrotic tissue resulting from advanced age-related macular degeneration [Figure 19-3]).

Age is directly correlated to a higher incidence of ERM and should be considered as an important risk factor.

Depending on the symptom severity and ERM biomicroscopic appearance, 2 different conditions are usually distinguished: the macular cellophane and the pucker.

The first one is a mild expression of the disease which is characterized by nonsignificant or absent symptoms, usually identified by chance during a common ocular visit. At the fundus ophthalmoscopic examinations, cellophane appears as a nonhomogeneous enhanced reflex on the

Huang D, Duker JS, Fujimoto JG, Lumbroso B, Schuman JS, Weinreb RN.
Imaging the Eye from Front to Back with RTVue Fourier-Domain Optical Coherence Tomography (pp 175-178).

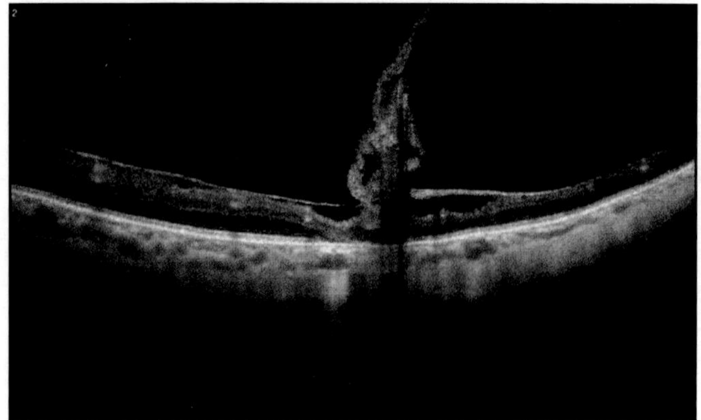

Figure 19-2. A case of proliferative vitreoretinopathy in proliferative diabetic retinopathy. The ERM, detectable above the retinal profile, appears thicker and nonhomogeneous at the margins of the proliferation (clearly visible at the center of the image, merged with a thinned retinal tissue).

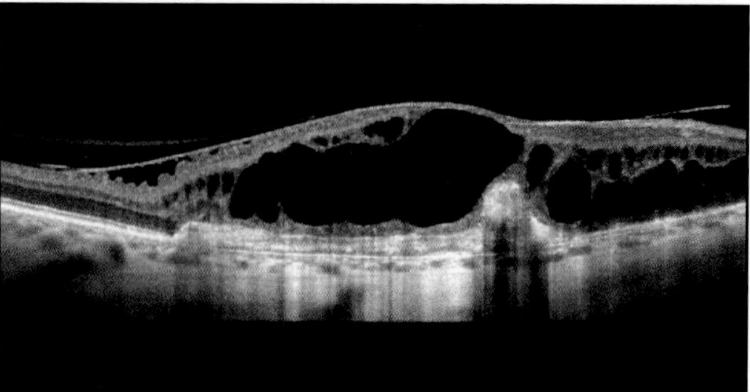

Figure 19-3. A case of advanced wet age-related macular degeneration. The retinal anatomy appears subverted by the presence of various giant cystic spaces involving the whole retinal thickness. The ERM can be easily distinguished on the left side of the image, where it appears separated from the wrinkled retinal surface. Upon the ERM, the vitreous appears partially detached.

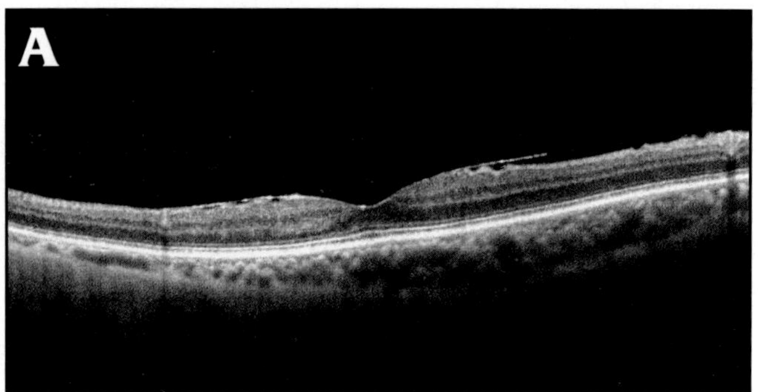
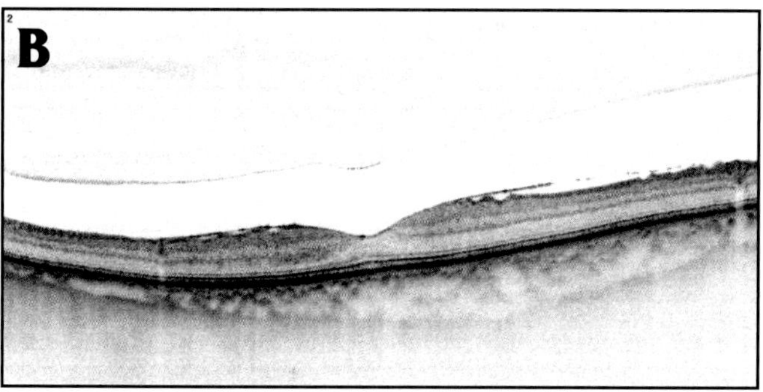

Figure 19-4. As case of ERM with conserved foveal profile. The membrane was incompletely adherent to the retinal surface, with multiple traction points. (A) In the normal gray scale image, the detached posterior hyaloid is detectable on the left part of the scan, but is not clear on the right side. (B) The inverted grayscale image improved the visualization of the posterior hyaloid (both the hyaloid face and internal hyaloid scattering) and the diagnosis of a complete PVD.

posterior pole that is usually idiopathic and frequently bilateral.

The macular pucker, on the contrary, is a more severe condition constantly defined by a decreased visual acuity and quality. The membrane at this stage is thicker, tenaciously adherent to the retina, and frequently contracted; this causes wrinkles on the retinal surface, distortion of the retinal vessels course, and sometimes the development of a cystoid macular edema.

Usually the retinal thickening caused by ERM contraction brings a decrease of visual acuity. However the central retinal thickness cannot be directly related to the visual loss. In fact, there are many other elements characterizing this pathological condition that can alter image perception and reduce the quality of vision (eg, a chronic macular edema caused by the ERM can damage photoreceptors or other retinal layers). Moreover, ERM contraction can cause the distortion of the physiological retinal surface producing metamorphopsia.

On the contrary, a patient affected by a lamellar macular hole, associated with an ERM and a doubled central retinal thickness, could maintain an optimal visual acuity if the photoreceptor complex is conserved.

INTERPRETATION OF OPTICAL COHERENCE TOMOGRAPHY

On an OCT cross-section, the ERM appears as a hyper-reflective line upon the retinal surface. Depending on the degree of adherence of this structure to the vitreous and the beneath inner retinal layers, it can be difficult to distinguish it from the surrounding elements.

In a patient with no OCT evidence of PVD it can be difficult, or even impossible, to differentiate an adherent ERM from a thickened posterior hyaloid.

Otherwise, when the PVD has occurred and hyaloid is clearly visible, an early ERM on a still normal retinal profile has to be distinguished from the nerve fiber layer, which is also highly reflective. In these ambiguous situations, it can be useful to shift the image to gray scale mode, or even invert image gray scale, in order to increase the scan details (Figure 19-4).

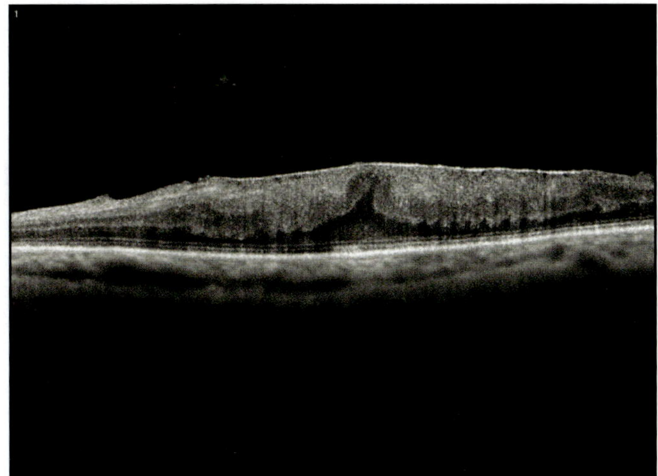

Figure 19-5. A typical case of ERM globally adherent to the retinal profile, with loss of foveal depression. The physiological layers, architecture, and stratification appear almost conserved in spite of the impressive retinal thickening. Rare microcysts are detectable, with no evidence of tissue loss.

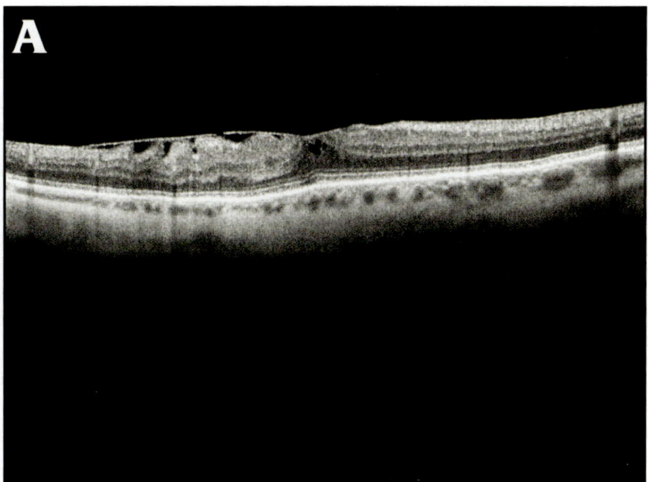

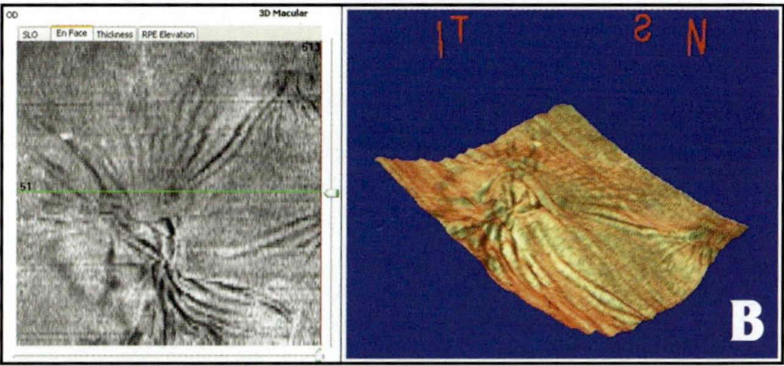

Figure 19-6. (A) Section across the fovea. An ERM is clearly visible along the whole retinal profile. On the right side of the image, the ERM appears tightly adherent to the tissue beneath. On the contrary, on the left side, there are a lot of discontinuities along the interface between the membrane and the retina. The retinal profile shows a series of humps whose heights correspond to the adherence points; all the beneath layers, except for the RPE, undergo a parallel wavy distortion following these alterations. In the foveal region, which appears almost rectilinear, little cystic formations can be appreciated. (B) *En face* view showing the distribution of inner retinal folds in the macula.

ERM can even present a discontinuous adherence to the retina showing areas of cleavage alternating with contact zones; this pattern is typically related to an incomplete PVD.

With the evolution of the membrane from the stage of cellophane to the pucker one, the thickness of the ERM increases, and this is accompanied by an enhancement of its reflectivity in OCT image. The consequences of this process are a contraction of the membrane, followed by a stretching of the retinal layers (especially the inner ones) along the traction force lines (clearly detectable with OCT analysis).

When the ERM is completely adherent to the retinal profile, its traction is equally spread on the beneath tissue and this results in a global retinal thickening (Figure 19-5) accompanied by diffuse wrinkling of its surface; otherwise, in case of incomplete adhesion, this effect appears localized, or more evident, close to the anchor points (Figure 19-6).

In long-standing puckers, the contraction of the ERM can completely subvert the retinal anatomy. This results in a loss of the physiological layers stratification, accompanied by a very important decrease of the visual acuity (Figure 19-7).

Persisting retinal thickening brings to the development of little, dark, almost round formations detectable with the OCT scans, which correspond to intraretinal cystic spaces; these are usually at first small and localized in the inner layers. However, with the progression of the disease, they can increase in dimension and extend through the whole retinal layers, or merge themselves in the foveal region (Figure 19-8) sometimes resulting in the formation of a macular hole.

In the advanced stages of pucker, ERM traction can rarely cause a detachment of the neuroretina that can be observed in OCT images as a dark band between the RPE and the outer retinal layers.

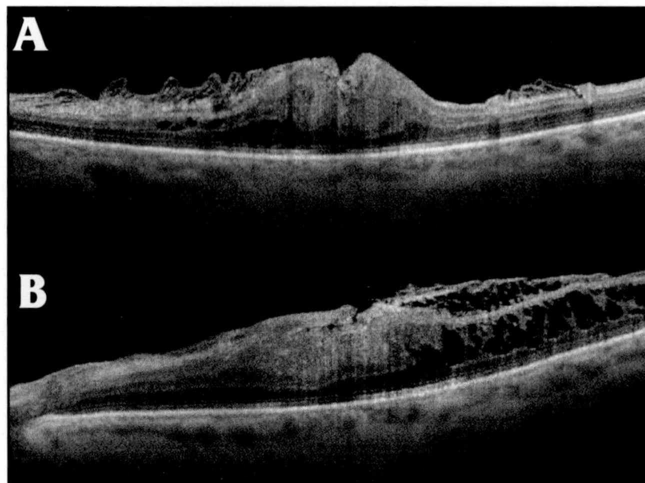
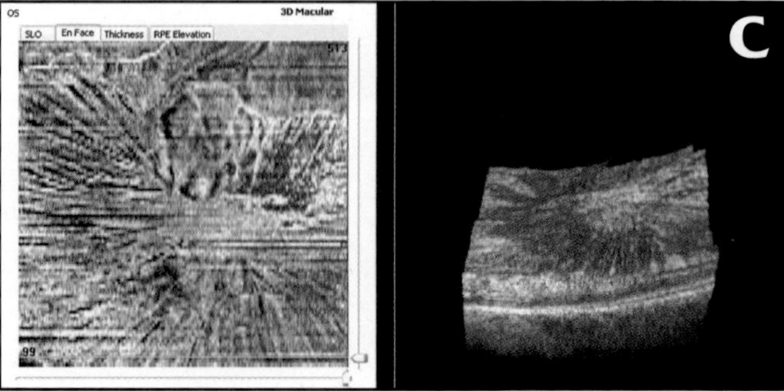

Figure 19-7. This is a case of a patient affected by a very long-standing pucker. (A) Cross line/horizontal scan. The retinal architecture was disrupted. In the thickened foveal region, retinal layers appear fused and indistinguishable. On the left side of the fovea, intraretinal cystic spaces can be detected (mainly in the external layers). A superficial retinal fraying caused by the ERM contraction is visible along the profile (especially on left side). (B) Cross line/vertical scan. Fraying of layers involves the whole retinal thickness. (C) *En face*/3D view. Wrinkled, distorted retinal surface appearance rendered through the instrument elaboration. This overview is useful in order to appreciate the spreading of the alterations to the whole posterior pole.

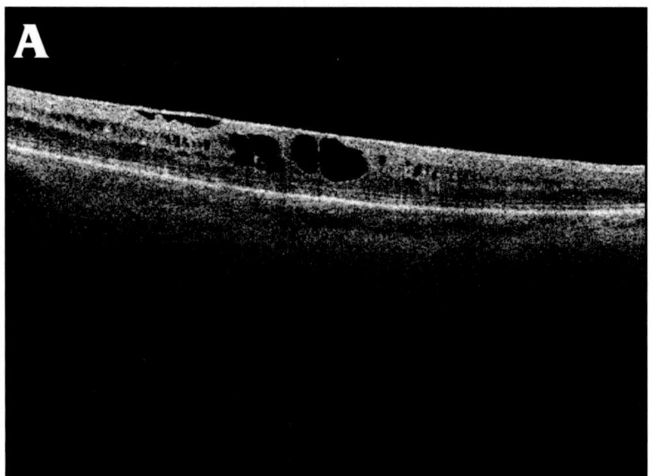
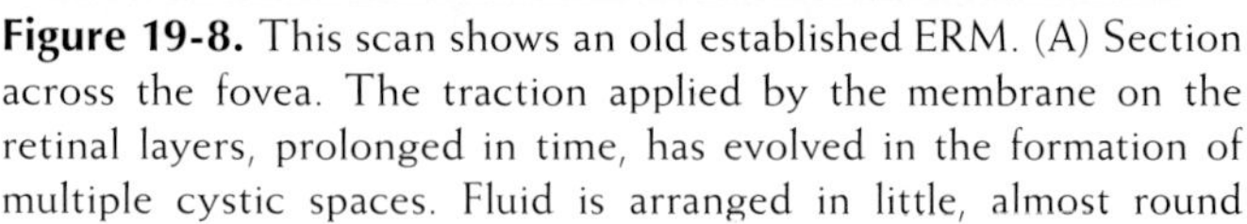
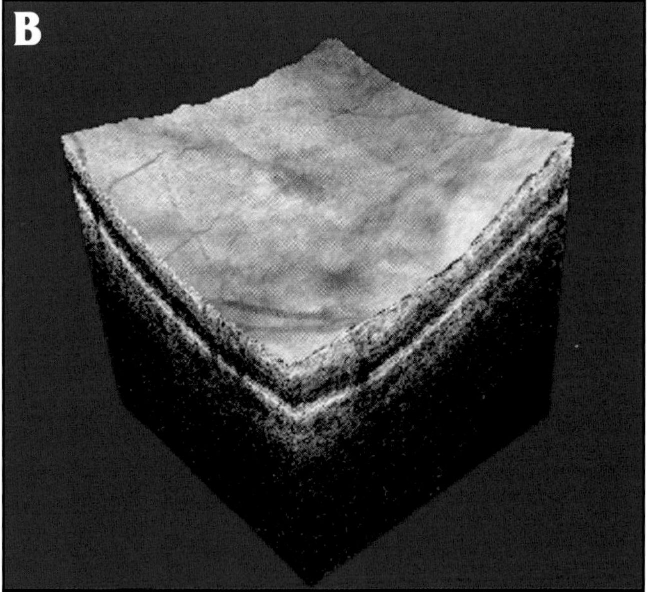

Figure 19-8. This scan shows an old established ERM. (A) Section across the fovea. The traction applied by the membrane on the retinal layers, prolonged in time, has evolved in the formation of multiple cystic spaces. Fluid is arranged in little, almost round shapes within the whole retinal layers. At the center of the image 4 big cysts, involving the foveal region, completely alter the tissue architecture. The physiological foveal depression was obliterated by the ERM. On the left side of the image a nonadherent zone between the membrane and the inner retinal layers is clearly visible. (B) A 3D view showing the loss of foveal depression and the surface topography of the ERM.

Inherited Retinal Diseases

Motokazu Tsujikawa, MD, PhD

INTRODUCTION

The neurosensory retina is thought to be a relatively simple model of a vertebrate central nervous system. It contains 6 different neurons and 1 glia, and each cell is characterized by a unique morphology, position, and function. Even their positions, which also correspond to their function, are important. The visual signal starts at the outermost layer (most apical side) of neurosensory retina (photoreceptor outer segment) and transmits to the second and third neurons, which lie in a more inner part of the retina (toward basal). The position of each cell is strictly controlled, so the retina appears to have a laminar pattern.

The inherited retinal disease means the genetic disorder in retina in term, so it should appear in every 7 different cells. However, except for some diseases, the most inherited retinal diseases affect mainly photoreceptor cells. So the term of inherited retinal diseases is sometimes equal to inherited photoreceptor diseases. It is an important feature of inherited retinal disease.

Each photoreceptor cell has unique cell morphology. The cell has column shape and shows clear apico-basal polarity. The most prominent feature is the presence of the outer segment (OS), which is a derivative of cilia. The OS plays a central function in the photoreceptor cell, as a photon detector. It also requires high metabolism of visual pigments. A salamander rod OS contains greater than 10^9 rhodopsin molecules,[1] and the most distal part of the OS is eliminated by retinal pigment epithelium (RPE) cells, while new OS membranes are born at the OS base. The entire OS is renewed every 10 days.[2] So, under an unhealthy condition of photorecep-

tor, the OS is sometimes affected first. The length of the OS becomes shorter as the photoreceptor degenerates. It results in the disturbance of the retinal layer structure, especially in its outer part.

Because of the features described previously, the evaluation of the layer structure of retina and photoreceptor morphology is important. For this purpose, a cross-section scan of the retina is the best way. In fact, in an *in vitro* analysis of model animals, the thickness of the OS or outer nuclear layer (ONL) is an ordinary analysis of photoreceptor degeneration. However, before the development of OCT, we had no way to obtain a cross-sectional image of living patients in an outpatient clinic. The invention of first-generation, time-domain OCT enabled us to obtain cross-sections of living patients, however, its resolution was not enough to describe the details of the OS. On the other hand, RTVue, a spectral-domain OCT, has enough resolution for describing IS/OS (inner/outer segments) and OLM lines.

RETINITIS PIGMENTOSA

Retinitis pigmentosa (RP) is a set of inherited retinal diseases that lead to blindness. It is the most common inherited retinal disease, 1 in every 4000 people are affected. The obvious features of the disease are late onset and slow progression. Usually, patients have no prominent symptoms in their childhood, however, because of progressive photoreceptor cell degeneration, their vision is gradually lost during adolescence and adulthood.[3] Diagnosing RP is not difficult because of an obvious fundus appearance including bone spicules

Huang D, Duker JS, Fujimoto JG, Lumbroso B, Schuman JS, Weinreb RN.
Imaging the Eye from Front to Back with RTVue Fourier-Domain Optical Coherence Tomography (pp 179-184).

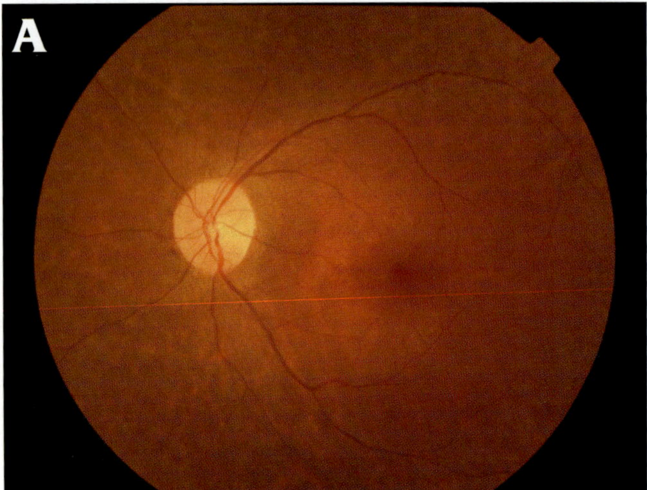

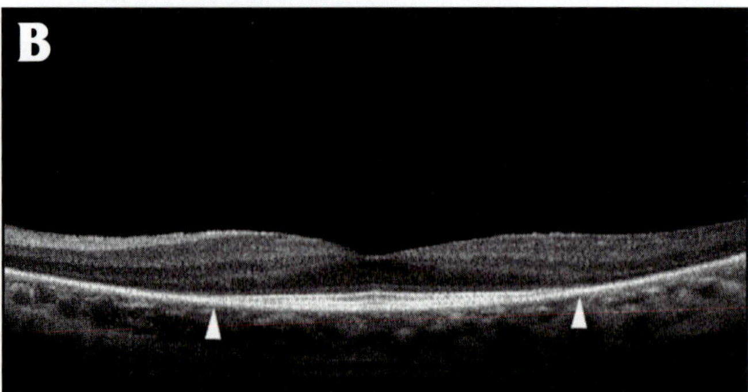

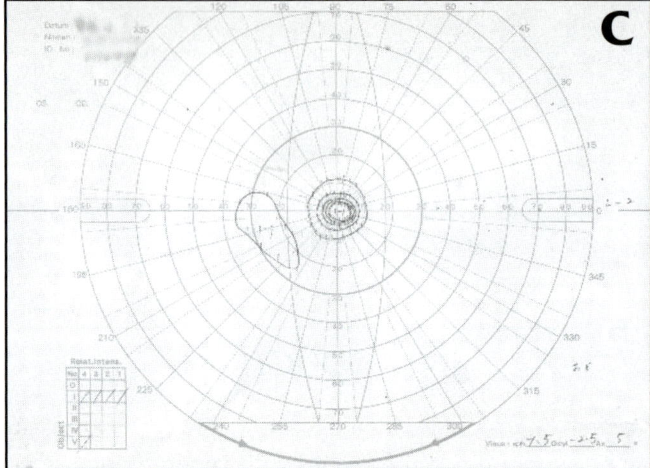

Figure 20-1. Retinitis pigmentosa, moderate case. (A) Fundus photograph. (B) RTVue image. (C) Goldmann perimetry.

and severe abnormality on electroretinogram (ERG). The basic problem of RP is making an assessment of the progress of the disease. The RP first invents rod photoreceptors, which are placed on the peripheral retina. So the patient first loses peripheral visual fields. It might be inconvenient for the patient, but does not cause severe problems because the central visual field still functions and visual acuity is still preserved. However, at the end stage of the disease, the residual function will be lost, and patients will not be able to do social activities by themselves. So, to assess the central vision is important in an RP clinic. However, to date, it is done by subjective experiments, including visual acuity and visual field. RTVue provides objective results corresponding to retinal morphology, which will be important to evaluate RP prognoses of these patients.

- Inherited trait: autosomal dominant, recessive, X-linked recessive/etc
- Disease locus (responsible gene): many[3]

Case 1

A 48-year-old male with a best-corrected visual acuity (BCVA) of 0.8.

1. Fundus photograph of RP patients. Posterior pole shows relatively normal appearance, whereas peripheral showed mottled RPE appearance (Figure 20-1A).

2. RTVue clearly showed IS/OS and OLM lines only in central retina (between arrowheads). It corresponds to relatively good visual acuity. However, the area is much smaller than normal appearance on fundus photograph. ONL was not clear in more peripheral area (Figure 20-1B).

3. Perimetry analysis revealed that the patient possessed only 10 degrees of visual field, which cor-

responded to RTVue analysis not to fundus appearance (Figure 20-1C).

Case 2

A 44-year-old male with a BCVA of 0.3.

1. Fundus photograph. Posterior pole showed relatively normal appearance compared with peripheral retina. However, the reflex has almost disappeared and patchy depigmentation was observed. At the peripheral area, bone spicules were observed (Figure 20-2A).

2. ONL at macular area is quite thin compared with Case 1. Inner nuclear layer (INL) and ganglion cell layer were preserved even in peripheral retina (Figure 20-2B).

3. The patient could not detect I-4 target in Goldmann perimetry (Figure 20-2C).

CONE DYSTROPHY

Cone dystrophy affects predominantly cone photoreceptors, whereas RP affects rod photoreceptors. So, clinical features appear in a central visual loss and color blindness. Unlike RP, peripheral vision loss and night blindness are rare symptoms. The common macular

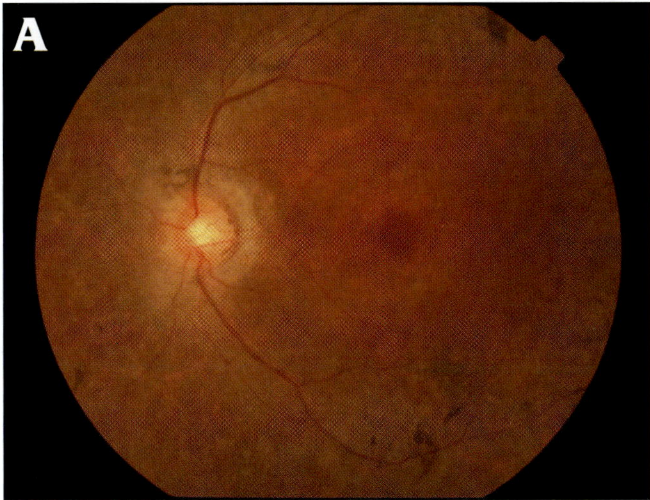

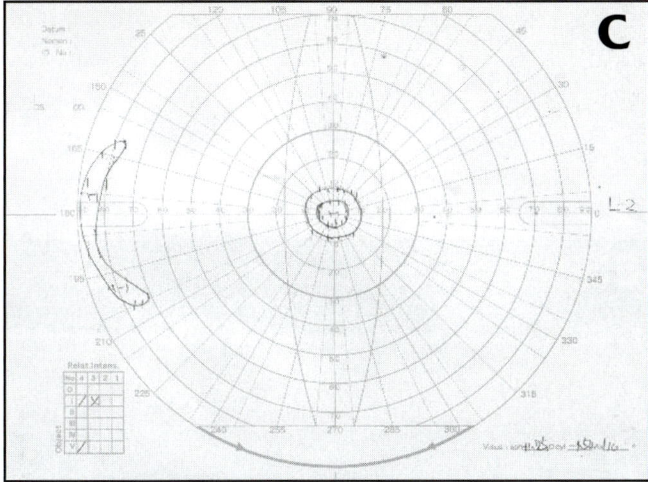

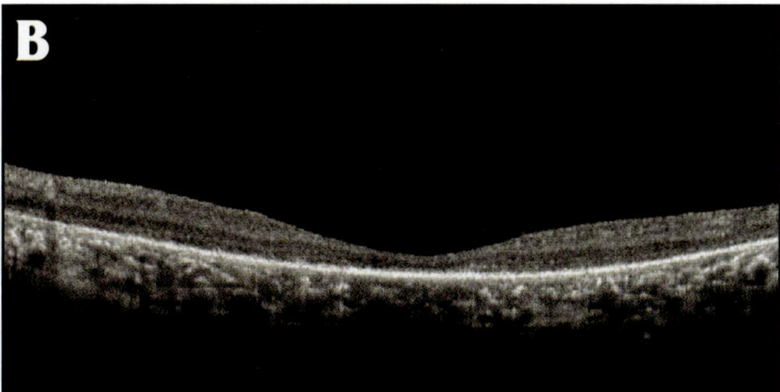

Figure 20-2. Retinitis pigmentosa, severe case. (A) Fundus photograph. (B) RTVue image. (C) Goldmann perimetry.

STARGARDT'S DISEASE

Stargardt's disease is one of the most common inherited macular dystrophies in young people. The appearance is a progressive atrophic macular area surrounded by some or many yellowish, ill-defined flecks.[4,5] After the onset, visual acuity rapidly deteriorates to under 20/200 in most cases. However, in some cases, the BCVA is reserved even in their late stage.

Sometimes the usage of the term *Stargardt's disease* is misunderstood. Many ophthalmologists do not restrict the diagnosis of Stargardt's disease to patients with fluorescein angiographic evidence of choroidal silence. I usually use this term when the patients show atrophic macular dystrophy with flecks, relatively normal ERG (except for the cases in which the entire posterior pole has hardly degenerated), and choroidal silence in FA. So, OCT is not important in diagnosing Stargardt's disease. However, in some cases with good visual acuity, photoreceptor survival at fovea region is clearly observed.

- Inherited trait: autosomal recessive
- Disease locus (responsible gene): ABCA4 or CNGB3 on 1p21-13 (Stargardt disease 1),[6] ELOVL4 on 6p14 (Stargardt disease 3)[7]; another loci was reported on chromosome 4

Case 4

A 19-year-old boy with a BCVA of 0.1.
1. Fundus photograph showed beaten-bronze central atrophy at macular and wide spread flecks on posterior pole (Figure 20-4A).
2. RTVue showed extremely thin ONL among posterior pole, especially at fovea. None of the normal retina structures were observed (Figure 20-4B).

Case 5

A 56-year-old male with a BCVA of 0.7.
1. Fundus photograph showed flecks and pigment deposition on posterior pole (Figure 20-5A).

lesion is a bull's eye appearance. However, the macular lesion shows variation; in some cases, we could not find any abnormality in ophthalmic scope experiment. The findings of ERG were distinctive, photopic shows severe abnormality and scotopic shows normal.

- Inherited trait: autosomal dominant, recessive, X-linked recessive/etc
- Disease locus (responsible gene): many

Case 3

A 49-year-old male with a BCVA of 0.1.
1. Fundus photograph. The case showed macular atrophy like a bull's eye appearance (Figure 20-3A).
2. RTVue showed that ONL disappeared in the lesion. Peripheral retina shows normal morphology (Figure 20-3B).

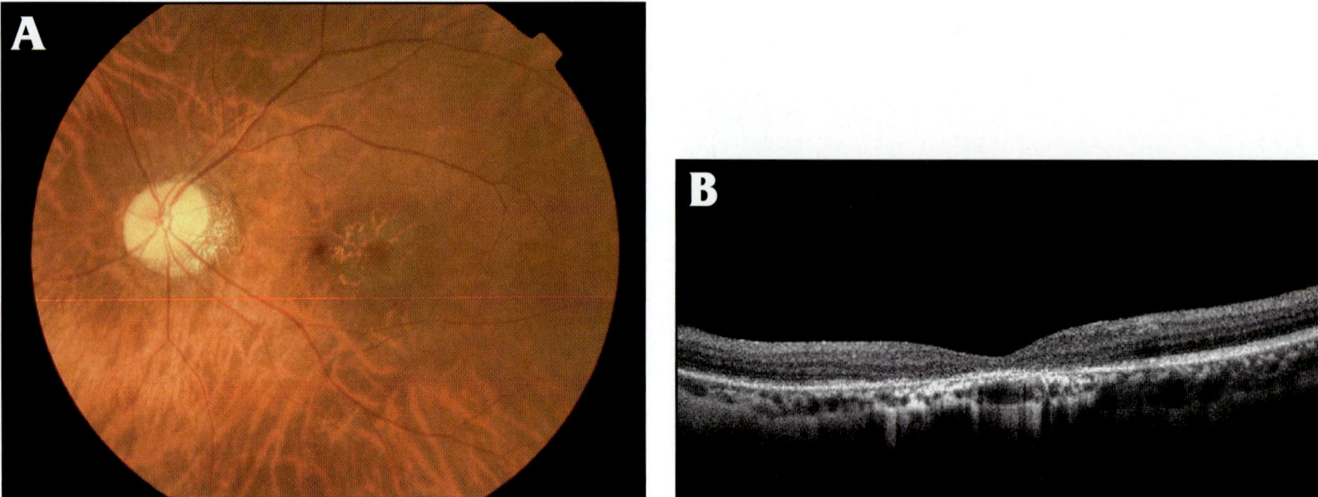

Figure 20-3. Cone dystrophy. (A) Fundus photograph. (B) RTVue image.

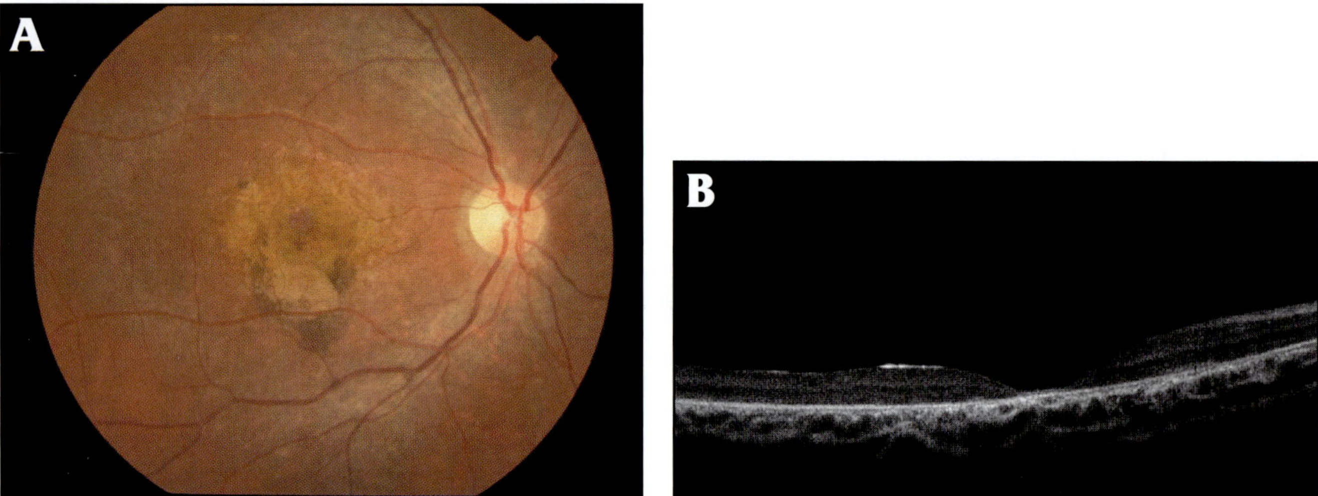

Figure 20-4. Stargardt's disease, severe case. (A) Fundus photograph. (B) RTVue image.

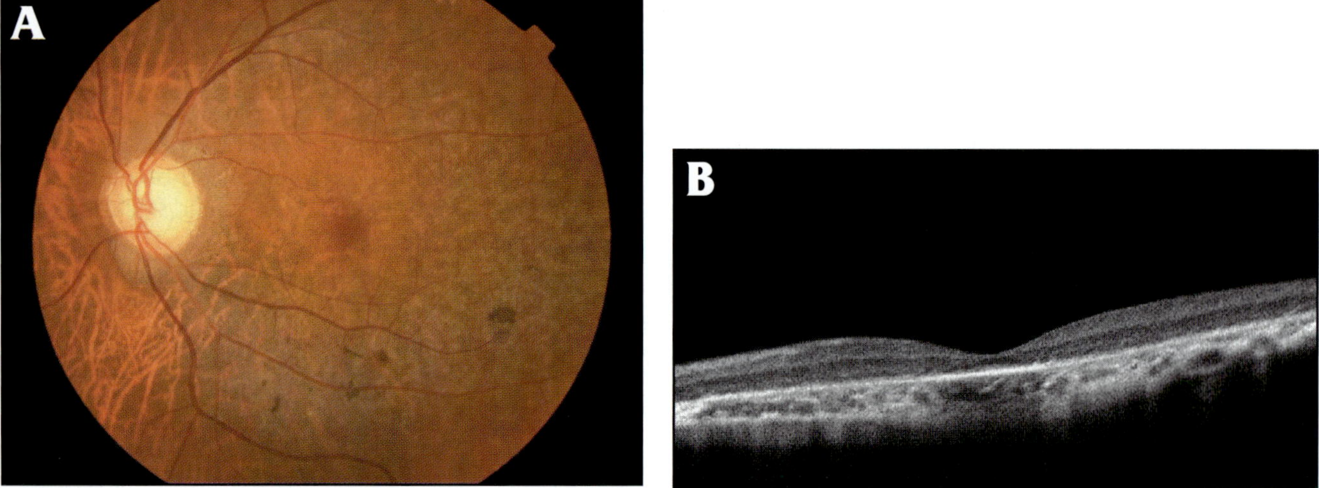

Figure 20-5. Stargardt's disease, moderate case. (A) Fundus photograph. (B) RTVue image.

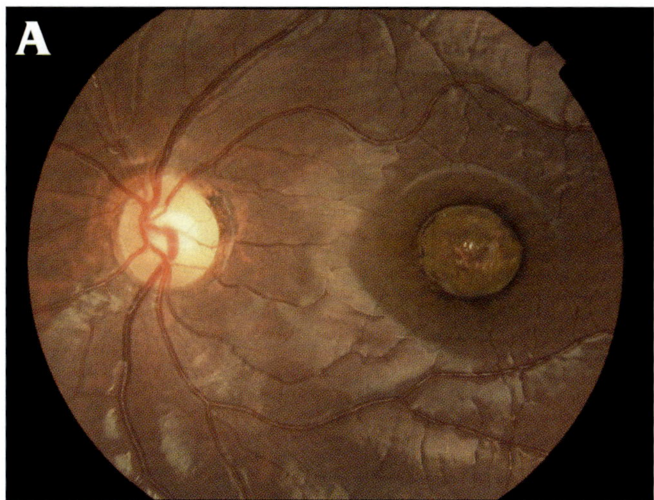

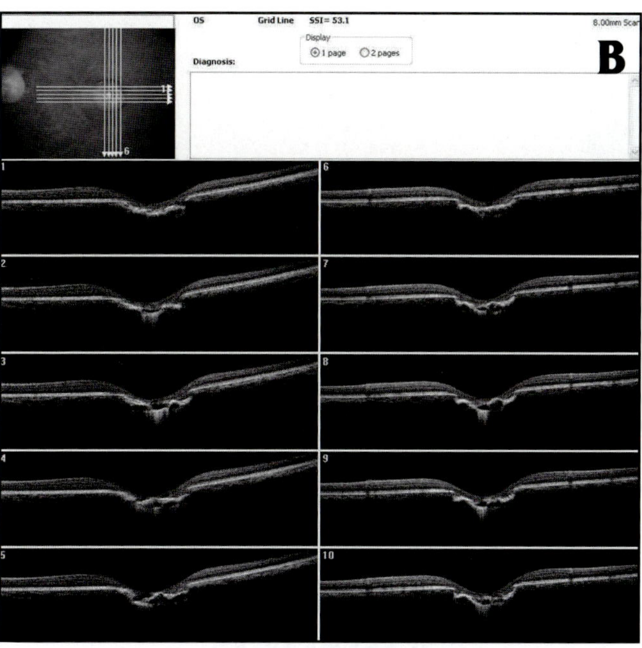

Figure 20-6. Vitelliform macular dystrophy (best macular dystrophy). (A) Fundus photograph. (B) RTVue image.

2. On RTVue image, only fovea region, IS/OS, and OLM lines were observed. Surrounding macular area did not possess the line, and retinal layer structure was disturbed (Figure 20-5B).

VITELLIFORM MACULAR DYSTROPHY (BEST MACULAR DYSTROPHY)

Vitelliform macular dystrophy, usually known as Best disease, is another form of inherited macular dystrophy. This appears in macular lesion called vitelliform, which resembles the yolk of an egg,[8] a yellow-orange macular cyst up to 5-disc diameters that is usually bilateral and causes mild visual loss. However, the disease is progressive and the cyst is absorbed, leaving large RPE atrophy and central visual loss (under 20/200). Age of onset is usually first or second decade. Adult onset vitelliform macular dystrophy is thought to be a distinct form of the disease.

- Inherited trait: autosomal dominant
- Disease locus (responsible gene): BEST1 (11q13),[9] however, locus heterogeneity was reported

Case 6

A 9-year-old boy who suffered from visual acuity loss. His BCVA is 0.5.
1. Fundus photograph showed vitelliform deposition at the macula. It had started being absorbed and did not show a cystic appearance (Figure 20-6A).

2. RTVue showed severe morphological abnormality in the lesion. The neurosensory retina was thin and the choroid dipped toward the sclera; the signal was highly disturbed. In the center,[6] structure of the neurosensory retina was not observed (Figure 20-6B).

X-LINKED RETINOSCHISIS

Retinoschisis is the splitting of the neurosensory retina, so it does not directly imply photoreceptors. However, the disturbance of the layer structure of the retina is easily observed by OCT. The inherited trace is X-linked recessive, so it affects the male population.[10] The macular lesion sometimes shows a radiate placation formed by small folds of the inner retina (radiating retinal cysts). About half of the patients show retinoschiesis peripheral schisis, in addition to maclar schisis.

- Inherited trait: X-linked recessive
- Disease locus (responsible gene): RS1[11]

Case 7

A 20-year-old male with a BCVA of 0.4.
1. Fundus photograph showed abnormal macular reflex. In this case, it did not show radiate placation (Figure 20-7A).

2. RTVue revealed the detail of retinal structure abnormalities. The schisis was observed between INL to ONL. At the macular region, serous retinal detachment was also observed (Figure 20-7B).

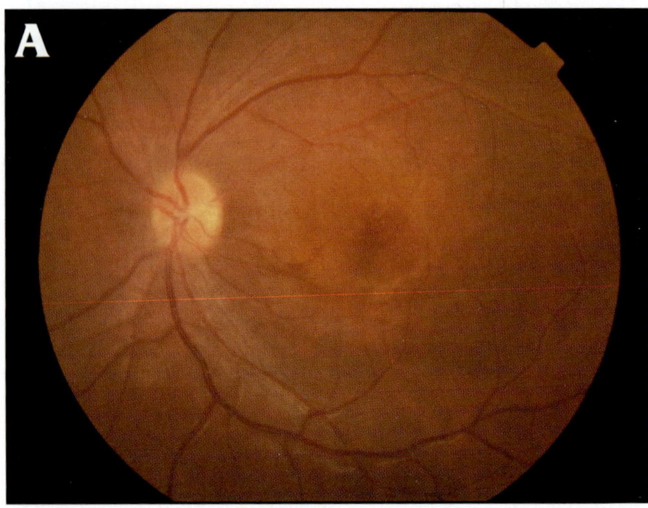

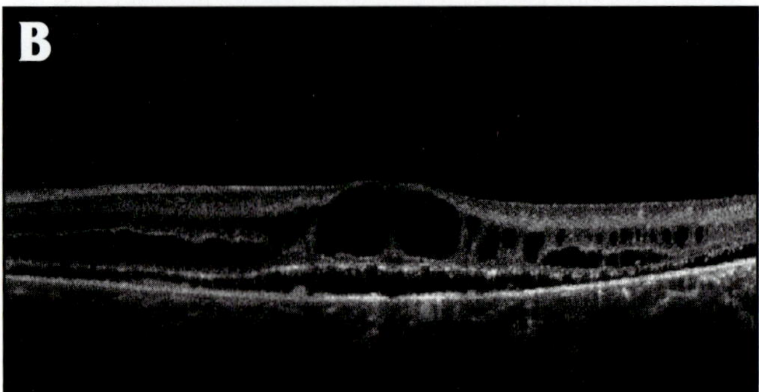

Figure 20-7. X-linked retinoschisis. (A) Fundus photograph. (B) RTVue image.

REFERENCES

1. Pugh EN, Lamb TD. Phototransduction in vertebrate rods and cones. *Handbook of Biological Physics.* 2000:183-255.
2. Young RW. The renewal of photoreceptor cell outer segments. *J Cell Biol.* 1967;33:61-72.
3. Hartong DT, Berson EL, Dryja TP. Retinitis pigmentosa. *Lancet.* 2006;368:1795-1809.
4. Stargardt K. Uber familiare, progressive degeneration in der maculagegend des auges. *Albrecht von Graefe's Archiv furr Ophthalmologie.* 1909;71:534-550.
5. Juan M. Jimenez-Sierra TEO, van Boemel GB. *Inherited Retinal Diseases.* St. Louis, MO: Mosby; 1989.
6. Allikmets R, Singh N, Sun H, et al. A photoreceptor cell-specific ATP-binding transporter gene (ABCR) is mutated in recessive Stargardt macular dystrophy. *Nat Genet.* 1997;15:236-246.
7. Zhang K, Kniazeva M, Han M, et al. A 5-bp deletion in ELOVL4 is associated with two related forms of autosomal dominant macular dystrophy. *Nat Genet.* 2001;27:89-93.
8. Braley AE, Spivey BE. Hereditary vitelline macular degeneration. A clinical and functional evaluation of a new pedigree with variable expressivity and dominant inheritance. *Arch Ophthalmol.* 1964;72:743-762.
9. Petrukhin K, Koisti MJ, Bakall B, et al. Identification of the gene responsible for best macular dystrophy. *Nat Genet.* 1998;19:241-247.
10. Gieser EP, Falls HF. Hereditary retinoschisis. *Am J Ophthalmol.* 1961;51:1193-1200.
11. Sauer CG, Gehrig A, Warneke-Wittstock R, et al. Positional cloning of the gene associated with X-linked juvenile retinoschisis. *Nat Genet.* 1997;17:164-170.

21 Cystoid Macular Edema

Bruno Lumbroso, MD; Marco Rispoli, MD; and
Maria Cristina Savastano, MD

Cystoid macular edema (CME) is frequently seen in retinal diseases as different as diabetic retinopathy, complications of ocular surgery, retinal vein occlusion, uveitis, and epiretinal membrane. Several pathogenic mechanisms occur in the development of macular edema (hypoxia, altered blood flow, retinal ischemia, inflammation, and retinal traction), but the exact process continues still to be not fully understood. Spectral-domain OCT is widely used in the diagnosis and management of CME and allows the user to quantify the evolution of macular edema without or after treatment. OCT is a wonderful tool to study *in vivo* intraretinal edema (Table 21-1).

Diffuse retinal edema is manifested on OCT scans as a diffuse decrease in retinal reflectivity due to a reduced density of optical backscattering within the retina.

OCT cross-sectional (sagittal) sections of CME show areas of low intraretinal reflectivity, which correspond to intraretinal fluid accumulation in cystic spaces. These cystic spaces form rounded low reflective areas found in the retinal layers, arranged in various patterns. Spectral-domain OCT has a high scan resolution, but not enough to determine whether the cystic swelling is intracellular or extracellular.

The possibility to use *en face* C-scan allows us to highlight other aspects and patterns of CME as seen in frontal sections.

Fluorescein angiography shows, in these cases, vascular leakage and different patterns of fluid accumulation in honeycomb or petaloid spaces, but it does not provide a quantitative assessment.

Monitoring quantitative retinal thickness with OCT is useful in assessing efficacy of treatment and is a valuable tool in deciding the opportunity of retreatment. Macular thickness tends to correlate inversely with visual acuity in patients with macular edema, although that is not always the case.

MÜLLER CELLS AND CYSTIC MACULAR EDEMA

Müller cells play a central role in supporting neuronal activity through blood-retinal integrity barrier and homeostasis regulation.[1-3] The disruption of their metabolism has been implicated in the production of retinal edema. In normal conditions, they mediate the fluid absorption into the blood by a co-transport of water and osmolytes, especially potassium ions.[4]

Several pathogenic mechanisms occur in the development of macular edema (hypoxia, altered blood flow, retinal ischemia, and inflammation), but the exact process continues still not fully understood.

Vasoactive factors induce macula edema (eg, prostaglandins, interleukins, tumor necrosis factor) although vascular endothelial growth factor (VEGF) is the major cytokine involved.[5,6] Through proper stimuli, many cells are able to produce VEGF, not only Müller cells and pericytes; also retinal pigment epithelium (RPE) cells, which are express VEGF receptors (VEGR-R).[7]

The maintenance of the blood-retinal barrier is a complex process involving several factors. The function of 2 major components, the outer and inner barriers are interrelated, leading to proper retina function.

Huang D, Duker JS, Fujimoto JG, Lumbroso B, Schuman JS, Weinreb RN.
Imaging the Eye from Front to Back with RTVue Fourier-Domain Optical Coherence Tomography (pp 185-194).

Table 21-1

CAUSES OF CYSTOID MACULAR EDEMA

Diabetic Retinopathy	Uveitis	Macular Dystrophies
Subretinal New Vessels	Choroiditis	Vogt-Koyanagi Harada
Complications of Ocular Surgery	Vitreoretinal Traction	Sympathetic Ophthalmia
Retinal Vein Occlusion	Epiretinal Membrane	
Chronic Pigment Epitheliopathies	Retinitis Pigmentosa	

Although some clinical studies suggest that the inner barrier is the primary site involved in vascular leakage of macular edema, many experimental researches confirm the important role of the outer barrier in the genesis of edema.[8,9] Some authors described the osmotic swelling of Müller cells after interruption of the inner blood-retinal barrier.[10,11] A study of the retinal morphology by Yanoff et al[12] describes Müller cell alterations. In light and electron-microscopic examination of eyes with CME, the authors found intracytoplasmic swelling of the Müller cells to be the anatomic basis of central macular edema. Intercellular (extracellular) fluid accumulation was interpreted as late, end-stage results of prolonged and excessive intracellular edema, cell death, and disruption.

Others theories are based on the damage of RPE, responsible for the integrity of the outer hematoretinal barrier. The break down of the outer hematoretinal barrier causes alteration of the RPE pump, producing abnormal accumulation of fluid in the *extracellular* spaces of the intraretinal layers.[13]

CYSTOID PATTERNS AND MACULAR AND RETINAL INFRASTRUCTURE FRAMEWORK

The shape, dimensions, and extension of the cystic cavities depend on the framework of retinal structures and layers in the macula that directly influence the cystoid patterns.

The Henle fibers structure, seen frontally, is radial, star-shaped, and centered on the foveola (Figure 21-1).

At the center of the macula, the Müller cells form a truncated inverted cone, described by Gass and Yamada,[14,15] shaped as a plug that binds together the receptor cells in the foveola. According to Gass,

> The Müller cell cone is the primary structural support for the fovea. The Müller cell cone serves as a plug to bind together the receptor cells in the foveola. Without this plug of glial cells, the retinal receptor cell layer

with its thin layer of horizontally radiating nerve fibers would be highly susceptible to disruption at the umbo and hole formation under a variety of circumstances, including sensory retinal detachment, minor trauma, cystoid macular edema, and macular degeneration.[14]

Gass hypothesized that the extension of a choroidal neovascularization into the capillary-free zone in the macula may disrupt the photoreceptor-external limiting membrane complex and lead to CME. Congenital abnormalities affecting the Müller cell cone are also responsible for the pathognomonic biomicroscopic picture of foveomacular schisis.

On the edge of the fovea, the Müller cells are radially disposed, following Henle fibers radial pattern (Figure 21-2).

Müller cells and Henle fibers form a framework that is important in shaping the cystoid spaces patterns. In the fovea, Henle's fibers disposition leads to the characteristic star-like appearance of degenerative processes of the macula and juvenile retinoschisis.

Yanoff microscopic studies[12] on CME seem to show that cystic spaces found by OCT are the result of swelling of Müller cells and that the extent of this swelling is influenced by and depends on the surrounding retinal structures and layers. He writes "Müller cells are arranged radially and not horizontally. This may explain why, using HD OCT imaging, cystic spaces, swelling, or both were not confined to a single layer in each leakage type."

VERTICAL AND HORIZONTAL STRUCTURES

The retina consists of a complex framework of fibers, capillaries, and cells organized in structures that effectively form barriers. Retinal framework is sustained by vertical structures and horizontal structures.

Pathological process extension is blocked vertically and horizontally by anatomical barriers. The retina is divided by different horizontal and vertical structures

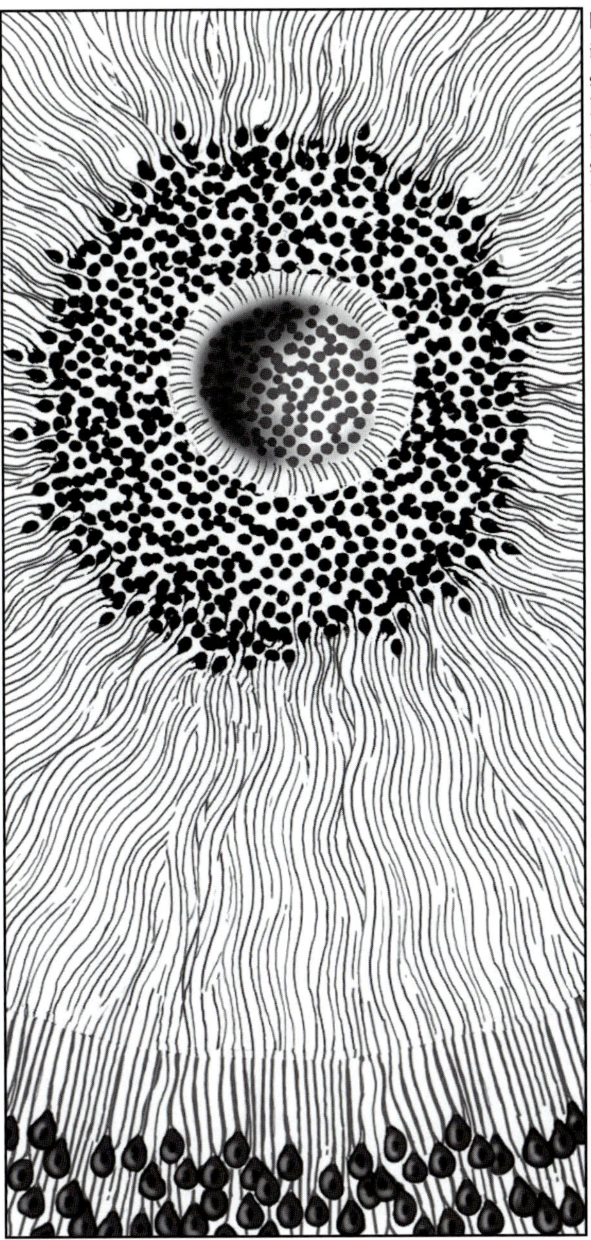

Figure 21-1. Frontal section of the fovea. Henle fiber radial structure in a frontal section of the retina in the foveal area. In the center can be seen foveal cones. Cone nuclei are placed around it. From it, the radial Henle fibers connect the central cone nuclei to bipolar cells nuclei in the periphery. (Drawing reconstructed from the description and histological sections by Vrabec and Polyak cited by Duke Elder. Illustration by Ms. D. Piccioli.)

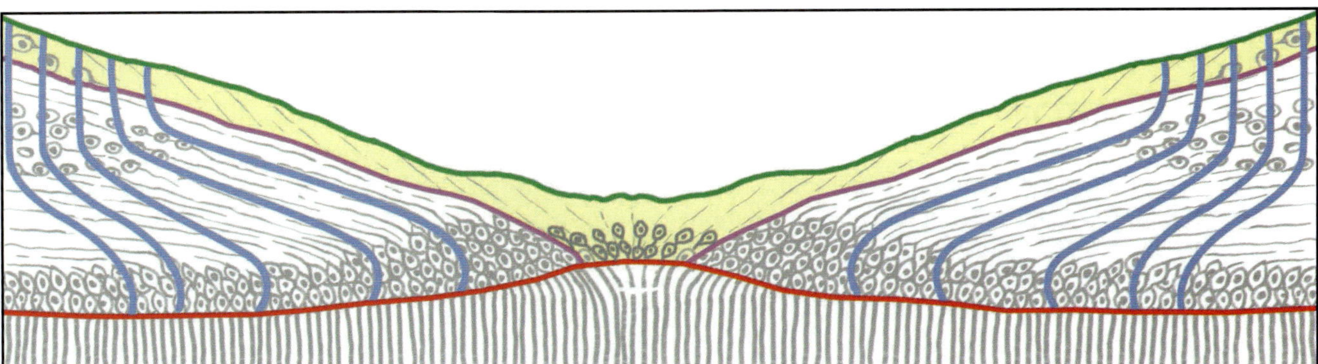

Figure 21-2. Cross-section of the fovea. (Drawing of the Müller cell cone as described by Yamada and Gass.) Müller cell cone is painted in yellow and forms a truncated inverted cone limited by inner and outer limiting membranes. Müller cells are drawn (in blue) according to the description and histological sections by Polyak cited by Duke Elder. They are located between the Henle fibers.

that explain the localization; dimensions; and shape of exudates, hemorrhages, and cystoid cavities that we can analyze with OCT.

In the extramacular area, the retinal framework is simpler. Müller cells unite the inner limiting membrane to the outer limiting membrane, crossing the 2 plexiform layers.

Vertical Structures

Müller cells unite the inner limiting membrane to the outer limiting membrane. Other vertical structures are the chains formed by photoreceptors cells linked to the bipolar cells and to ganglion cells.

Horizontal Structures

The inner limiting membrane and the outer limiting membrane are membranes that can be recognized on histological sections of the retina. The inner limiting membrane is the basement membrane of the Müller cells.

The inner (IPL) and outer plexiform layers (OPL) are formed by horizontal and amacrine cells and by a net of intercellular synaptic connections between bipolar cells and the photoreceptors (on one side) and the ganglion cells (on the other side). They are not true histological membranes, but they effectively form barriers even if they are less strong and resistant than the inner and the outer limiting membranes. Due to this complex intertwining of fibers, they act as horizontal barriers against the spread of fluid throughout the retina, giving their shape to the cystoid cavities. The IPL is stronger than the OPL.

These vertical and horizontal structures limit expansion of pathological lesions, ie, exudates, hemorrhages, and edema. Cystoid edema appears on OCT as round cavities and holes limited by the tissue barriers (Figure 21-3).

OPTICAL COHERENCE TOMOGRAPHY PATTERNS OF CYSTOID EDEMA

At least 2 pathogenic mechanisms occur in the development of macular edema. The vascular mechanism is related to the disruption of the blood-retinal barrier; another mechanism is the vitreomacular traction that pulls the retina forward and modifies the retinal vascular system, leading to the formation of CME.

Cystoid cavity morphology may help define the stage in the evolution of the edema.

EARLY CYSTOID EDEMA

At the cystoid edema onset, we can see one or more rounded edema cells mainly in the inner (INL) and outer nerve layers (ONL). Later the retinal cavities are small, numerous, and separated by thick septa or walls. Their dividing walls are regular, smooth, and rounded. The cysts are mainly seen in the ONL, INL, and IPL (Figure 21-4).

ADVANCED CYSTOID EDEMA

During the edema evolution, whatever the etiology, the small cavities coalesce, cystoid cells merge, their dimensions increase, the walls get thinner; but the cystoid cavities shape remain smooth, regular, globular, and spherical, maintaining a rounded shape.

Müller cell necrosis leads to larger cavities, first in the INL and IPL, and later in the ONL and OPL. Cystoid edema begins in the nuclear layers and then extends vertically to full retinal thickness, forming large chambers or crypts. The IPL yields to the cells expansion, allowing vertical merging between cells in the INL and the ONL. The largest cystic spaces are found centrally; forming 1 or more bulging chambers. *Localization of the larger cells at the fovea and their vertical shape is induced by the arrangement of the Müller fibers and by the foveal cone described by Gass.*[14] The petaloid pattern is caused by the Henle fibers structure. At the onset the fovea is not affected, but later a central cell may appear and lead to a decrease of visual acuity.

Smaller cystic spaces are found in the periphery of the macula.

The longer the evolution, the larger the cystoid cavities. Their partitions gradually disappear, leading to larger and more irregular cells. Finally we have no more clusters of spherical cells, but rather large vaults sustained by vertical pillars of residual tissue similar to columns in a cathedral or tree trunks in a forest.

According to Bolz et al,[16] in OCT

> The largest cystic spaces are found centrally; smaller cystic spaces are seen parafoveally and extrafoveally. This aspect gives a major thickness to the macular centre, could be the result of the characteristic distribution on the retinal structures within the fovea. In the centre, the absence of the ganglion cell layer, inner plexiform layer, and inner nuclear layer seems to be a precondition for the formation of cystic spaces larger than in the surrounding macular fields. This indicates that a measurement of retinal thickness in the centre reflects the size of a few or a single central cyst rather than an affection of the

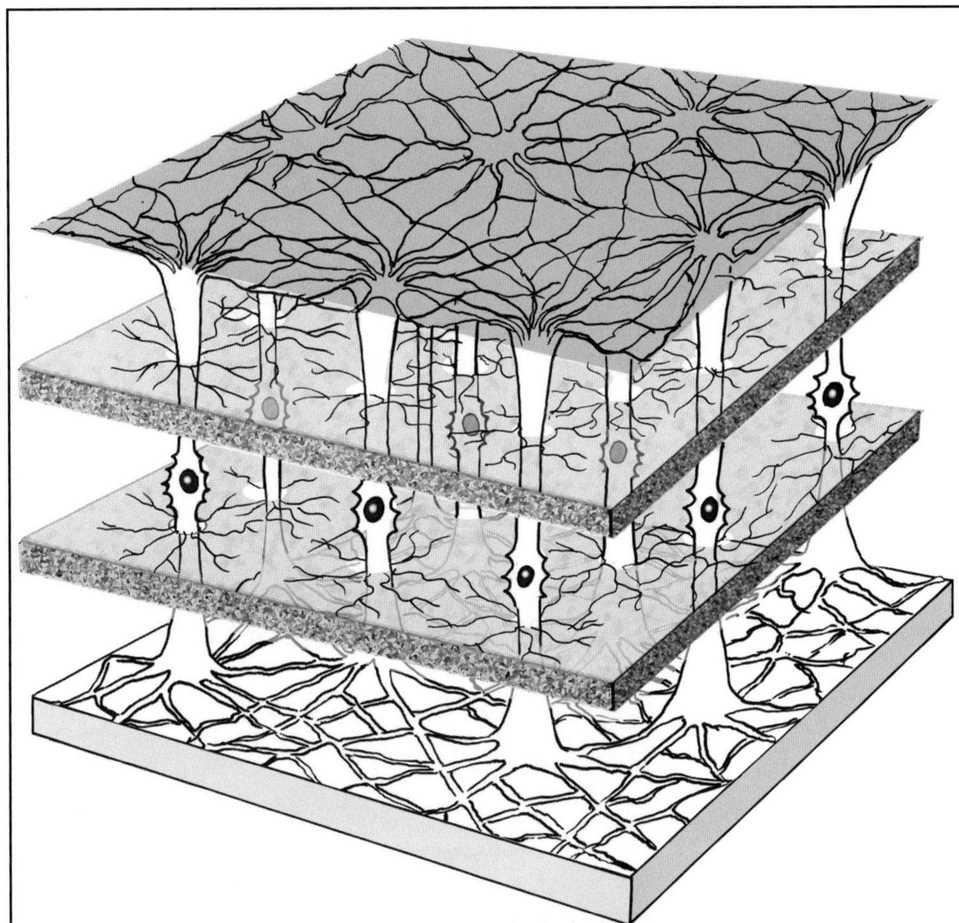

Figure 21-3. Retinal framework structure. Müller cells, inner and outer limiting membranes, IPL, and OPL form the macular framework structure. Vertical structures: Müller cells unite the inner limiting membrane to the outer limiting membrane. Other vertical structures are the chains formed by photoreceptors cells linked to the bipolar cells and to ganglion cells. Horizontal structures: the inner limiting membrane and the outer limiting membrane are membranes seen in histology. The inner limiting membrane is the basement membrane of the Müller cells. The INL and OPL are formed by horizontal and amacrine cells, by short horizontal processes from the Müller cells, and by a net of intercellular synaptic connections between bipolar cells and the photoreceptors (on one side) the ganglion cells (on the other side). They are not histological membranes, but effectively form barriers even if they are less strong and resistant than true membranes. The IPL is stronger than the OPL. These vertical and horizontal structures limit expansion of exudates, hemorrhages, and edema. (Illustration by Ms. D. Piccioli.)

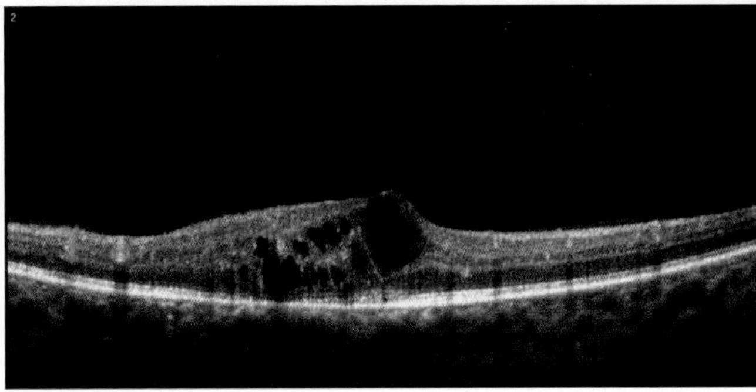

Figure 21-4. Early cystoid edema. In this case of venous branch occlusion, the cavities are mainly seen in the INL. Retinal cavities are small, numerous, and separated by thick septa or walls. Their walls and dividing lines are regular and rounded. Some cells are also seen in the ONL. One bigger ovalar vertical cell intersects vertically the full retina.

entire retina. These morphologic findings have implications for follow-up data during therapy (Figures 21-5 and 21-6).

REGRESSED CYSTOID EDEMA

After a long evolution, with moments of progression followed by regressions and periods of stabilization, cystoid cells and cavities tend to decrease in dimension without disappearing. Their walls are less regular, sometimes concave, bulging toward the cavity center. *Cavities loose their globular or rounded aspect but take a vaguely triangular shape, with sharp angles and knees (Figure 21-7).* Their walls are thinner and make angles and bifurcations. Sometimes they are triangle-shaped with a vertical wall toward the fovea and a sharp angle toward the retinal periphery. The wall profiles are no more smooth, but are irregular and granulated. Sometimes the older edema cavities are seen as thin intraretinal fluid detachments between nuclear and plexiform layers.

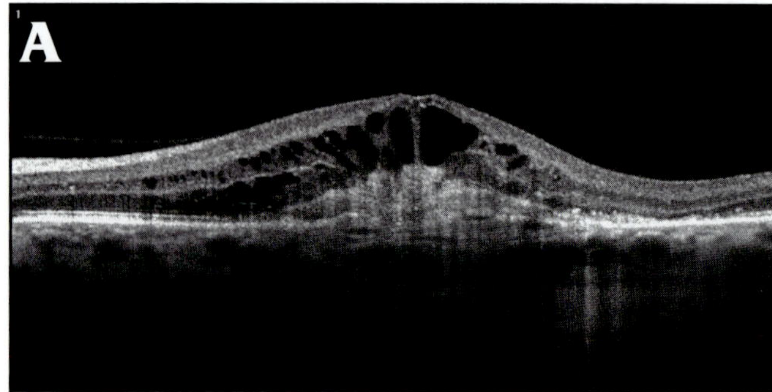

Figure 21-5. Advanced cystoid edema. In this case of choroidal neovascularization, the largest cystic spaces are found centrally, forming some bulging chambers above the neovascular membrane. We do not see spherical cell clusters, but large vaults sustained by vertical pillars of residual tissue as columns in a cathedral or tree trunks in a forest. The longer the evolution, the larger the cystoid cavities. The localization of the larger cells at the fovea, and their vertical shape is induced by the vertical arrangement of the Müller fibers and the foveal cone described by Gass.[14] Smaller cystic spaces are found in the periphery of the macula.

Cystoid Edema in Choroiditis and Uveitis

During the evolution of inflammatory diseases we can sometimes see bigger cystoid cells and cavities (Figure 21-8). Their walls are thicker than normally seen and highly reflective, frequently with granular irregularities. These cavities are globular or rounded with high reflectivity content.

Tractional Cystoid Macular Edema, Epiretinal Membrane

Transversal retinal traction induces the formation of small retinal folds. OCT examination shows hyperreflective epiretinal membrane on the surface of the retina as a thin band of increased signal. The membrane is separated from the inner retinal surface in small multiple areas, enhancing visibility. The retina underlying this membrane is diffusely thickened with loss of the normal fovea contour. Some irregular cells of cystoid edema are seen below the retinal surface and in the INL and ONL. The retinal thickness map shows the increased thickness of the retina.

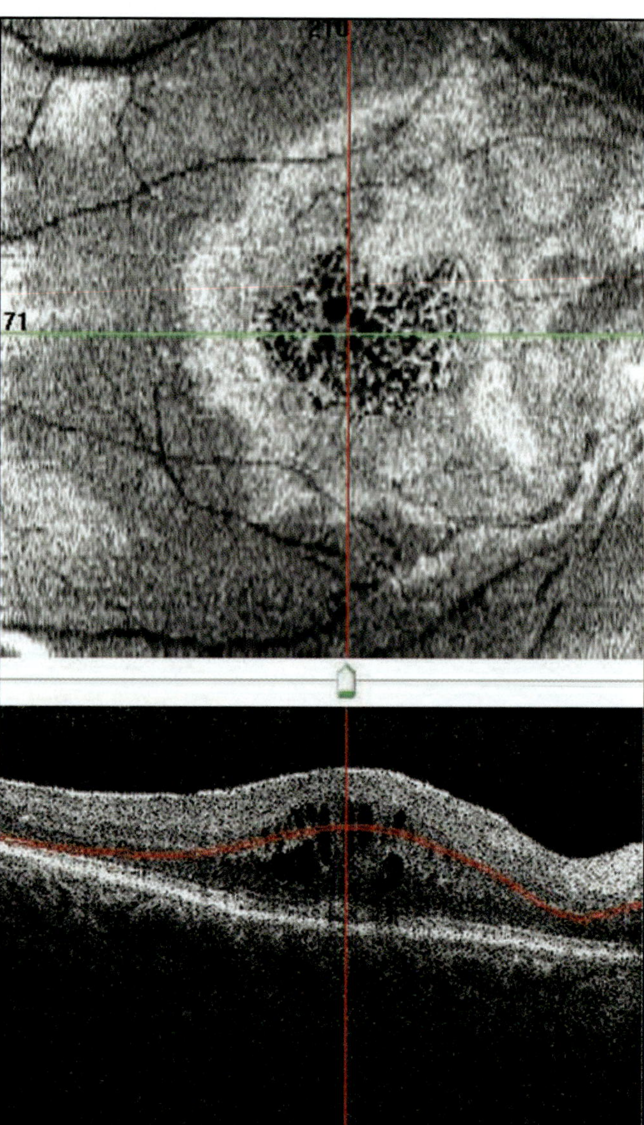

Figure 21-6. Advanced cystoid edema. In a case of postsurgical cystoid edema B-scan, we see few central large regular cystoid spaces surrounded by many smaller cysts. *En face* C-scan shows another aspect of these central large cystoid cavities surrounded by smaller cysts. The shape of the cavities seen on *en face* scan is more irregular than seen on sagittal (anterior-posterior scans). The cavities are divided by septa or columns. B-scans give us a 2D view of the retina. A third dimension is given by frontal *en face* scans. RTVue *en face* scans are a great improvement on flat C-scans because they give a retinal section that follows the shape of the retinal layer.

Vitreomacular Traction and Cystoid Macular Edema

Vitreomacular traction pulls the retina forward and leads to the formation of CME. OCT cross-section scans show attachment of the posterior hyaloid directly to the fovea. The tractions are seen as 2 reflective lines in the vitreous. Increased linear reflectivity along the anterior

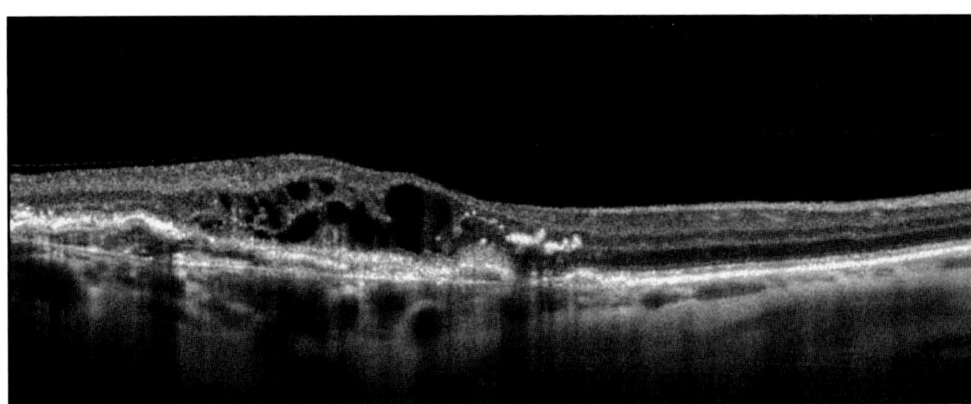

Figure 21-7. Regressed cystoid edema. Age-related macular degeneration/choroidal neovascularization treated by antiangiogenic intravitreal injections. Cystoid cells and cavities have smaller dimensions than in the active edema. Their walls are thinner and irregular, sometimes bulging toward the cavity center, and the cavities take a vaguely triangular or rectangular shape.

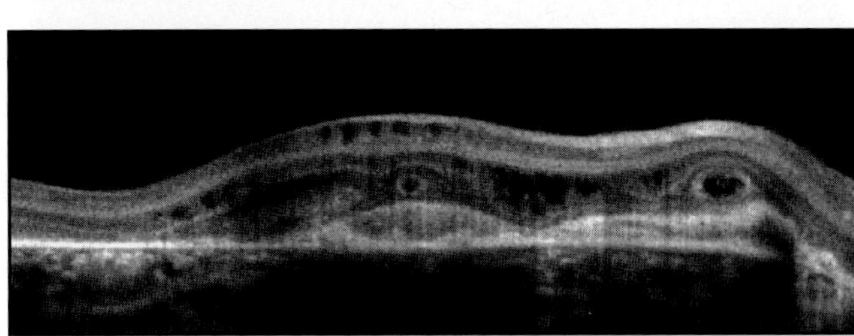

Figure 21-8. Cystoid edema in choroiditis. In a case of choroiditis the cavity walls are thick and highly reflective with granular irregularities. These cavities are globular or rounded with a highly reflective content.

retina represents epiretinal membrane formation. The tractional forces from the vitreous lead to edema formation. CME is visible as large, hyporeflective, rounded spaces in the fovea with increased foveal thickness.

MACULAR HOLE AND CYSTOID MACULAR EDEMA

Full-thickness macular hole has distinct features on *en face* OCT C-scan images. Typically the hole is surrounded by 2 or more rings of cysts of various sizes. The hole appears as a nonreflective circle surrounded by a less reflective layer.

RETINOSCHISIS

One of the differential diagnoses of CME is X-linked retinoschisis. *En face* OCT shows areas of harmonious, regular, uniform cystoid cells corresponding to schisis cavities in the INL and ONL and in the IPL. Frequently the OCT images are star-shaped or stellate, following the structure of Henle fibers and Müller cells. The walls form parallel harmonious curves centering the fovea (Table 21-2 and Figures 21-9 through 21-11).

CONCLUSION

Spectral-domain OCT is widely used in the diagnosis and management of CME and to study and quantify its evolution. It is a wonderful tool to study *in vivo* intraretinal edema.

OCT cross-sectional (sagittal) sections of CME show the classical aspects of the cystic spaces. The use of RTVue *en face* C-scans adapted to the retinal concavity allows us to highlight other aspects and patterns of CME corresponding to the structure of the fovea and posterior pole retina.

REFERENCES

1. Bringmann A, Pannicke T, Grosche J, et al. Müller cells in the healthy and diseased retina. *Prog Retin Eye Res.* 2006;25:397-424.
2. Tout S, Chan-Ling T, Holländer H, Stone J. The role of Müller cells in the formation of the blood-retinal barrier. *Neuroscience.* 1993;55:291-301.
3. Bringmann A, Pannicke T, Biedermann B, et al. Role of retinal glial cells in neurotransmitter uptake and metabolism. *Neurochem Int.* 2009;54:143-160.
4. Nagelhus EA, Horio Y, Inanobe A, et al. Immunogold evidence suggests that coupling of K+ siphoning and water transport in rat retinal Müller cells is mediated by a coenrichment of Kir4.1 and AQP4 in specific membrane domains. *Glia.* 1999;26:47-54.
5. Aiello LP, Bursell SE, Clermont A, et al. Vascular endothelial growth factor-induced retinal permeability is mediated by protein kinase C in vivo and suppressed by an orally effective beta-isoform-selective inhibitor. *Diabetes.* 1997;46:1473-1480.

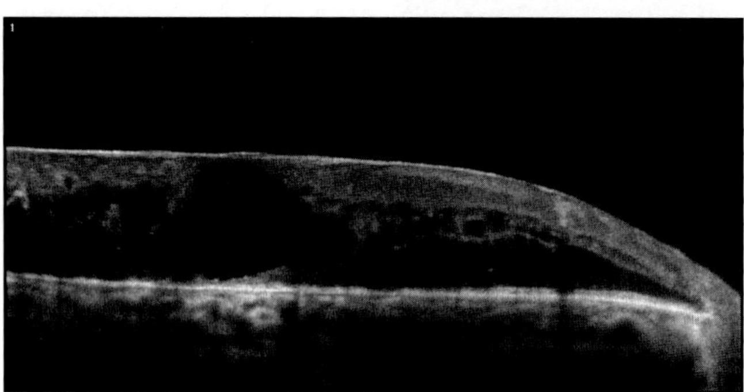

Figure 21-9. Epiretinal membrane, tractional CME. Epiretinal membrane on the surface of the retina is a thin band of increased signal. The membrane is adherent to the inner retinal surface. The retina underlying this membrane is diffusely thickened with loss of the normal fovea contour. Irregular cavities of cystoid edema are found below the retinal surface and mainly in the INL and ONL. We see large chambers with vertical pillars of residual tissue.

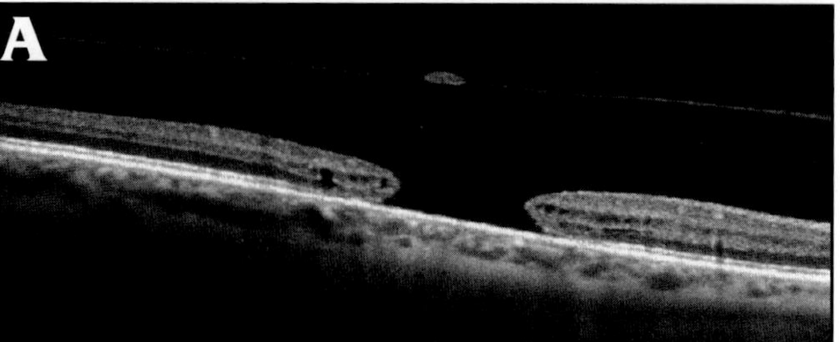

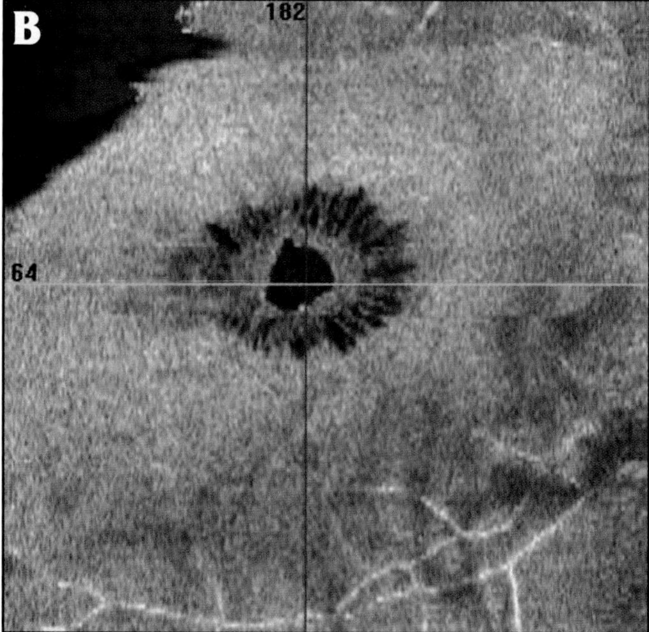

Figure 21-10. Full-thickness macular hole and CME. Full-thickness macular hole has distinct features on *en face* OCT C-scan images. Typically the hole is surrounded by 2 concentric rings of cysts of various sizes. The hole appears as a nonreflective circle surrounded by small regular rounded cavities. On cross-section scan it is difficult to assess the localization and importance of the edema pseudo cells.

6. Shimada H, Akaza E, Yuzawa M, Kawashima M. Concentration gradient of vascular endothelial growth factor in the vitreous of eyes with diabetic macular edema. *Invest Ophthalmol Vis Sci.* 2009;50(6):2953-2955.

7. Ohno-Matsui K, Yoshida T, Uetama T, Mochizuki, M, Morita I. Vascular endothelial growth factor upregulates pigment epithelium-derived factor expression via VEGFR-1 in human retinal pigment epithelial cells. *Biochem Biophys Res Commun.* 2003;303:962-967.

8. Sander B, Larsen M, Moldow B, Lund-Anderson H. Diabetic macular edema: Passive and active transport of fluorescein through the blood-retina barrier. *Invest Ophthalmol Vis Sci.* 2001;42:433-438.

9. Tso MO, Cunha-Vaz JG, Shih CY, Jones CW. Clinicopathologic study of blood-retinal barrier in experimental diabetes mellitus. *Arch Ophthalmol.* 1980;98:2032-2040.

10. Bringmann A, Reichenbach A, Wiedemann P. Pathomechanisms of cystoid macular edema. *Ophthalmic Res.* 2004;36:241-249.

11. Hirrlinger PG, Wurm A, Hirrlinger J, Bringmann A, Reichenbach A. Osmotic swelling characteristics of glial cells in the murine hippocampus, cerebellum, and retina in situ. *J Neurochem.* 2008;105:1405-1417.

12. Yanoff M, Fine BS, Brucker AJ, Eagle RC Jr. Pathology of human cystoid macular edema. *Surv Ophthalmol.* 1984;28(Suppl):505-511.

13. Cunha-Vaz JG, Shakib M, Ashton N. Studies on the permeability of the blood-retinal barrier. I. On the existence, development and site of blood-retinal barrier. *Br J Ophthamol.* 1966;50:441-453.

14. Gass JDM. Muller cell cone, an overlooked part of the anatomy of the fovea centralis. *Arch Ophthalmol.* 1999;117:821-823.

15. Yamada E. Some structural features of the fovea centralis in the human retina. *Arch Ophthalmol.* 1969;82:151-159.

16. Bolz M, Ritter M, Schneider M, et al. A systematic correlation of angiography and high-resolution optical coherence tomography in diabetic macular edema. *Ophthalmology.* 2009;116:66-72.

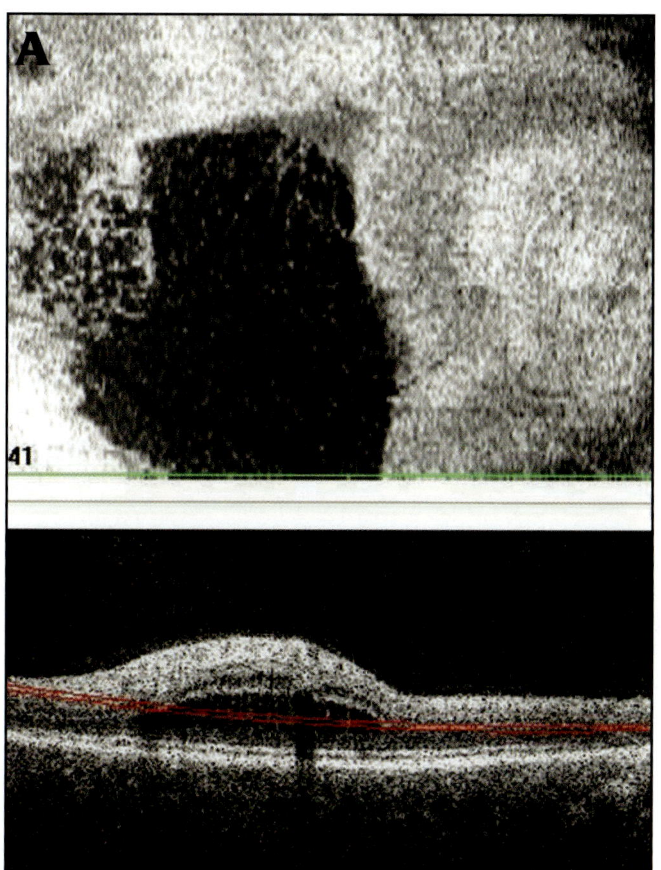

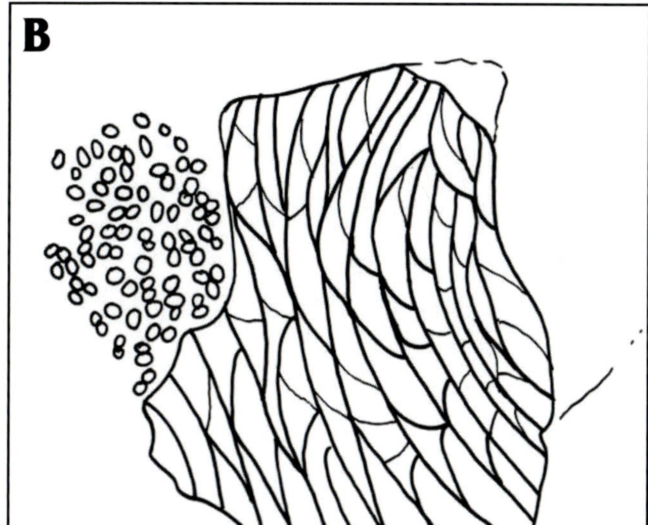

Figure 21-11. Retinoschisis and CME, cross-section scan and *en face* scan. In this young patient, retinoschisis and cystoid edema coexist. We can see an area of harmonious, regular, uniform cells corresponding to schisis cavities in the ONL. The walls of these cavities appear to form parallel harmonious curves centering on the fovea. The fovea is outside the retinoschisis area. Close to the retinoschisis zone, CME is visible as small, hyporeflective, rounded, less regular spaces located in the INL.

22 Optical Coherence Tomography in Diabetic Retinopathy

Tarek S. Hemeida, MD and Amani A. Fawzi, MD

Following the first *ex vivo* OCT imaging in 1991, *in vivo* OCT imaging studies of the human retina were performed in 1993.[1-3] Since then, OCT has become an indispensable tool in the diagnosis and management of diabetic patients, especially with recent advances in pharmacologic therapies.[4,5]

PATHOGENESIS OF DIABETIC RETINOPATHY AND DIABETIC MACULAR EDEMA

Hyperglycemia is a major risk factor for development of diabetic retinopathy. It leads to high intracellular levels of glucose, formation of free radicals (oxidative stress), and protein kinase C activation.[6] Chronic hyperglycemia also leads to formation of advanced glycation end products (AGEs), which may be the inciting event for diabetic retinopathy and maculopathy.[7] Accumulation of AGEs in the vitreous and vitreoretinal interface is associated with neurovascular injury seen in diabetic retinopathy. Other factors such as hypoxia, altered blood flow, retinal ischemia, and inflammation are also associated with the progression of diabetic macular edema (DME). Inflammatory processes, as well as increased vascular endothelial growth factor (VEGF) levels, endothelial dysfunction, leukocyte adhesion, decreased pigment epithelium derived factor (PEDF) levels, and increased protein kinase C production, combine to lead to breakdown of the blood-retinal barrier and increased vascular permeability within the diabetic retinal vasculature.[7] In addition to increased vascular permeability, changes at the vitreomacular interface may contribute significantly to the progression of macular edema.[2-8]

CLINICAL ASSESSMENT OF DIABETIC MACULAR EDEMA

Since the publication of the Early Treatment Diabetic Retinopathy Study (ETDRS),[8] DME has been defined as diffuse or focal macular thickening, with or without hard exudates that can be complicated by the formation of intraretinal cystoid spaces. This thickening is judged clinically relevant when involving or threatening the macula (thickening or hard exudates plus adjacent thickening at <500 µm from the center of foveal avascular zone) and called clinically significant macular edema (CSME). Patients with CSME are deemed to be at great risk of visual loss if left without treatment.[9]

It has been stated that visual function in patients with DME is inversely correlated to central retinal thickening and not to capillary leakage, and that restoration of normal macular thickness may be followed by partial recovery of visual function.[10]

While massive or extensive edema is easily seen at the slit lamp, subtle CSME, as well as small intraretinal cystoid spaces or subtle epiretinal changes may be difficult to appreciate. Moreover, the evaluation of retinal thickness changes after medical or surgical interventions was generally relegated to clinical and subjective impression, until the introduction of objective OCT measurements.[11]

Huang D, Duker JS, Fujimoto JG, Lumbroso B, Schuman JS, Weinreb RN.
Imaging the Eye from Front to Back with RTVue Fourier-Domain Optical Coherence Tomography (pp 195-202).
© 2010 SLACK Incorporated.

With high-resolution, rapid cross-sectional scanning of the entire macular area, OCT can be instrumental in the diagnosis and follow-up of DME, allowing not only an objective measurement of overall macular thickness, but also cross-sectional detailed analysis of intraretinal structure and epiretinal traction.[12] Similar to its effect on various retinal diseases, the use of OCT in patients with diabetic retinopathy has added new insights into the various presentations of DME and their impact on visual function, as well as on the potential benefit of new therapies.[11]

Qualitative Optical Coherence Tomography Assessment of Diabetic Macular Edema

OCT studies of CSME, as defined by the ETDRS, have revealed 4 basic structural changes in the neurosensory retina: sponge-like retinal swelling (Type 1), cystoid macular edema (CME) (Type 2), serous retinal detachment (Type 3), and tractional retinal detachment (Type 4).[11,13-15] The presence of CME on OCT is associated with more severe reduction in visual acuity and poorer response to treatment than OCTs displaying sponge-like retinal thickening alone.[16] Improved characterization of DME on OCT may allow clinicians to determine when irreversible structural damage has occurred and, therefore, the point at which further treatment is unlikely to be of benefit.

Type 1: Sponge-Like Retinal Swelling

This is the most common configuration of DME; approximately 90% of patients with DME who are examined using OCT show evidence of sponge-like retinal thickening (Figures 22-1 and 22-2).[11,13] It is identified as increased retinal thickness with reduced intraretinal reflectivity. The retina resembles a sponge with the internal layers maintaining normal reflectivity, with areas of low reflectivity located in the outer layers. Sponge-like retinal thickening may occur in an isolated setting, where it may portend a poor prognosis or be seen in association with both CME and serous retinal detachment.[17]

Type 2: Cystoid Macular Edema

Round or oval hyporeflective areas on OCT are consistent with intraretinal cystoid space formation (Figure 22-3). This form of edema, isolated or in combination with other patterns, presents in approximately 50% of patients with DME.[11,13] Panozzo et al[11] further classified this type of DME according to the number of cystoid

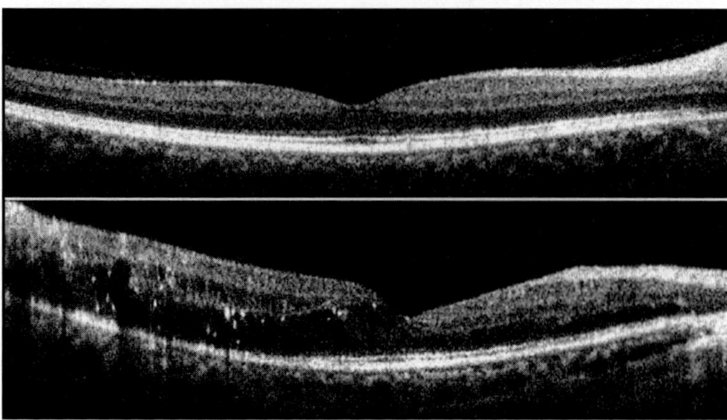

Figure 22-1. RTVue Fourier-domain OCT image, horizontal line centered at fovea, scan length = 8 mm. Top: normal eye. Bottom: CSME patient, left side of the scan shows outer retinal spongy thickening, loss of the definition of the IS/OS line, and multiple highly reflective deposits with relative preservation of the foveal contour.

spaces into Type 2A (presence of 1 or 2 small cysts in the inner plexiform layer; the normal macular profile is still visible but the central foveal thickness is >325 μm), Type 2B (petaloid pattern of intraretinal cysts; central foveal thickness >485 μm), and Type 2C (chronic cystoid edema, major intraretinal damage with cysts fused in large cavities resembling retinoschisis; central foveal thickness >405 μm).

Type 3: Serous Retinal Detachment

Although rare (15%), subfoveal shallow retinal detachment can be seen in some cases of DME.[11,13] Serous retinal detachment is seen on OCT as an optically clear/hyporeflective space between the outer layer of photoreceptors and the highly hyper-reflective retinal pigment epithelium (RPE) band. Serous retinal detachment is not easily discernible on clinical examination and does not appear to be associated with a specific pattern of leakage on fluorescein angiography.[18,19] The presence of a serous retinal detachment was initially thought to signal a severe stage of DME.[17] However, this is in contrast to recent reports that document serous retinal detachment in eyes where the central retinal thickness above the detachment is within the normal range.[20] Furthermore, serous retinal detachment may resolve despite worsening of DME, and the presence of this feature on OCT does not seem to correlate with a poor prognosis.[20,21]

Type 4: Tractional Retinal Detachment

In some cases, DME can be associated with changes of the internal retinal surface, characterized by diffuse or focal traction by the thickened, taut posterior hyaloid, thickened ILM; or by epiretinal membranes (Figure 22-4).[15,22] The presence of tangential or anterior-posterior

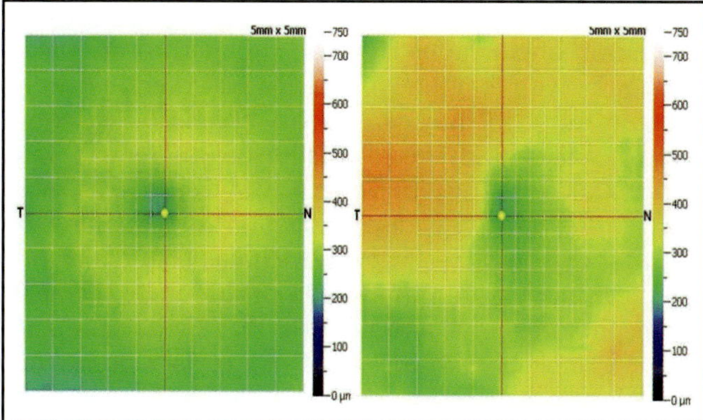

Figure 22-2. Corresponding total retinal thickness map created by RTVue Fourier-domain OCT based on MM5 scan. Left: normal eye from Figure 22-1. Right: CSME patient from Figure 22-1.

forces can be demonstrated by OCT and may help in directing the patient to surgery as the most appropriate therapeutic intervention.[11] There is increasing evidence that these forces may play a role in the determination of edema.[15,23] Nasrallah and associates were among the first to suggest that the vitreous may play an important role in the pathogenesis of DME.[24-26] They found that diabetic patients, 60 years of age or older, with macular edema had significantly higher prevalence of attached posterior vitreous than diabetics without macular edema. There are several reports of resolution of DME following either spontaneous or surgically induced posterior vitreous detachment (PVD).[27,28] Evaluation of the vitreomacular interface is an essential component in the work up of in patients with DME, particularly when retinal thickening is present on clinical examination without fluorescein angiographic evidence of significant leakage or ischemia.[29]

Quantitative Optical Coherence Tomography Assessment of Diabetic Macular Edema

Using Time-Domain Optical Coherence Tomography

Hee et al analyzed 73 normal eyes from 41 healthy volunteers (mean age, 38 years, range 23 to 79 years).[30] The average thickness (mean ± SD) in an area of 500 μm in diameter centered in the fovea, was 174 ± 18 μm and never exceeded 216 μm, while central foveal thickness was 152 ± 21 μm. Their data have been further confirmed by different authors.[11,13]

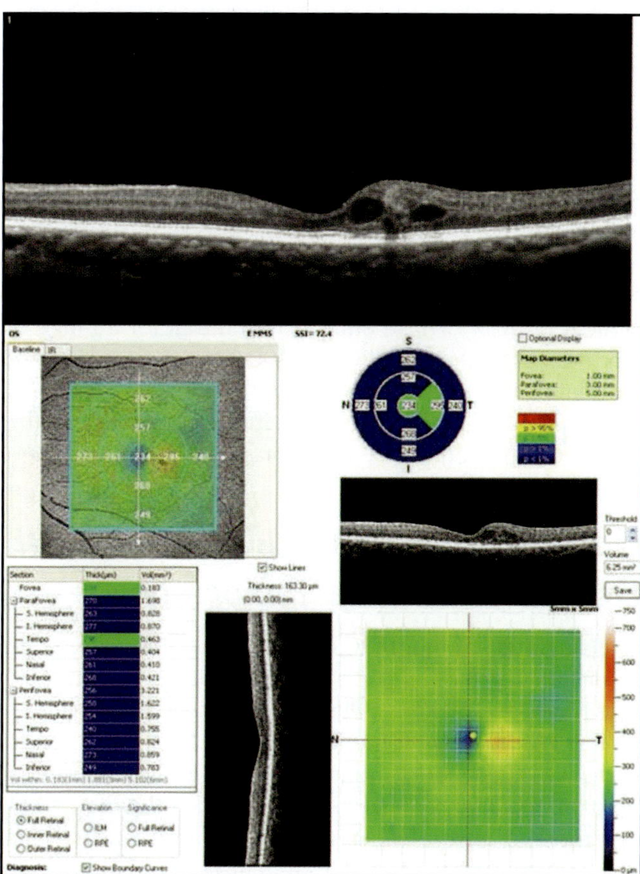

Figure 22-3. High-definition line scan (16x averaged lines) through the fovea illustrating an example of Type 2A cystoid maculopathy, with only 2 cysts seen in the OPL. Bottom map illustrates the focal location of those cysts, and the relative preservation of the foveal contour with their effect on the temporal parafovea. In this particular patient the maximum thickening does not exceed 300 μm, which makes this probable subclinical thickening.

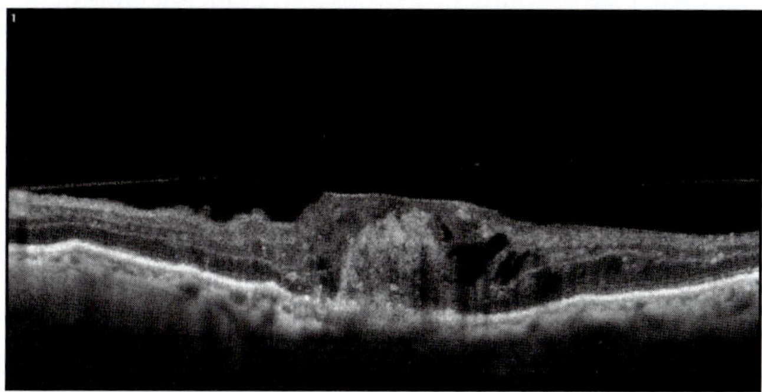

Figure 22-4. High-definition OCT of the fovea demonstrating vitreous traction, taut posterior hyaloid, and subfoveal cholesterol. This patient had a chronic tractional component contributing to her DME (focal vitreous traction, taut posterior hyaloids, and epiretinal membrane formation). The presence of organized subfoveal cholesterol (highly reflective deposits above the RPE) disrupting the outer retinal integrity including the IS/OS junction and ONL, portends a very poor prognosis.

Central Foveal Thickness in Diabetic Macular Edema

Panozzo et al analyzed the OCT macular thickness in 136 eyes with diabetic retinopathy.[11] Similar to other authors, they considered the diagnosis of DME in patients with central foveal thickness of 200 µm or more.[11,13,30] DME that is detected on OCT, but not on biomicroscopic examination has been termed "subclinical DME."[31] These authors found that for a range of mild foveal thickening (200 to 300 µm), there was poor correlation between biomicroscopic detection of foveal thickening and OCT detection, further highlighting the improved detection of macular thickening as determined by OCT (see Figure 22-3).

Grading Diabetic Macular Edema with Optical Coherence Tomography

Using time-domain OCT, Sadda et al[32] developed an algorithm for Stratus (Carl Zeiss Meditec, Dublin, CA) Fast Macular Thickness Map (FMTM) scan and Macular Grid 5 (MG5) to automatically detect DME. Their software was designed to grade DME as CSME or nonsignificant macular edema. An automatic algorithm was designed to detect the retinal boundaries and extract a retinal thickness map. The retinal thickness map was then compared with a normative database to identify presumed areas of retinal edema and to establish a grading system of CSME. The analysis algorithm seems to be accurate and repeatable. Automated detection of CSME by the MG5 analysis correlated well with the clinical grading and standard OCT analysis (ie, FMTM).[32]

Movement artifacts are known to plague Stratus time-domain OCT, which has a scanning speed of 400 A-scans per second. On that system, both FMTM and MG5 will take 1.92 seconds. Saccades and microsaccades can reduce the accuracy and repeatability of the thickness map, since this device does not track fixation and retinal location during scanning.

The RTVue Fourier-domain OCT allows a scanning speed of 26,000 A-scans per second, 65 times faster than Stratus time-domain OCT. It also doubles the axial resolution to 5 µm. It can provide high-resolution cross-section images by averaging multiple frames (see Figure 22-1). It also provides a MM5 scan which can be done in 0.78 seconds and provides more detailed retinal thickness maps (see Figures 22-2 and 22-3).

MACULAR ISCHEMIA

Clinical detection of macular ischemia is challenging, especially in the presence of CME. In diabetic patients with macular capillary nonperfusion, Unoki et al[33] stud-ied OCT scans and correlated the finding to areas of fluorescein angiographic capillary nonperfusion. These authors showed that the inner retina was predominantly disorganized in the area of capillary nonperfusion, compared to adjacent, perfused locations of the retina. They demonstrated that the outer retinal thickness was relatively preserved, with disorganization and thinning of the inner retina. Interestingly, they characterized a highly reflective band within the region of the outer segment of the photoreceptor and the RPE that specifically and focally coincides with the areas of capillary nonperfusion.[33]

OPTICAL COHERENCE TOMOGRAPHY-GUIDED TREATMENT OF DIABETIC MACULAR EDEMA

Laser photocoagulation remains the mainstay of therapy for DME.[9,34] Pharmacotherapies such as corticosteroids or anti-VEGF agents are usually reserved for cases of DME that are refractory to laser treatment. In more severe cases, or in cases where clinical examination and OCT reveal evidence of tractional components to the macular thickening, surgical intervention in the form of pars plana vitrectomy with or without peeling of the tractional membranes is often considered. The effect of each treatment has to be evaluated in light of the changes in visual acuity, as well as an objective assessment of the retinal structural changes. Fourier-domain OCT allows an objective and reproducible measurement of retinal thickness as well as detailed, high-resolution maps of tractional and structural retinal changes, which has made it an attractive tool to monitor the results of surgical and medical interventions in recent literature (Figures 22-5 and 22-6).[21,29,35,36]

FUTURE DIRECTIONS

Software Improvements

Similar to the application on time-domain OCT,[32] future software improvements in RTVue Fourier-domain OCT may potentially explore an automated DME classification method. One possible approach can compare a given patient's MM5 scan against a normative database, yielding a probability index for his or her overall as well as his or her focal macular thickness. This is approach is somewhat similar to probability indices on a visual field test, or a retinal nerve fiber layer scan.

Automated quantification of individual retinal layers, serous retinal detachment, or CME may be of use

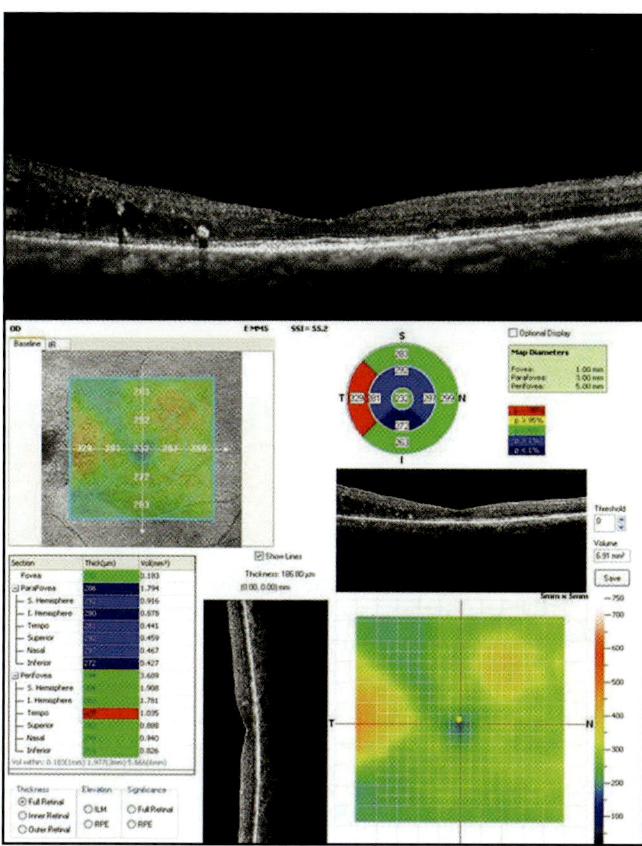

Figure 22-5. Top: shows a horizontal high-definition, frame-averaged scan through the fovea, which highlights the temporal thickening and presence of highly reflective lipid deposits intraretinally with preservation of the foveal contour. The map (bottom) highlights the extent of the macular edema, its maximum thickness, and relative proximity to the central fovea (ETDRS map overlaid on the fundus reconstruction). This information is very helpful not only in planning the laser treatment, but also in illustrating the disease process to the patient, as an educational tool.

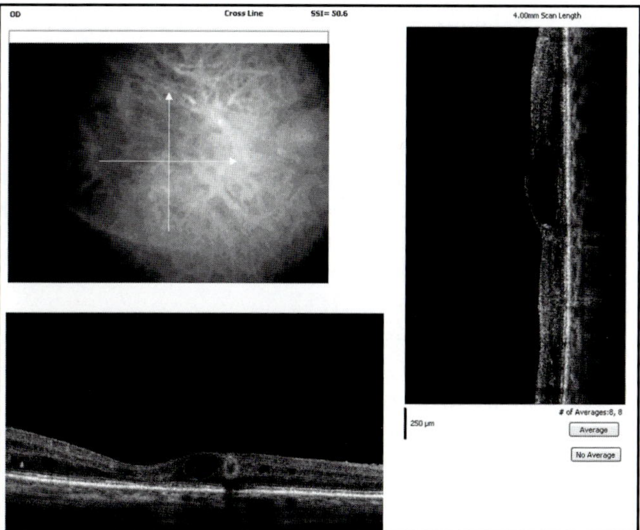

Figure 22-6. Cross-hair scan. Each cross-sectional image is an average of 8 frames. The horizontal scan demonstrates a large microaneurysm in the OPL, adjacent to a large foveal cyst. The microaneurysm can be differentiated from the cyst due to its thick hyper-reflective wall, compared to the thin wall of the cysts.

in patients with DME, but is not possible with currently available commercial software. Assessment of other, more novel, parameters encoded in OCT images (eg, signal intensity) may provide additional prognostic information for patients with DME.[28,37]

Doppler Optical Coherence Tomography for Measuring Retinal Blood Flow and Flow Velocity

Ophthalmologists currently do not have a standardized reproducible and clinically applicable tool to measure retinal blood flow *in vivo*. Fluorescein angiography is used in clinical practice to evaluate retinal blood flow. It is a relatively invasive test that involves the injection of an intravenous dye to show areas of capillary occlusion, new blood vessel formation, and arteriovenous shunting. It does not allow for accurate quantitative measurements of retinal blood flow. Using ultrasound color Doppler imaging (CDI), previous authors have shown decreased blood velocity in diabetic patients with proliferative diabetic retinopathy,[38] as well as in patients with nonproliferative disease, which was suggestive of increased resistance in peripheral microvasculature.[39,40] In contrast, other authors have identified higher blood flow velocity in the central retinal vein than the artery in a group of patients with nonproliferative diabetic retinopathy.[41] These controversies show the limitations of CDI measurements, which are limited to larger retrobulbar blood vessels, and are unable to quantify total retinal blood flow. Knowledge of the diameter of the blood vessels is needed in order to permit quantification of blood flow; however, CDI technology cannot accurately measure vessel diameter. Laser Doppler flowmetry is another optical method used to measure retinal hemodynamics. In a group of well-controlled Type 1 diabetic patients, Lorenzi et al showed normal retinal blood flow, compared to controls.[42] Unlike laser Doppler flow, which cannot spatially resolve the origin of the Doppler signal in retinal tissue, OCT Doppler permits depth localization with unprecedented axial resolution. Scanning laser Doppler is limited by the confocal aperture, which limits the axial resolution of that technology to 300 µm.[43] Although these flow or velocity measurement techniques were able to demonstrate the difference between diabetic and normal eyes as groups, they are not accurate enough to diagnose abnormality in individual eyes.

Recently, human total retinal blood flow measurements using Doppler Fourier-domain OCT have become possible.[44,45] As outlined in Chapter 10, this approach allows calculation of the total retinal venous flow by obtaining measurements of blood velocity and vessel diameter in all the retinal veins around the optic nerve. In a pilot study of diabetic patients, Wang et al[46] have shown that retinal blood flow in a diabetic patient without retinopathy was normal, while patients with treated proliferative diabetic retinopathy had blood flows that were lower than normal, similar to the findings of others using other blood flow measurement approaches and technologies.[47-50]

Further patient-based studies and large-scale data collection are underway to determine the reliability and reproducibility of Doppler OCT as an objective tool for evaluating retinal blood flow in diabetics. Furthermore, systematic prospective research studies will determine the role that can be played by Doppler OCT in the clinical management of diabetic patients.

REFERENCES

1. Huang D, Swanson EA, Lin CP, et al. Optical coherence tomography. *Science.* 1991;254(5035):1178-1181.
2. Fercher AF, Hitzenberger CK, Drexler W, Kamp G, Sattmann H. In-vivo optical coherence tomography. *Am J Ophthalmol.* 1993;116(1):113-115.
3. Swanson EA, Izatt JA, Hee MR, et al. In-vivo retinal imaging by optical coherence tomography. *Optics Letters.* 1993;18(21):1864-1866.
4. Mohamed Q, Gillies MC, Wong TY. Management of diabetic retinopathy: A systematic review. *JAMA.* 2007;298:902-916.
5. Massin P, Girach A, Erginay A, et al. Optical coherence tomography: A key to the future management of patients with diabetic macular oedema. *Acta Ophthalmol.* 2006;84:466-474.
6. Witmer AN, Vrensen GF, VanNoorden CJ, Schlingemann RO. Vascular endothelial growth factors and angiogenesis in eye disease. *Prog Retin Eye Res.* 2003;22(1):1-29.
7. Bhagat N, Grigorian R, Tutela A, Zarbin M. Diabetic macular edema: Pathogenesis and treatment. *Surv Ophthalmol.* 2009;54(1):1-32.
8. Early Treatment Diabetic Retinopathy Study Group. Early Treatment Diabetic Retinopathy Study design and baseline patient characteristics. ETDRS report number 7. *Ophthalmology.* 1991;98:741-756.
9. Early Treatment Diabetic Retinopathy Study Research Group. Early photocoagulation for diabetic retinopathy. ETDRS report number 9. *Ophthalmology.* 1991;98:766-785.
10. Nussenblatt RB, Kaufman SC, Palestina AG, et al. Macular thickening and visual acuity. Measurement in patients with cystoid macular edema. *Ophthalmology.* 1987;94:1134-1139.
11. Panozzo G, Gusson E, Parolini B, Mercanti A. Role of OCT in the diagnosis and follow up of diabetic macular edema. *Sem Ophthalmol.* 2003;18:74-81.
12. Hee MR, Izatt JA, Swanson EA, et al. Optical coherence tomography of the human retina. *Arch Ophthalmol.* 1995;113:325-332.
13. Otani T, Kishi S, Maruyama Y. Patterns of diabetic macular edema with optical coherence tomography. *Am J Ophthalmol.* 1999;127:688-693.
14. Yamamoto S, Yamamoto T, Hayashi M, et al. Morphological and functional analyses of diabetic macular edema by optical coherence tomography and multifocal electroretinograms. *Graefes Arch Clin Ophthalmol.* 2001;239:96-101.
15. Kaiser PK, Riemann CD, Sears JE, et al. Macular traction detachment and diabetic macular edema associated with posterior hyaloidal traction. *Am J Ophthalmol.* 2001;131:44-49.
16. Brasil OF, Smith SD, Galor A, et al. Predictive factors for short-term visual outcome after intravitreal triamcinolone acetonide injection for diabetic macular oedema: An optical coherence tomography study. *Br J Ophthalmol.* 2007;91:761-765.
17. Alkuraya H, Kangave D, Abu El-Asrar AM. The correlation between optical coherence tomographic features and severity of retinopathy, macular thickness and visual acuity in diabetic macular edema. *Int Ophthalmol.* 2005;26:93-99.
18. Soliman W, Sander B, Jørgensen TM. Enhanced optical coherence patterns of diabetic macular oedema and their correlation with the pathophysiology. *Acta Ophthalmol.* 2007;85:613-617.
19. Soliman W, Sander B, Hasler PW, et al. Correlation between intraretinal changes in diabetic macular oedema seen in fluorescein angiography and optical coherence tomography. *Acta Ophthalmol.* 2008;86:34-39.
20. Gaucher D, Sebah C, Erginay A, et al. Optical coherence tomography features during the evolution of serous retinal detachment in patients with diabetic macular edema. *Am J Ophthalmol.* 2008;145:289-296.
21. Ozdemir H, Karacorlu M, Karacorlu SA. Regression of serous macular detachment after intravitreal triamcinolone acetonide in patients with diabetic macular edema. *Am J Ophthalmol.* 2005;140:251-255.
22. Yamamoto T, Akabane N, Takeuchi S. Vitrectomy for diabetic macular edema: the role of posterior vitreous detachment and epimacular membrane. *Am J Ophthalmol.* 2001;132:369-377.
23. Ikeda T, Sato K, Katano T, Hyashi Y. Attached posterior hyaloid membrane and the pathogenesis of honeycombed cystoid macular edema in patients with diabetes. *Am J Ophthalmol.* 1999;127(4):478-479.
24. Nasrallah FP, Alex JE, Van Coppenolle F, et al. The role of the vitreous in diabetic macular edema. *Ophthalmology.* 1988;95:1335-1339.
25. Nasrallah FP, Van de Velde F, Jalkh AE, et al. Importance of the vitreous in young diabetics with macular edema. *Ophthalmology.* 1989;96(10):1511-1516, discussion 1516-1517.
26. Sebag J. Anatomy and pathology of the vitreo-retinal interface. *Eye.* 1992;6(Pt6):541-552.
27. Hikichi T, Fujio N, Akiba J, et al. Association between the short-term natural history of diabetic macular edema and the vitreomacular relationship in type II diabetes mellitus. *Ophthalmology.* 1997;104(3):473-478.
28. Tachi N, Ogino N. Vitrectomy for diffuse macular edema in cases of diabetic retinopathy. *Am J Ophthalmol.* 1996;8:258-260.
29. Ghazi NG, Ciralsky JB, Shah SM, et al. Optical coherence tomography findings in persistent diabetic macular edema: The vitreomacular interface. *Am J Ophthalmol.* 2007;144:747-754.
30. Hee MR, Puliafito CA, Duker JS, et al. Topography of diabetic macular edema with optical coherence tomography. *Ophthalmology.* 1998;105:360-370.
31. Brown JC, Solomon SD, Bressler SB, et al. Detection of diabetic foveal edema: Contact lens biomicroscopy compared with optical coherence tomography. *Arch Ophthalmol.* 2004;122:330-335.
32. Sadda SR, Tan O, Walsh AC, et al. Automated detection of clinically significant macular edema by grid scanning optical coherence tomography. *Ophthalmology.* 2006;113:1187,e1-12.
33. Unoki N, Nishijima K, Sakamoto A, et al. Retinal sensitivity loss and structural disturbance in areas of capillary nonperfusion of eyes with diabetic retinopathy. *Am J Ophthalmol.* 2007;144:755-760.
34. Mohamed Q, Gillies MC, Wong TY. Management of diabetic retinopathy: A systematic review. *JAMA.* 2007;298:902-916.
35. Otani T, Kishi S. Tomographic assessment of vitreous surgery for diabetic macular edema. *Am J Ophthalmol.* 2000;129(4):487-494.
36. Gibran SK, Khan K, Jungkim S, et al. Optical coherence tomographic pattern may predict visual outcome after intravitreal triamcinolone for diabetic macular edema. *Ophthalmology.* 2007;114:890-894.

37. Barthelmes D, Sutter FK, Gillies MC. Differential optical densities of intraretinal spaces. *Invest Ophthalmol Vis Sci.* 2008;49:3529-3534.

38. Mendivil A, Cuartero V, Mendivil MP. Ocular blood flow velocities in patients with proliferative diabetic retinopathy and healthy volunteers: A prospective study. *Br J Ophthalmol.* 1995;79:413.

39. Guven D, Ozdemir H, Hasanreisoglu B. Hemodynamic alterations in diabetic retinopathy. *Ophthalmology.* 1996;103:1245-1249.

40. MacKinnon JR, McKillop G, O'Brien C, et al. Colour Doppler imaging of the ocular circulation in diabetic retinopathy. *Acta Ophthalmologica Scandinavica.* 2000;78:386-389.

41. Fujioka S, Karashima K, Nishikawa N, Saito Y. Correlation between higher blood flow velocity in the central retinal vein than in the central retinal artery and severity of nonproliferative diabetic retinopathy. *Jpn J Ophthalmol.* 2006;50:312-317.

42. Lorenzi M, Feke GT, Cagliero E, et al. Retinal haemodynamics in individuals with well-controlled type 1 diabetes. *Diabetologia.* 2008;51:361-364.

43. Yazdanfar S, Rollins AM, Izatt JA. In vivo imaging of human retinal flow dynamics by color Doppler optical coherence tomography. *Arch Ophthalmol.* 2003;121:235.

44. Wang Y, Bower BA, Izatt JA, Tan O, Huang D. Retinal blood flow measurement by circumpapillary Fourier domain Doppler optical coherence tomography. *J Biomed Opt.* 2008;13(6):064003.

45. Wang Y, Lu A, Gil-Flamer J, Tan O, Izatt JA, Huang D. Measurement of total blood flow in the normal human retina using Doppler Fourier-domain optical coherence tomography. *Br J Ophthalmol.* 2009;93(5):634-7, Epub 2009 Jan 23.

46. Wang Y, Fawzi A, Tan O, Gil-Flamer J, Huang D. Retinal blood flow detection in diabetic patients by Doppler Fourier domain optical coherence tomography. *Optics Express.* 2009;17(5):4061-4073.

47. Oswald B, Vilser W, Oswald H, et al. Measurement of flow physiologic parameters of retinal blood circulation in type 1 and 2 diabetics before and after photocoagulation. *Graefes Arch Clin Exp Ophthalmol.* 1985;223:154-157.

48. Grunwald JE, Riva CE, Brucker AJ, Sinclair SH, Petrig BL. Effect of panretinal photocoagulation on retinal blood flow in proliferative diabetic retinopathy. *Ophthalmology.* 1986;93:590-595.

49. Grunwald JE, Riva CE, Sinclair SH, Brucker AJ, Petrig BL. Laser Doppler velocimetry study of retinal circulation in diabetes mellitus. *Arch Ophthalmol.* 1986;104:991-996.

50. Grunwald JE, Brucker AJ, Petrig BL, Riva CE. Retinal blood flow regulation and the clinical response to panretinal photocoagulation in proliferative diabetic retinopathy. *Ophthalmology.* 1989;96:1518-1522.

23 Vitreomacular Traction Syndrome

Andre J. Witkin, MD and Jay S. Duker, MD

As a person ages, the vitreous liquefies and detaches from the retina. Posteriorly, the vitreous is bound by a thin membrane of condensed collagen, termed the posterior hyaloid. Typically, the vitreous is most adherent to the vitreous base, which encompasses the ora serrata; less firm adhesions are present at the optic nerve and in the macula. In some individuals, adhesions between the vitreous and the macula are abnormally firm, and as the vitreous begins to separate from the retina, tractional forces are generated in areas of vitreomacular adhesion. Vitreomacular traction is the generating force behind a number of disease entities. Full- or partial-thickness macular holes may be formed as vitreomacular traction pulls on the central macula. Subretinal fluid and/or macular edema may also develop; this entity is most commonly referred to as vitreomacular traction syndrome.

With the advent of OCT, the diagnosis of vitreomacular traction was revolutionized, as a detached posterior hyaloid is often invisible to the examiner but is easily visible using OCT imaging. On OCT, a partially detached posterior hyaloid appears as a thin, moderately reflective line anterior to the neural retina. In vitreomacular traction syndrome, this line can be seen bowing posteriorly and attaching to the macula at the site of adhesion, causing increased macular thickness, cystoid macular edema, or appearance of subretinal fluid (Figures 23-1 and 23-2).

Often, an associated epiretinal membrane may be visualized on OCT as a highly reflective line just anterior to the nerve fiber layer, which is usually adherent to the retina in numerous areas (Figure 23-3). In some cases, the epiretinal membrane may in fact extend from the retinal surface anteriorly onto the attached hyaloid face.

Occasionally, vitreomacular attachment may be present on OCT imaging without symptomatic change in vision. OCT has demonstrated that during the progression of normal posterior vitreous detachment, the posterior hyaloid first separates from the temporal retina, then between the fovea and optic nerve, followed by the fovea, and then the hyaloid finally detaches from the optic nerve—creating the classic Weiss ring visible on biomicroscopy. Vitreofoveal attachment is therefore a common finding in OCT imaging. In some cases, subtle changes in the macular contour may be found, which are sometimes associated with metamorphopsia and other visual changes (Figure 23-4).

The clinical course of vitreomacular traction syndrome is varied. Many patients have only mild symptoms, while others note more severe visual disability. Some patients may have a spontaneous resolution of traction as the posterior hyaloid completes its separation from the macula. In other cases, adherence between the vitreous and macula remains, and vitrectomy surgery may be warranted to peel the posterior hyaloid (with or without an associated epiretinal membrane) from the macula.

The increased axial resolution and high scan density of the RTVue allows the observer to view adhesions of the vitreous to the retina in 3D. Small abnormalities within the neural retina, such as disruptions in the photoreceptor inner/outer segment (IS/OS) junction layer, may be visualized as a result of tractional forces generated on the anterior surface of the macula. Surgical planning may be aided by visualization of hyaloid and epiretinal membranes in 3D. Macular maps using the EMM5 protocol may be used to define statistically significant macular thickening and follow thickness over

Huang D, Duker JS, Fujimoto JG, Lumbroso B, Schuman JS, Weinreb RN.
Imaging the Eye from Front to Back with RTVue Fourier-Domain Optical Coherence Tomography (pp 203-206).

Figure 23-1. Color scale line scan from an earlier version of the RTVue software. It is a case of vitreomacular traction syndrome in an eye with 20/50 vision. Note the moderately reflective line of posterior hyaloid, partially separated from the macula while remaining firmly attached at the fovea. The fovea is thickened and there is a small area of elevation of the retina from the RPE, most obviously visualized as a drawbridge-like lifting of the highly reflective photoreceptor IS/OS junction line.

Figure 23-2. Black-on-white cross scan from the RTVue. This is a case of vitreomacular traction in an eye with 20/40 vision. Again, the moderately reflective line of posterior hyaloid is partially separated from the macula while remaining firmly attached at the fovea. In this eye, there appears to be a lamellar separation between the inner and outer foveal layers. Note that this color scheme highlights higher reflectivity in darker color; the posterior hyaloid may be easier to visualize using this black-on-white scheme.

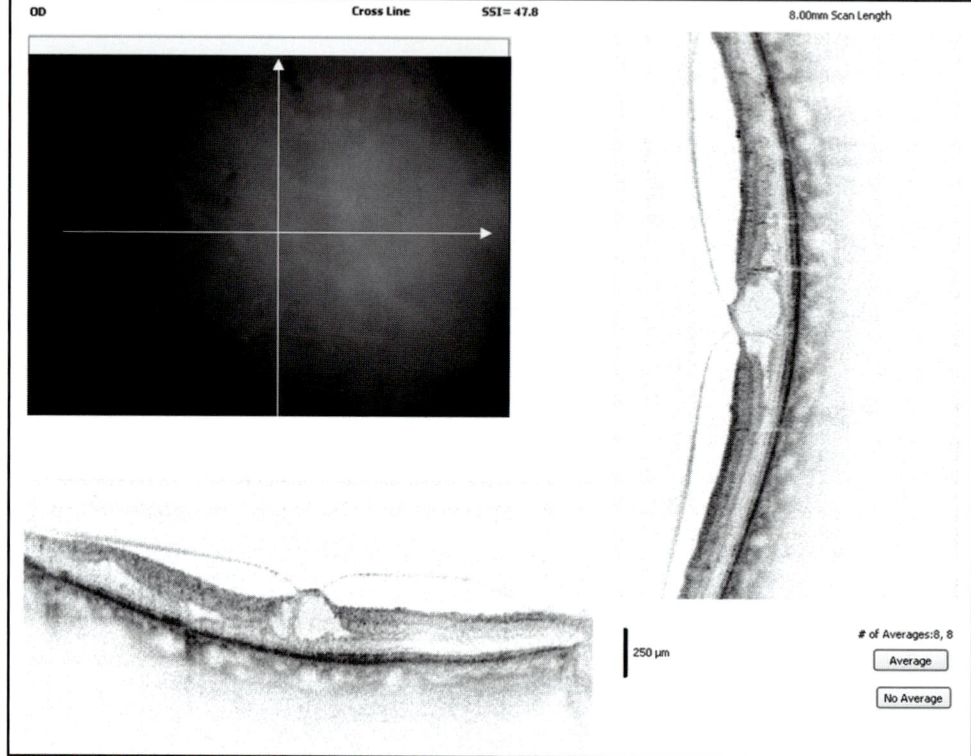

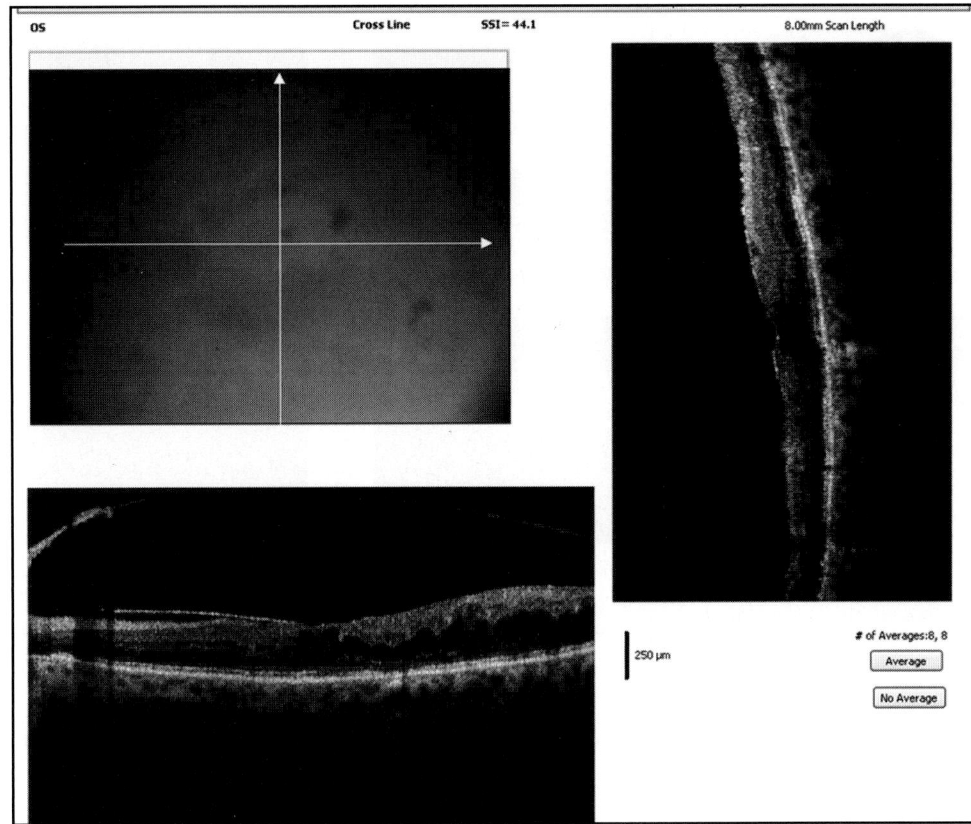

Figure 23-3. White-on-black cross scan from the RTVue. This is a case of vitreomacular traction in an eye with 20/60 vision. There is a schisis, or split, within the posterior hyaloid, with part of it detached from and bowing over the macula, and part still attached to the fovea (evident on the horizontal image). A thin epiretinal membrane is also present and tightly adherent to the macula (evident on the vertical image). There are also obvious intraretinal cystic spaces of fluid. This white-on-black scheme may help visualize some smaller intraretinal details compared to the chromatic image color scheme as demonstrated in Figure 23-1.

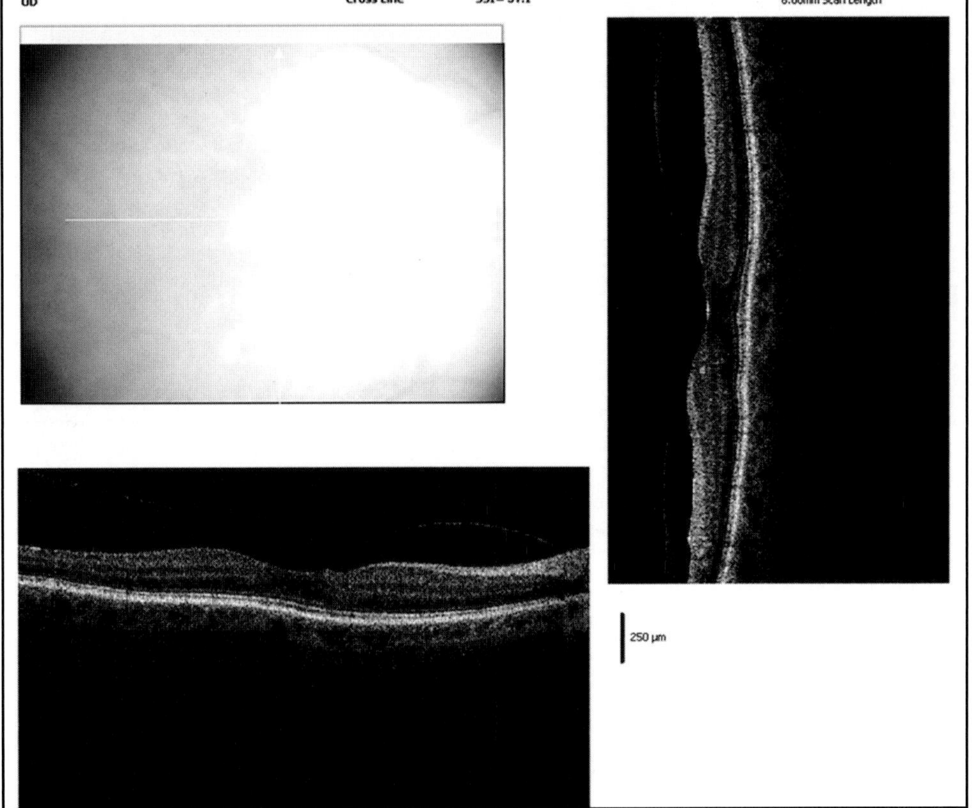

Figure 23-4. White-on-black cross scan from the RTVue. This image is from the fellow eye of the eye imaged in Figure 23-1. Vision is 20/20 and the patient is asymptomatic in this eye. In this case, the posterior hyaloid remains attached to the fovea; in locations of attachment, there are small irregularities in the foveal contour. This may be a stage in the normal process of posterior detachment, or may represent a milder form of abnormal vitreomacular adhesions.

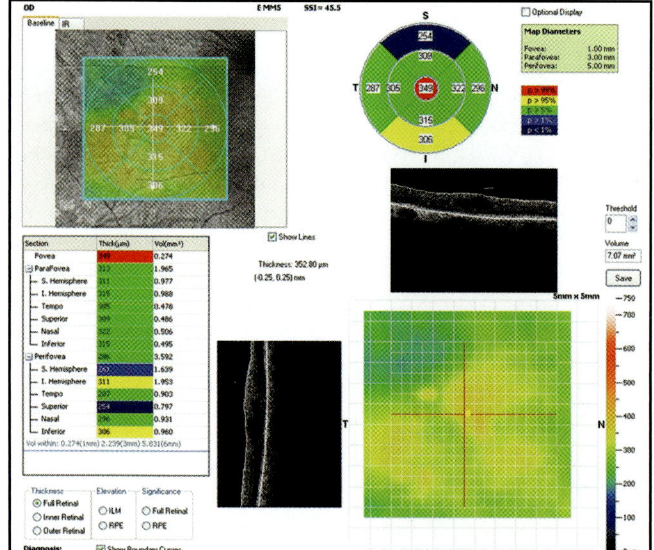

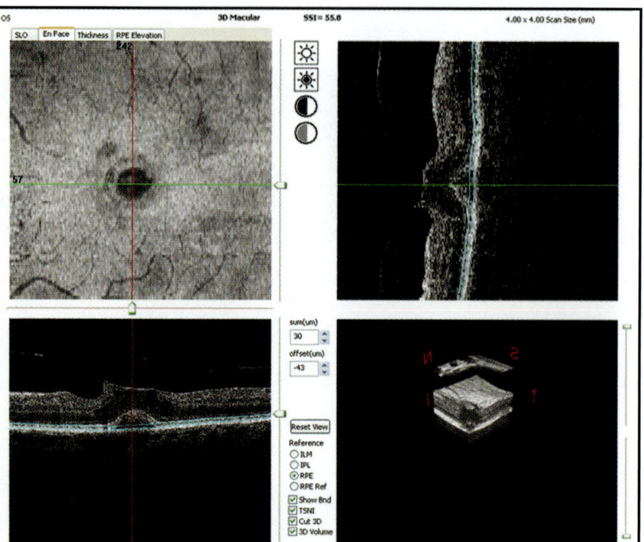

Figure 23-5. EMM5 scan from the same eye as shown in Figure 23-2. Note a statistically significant area of thickening in the foveal circle of the ETDRS map (highlighted in red), while other areas in the map remain of relatively normal thickness. A detailed macular thickness map is also shown, overlaid on an *en face* fundus reconstruction map.

Figure 23-6. Macular scan from the same eye as shown in Figure 23-1. An *en face* slice is created in the same contour as the RPE line, and moved just anterior to the RPE into an area of subretinal fluid. The size of the small circular foveal detachment may then be visualized on the *en face* SLO OCT image.

time either to monitor disease progression or return to normal macular thickness after medical or surgical treatment (Figures 23-5 and 23-6).

Notably, the posterior hyaloid is often mistaken as the inner retinal border by OCT thickness measurement software and seen as falsely elevated areas of macular thickness within OCT mapping displays. On the EMM5 display, these artifacts are often seen as "spokes" of falsely elevated macular thickness (Figure 23-7).

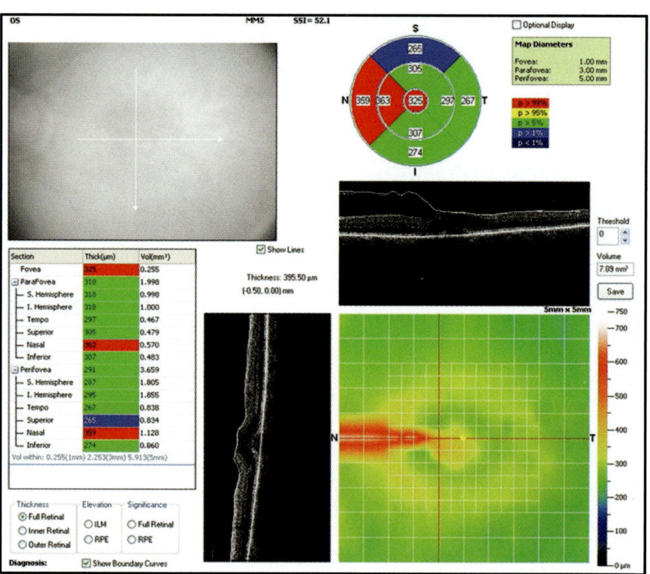

Figure 23-7. MM5 scan from the same eye as shown in Figure 23-1, using an older version of RTVue software. Note a spoke of increased thickness (red) within the macular thickness map, which corresponds to an artifactual measurement of the posterior hyaloid as the inner retinal border. This posterior hyaloid artifact is common in vitreomacular traction syndrome, and care must be taken when using macular thickness measurements in these patients.

SUGGESTED READING

Chang LK, Fine HF, Spaide RF, Koizumi H, Grossniklaus HE. Ultrastructural correlation of spectral-domain optical coherence tomographic findings in vitreomacular traction syndrome. *Am J Ophthalmol.* 2008;146:121-127.

Jaffe NS. Vitreous traction at the posterior pole of the fundus due to alterations in the vitreous posterior. *Trans Am Acad Ophthalmol Otolaryngol.* 1967;71:642-652.

Johnson MW. Tractional cystoid macular edema: A subtle variant of the vitreomacular traction syndrome. *Am J Ophthalmol.* 2005;140:184-192.

Maumenee AE. Further advances in the study of the macula. *Arch Ophthalmol.* 1967;78:151-165.

McDonald HR, Johnson RN, Schatz H. Surgical results in the vitreomacular traction syndrome. *Ophthalmology.* 1994;101:1397-1403.

Smiddy WE. Vitreomacular traction syndrome. In: Yanoff M, Duker JS, eds. *Ophthalmology.* 2nd ed. St. Louis, MO: CV Mosby; 2004:951-955.

Smiddy WE, Michels RG, Glaser BM, deBustros S. Vitrectomy for macular traction caused by incomplete vitreous separation. *Arch Ophthalmol.* 1988;106:624-628.

Smiddy WE, Michels RG, Green WR. Morphology, pathology, and surgery of idiopathic vitreoretinal macular disorders. *Retina.* 1990;10:288-296.

Uchino E, Uemura A, Ohba N. Initial stages of posterior vitreous detachment in healthy eyes of older persons evaluated by optical coherence tomography. *Arch Ophthalmol.* 2001;119:1475-1479.

Witkin AJ, Ko TH, Fujimoto JG, et al. Vitreofoveal attachment causing metamorphopsia: an ultrahigh-resolution optical coherence tomography finding. *Retina.* 2006;26(9):1085-1087.

24 Posterior Uveitis

Nobuyuki Ohguro, MD

INTRODUCTION

Intraocular inflammation often increases vascular permeability and results in retinal edema. The migration of inflammatory cells into the eye also can lead to morphologic changes in the retina. In chronic inflammation, fibrosis or atrophy may replace the pathological lesion. The inflammatory process often is associated with dynamic changes from the onset to the late phase of the pathology.

Fluorescein angiography may be the most widely used tool for detecting intraocular inflammation, especially in the posterior segment. However, this examination is inadequate for evaluating the depth of retinal and subretinal lesions. Fluorescein angiography is also invasive, although the associated risk is not high and the adverse effects are not so serious. OCT has been a widely accepted retinal imaging technique and has become a powerful tool for identifying pathological retinal lesions, especially in the posterior pole.[1]

Many kinds of diseases, including systemic autoimmune diseases, infectious diseases, and malignancies, can cause posterior uveitis. Whatever the cause, posterior uveitis can result in retinal morphologic damage that can best be seen in cross-sectional imaging. Therefore, a combination planar evaluation using fluorescein angiography and cross-sectional OCT images provides useful information about localization of the inflammation, disease activity, and diagnosis.

CASE 1: MACULAR EDEMA DUE TO IDIOPATHIC POSTERIOR UVEITIS

Because macular edema is the leading cause of decreased vision in patients with uveitis,[2] evaluation and quantification of macular edema is essential to managing patients with posterior uveitis.

A 22-year-old woman complained of blurred vision in her right eye in which the best-corrected visual acuity (BCVA) was 12/20. The left eye was normal (Figure 24-1).

CASE 2: ACUTE PHASE OF VOGT-KOYANAGI-HARADA DISEASE

Vogt-Koyanagi-Harada (VKH) disease is a granulomatous inflammatory disorder that affects the eyes, auditory system, meninges, and skin. The ocular manifestations include uveitis associated with exudative retinal detachment, iridocyclitis, diffuse choroidal thickening, and hyperemia of the optic disc. Depigmentation, including a sunset-glow fundus, vitiligo, and poliosis, is a typical clinical feature in the late phase.

A 55-year-old woman complained of bilateral decreased vision. The BCVA was 8/20 in the right eye and 12/20 in the left eye (Figure 24-2).

Huang D, Duker JS, Fujimoto JG, Lumbroso B, Schuman JS, Weinreb RN.
Imaging the Eye from Front to Back with RTVue Fourier-Domain Optical Coherence Tomography (pp 207-212).

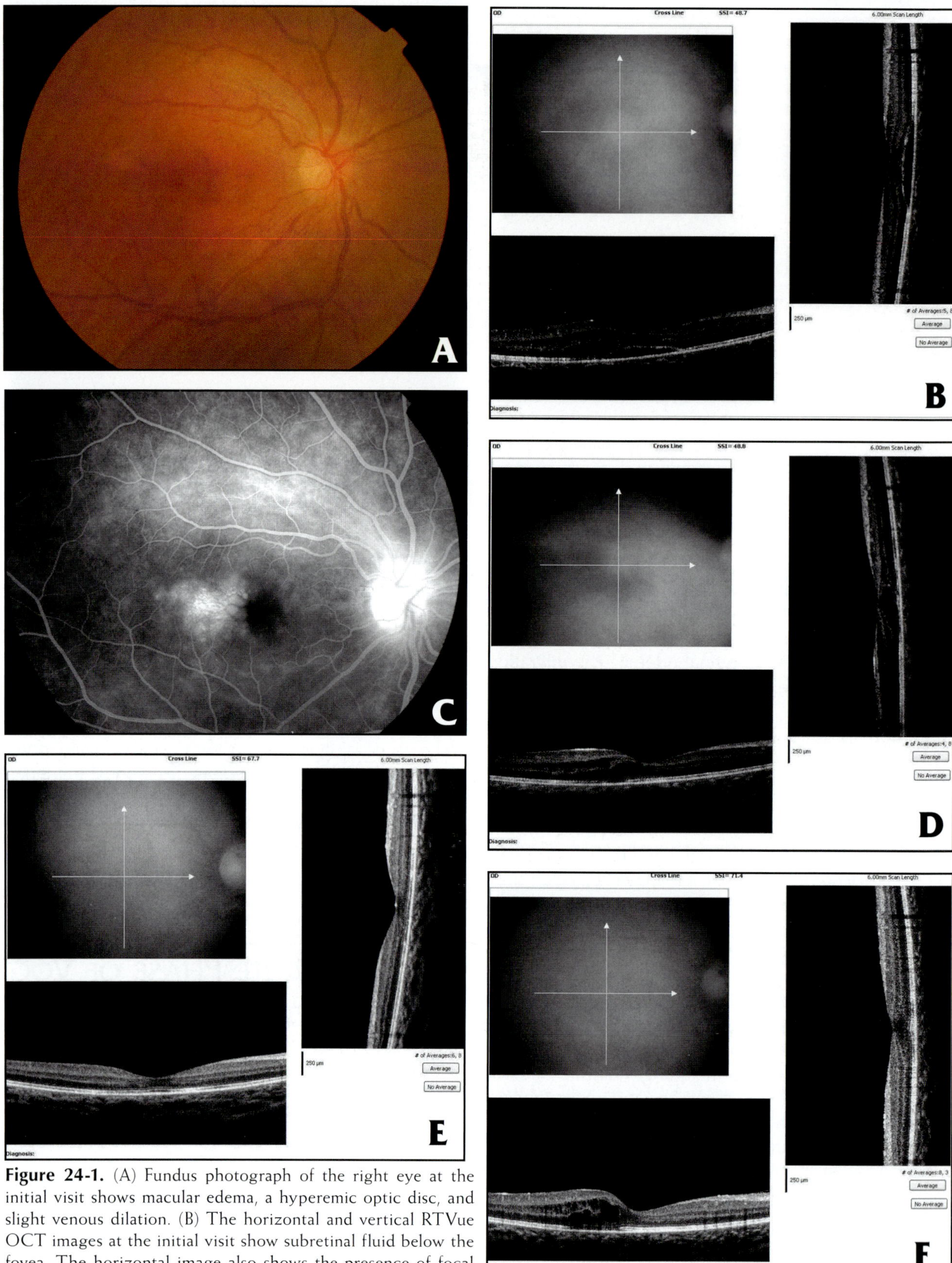

Figure 24-1. (A) Fundus photograph of the right eye at the initial visit shows macular edema, a hyperemic optic disc, and slight venous dilation. (B) The horizontal and vertical RTVue OCT images at the initial visit show subretinal fluid below the fovea. The horizontal image also shows the presence of focal edema. (C) Despite treatment with systemic prednisolone (30 mg) for 1 week, subsequent fluorescein angiography shows that cystoid macular edema (CME) and a hyperfluorescent optic disc are still present. (D) RTVue OCT images obtained on the same day as fluorescein angiography show several persistent intraretinal cysts. (E) One month later, the inflammation has subsided and the RTVue OCT images appear almost normal. (F) When the systemic corticosteroid was tapered to 15 mg, the CME recurred. The horizontal RTVue OCT images show focal CME.

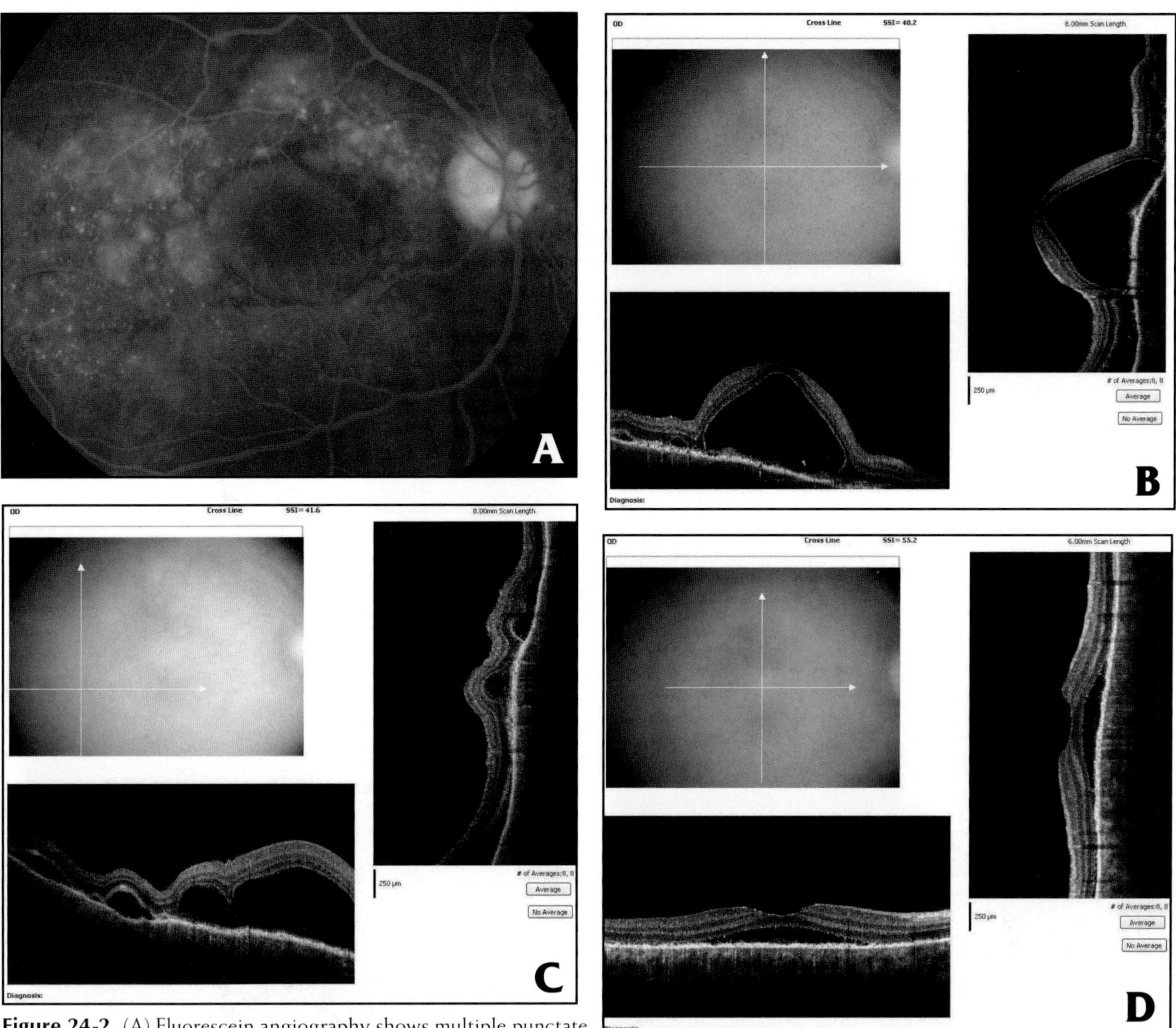

Figure 24-2. (A) Fluorescein angiography shows multiple punctate hyperfluorescent dots at the level of the RPE, and pooled fluorescence at the subretinal space. (B, C) RTVue OCT images show serous retinal detachments with septa that consist of a fibrinous membrane. The septa correspond to the dark line at the margin of the lobular hyperfluorescent areas on fluorescein angiography, as described previously.[3] (D) After corticosteroid pulse therapy (methylprednisolone 1g/d, 3 consecutive days), the serous retinal detachment dramatically improved, but residual subretinal fluid remains.

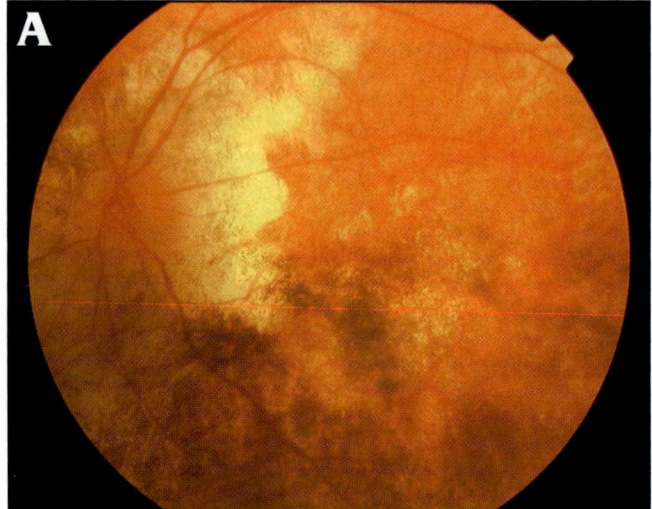

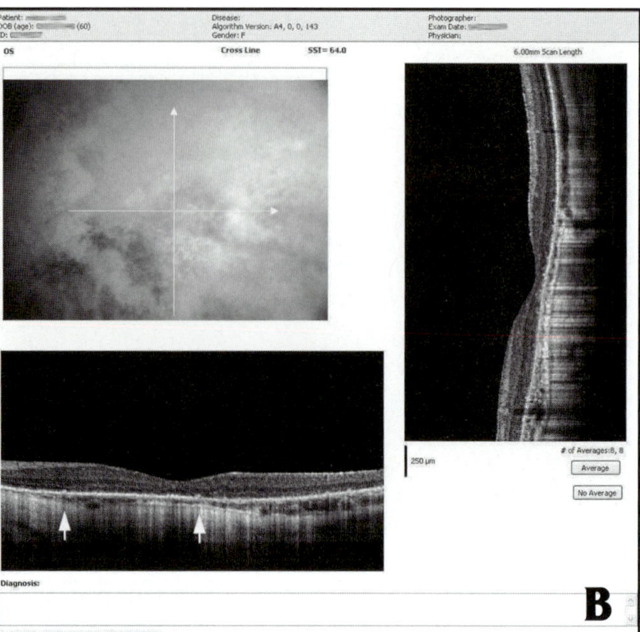

Figure 24-3. (A) Current fundus photograph of the right eye shows a sunset-glow fundus and severe chorioretinal scars and alterations of the RPE. (B) RTVue OCT images show that the structure of the retinal layer appears normal, except for an irregular IS/OS line and RPE hyperplasia (arrows).

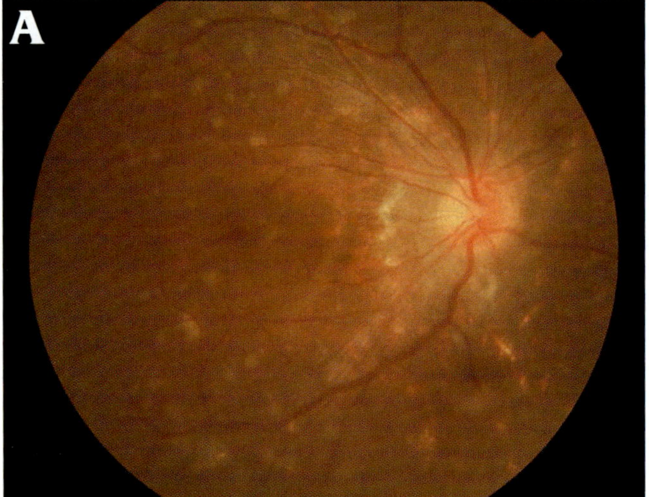

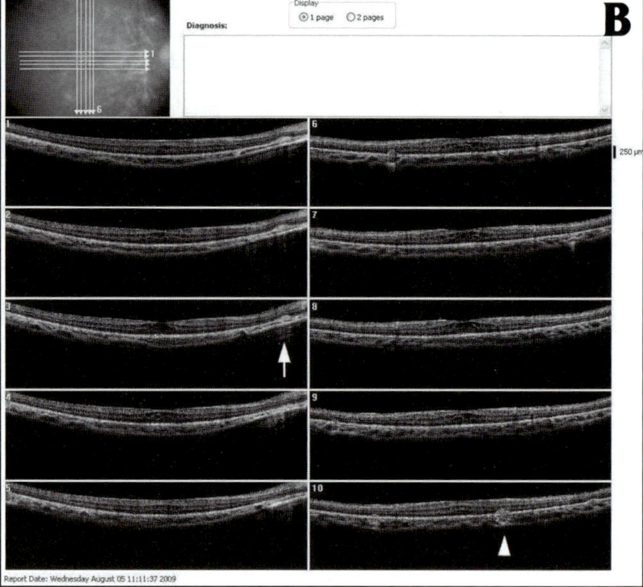

Figure 24-4. (A) Fundus photograph of the right eye shows the typical appearance of MFC. (B) In cross-sectional RTVue OCT image 3, typical transretinal hyper-reflectivity is seen (arrow).[4] In image 10, the typical appearance of disorganized RPE (arrowhead) is seen.[4]

CASE 3: LATE-PHASE VOGT-KOYANAGI-HARADA DISEASE

A 68-year-old woman developed VKH disease 20 years previously. The disease activity subsided soon after administration of a systemic corticosteroid. The BCVA was 12/20 in the right eye and 10/100 in the left eye (Figure 24-3).

CASE 4: MULTIFOCAL CHOROIDITIS

The primary clinical features of acute-phase multifocal choroiditis (MFC) are multiple round-to-oval, yellow-gray lesions at the level of the retinal pigment epithelium (RPE). These lesions subsequently change to scar tissue with proliferation of RPE and fibrosis, which often leads to choroidal neovascularization.

A 16-year-old girl complained of bilateral visual loss 1 year previously. The clinical appearance was compatible with MFC, which was refractory to a systemic steroid. We administered infliximab (intravenous infusion 8 mg/every 8 weeks), which had a substantial effect in this case (Figure 24-4).

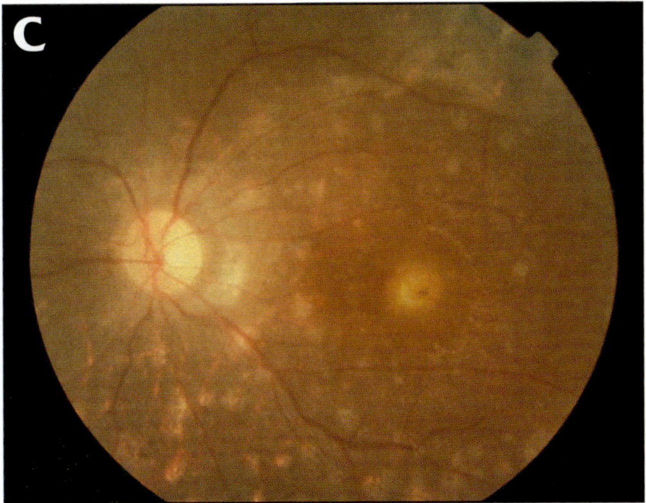

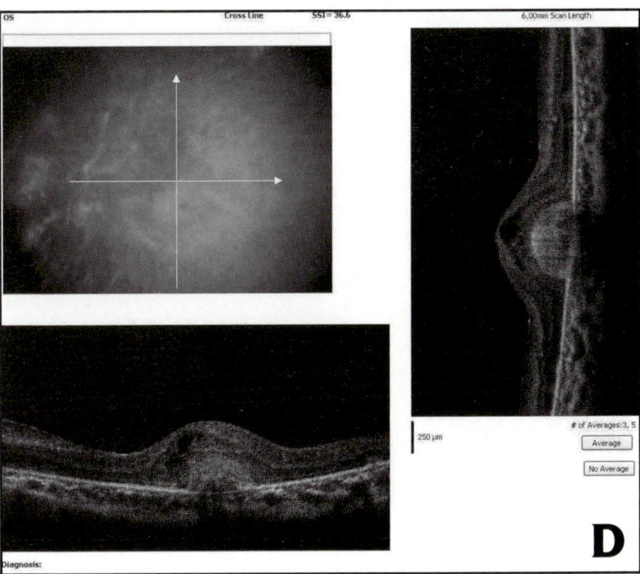

Figure 24-4. (C) A fundus photograph of the left eye suggests the presence of a choroidal neovascular membrane. (D) RTVue OCT images clearly show choroidal vascularization.

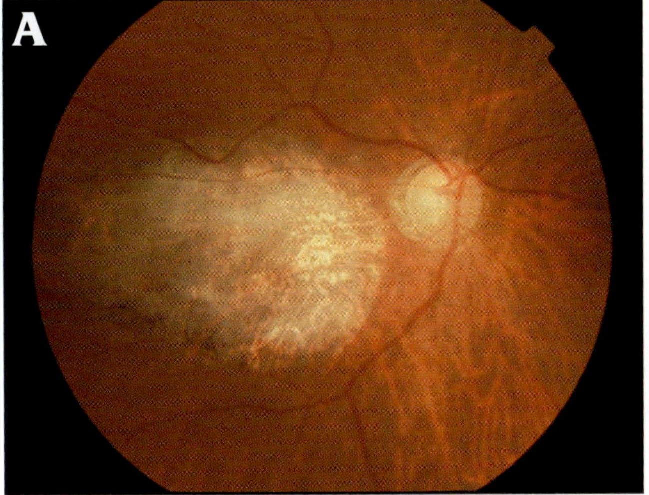

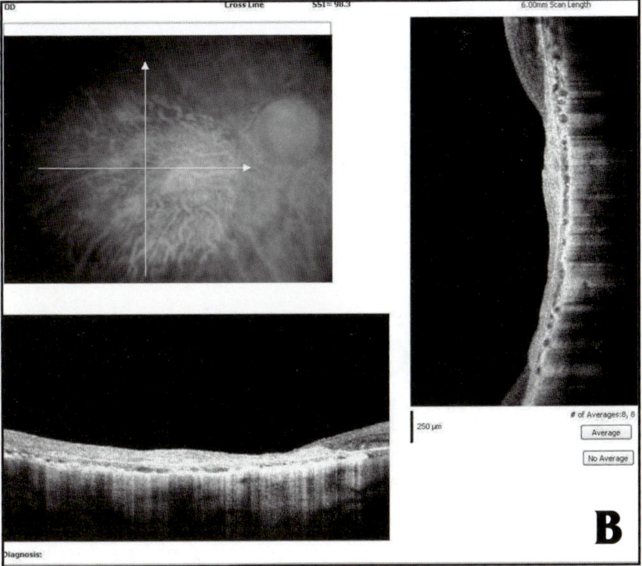

Figure 24-5. (A) In the atrophic area, bare sclera is partially observed. (B) An RTVue OCT image clearly shows scarring resulting from necrotizing retinitis and complete loss of neurosensory retinal tissue.

Case 5: Chorioretinal Scar in Acquired Ocular Toxoplasmosis

Acquired ocular toxoplasmosis generally is characterized by one unilateral active lesion without a previous chorioretinal scar. The typical fundus has a yellow-white exudate with ill-defined borders. A healed scar has well-defined borders with central chorioretinal atrophy and peripheral RPE hyperplasia.

A 75-year-old woman developed acquired ocular toxoplasmosis in her right eye 3 years previously. The ocular inflammation subsided with treatment that included a combination of an oral corticosteroid and antibiotics. The current BCVA in the right eye is 10/100 (Figure 24-5).

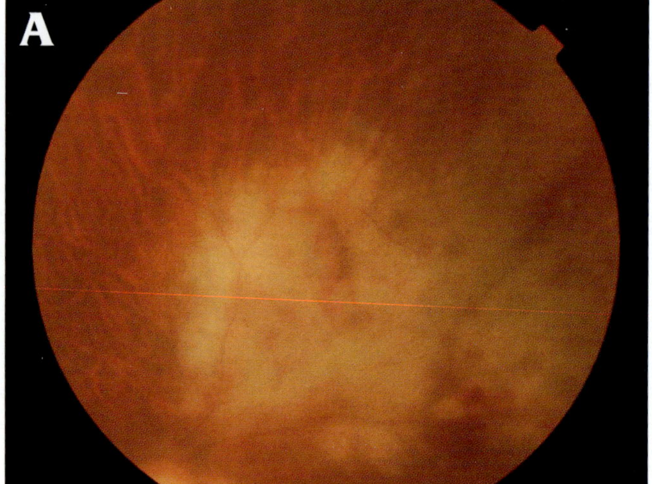

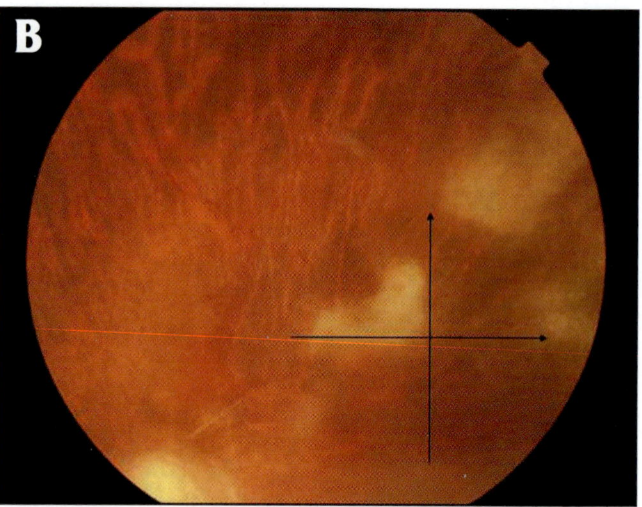

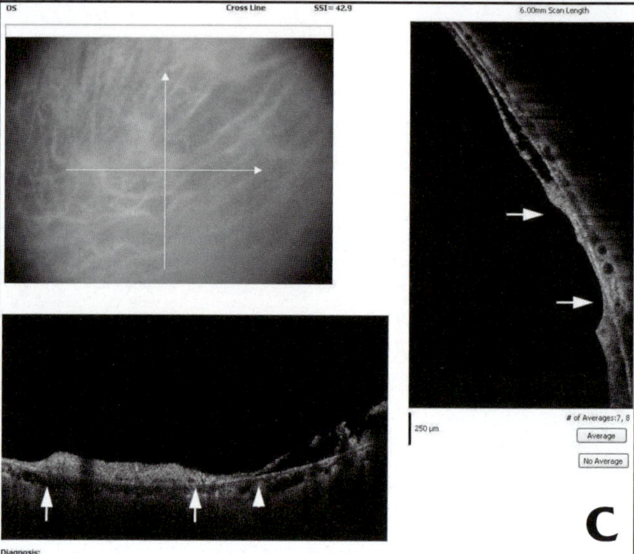

Figure 24-6. (A) Fundus photograph of the left eye at the initial visit shows fluffy, dense, white retinal infiltrations with multiple retinal hemorrhages. (B) One month later, the retinitis has subsided with residual atrophic lesions. Some lesions were still active. The arrows indicate the area scanned by RTVue OCT and seen in C. (C) RTVue OCT images clearly show atrophic lesions and a severely thinned retina (on the left of the left arrow and between the right arrow and the arrowhead in the horizontal image and between the arrows in the horizontal image). Focal retinal schisis also is seen on the right of the arrowhead in the horizontal image and above the upper arrow in the horizontal image. There is no normal retinal structure in the rest of the active area between the arrows in the horizontal image.

CASE 6: CYTOMEGALOVIRUS RETINITIS

Cytomegalovirus (CMV) in the retina usually occurs as an opportunistic infection in immunosuppressed individuals and tends to develop in patients who are the most immunosuppressed with an HIV infection. Typically, CMV retinitis is a slow-growing necrotizing retinitis that begins as a small white retinal infiltrate resembling a large cotton-wool spot. The affected area often has irregular borders and is surrounded by satellite lesions.

A 55-year-old man who had undergone chemotherapy for Hodgkin's disease complained of blurred vision in his left eye. Ophthalmic examination showed CMV retinitis. An anti-CMV drug was immediately started (Figure 24-6).

REFERENCES

1. Huang D, Swanson EA, Lin CP, et al. Optical coherence tomography. *Science*. 1991;254:1178-1181.
2. Malinowski SM, Pulido JS, Folk JC. Long-term visual outcome and complications in patients with uveitis. *Ophthalmology*. 1993;100:818-825.
3. Yamaguchi Y, Otani T, Kishi S. Tomographic features of serous retinal detachment with multilobular dye pooling in acute Vogt-Koyanagi-Harada disease. *Am J Ophthalmol*. 2007;144:260-265.
4. Gallagher MJ, Yilmaz T, Cervantes-Castaneda RA, et al. The characteristic features of optical coherence tomography in posterior uveitis. *Br J Ophthalmol*. 2007;91:1680-1685.

25 Diagnosis of Angle Closure

Hiroshi Ishikawa, MD

With the RTVue cornea adaptor modules (CAM), there is one scan type dedicated for angle scanning called literally "angle." This scan pattern performs 16 repeated linear scannings with 1024 A-scan lines along a straight line (1 to 3 mm for CAM-S, 2 to 6 mm for CAM-L), which takes 0.63 seconds (Figure 25-1). Please refer to Chapters 4 and 5 for more details about the RTVue anterior segment imaging.

As described in Chapter 5, the biggest limitation of the RTVue CAM is its poor penetration. Although the structures within the cornea (eg, Schlemm's canal) can be visualized very clearly, detailed angle structure is largely missing, especially the iris and the ciliary body. Since diagnosing a plateau iris configuration requires visualization of the ciliary sulcus region, it is impossible with RTVue CAM. Even just for an angle closure, careful angle scanning with a well-trained technician is needed to see the angle configuration at the vicinity of the scleral spur.

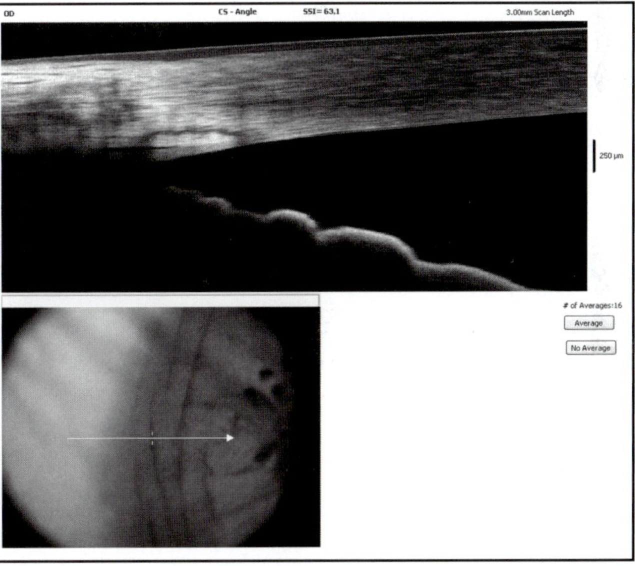

Figure 25-1. Angle scan (CAM-S). A single linear scan is repeated 16 times on the same location. Repeated scans are used to generate 1 averaged image so that speckle noise is reduced and signal-to-noise ratio improved. This sample image was obtained from 9 o'clock on the right eye of a healthy 37-year-old Caucasian female.

QUANTITATIVE ANALYSIS

An angle can be measured by specifying 3 points on an image (Figure 25-2). Most importantly for glaucoma application, an angle opening distance (AOD) and a trabecular-iris segment area (TISA) can also be measured at 500 and 750 µm anterior to the scleral spur with the software assistant.

AOD is first proposed by Pavlin et al for quantitative assessment of anterior chamber angle using ultrasound biomicroscopy (UBM).[1] Later, Ishikawa et al proposed angle recess area (ARA), which is an improved extension of AOD measurement made by measuring an area bounded by a roughly triangular space between the angle recess, corneal endothelium, iris surface, and a line perpendicular to a point on the corneal endothelium that is 750 µm anterior to the scleral spur.[2] Theoretically, ARA provides more accurate angle quantification as it

Huang D, Duker JS, Fujimoto JG, Lumbroso B, Schuman JS, Weinreb RN.
Imaging the Eye from Front to Back with RTVue Fourier-Domain Optical Coherence Tomography (pp 213-218).
© 2010 SLACK Incorporated.

reflects the variable undulation of the iris surface. Both measurements are well established and widely used for quantitative UBM analysis.[3,4]

With the anterior segment OCT, due to its lack of clear visualization of the angle recess and the iris surface near the scleral spur, ARA cannot be measured appropriately. Radhakrishnan et al proposed a further modified method, TISA, which does not require clear visualization of the angle recess, and is therefore more suitable for the anterior segment OCT application.[5] TISA-X is defined as an area bounded by a roughly trapezoidal shape between the scleral spur, a point on the corneal endothelium X µm anterior to the scleral spur, and the intersections of the anterior iris surface with perpendicular lines drawn from the first 2 landmarks. The analysis tools on the RTVue allow TISA-500 and TISA-750 measurement with the software assistant. After locating the scleral spur, a user specifies a point on the corneal endothelium anterior to the scleral spur (distance is automatically locked to either 500 or 750 µm) and 2 more points on the iris surface, where perpendicular lines from the scleral spur and another corneal point intersect. Although they are manually input, as the assisting software does not allow divergence from the perpendicular lines, the measurement consistency is reasonably maintained.

CASE EXAMPLES

Healthy Anterior Chamber Angle

This case involves a 35-year-old healthy Caucasian female. An angle scan visualized detailed anterior segment structures around the anterior chamber angle at 9 o'clock OD (Figure 25-3).

Narrow Angle

A 54-year-old Caucasian female has a narrow but open angle, visualized by angle scan at 9 o'clock OD (Figure 25-4). TISA was also measured on this case (Figure 25-5).

Closed Angle

A 48-year-old Caucasian male presented with partially closed angle, visualized by angle scan at 9 o'clock OS (Figure 25-6). TISA was also measured on this case (Figure 25-7).

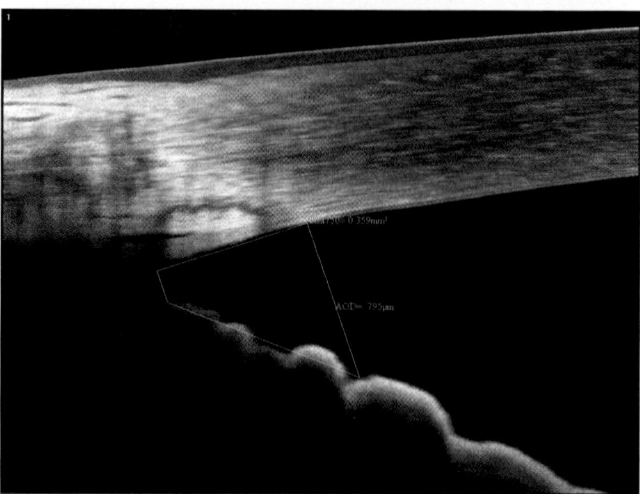

Figure 25-2. TISA measurement. RTVue software assists users to measure the TISA in a semi-automated fashion. First, a user locates the scleral spur location. Then the software displays a trapezoid with 4 movable anchors at each apex. User drags the anchors to the right locations one by one. The distance between the scleral spur and the corneal endothelium point is fixed to either 500 or 750 µm. The angle between the lines from the cornea to the iris is also fixed to the right angle. Therefore, the user can easily locate 4 points without fearing about the variability due to free hand (mouse) drawing. For this healthy open angle, TISA-750 was 0.359 mm^2 and AOD-750 was 795 µm.

CONCLUSION

With the CAM options, RTVue provides an "all-in-one" solution to ocular imaging for the ophthalmology clinic. However, for glaucoma application, incapability of visualizing ciliary processes is a significant disadvantage. In addition, TISA and AOD measurements may be misleading in a case like the Closed Angle example.

REFERENCES

1. Pavlin CJ, Harasiewicz K, Foster FS. Ultrasound biomicroscopy of anterior segment structures in normal and glaucomatous eyes. *Am J Ophthalmol.* 1992;113:381-389.

2. Ishikawa H, Esaki K, Liebmann JM, Uji Y, Ritch R. Ultrasound biomicroscopy dark room provocative testing: a quantitative method for estimating anterior chamber angle width. *Jpn J Ophthalmol.* 1999;43:526-534.

3. Ishikawa H, Schuman JS. Anterior segment imaging: ultrasound biomicroscopy. *Ophthalmol Clin North Am.* 2004;17:7-20.

4. Ishikawa H, Liebmann JM, Ritch R. Quantitative assessment of the anterior segment using ultrasound biomicroscopy. *Curr Opin Ophthalmol.* 2000;11:133-139.

5. Radhakrishnan S, Goldsmith J, Huang D, et al. Comparison of optical coherence tomography and ultrasound biomicroscopy for detection of narrow anterior chamber angles. *Arch Ophthalmol.* 2005;123:1053-1059.

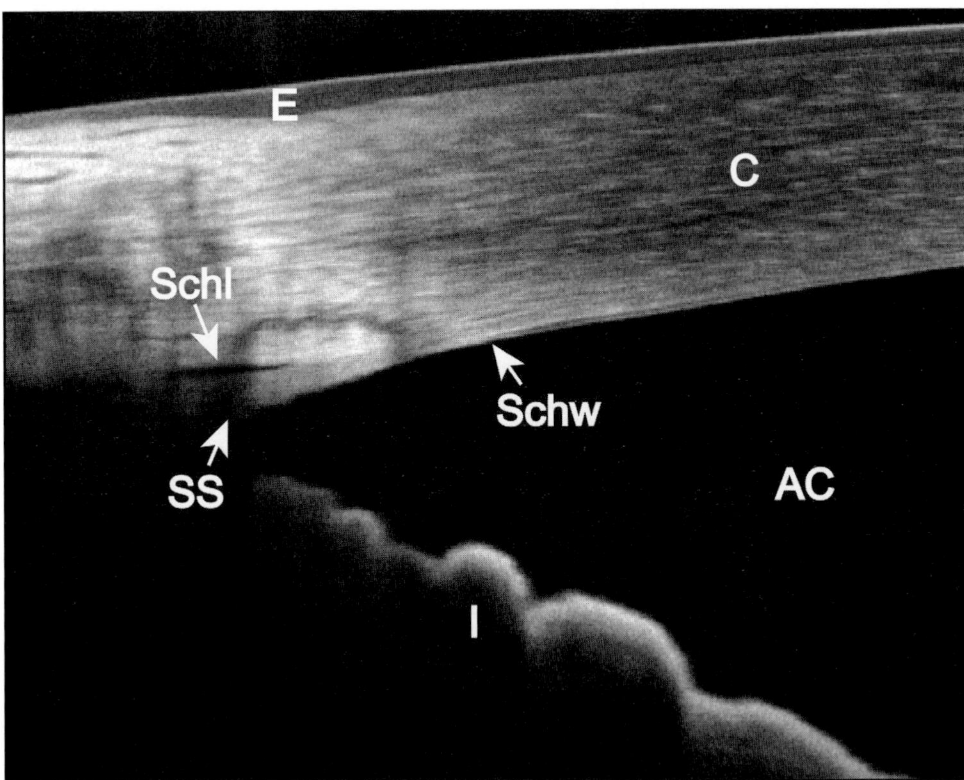

Figure 25-3. Healthy anterior chamber angle. A widely open angle is visualized together with the cornea (C), corneal epithelium (E), anterior chamber (AC), iris (I), scleral spur (SS), Schwalbe's line (Schw), and Schlemm's canal (Schl). Note that the signal level around the scleral spur is weak, and the iris surface is not visible in the angle recess. Also no structures behind the iris are visible.

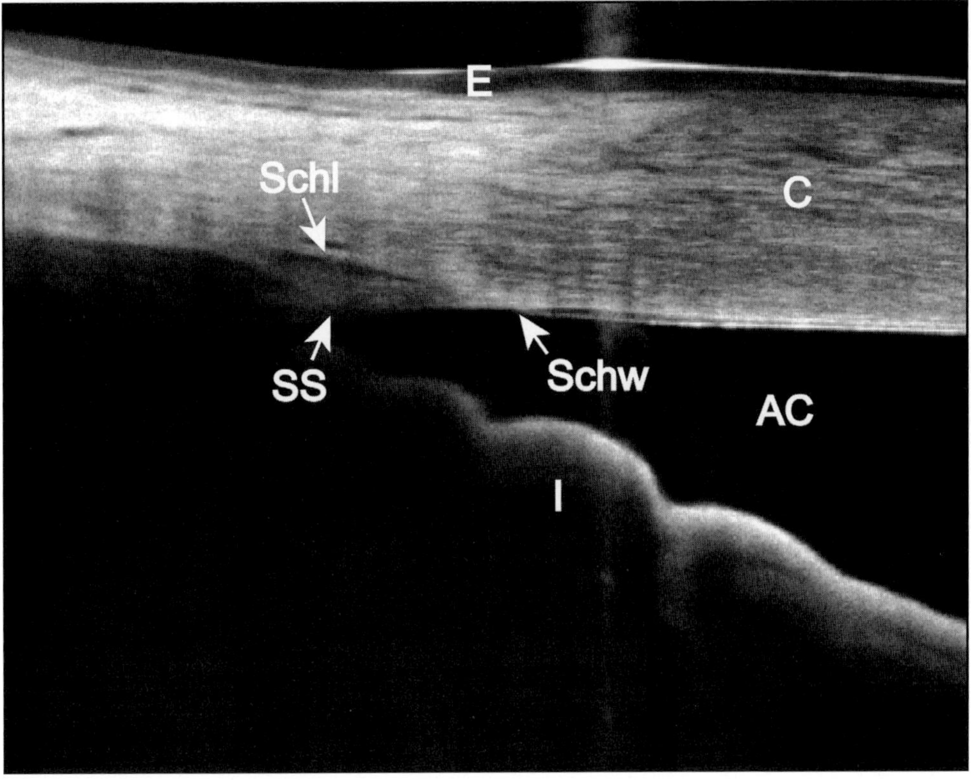

Figure 25-4. Narrow angle. The angle is open but narrow (see Figure 25-3). Labels are the same as Figure 25-3. None of the structures behind the iris are visible; therefore, detecting a plateau iris configuration is not possible.

Figure 25-5. TISA measurement on narrow angle. TISA-750 was 0.160 mm^2, and AOD-750 was 318 μm. Both measurements were less than half of the measurements on the healthy open angle (see Figure 25-2).

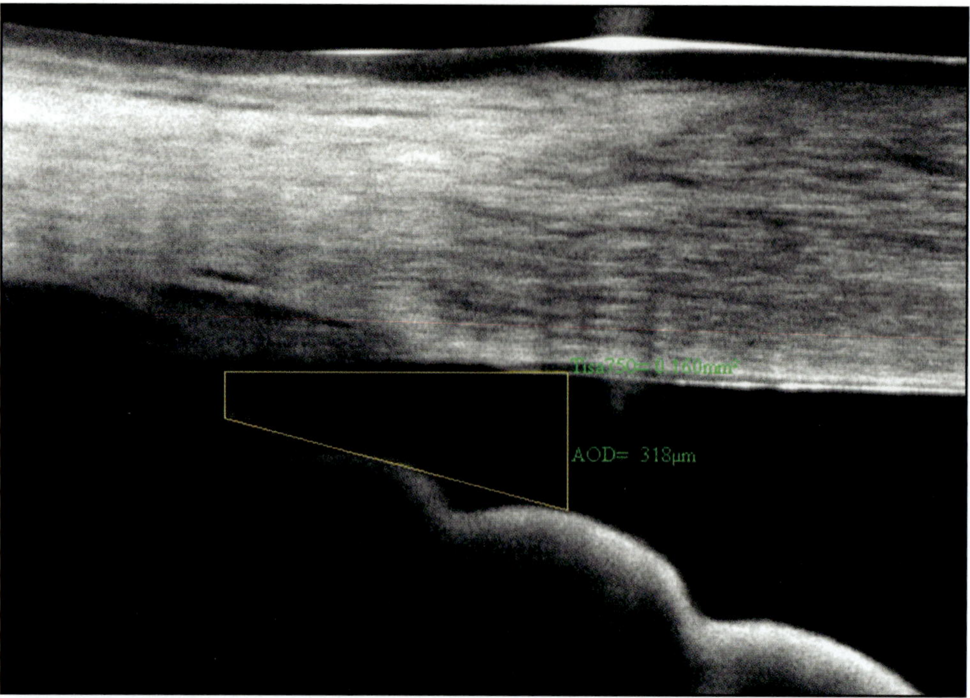

Figure 25-6. Partially closed angle. The angle shows an appositional closure in between the scleral spur and the Schwalbe's line. None of the structures behind the iris are visible. This type of partial angle closure often associates with plateau iris configuration. For the confirmation of a plateau iris configuration, a UBM test is necessary.

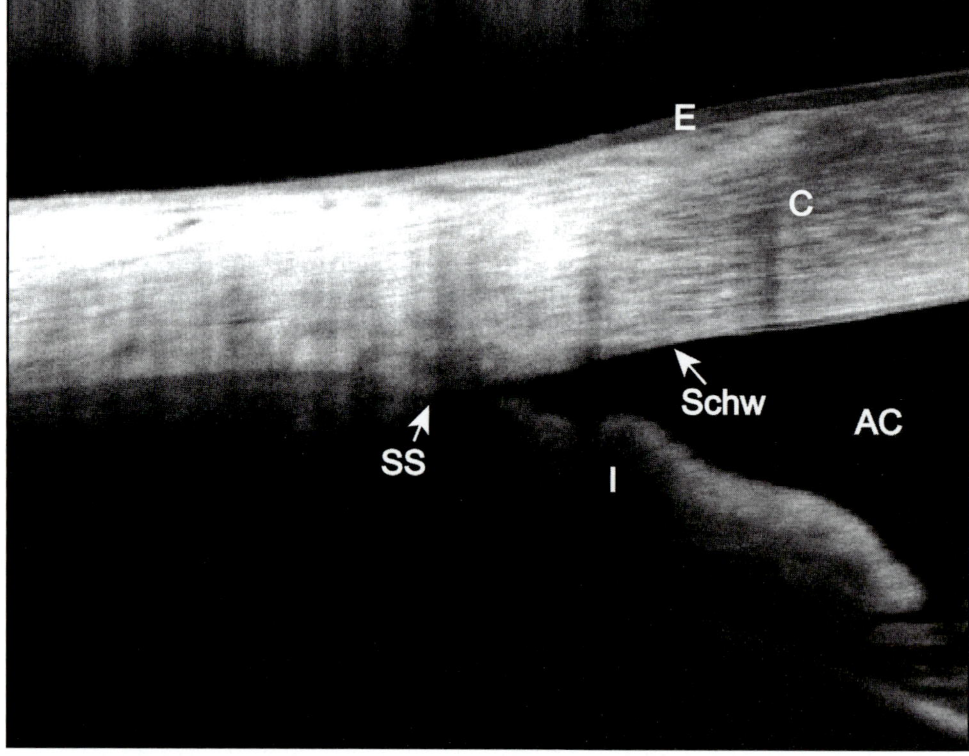

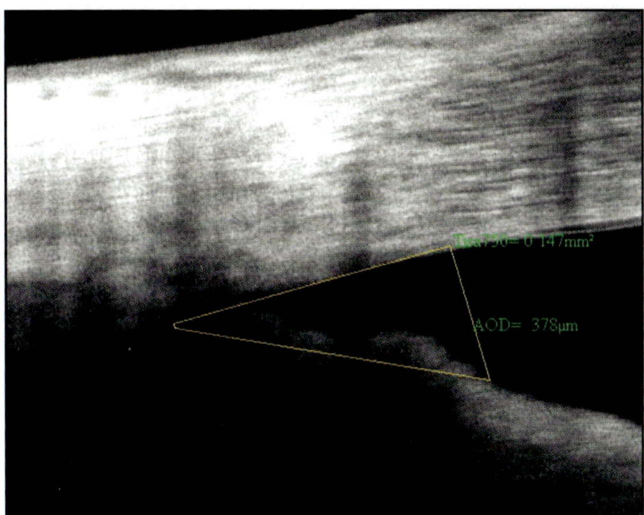

Figure 25-7. TISA measurement on partially closed angle. TISA-750 was 0.147 mm², and AOD-750 was 378 μm. Due to the straight line boundary along the iris surface, TISA-750 overestimated the actual angle opening because it did not take the irregular undulation of the iris surface into account. Interestingly, the AOD-750 was longer than the narrow angle case (see Figure 25-5). This implies that a simple distance measure may be misleading to quantify the angle opening.

26 Diagnosis of Glaucomatous Optic Neuropathy

Lindsey S. Folio; Gadi Wollstein, MD; Hiroshi Ishikawa, MD;
Larry Kagemann, MS BME; and Joel S. Schuman, MD, FACS

NORMAL EYES

Normal Nerve Fiber Layer, Macula, and Optic Nerve Head (Figures 26-1A through G)

A 45-year-old healthy man presented with visual acuity of 20/20, intraocular pressure (IOP) of 12 mmHg, and normal appearing anterior chamber with open angles. Fundus examination was normal and the optic nerve head (ONH) was normal in size had a small central cup with no evidence of retinal nerve fiber layer (NFL) defect (Figure 26-1A). Visual field was normal (Figure 26-1B).

Optical Coherence Tomography

The circumpapillary retinal NFL (Figure 26-1C) scan showed a normal retinal NFL thickness configuration (center) with thickest retinal NFL at the superior and inferior regions and thinnest at the nasal and temporal regions (double hump configuration). Comparison of the measurements to the RTVue normative database showed that the retinal NFL profile, global, and sectoral measurements were all within normal limits (bottom, green background). The NHM4 (nerve head map 4 mm) images (Figure 26-1D) also confirmed normal retinal NFL thickness in the 3.4 mm diameter circular region surrounding the ONH (right middle and bottom) and normal cupping of the ONH. The cross line image (Figure 26-1E) showed a vertical scan through the fovea that demonstrated symmetrical features above and below the fovea with no observable thinning in the macular region. This was confirmed by the MM5 (Figure 26-1F) scan of the full retinal thickness in the macular region that showed typical perifoveal circular thickening surrounding the foveal pit (top). Comparison of the sectoral measurements of full retinal thickness to the machine's normative data showed all measurements to be within normal limits (bottom right). The probability map for difference of the full retinal thickness from normal (bottom left) demonstrated a few minimally significant deviating points (blue) with no clear pathological relevance; this suggested either physiologic anomaly, scan artifact, or both. The ganglion cell complex (GCC) scan (Figure 26-1G) of the macula showed the thickness (bottom) and the deviation from normal (top) maps of the macular retinal NFL and ganglion cell layers. The map typically shows that the tissue is thinnest in the fovea and toward the periphery of the macula. The ganglion cell complex (GCC) parameters including overall GCC average, superior and inferior hemispheric average, and focal and general loss volume, all appeared in the normal region (see Figure 26-1G).

Huang D, Duker JS, Fujimoto JG, Lumbroso B, Schuman JS, Weinreb RN.
Imaging the Eye from Front to Back with RTVue Fourier-Domain Optical Coherence Tomography (pp 219-244).
© 2010 SLACK Incorporated.

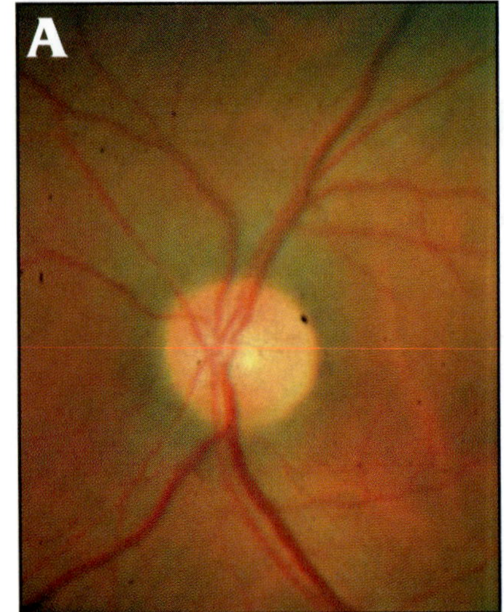

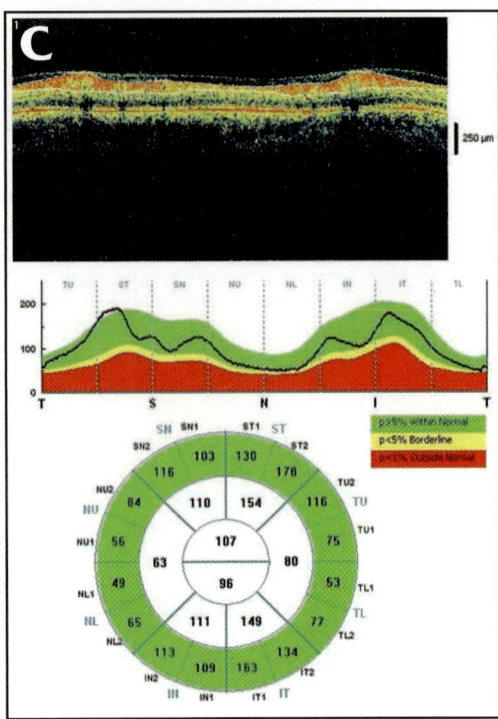

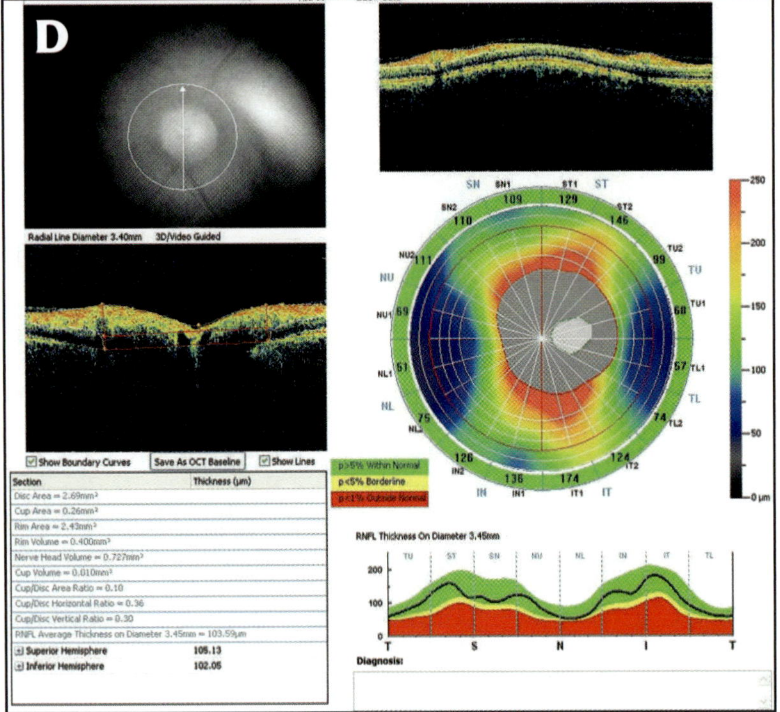

Section	Thickness (µm)
Disc Area = 2.69mm²	
Cup Area = 0.26mm²	
Rim Area = 2.43mm²	
Rim Volume = 0.400mm³	
Nerve Head Volume = 0.727mm³	
Cup Volume = 0.010mm³	
Cup/Disc Area Ratio = 0.10	
Cup/Disc Horizontal Ratio = 0.36	
Cup/Disc Vertical Ratio = 0.30	
RNFL Average Thickness on Diameter 3.45mm = 103.59µm	
Superior Hemisphere	105.13
Inferior Hemisphere	102.05

Diagnosis:

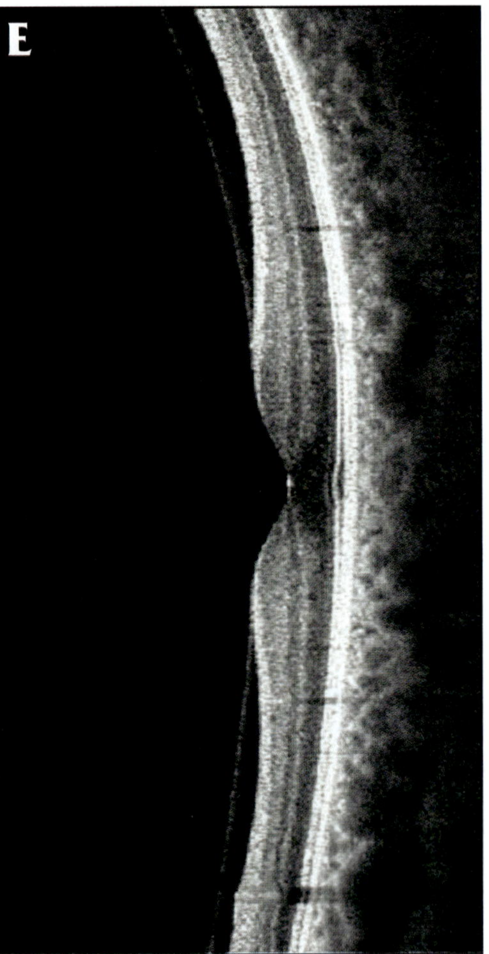

Figures 26-1A through E.

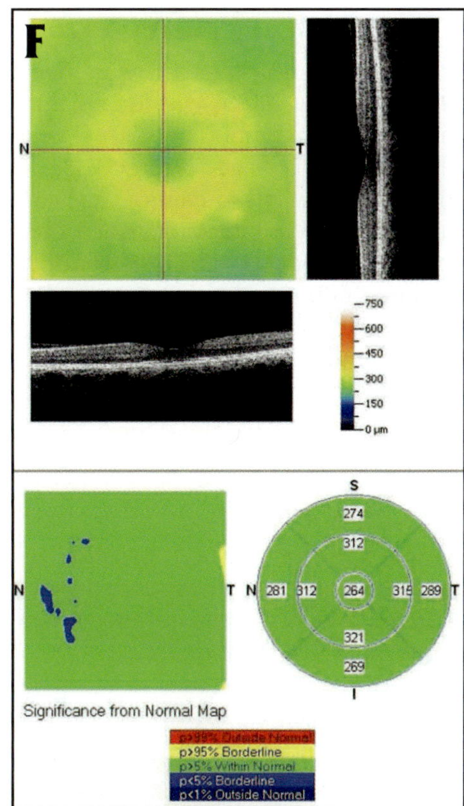

Figures 26-1F through G.

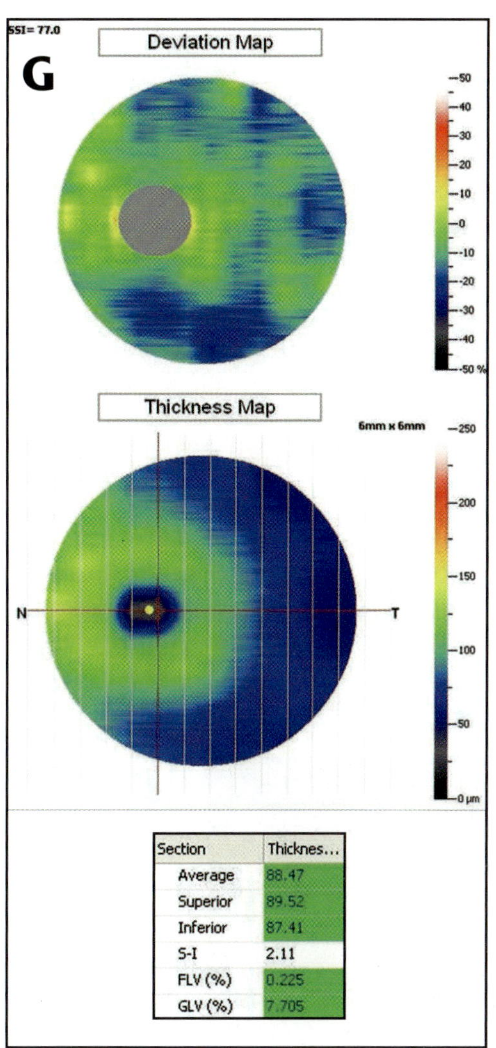

Section	Thicknes...
Average	88.47
Superior	89.52
Inferior	87.41
S-I	2.11
FLV (%)	0.225
GLV (%)	7.705

STRUCTURE-FUNCTION CORRESPONDENCE

Nerve Fiber Layer and Macula Region Structural Analysis Agreed with Visual Field Damage (Figures 26-2A through G)

A 48-year-old woman with normal tension glaucoma in the right eye was treated with selective laser trabeculoplasty in the past and currently with multiple topical medications. Ocular examination revealed a visual acuity of 20/25, IOP = 16 mmHg, with a normal appearing anterior segment and open angle. Central corneal thickness was within normal limits. Fundus examination showed posterior vitreous detachment and small central cupping of the ONH with inferotemporal crescent of peripapillary atrophy (Figure 26-2A). Visual field demonstrated reproducible superior nasal scotoma (Figure 26-2B).

Optical Coherence Tomography

Retinal NFL thickness measurements with both retinal NFL scan and NHM4 was focally thin (red background) in the inferior temporal region (Figures 26-2C and D). The cross line image (Figure 26-2E) also indicated visible thin tissue in the inferior macular region that could be most easily appreciated when compared to the corresponding location in the superior macula. The MM5 (Figure 26-2F) and GCC (Figure 26-2G) scans highlighted the thin inferior temporal region (blue and black background). This was most obvious on the GCC map, indicated by the outside normal thickness limits in the inferior hemisphere with a thickness of 72.63 µm shown in red (see Figure 26-2G).

Comment

Spectral-domain OCT provided evidence of structural damage that corresponded precisely to the visual field abnormality in the absence of any detectable glaucomatous ONH abnormality.

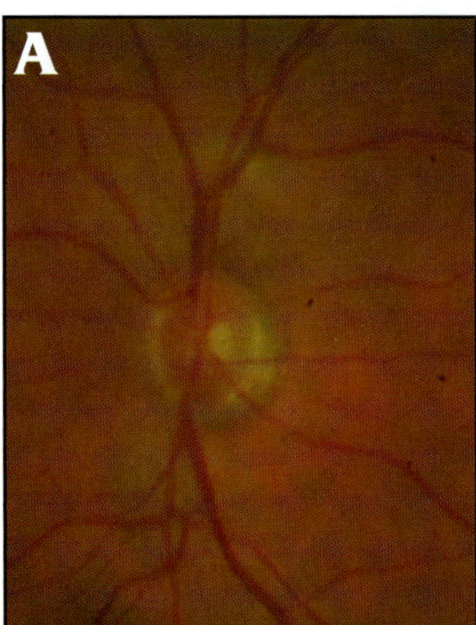

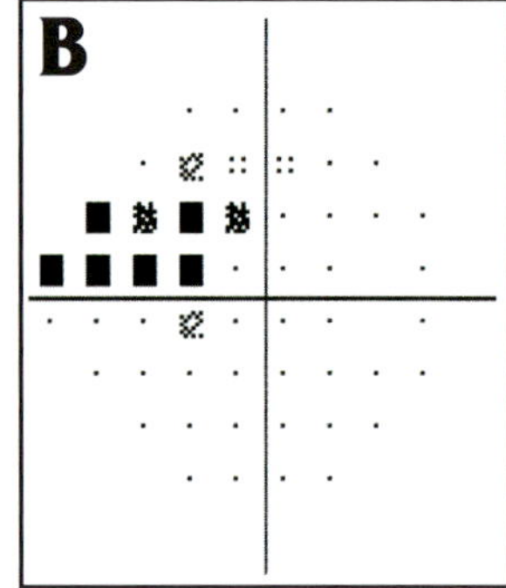

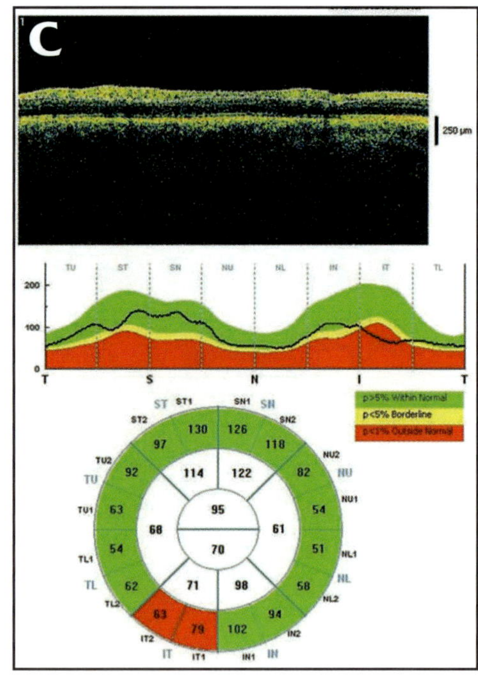

Figures 26-2A through C.

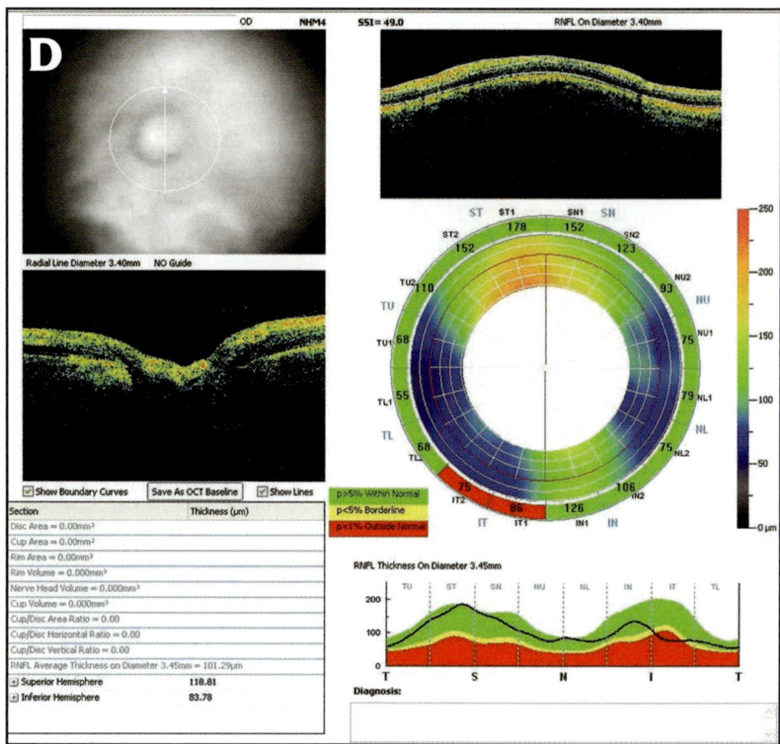

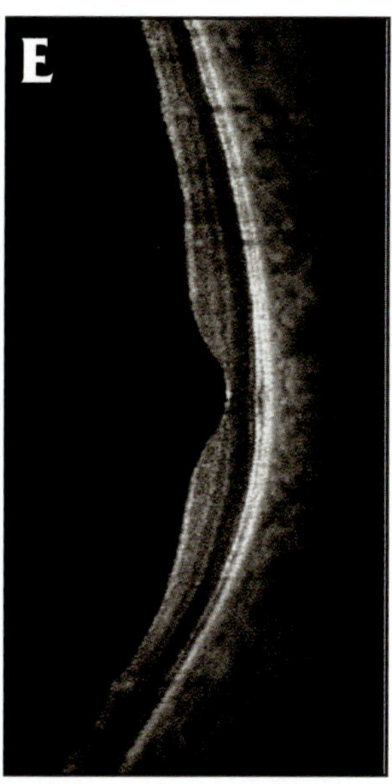

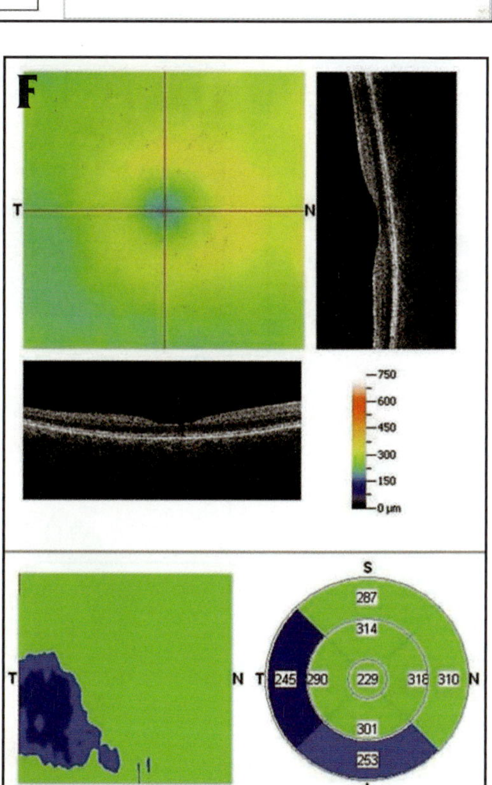

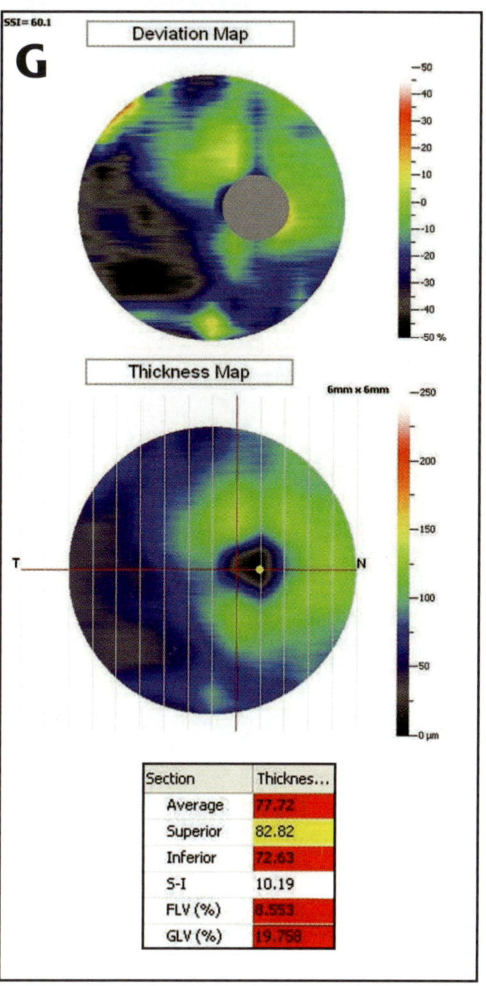

Figures 26-2D through G.

Structural Damage Correlated to Functional Damage and Imaging Shows Signs of a New Possible Defect (Figures 26-3A through H)

A 55-year-old woman with normal tension glaucoma was treated with oral and topical medications. On examination, visual acuity was 20/20, IOP = 11 mmHg, there was a normal anterior segment with open angle, and central corneal thickness was normal by ultrasound pachymetry. The posterior segment examination showed a large ONH cup with a neuroretinal rim thin to the disc margin at the inferior temporal region (Figure 26-3A). Red-free photography showed a wedge defect of the retinal NFL between 4:00 to 6:30 (Figure 26-3B). Visual field showed reproducible superior central and nasal scotomas with a possible inferior nasal step (Figure 26-3C).

Optical Coherence Tomography

The retinal NFL (Figure 26-3D) and NHM4 (Figure 26-3E) scans showed thin inferior circumpapillary retinal NFL and borderline thin (yellow background) superiorly. The cross line image (Figure 26-3F) of the macula showed thin retinal NFL in the entire inferior region. This was most obvious when compared to the superior macular region. The MM5 (Figure 26-3G) showed an advanced abnormality in the entire inferior macula, visible in all quantitative displays. In addition, an early arcuate thin area was seen superiorly on the deviation map (bottom left). The substantial structural damage was also demonstrated by the GCC (Figure 26-3H) analysis, especially in the inferior macula with an outside normal limits thickness of 65.61 μm in this region.

Comments

The structural analysis obtained by spectral-domain OCT verified the functional abnormalities observed in the visual field. Additionally, the retinal NFL, NHM4, and MM5 scans indicated an early defect in the superior region that corresponded with the possible visual field abnormality in the inferior nasal region. Preservation of the central 1 mm on the total retinal thickness macular scan in the presence of a central scotoma is something of a measurement artifact, as this area is almost entirely comprised by outer retina, and is relatively unaffected by glaucoma damage.

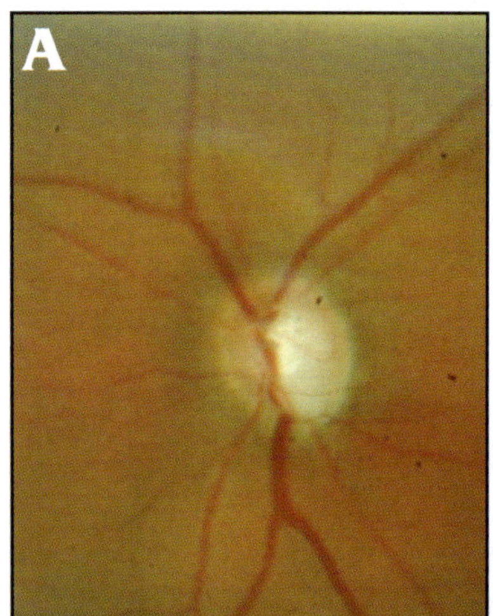

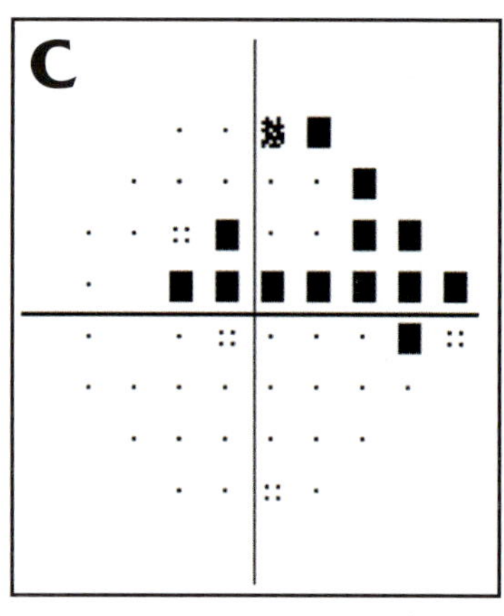

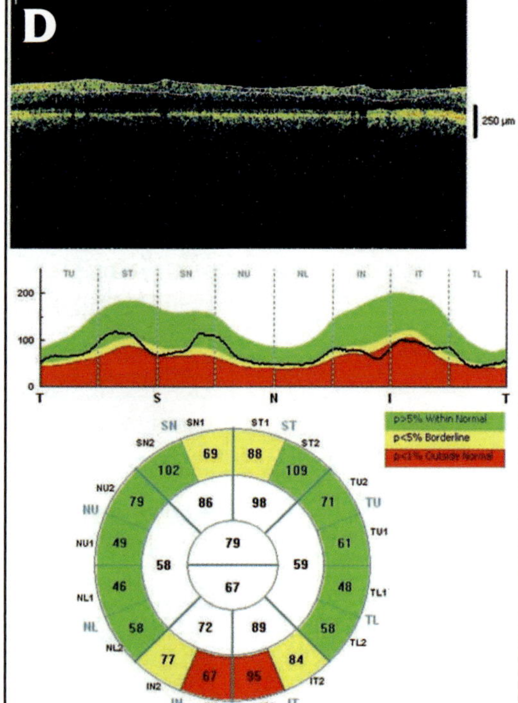

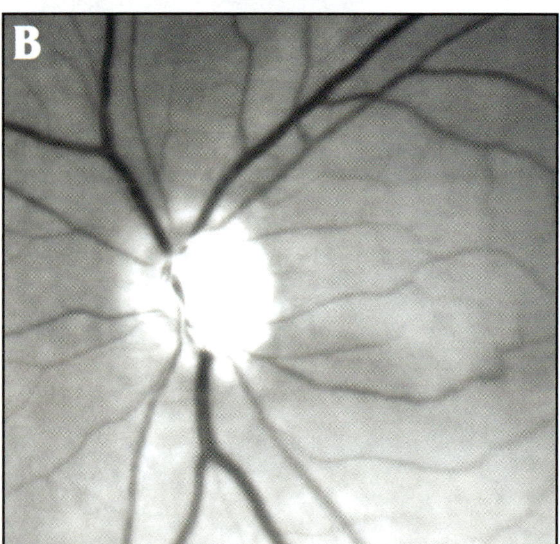

Figures 26-3A through D.

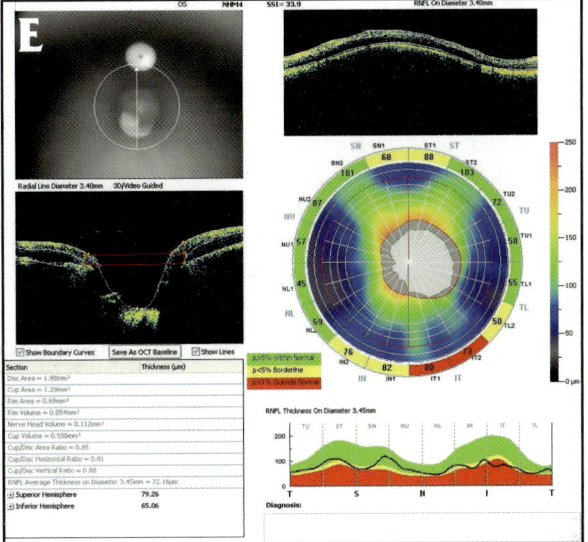

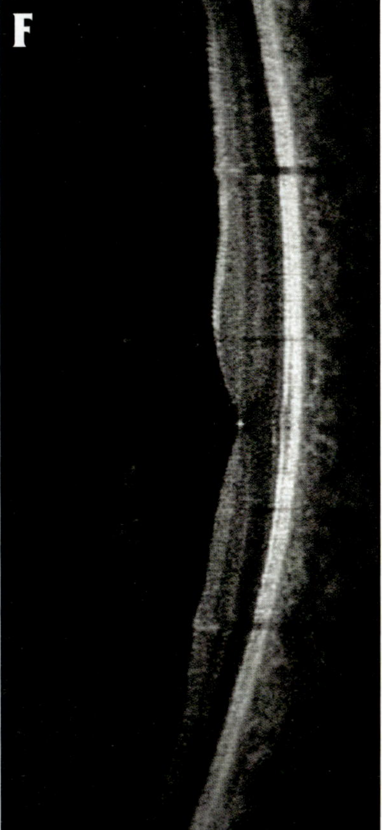

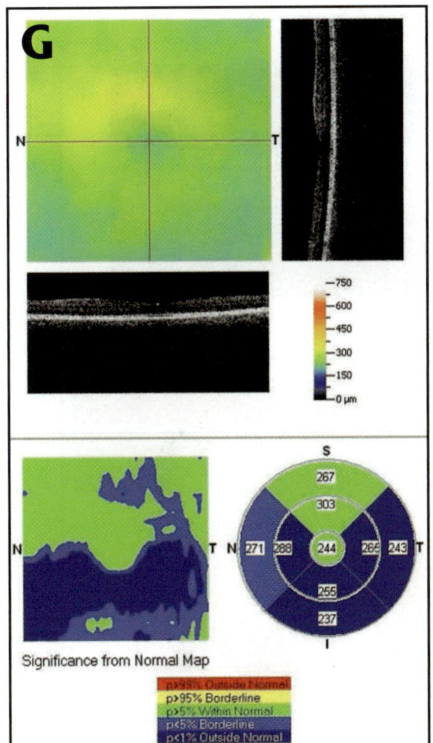

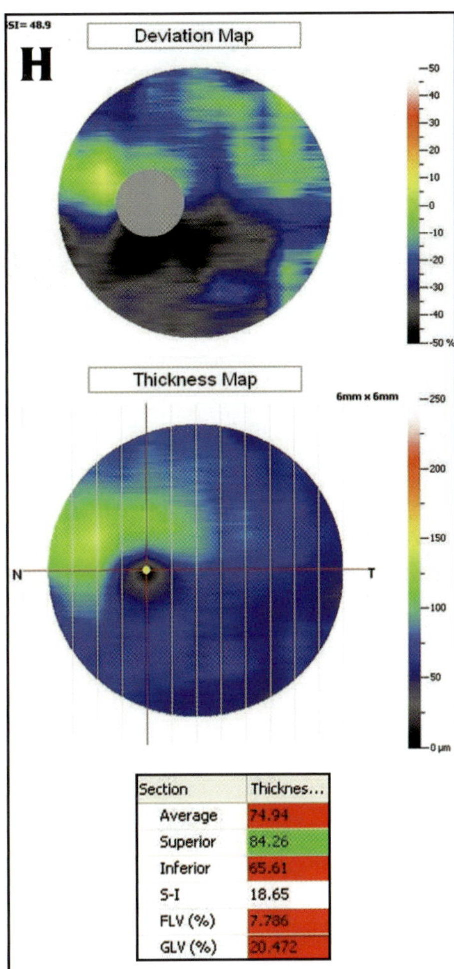

Figures 26-3E through H.

Structural Analysis Corresponded to Visual Field Abnormality in Advanced Glaucoma (Figures 26-4A through G)

A 63-year-old healthy woman with advanced primary open angle glaucoma in her right eye that was treated with selective laser trabeculoplasty and topical prostaglandins. Visual acuity was 20/30, IOP = 18 mmHg, the anterior segment appeared normal, and the anterior chamber angle was open. Fundus examination showed advanced cupping with a wide retinal NFL defect extending from 5:00 to 9:00 (Figure 26-4A). The visual field demonstrated a near total superior hemifield defect (Figure 26-4B).

Optical Coherence Tomography

The circumpapillary retinal NFL scan (Figure 26-4C) showed an obliteration of the inferior peak of the double hump configuration and borderline thinning in most of the remaining retinal NFL (center). The NHM4 scan (Figure 26-4D) showed thin retinal NFL across the entire scan (right) and a large ONH cup (left). The macular scans (Figures 26-4E and F) also indicated that the inferior region was thin, with involvement of the fovea and the inner perifoveal ring. The GCC image (Figure 26-4G) showed that the macula was thin throughout and that this was most pronounced in the outer region of the macula. All GCC parameters were found to be outside normal limits.

Comments

Substantial structural damage was observed with spectral-domain OCT in a subject with a corresponding advanced glaucomatous visual field defect.

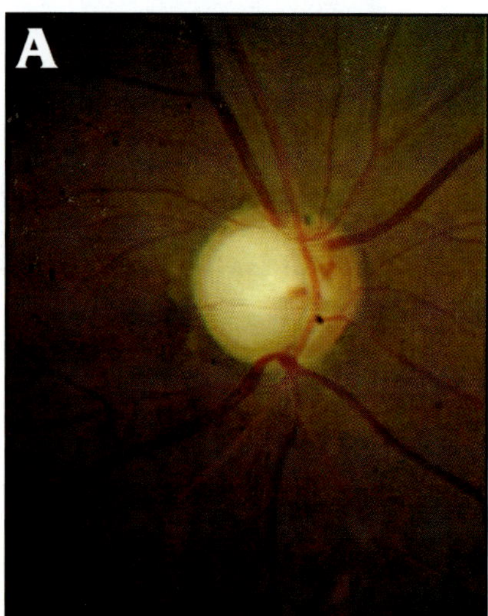
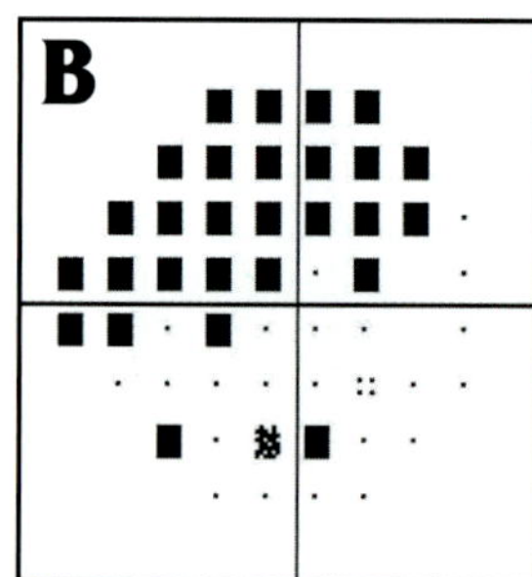
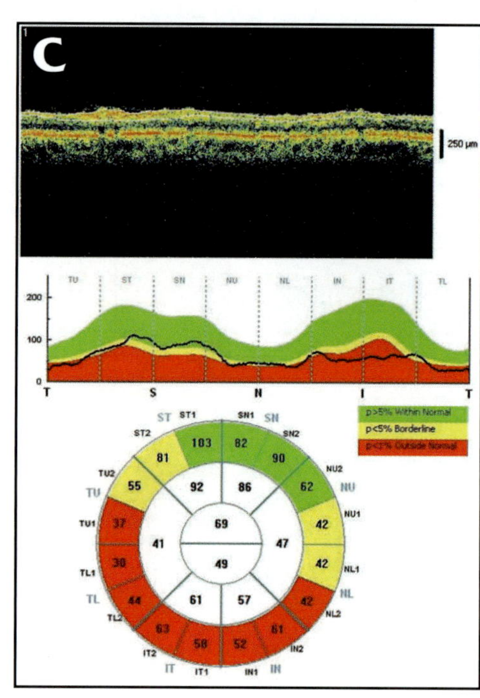

Figures 26-4A through C.

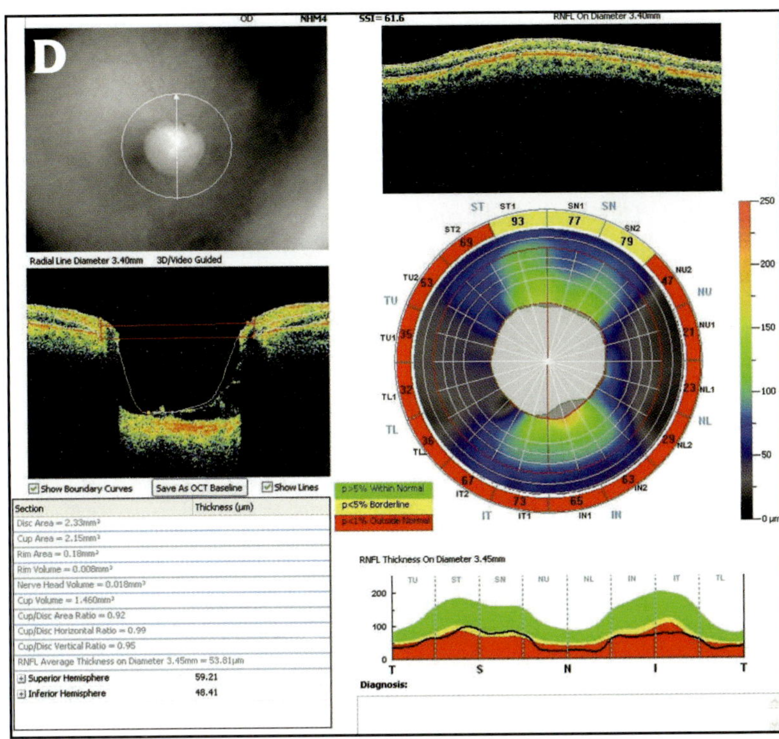

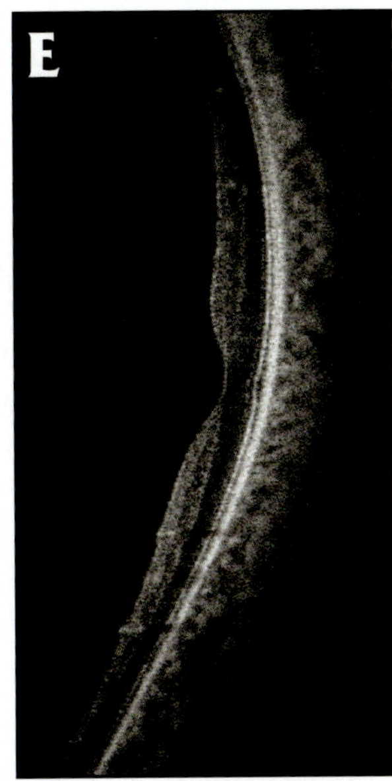

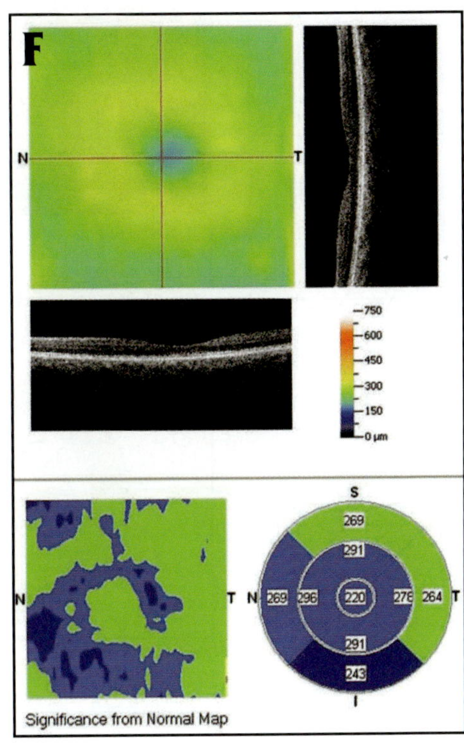

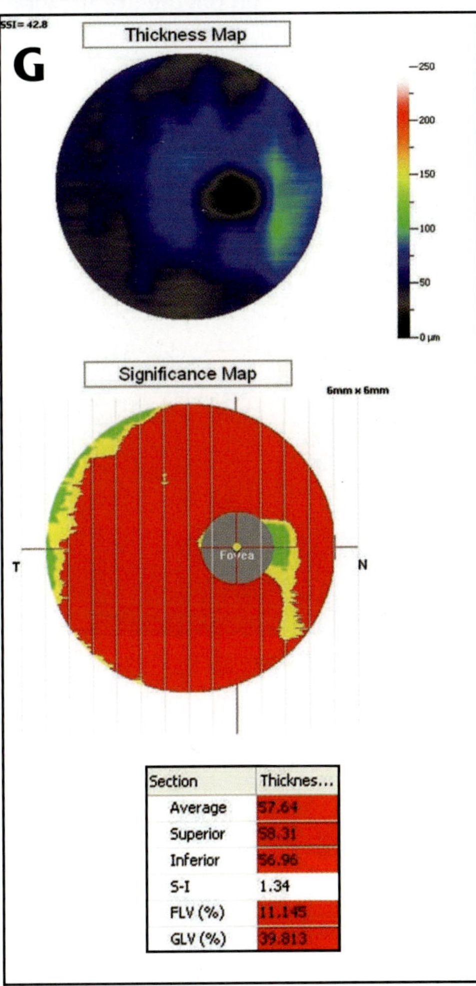

Figures 26-4D through G.

THE UTILITY OF IMAGING IN EARLY GLAUCOMA

Optical Coherence Tomography Supported Early Visual Field Abnormality (Figures 26-5A through G)

A 69-year-old woman had systemic hypertension and open angle glaucoma in her right eye. On examination, visual acuity was 20/30, IOP = 15 mmHg and showed a normal anterior chamber and open angle. Fundus examination revealed a large central cupping with no evidence of localized neuroretinal rim thinning or retinal NFL defect (Figure 26-5A). Visual field demonstrated a focal inferior scotoma (Figure 26-5B).

Optical Coherence Tomography

The retinal NFL scan (Figure 26-5C) indicated normal thickness throughout, but a well-centered NHM4 scan (Figure 26-5D) showed a focally thin superior region that corresponded to the defect seen in the visual field. The cross line image (Figure 26-5E) also indicated visible thin tissue in the superior macular region (upper part of the scan). The MM5 (Figure 26-5F) and the GCC (Figure 26-5G) showed an even more pronounced corresponding abnormality in the superior region of the macula. This was indicated by the borderline GCC superior hemisphere thickness of 79.28 µm. Also noted is the borderline global loss volume of 12.129%, which has been show to be an appropriate measure in glaucoma diagnosis.[1]

Comments

An early and localized scotoma as appeared in this patient might be considered questionable in clinical practice. OCT highlighted this thin region in both the retinal NFL and macular analysis, reinforcing the validity of the finding. In this case, the NHM4 was more informative than the retinal NFL scan; there are situations such as this where one scan pattern or the other may provide the most valuable clinical information.

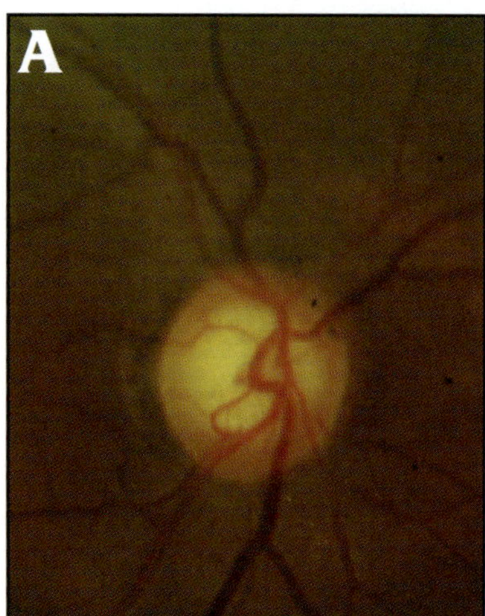

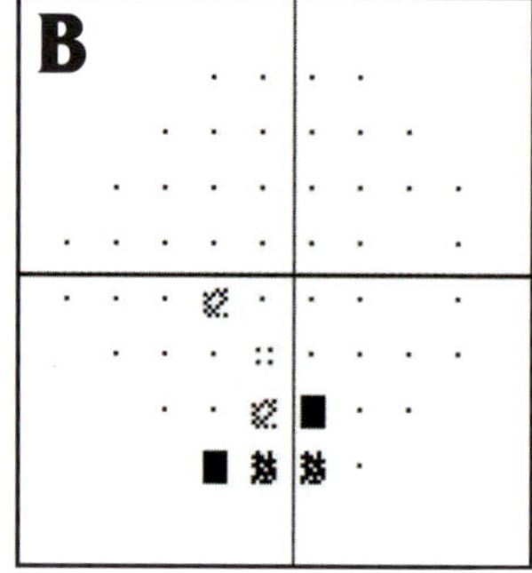

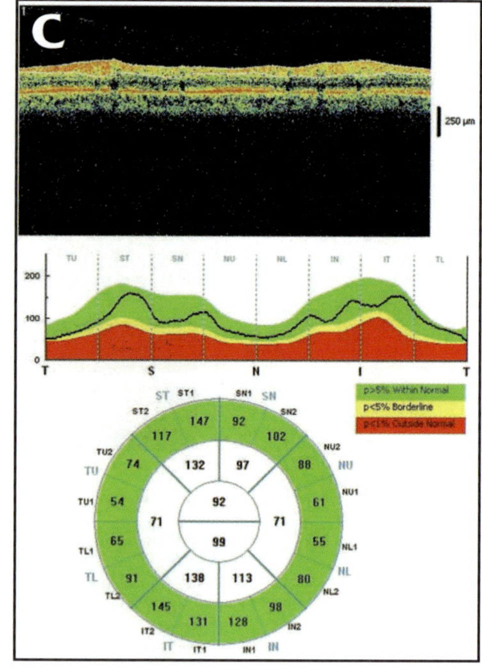

Figures 26-5A through C.

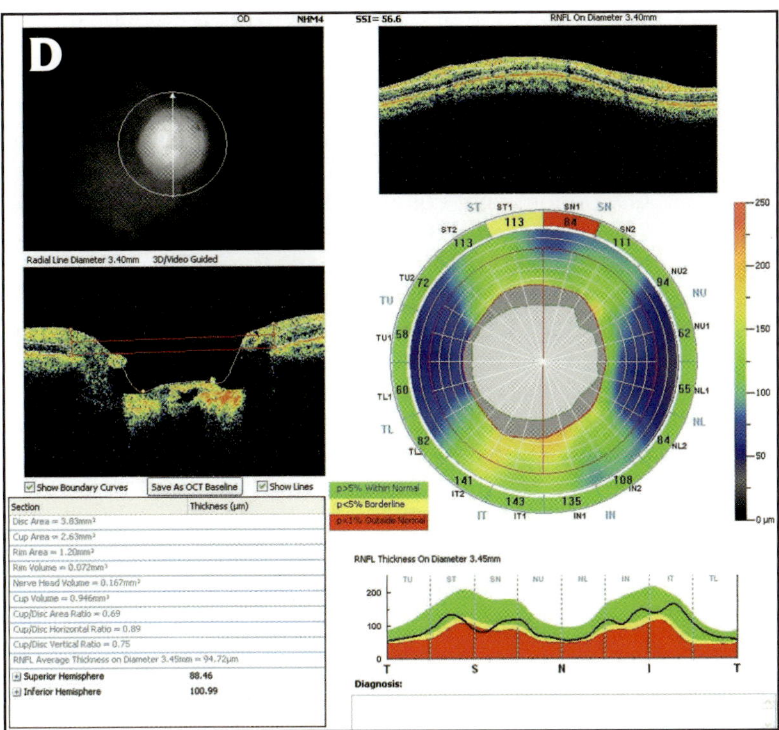

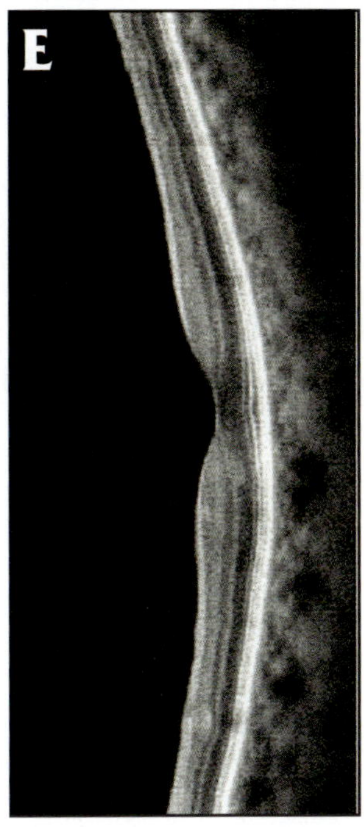

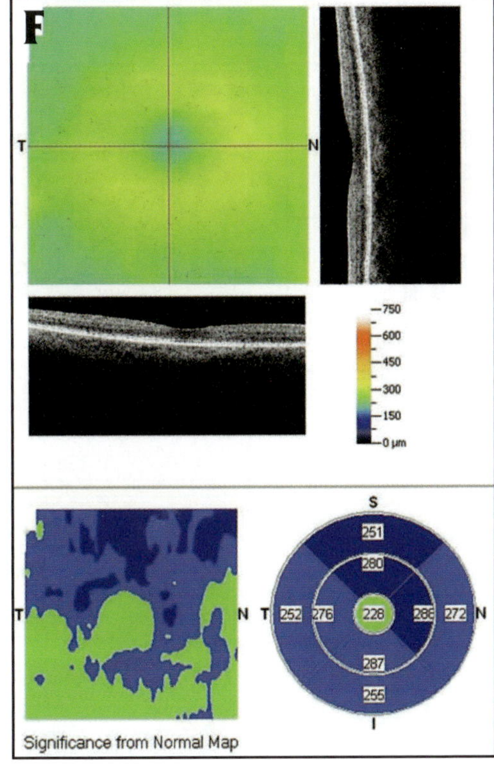

Figures 26-5D through G.

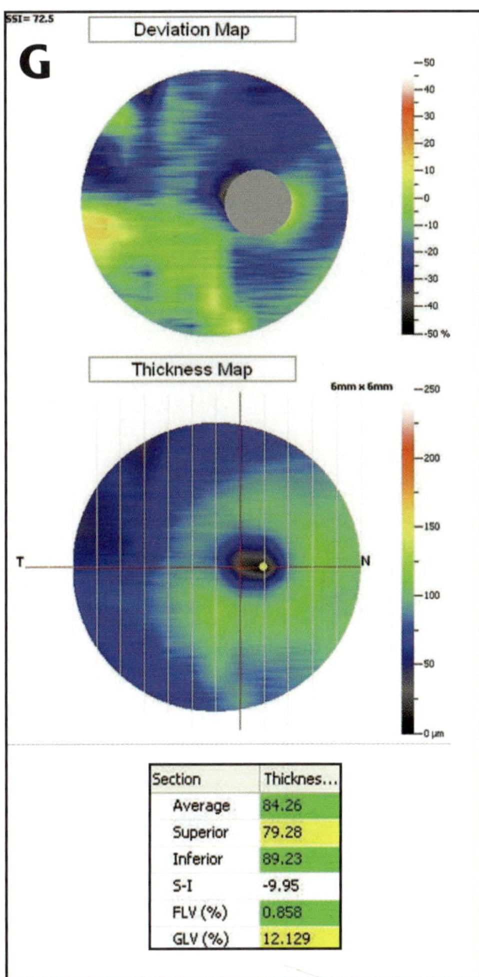

OVERALL STRUCTURAL DAMAGE SHOWN CLEARLY IN RETINAL NERVE FIBER LAYER PROFILE

Visual Field Abnormality Explained by Defect in Macula, Retinal Nerve Fiber Layer Assessment Indicated Additional Area of Structural Damage (Figures 26-6A through G)

A 64-year-old woman had systemic hypertension, history of breast malignancy, and iridocyclitis; she was diagnosed with primary open angle glaucoma in her right eye. Trabeculectomy with mitomycin C was performed in the eye, which previously had been treated only with timolol. On examination, the visual acuity was 20/25, IOP = 6 mmHg, there was an elevated and functional filtration bleb, the peripheral iridectomy was patent, and the angle was open. Posterior segment examination revealed a large central ONH (Figure 26-6A) cup with no evidence of a retinal NFL scotoma (Figure 26-6B).

Optical Coherence Tomography

Retinal NFL (Figure 26-6C) and NHM4 scans (Figure 26-6D) both demonstrated very thin retinal NFL throughout the scan as clearly indicated by the retinal NFL thickness profiles (see Figure 26-6C, center and Figure 26-6D, bottom right column). The cross line image (Figure 26-6E) showed a defect in the superior macular region starting from the location marked with the arrow. This was also apparent in the MM5 analysis (Figure 26-6F) by the dark blue region and clearly shown quantitatively in the GCC analysis (Figure 26-6G) by both an outside normal limits focal loss volume of 7.556% and global loss volume of 14.559%. A concurrent but lesser degree of thinning was also apparent in the inferior region.

Comments

All spectral-domain OCT scans indicated a thin retinal NFL, most pronounced in the superior region with lesser damage in the inferior region, corresponding to the visual field finding. This case illustrated the importance of the assessment of the retinal NFL profile; in this case that parameter provided the clearest indication of the overall magnitude of the structural damage.

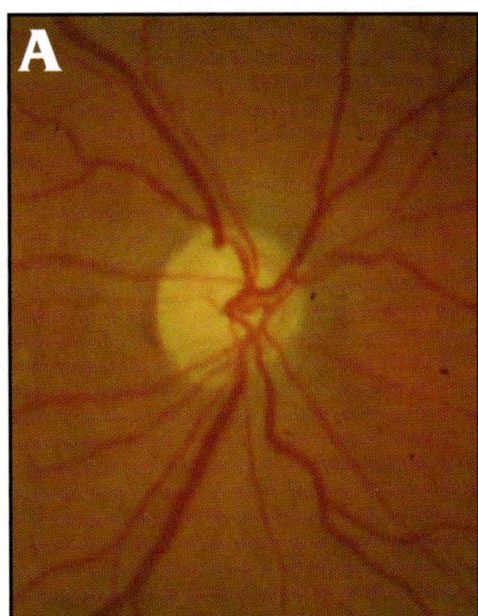

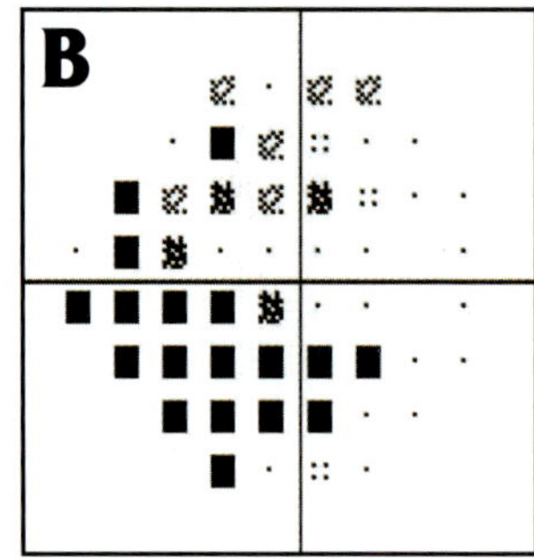

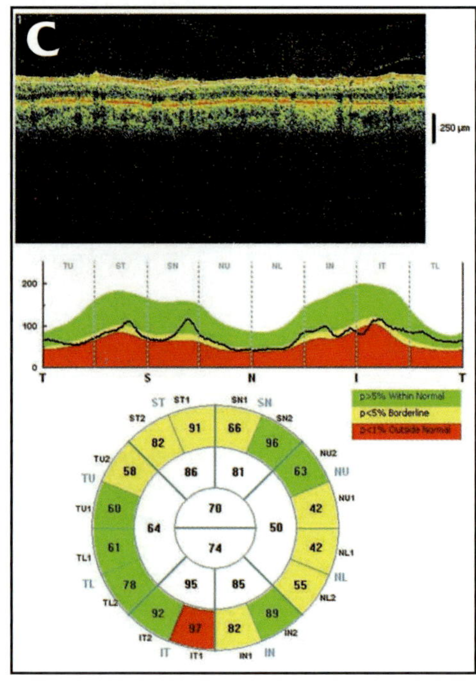

Figures 26-6A through C.

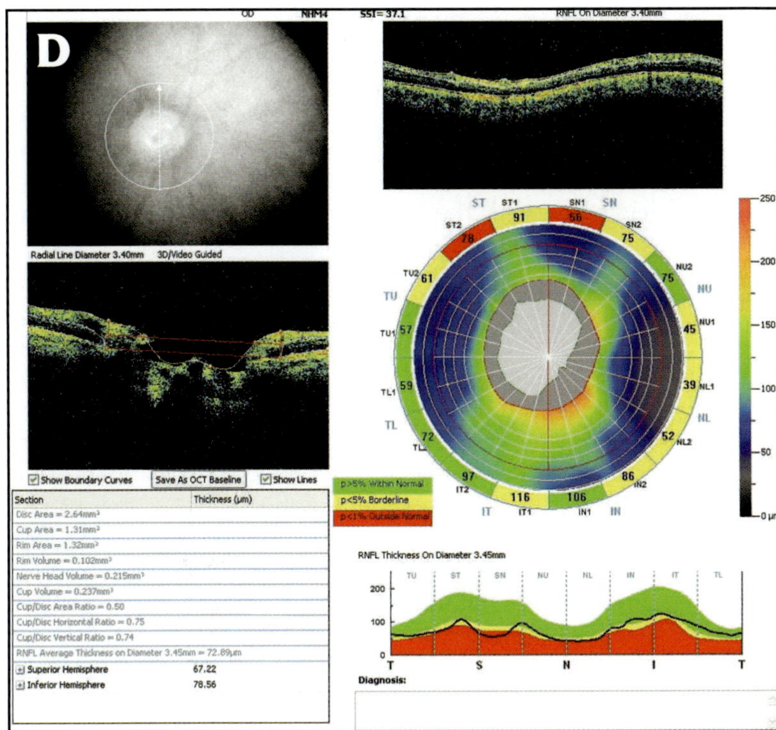

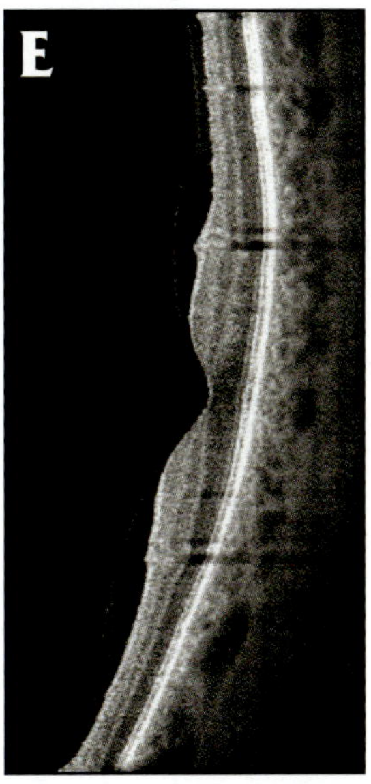

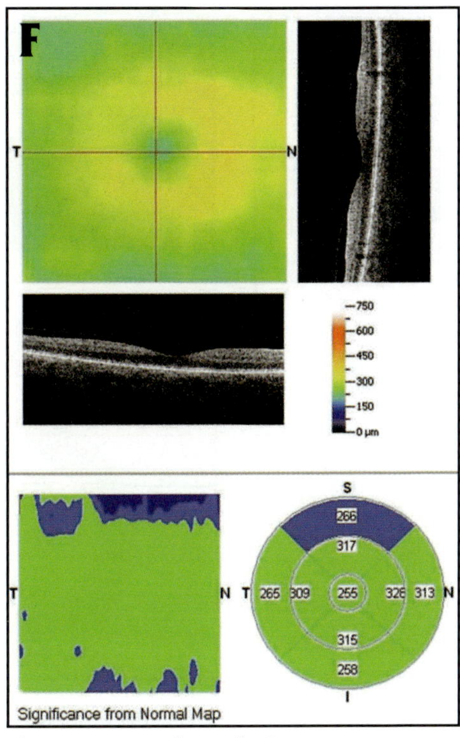

Figures 26-6D through G.

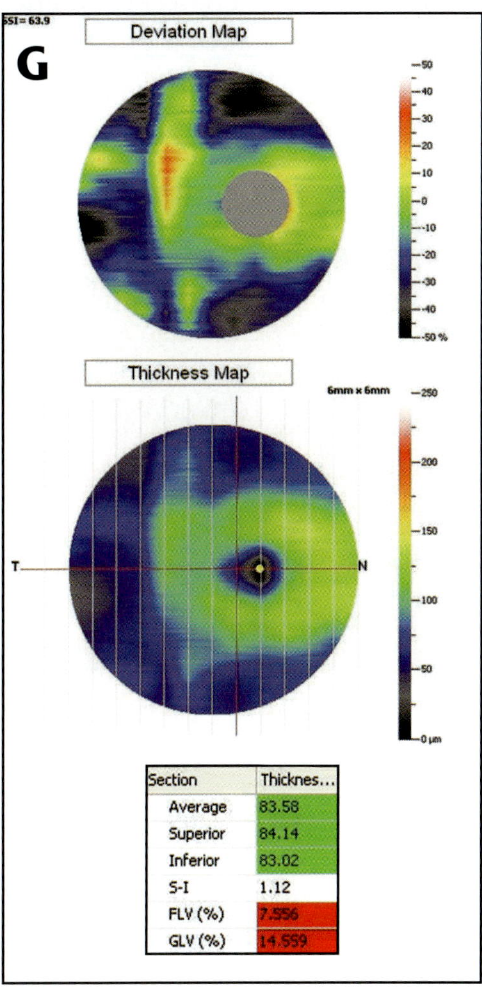

IMAGING DISCLOSED FUNCTIONAL ARTIFACT

Normal OCT Images Showed Visual Field Abnormality to be Artifact (Figures 26-7A through G)

A 58-year-old woman had systemic hypertension and glaucoma was suspected in her right eye. On examination, the visual acuity was 20/16, IOP = 14 mmHg, and there was a normal anterior segment and an open angle. The fundus examination showed a central excavation of the ONH with no evidence of localized neuroretinal rim thinning or NFL defect (Figure 26-7A). Visual field showed possible abnormalities in the inferior hemifield that varied in consecutive tests (Figure 26-7B).

Optical Coherence Tomography

All OCT images showed a structurally normal retinal NFL, ONH, and macular region. The retina was slightly thin in the MM5 (Figure 26-7F) scan in the inferior region, but this did not correspond to the visual field abnormality. Retinal NFL thickness was normal, as shown in Figure 26-7C. The NHM4 (Figure 26-7D) scan did not show large cupping, and revealed a normal retinal NFL thickness. The GCC (Figure 26-7G) scan did not agree with the MM5 analysis and showed normal thickness in the macula.

Comments

Although the visual field showed an abnormality in the inferior hemifield, OCT imaging disclosed no structural abnormality in that region. While total macular thickness showed inferior abnormality, this was not observed in the GCC map that is more specific to glaucomatous damage. Careful assessment of the NHM4 scan location (Figure 26-7E, top left column) discloses an upward shift that might artificially affect the measurements. While this displacement did not affect the retinal NFL measurements in this case, the scan location always needs to be considered when evaluating NHM4 scan results.

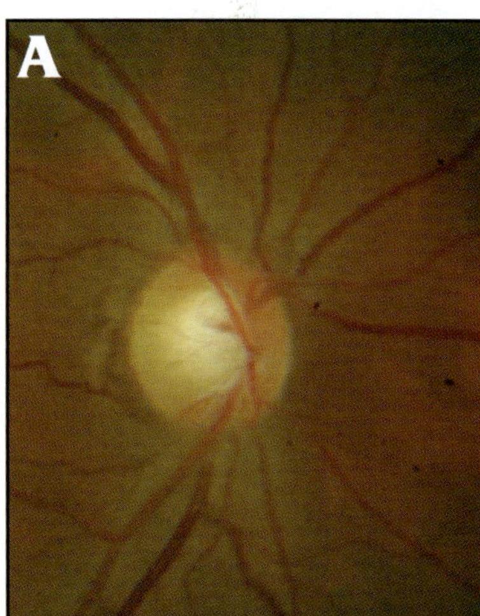

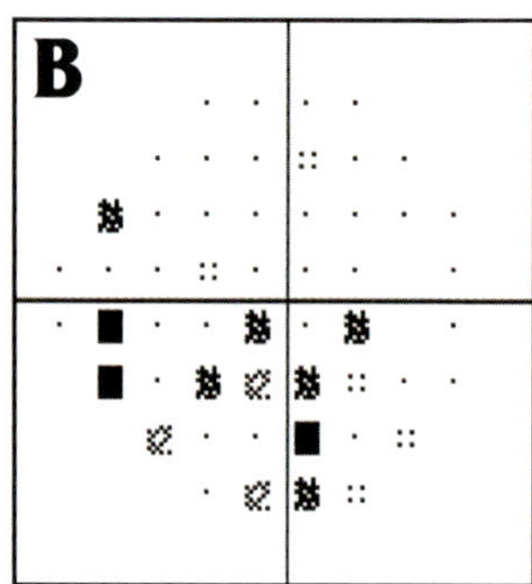

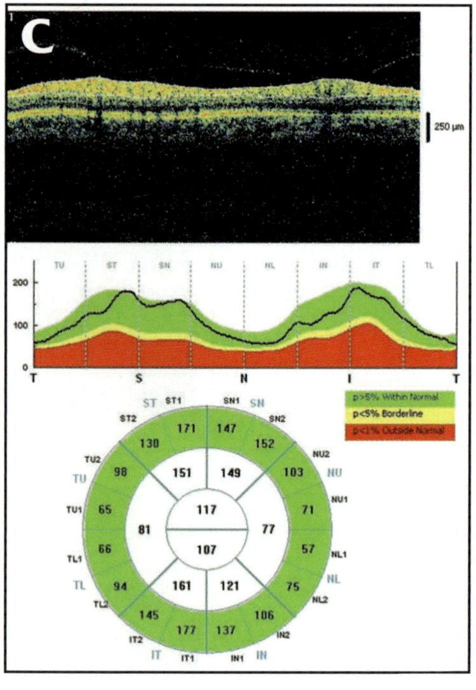

Figures 26-7A through C.

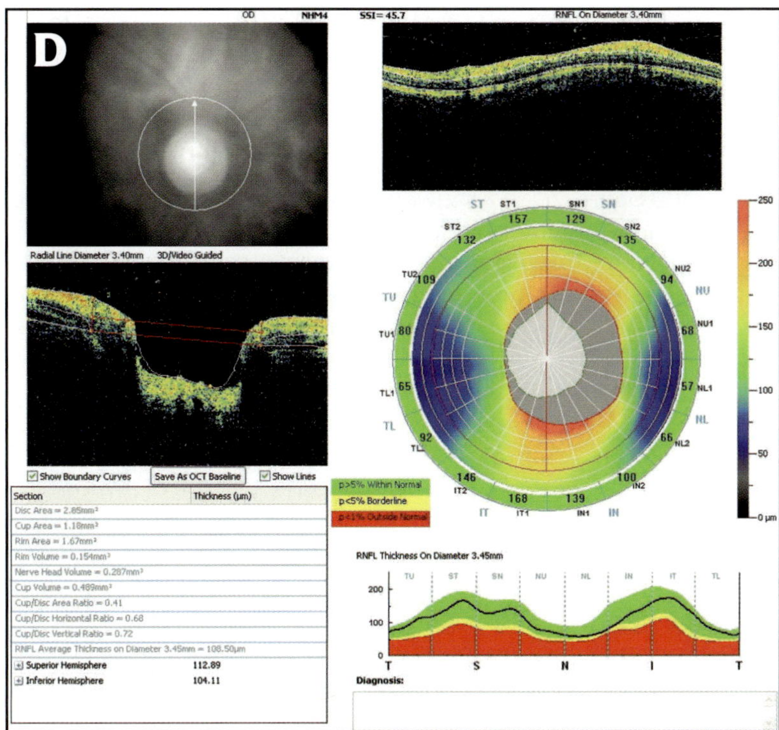

Section	Thickness (µm)
Disc Area = 2.85mm³	
Cup Area = 1.18mm³	
Rim Area = 1.67mm²	
Rim Volume = 0.154mm³	
Nerve Head Volume = 0.287mm³	
Cup Volume = 0.489mm³	
Cup/Disc Area Ratio = 0.41	
Cup/Disc Horizontal Ratio = 0.68	
Cup/Disc Vertical Ratio = 0.72	
RNFL Average Thickness on Diameter 3.45mm = 108.50µm	
Superior Hemisphere	112.89
Inferior Hemisphere	104.11

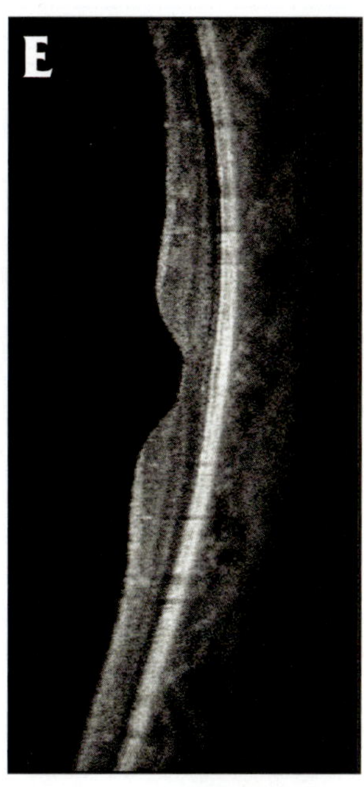

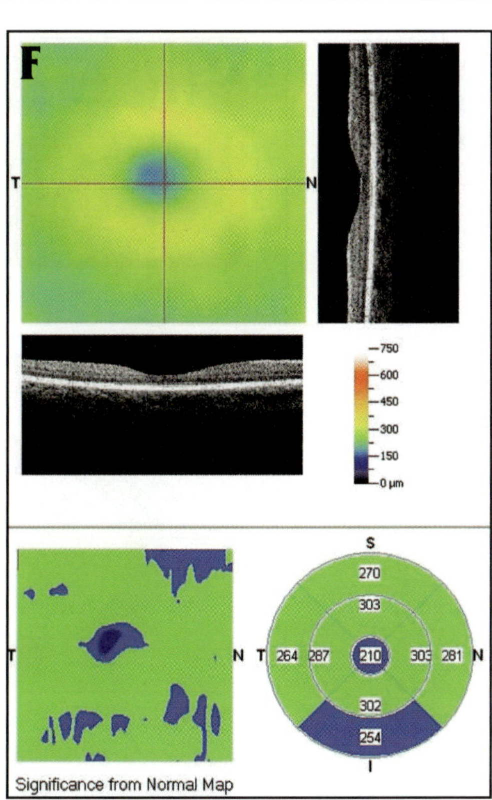

Figures 26-7D through G.

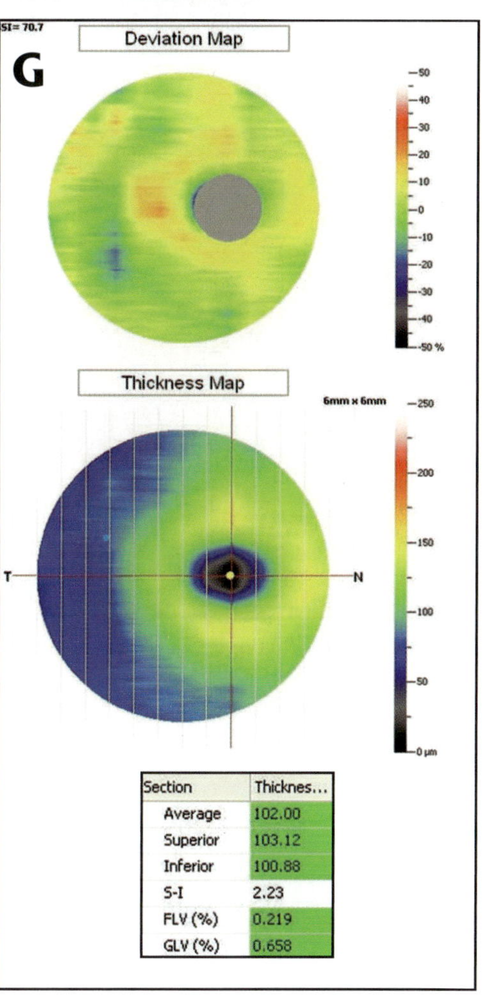

Section	Thicknes...
Average	102.00
Superior	103.12
Inferior	100.88
S-I	2.23
FLV (%)	0.219
GLV (%)	0.658

IMAGING SUGGESTS ACTUAL GLAUCOMA DAMAGE EXCEEDS VISUAL FIELD

Imaging Detected a Defect in the Superior Region of Retinal Nerve Fiber Layer Thickness (Figures 26-8A through G)

A 58-year-old man with atrial fibrillation and pigmentary glaucoma in his left eye was treated topically with multiple medications. On examination, visual acuity was 20/20 with IOP = 12 mmHg, normal anterior segment with open angle, and pigmentary deposits in the trabecular meshwork. Well-centered posterior chamber intraocular lens, large excavation of the ONH, peripapillary atrophy, and inferotemporal disc hemorrhage were noted after pupil dilation (Figure 26-8A). Visual field showed extensive scotoma in the superior hemifield (Figure 26-8B).

Optical Coherence Tomography

The structural damage in the inferior region was clearly evident through the retinal NFL (Figure 26-8C) and NHM4 (Figure 26-8D) scans with a focal superior defect observed with both scans. The cross line scan (Figure 26-8E) showed an extensively thin retinal NFL in the entire inferior macula. The MM5 (Figure 26-8F) and GCC (Figure 26-8G) analysis also showed thin tissue in this area with localized thickening in the fovea. The GCC map demonstrated a thin superior macula but to a lesser degree than the thin inferior region with an abnormal thickness of 68.59 μm.

Comments

Structural damage was observed in areas corresponding to the extensive superior visual field damage. In addition, structural assessment indicated damage to areas that currently showed limited visual field involvement (inferior). Disc hemorrhages are not detected by OCT and at the time of imaging there is no indication for corresponding structural damage.

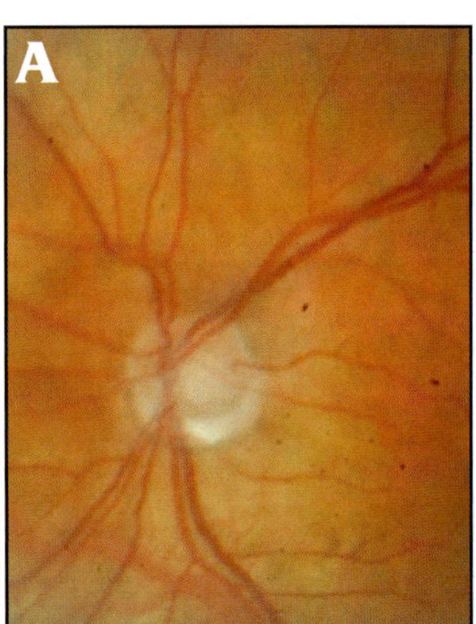

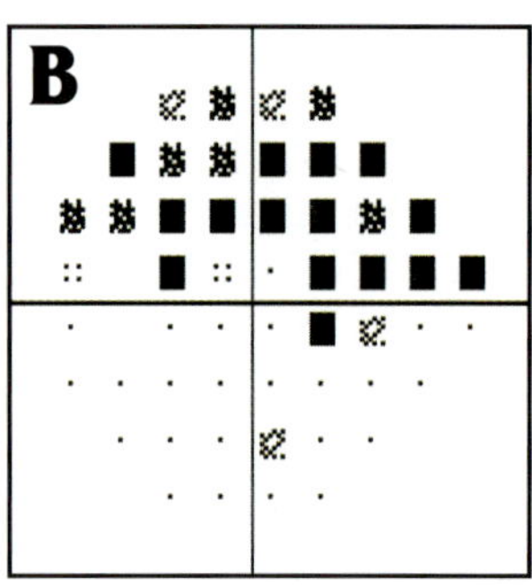

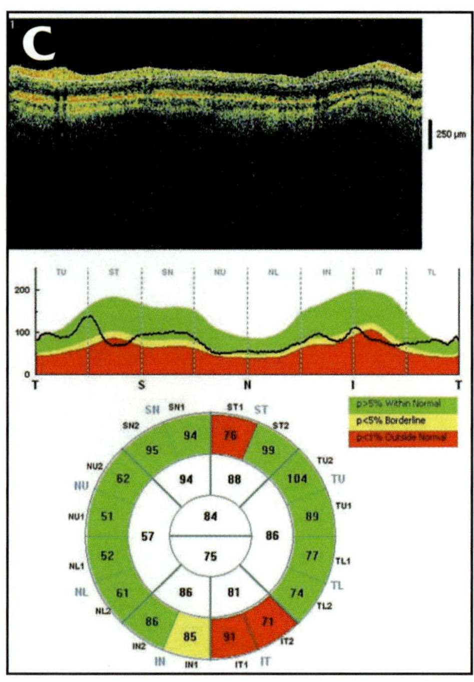

Figures 26-8A through C.

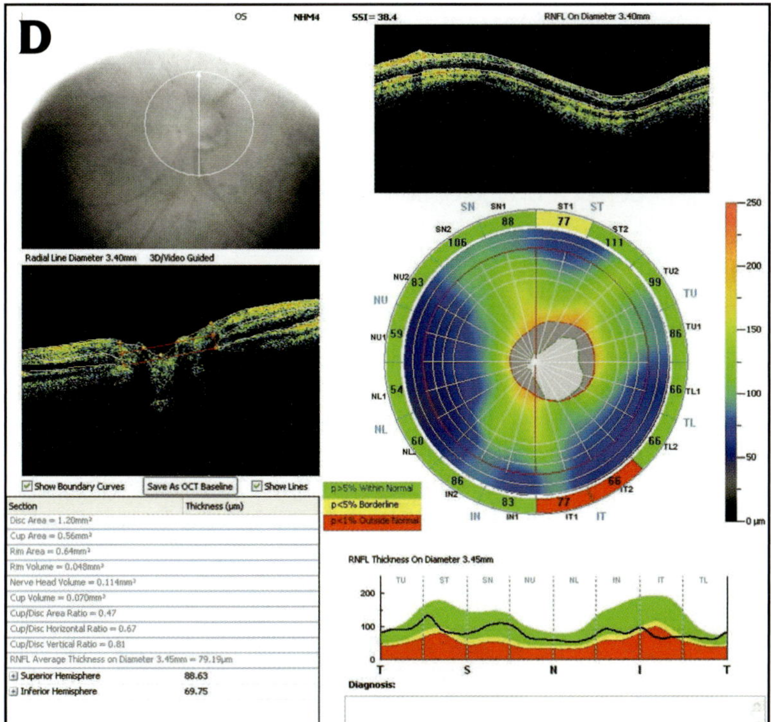
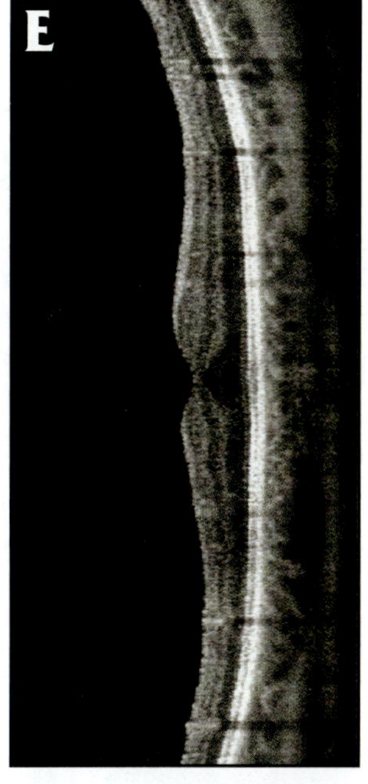
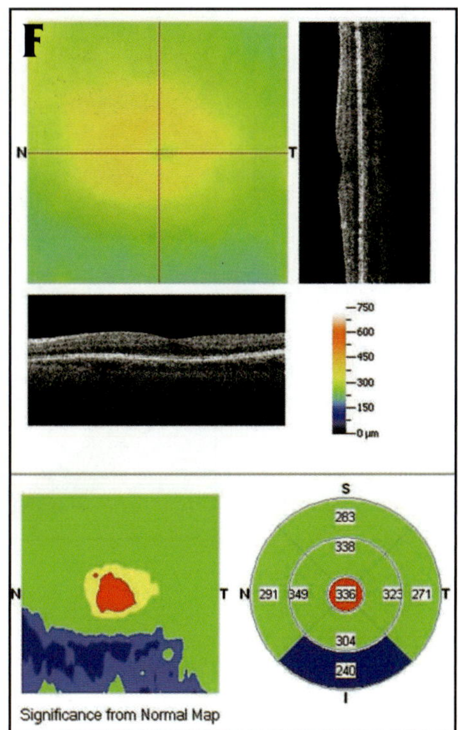
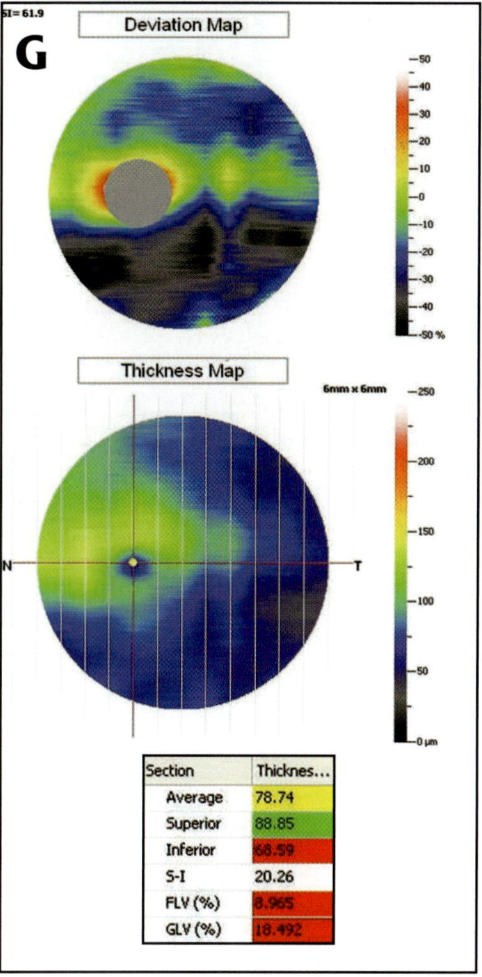

Figures 26-8D through G.

ABNORMAL STRUCTURE WITH NORMAL FUNCTION MAY SUGGEST FUTURE FUNCTIONAL LOSS

Imaging Detected Localized Retinal Nerve Fiber Layer Defect Not Shown in Macula Scans (Figures 26-9A through G)

A 58-year-old healthy woman had suspected glaucoma in her left eye. On examination, visual acuity was 20/20, IOP = 12 mmHg, with normal anterior chamber and open angle. The fundus examination revealed large cupping and peripapillary atrophy (Figure 26-9A). Visual field was interpreted as within normal limits (Figure 26-9B).

Optical Coherence Tomography

Both the retinal NFL (Figure 26-9C) and NHM4 (Figure 26-9D) scans were focally thin inferiorly, which was undetected by the visual field or by clinical examination for retinal NFL defect. The macula images did not detect any type of abnormalities (Figures 26-9E through G).

Comments

The localized retinal NFL defect might indicate a pre-perimetric stage, and inferior visual field area should be closely monitored for any functional abnormalities in future visits. The occurrence of the damage in the peripapillary region without any macular involvement might hint that the initial glaucomatous damage, in this case, originated at the level of the ONH and did not yet reach the macula region.

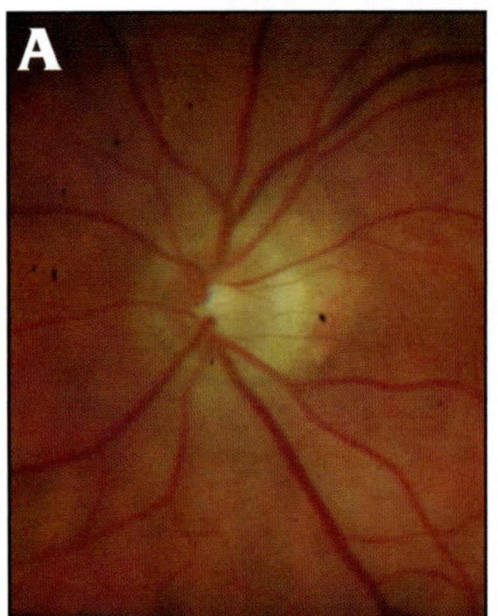

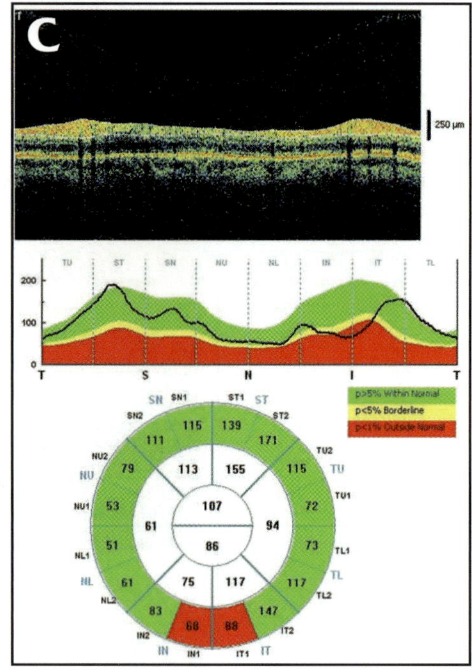

Figures 26-9A through C.

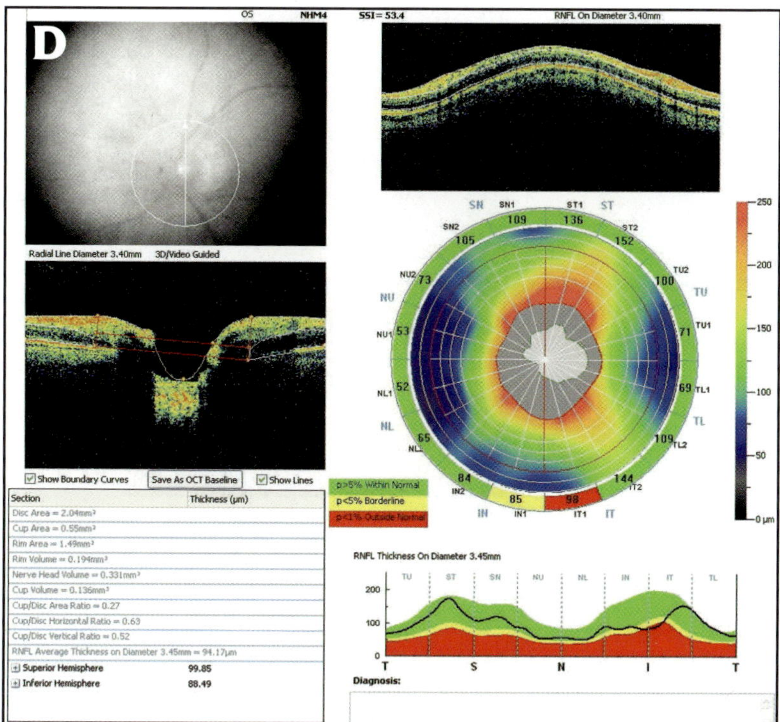

Section	Thickness (μm)
Disc Area = 2.04mm²	
Cup Area = 0.55mm²	
Rim Area = 1.49mm²	
Rim Volume = 0.194mm³	
Nerve Head Volume = 0.331mm³	
Cup Volume = 0.136mm³	
Cup/Disc Area Ratio = 0.27	
Cup/Disc Horizontal Ratio = 0.63	
Cup/Disc Vertical Ratio = 0.52	
RNFL Average Thickness on Diameter 3.45mm = 94.17μm	
Superior Hemisphere	99.85
Inferior Hemisphere	88.49

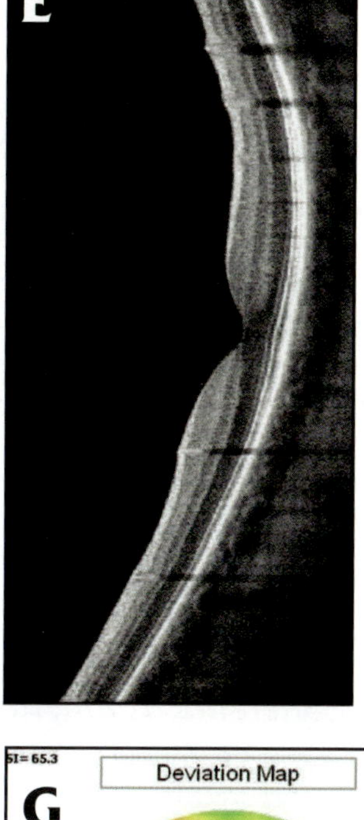

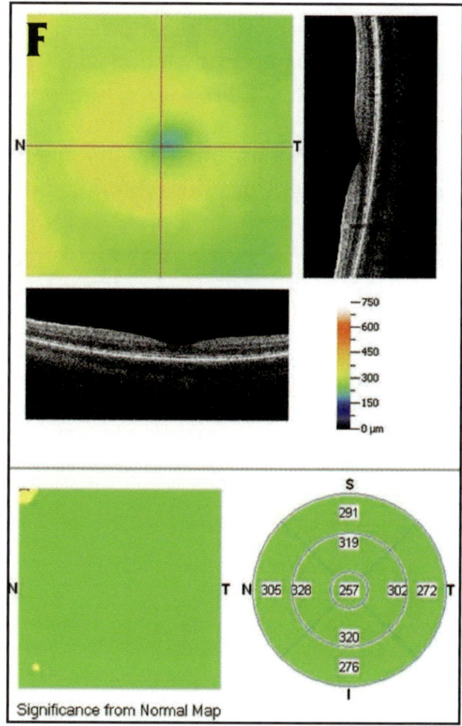

Figures 26-9D through G.

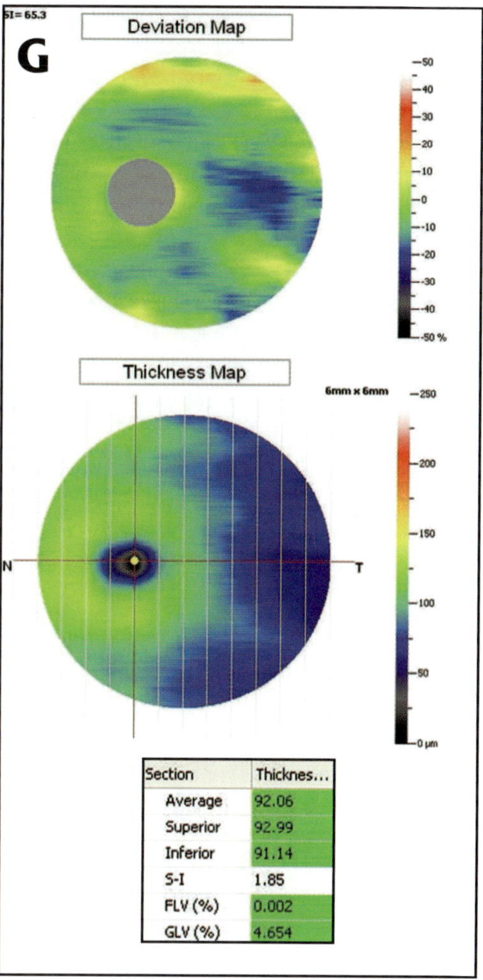

Section	Thicknes...
Average	92.06
Superior	92.99
Inferior	91.14
S-I	1.85
FLV (%)	0.002
GLV (%)	4.654

LONGITUDINAL ANALYSIS

Progression in Structure Found Before Functional Abnormalities (Figures 26-10A through F)

A 68-year-old woman with hypothyroidism and suspected glaucoma in her left eye. On examination, the visual acuity was 20/25, IOP = 13 mmHg with normal anterior segment and open angle. Fundus examination revealed large cupping with thinning of the neuroretinal rim in the inferotemporal region (Figure 26-10A). During the period of follow-up there were marked fluctuations in the visual field in the superior hemifield (Figure 26-10B). Although the patient was treated with topical medications and selective laser trabeculoplasty a reproducible scotoma was noted in the superior hemifield by the time of her most recent visit.

Optical Coherence Tomography

Since the beginning of the follow-up both the retinal NFL (Figure 26-10C) and NHM4 (Figure 26-10D) scans showed markedly thin retinal NFL, most pronounced in the inferior region and with lesser damage in the superior region. The MM5 (Figure 26-10E) analysis showed a substantially thin inferior region, with early changes in the superior macula appearing in the most recent visit. The GCC (Figure 26-10F) indicated a very prominently thin inferior region of the macula with gradual progression of the damage in both superior and inferior regions as indicated by the figure, GCC thickness values, and percent volume lost values. We see the global loss volume percent of the GCC start at 14.715% in 2006 and reach 17.305% in 2008.

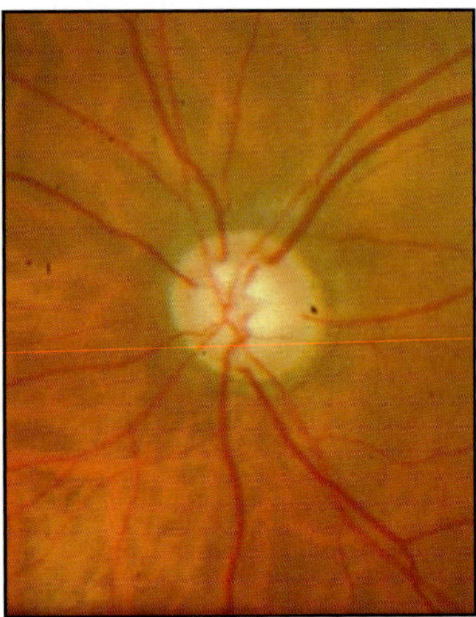

Figure 26-10A.

Comments

The fluctuating visual field at the beginning of the follow-up and especially when it returned to a near normal field (see Figure 26-10B, June 2007) may lead to erroneous clinical diagnosis. However, the OCT demonstrated repeated advanced glaucomatous damage through the entire follow-up. The presence of superior retinal NFL damage and the consequent progression of the corresponding area in the macula should be closely monitored for future inferior functional abnormalities.

REFERENCES

1. Tan O, Chopra V, Lu ATH, et al. Detection of macular ganglion cell loss in glaucoma by Fourier-domain optical coherence tomography. *Ophthalmology.* In press.

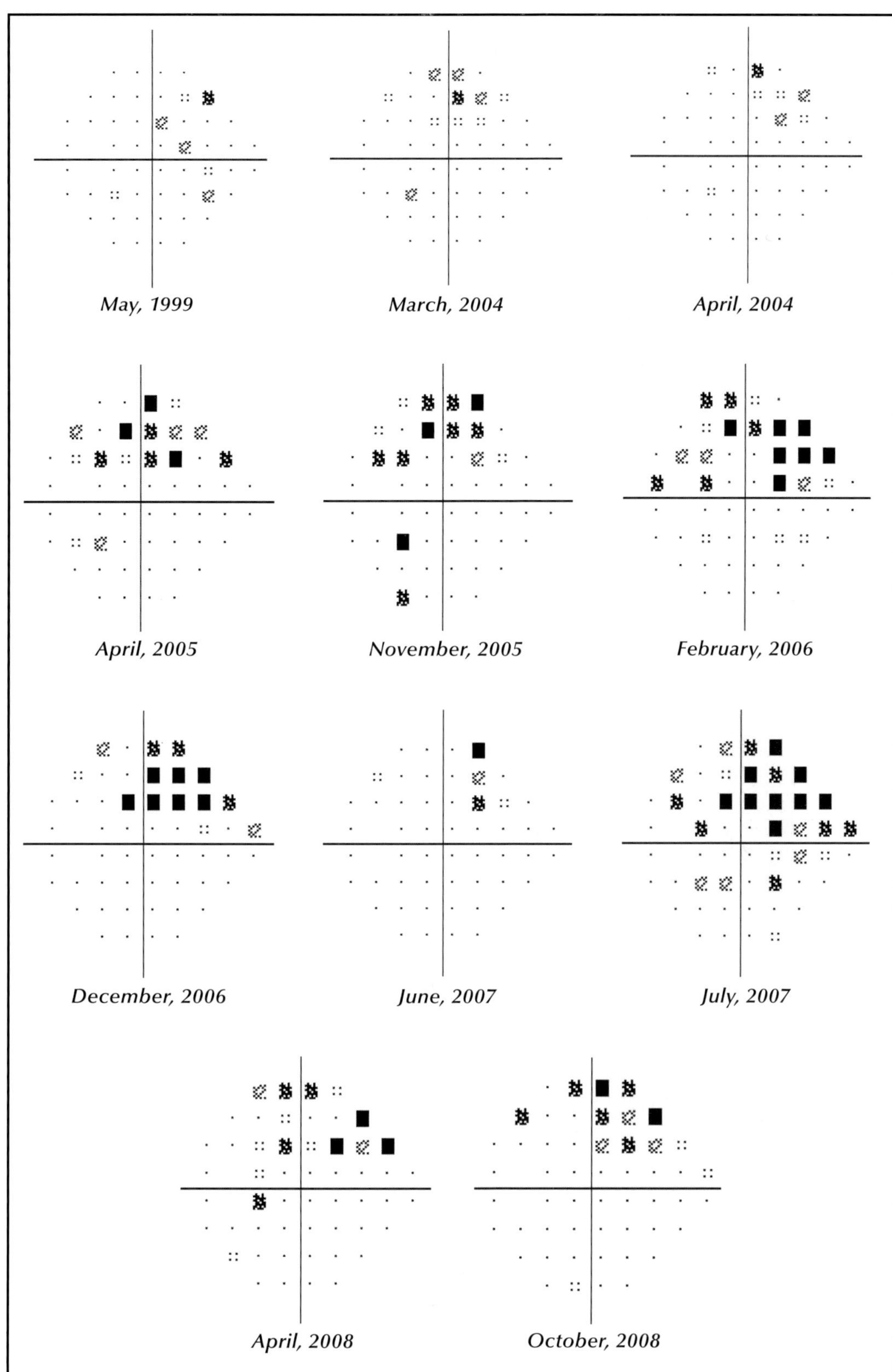

Figure 26-10B.

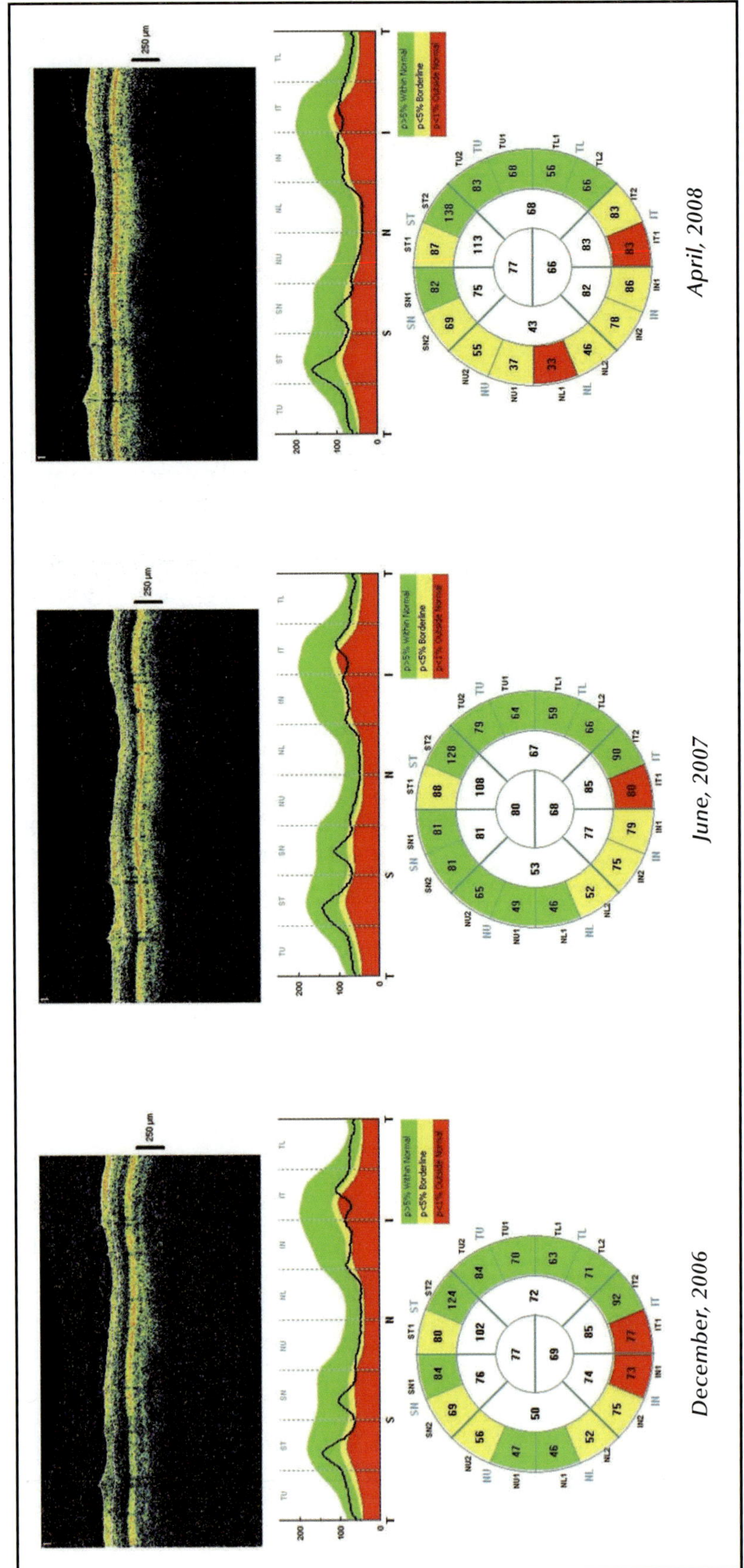

Figure 26-10C.

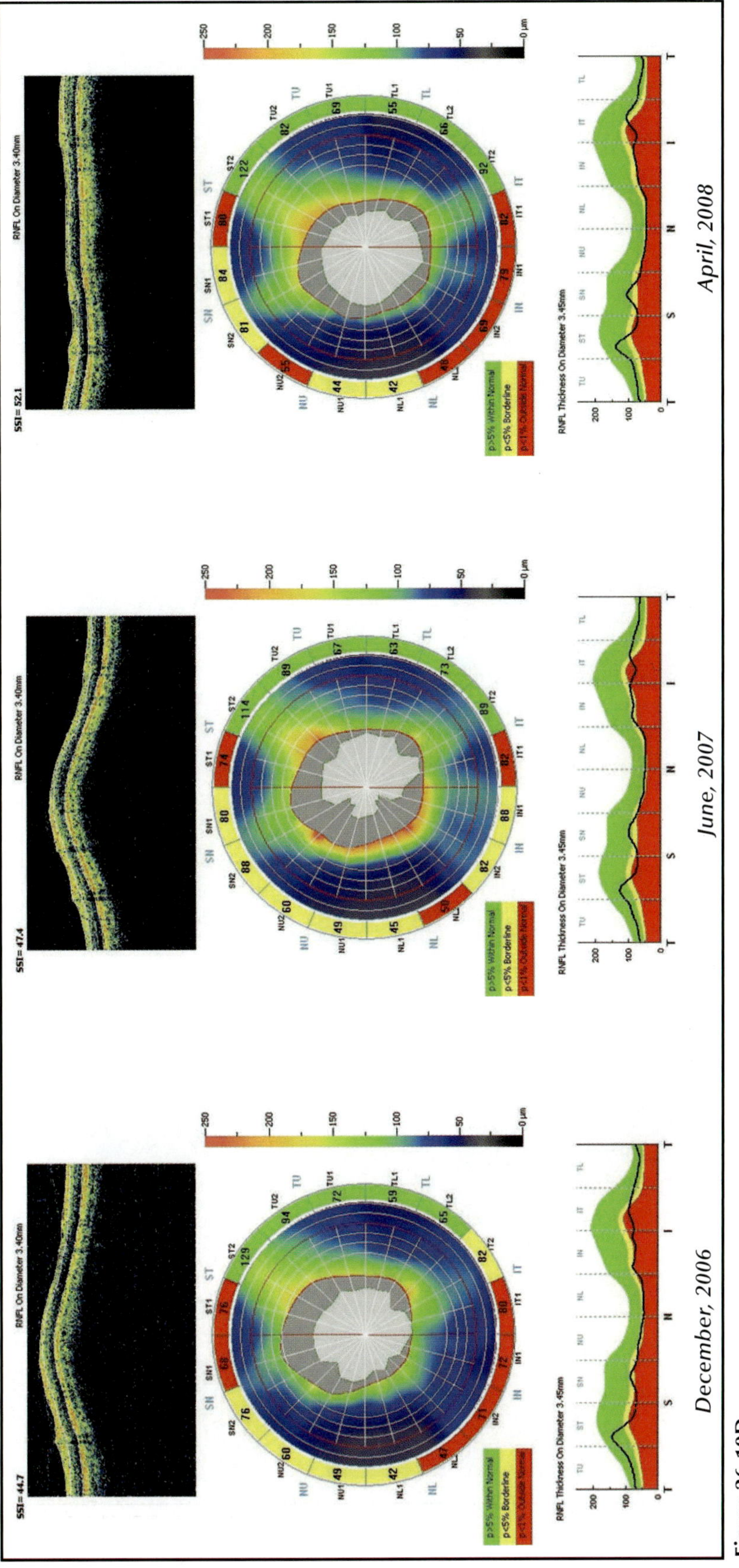

Figure 26-10D.

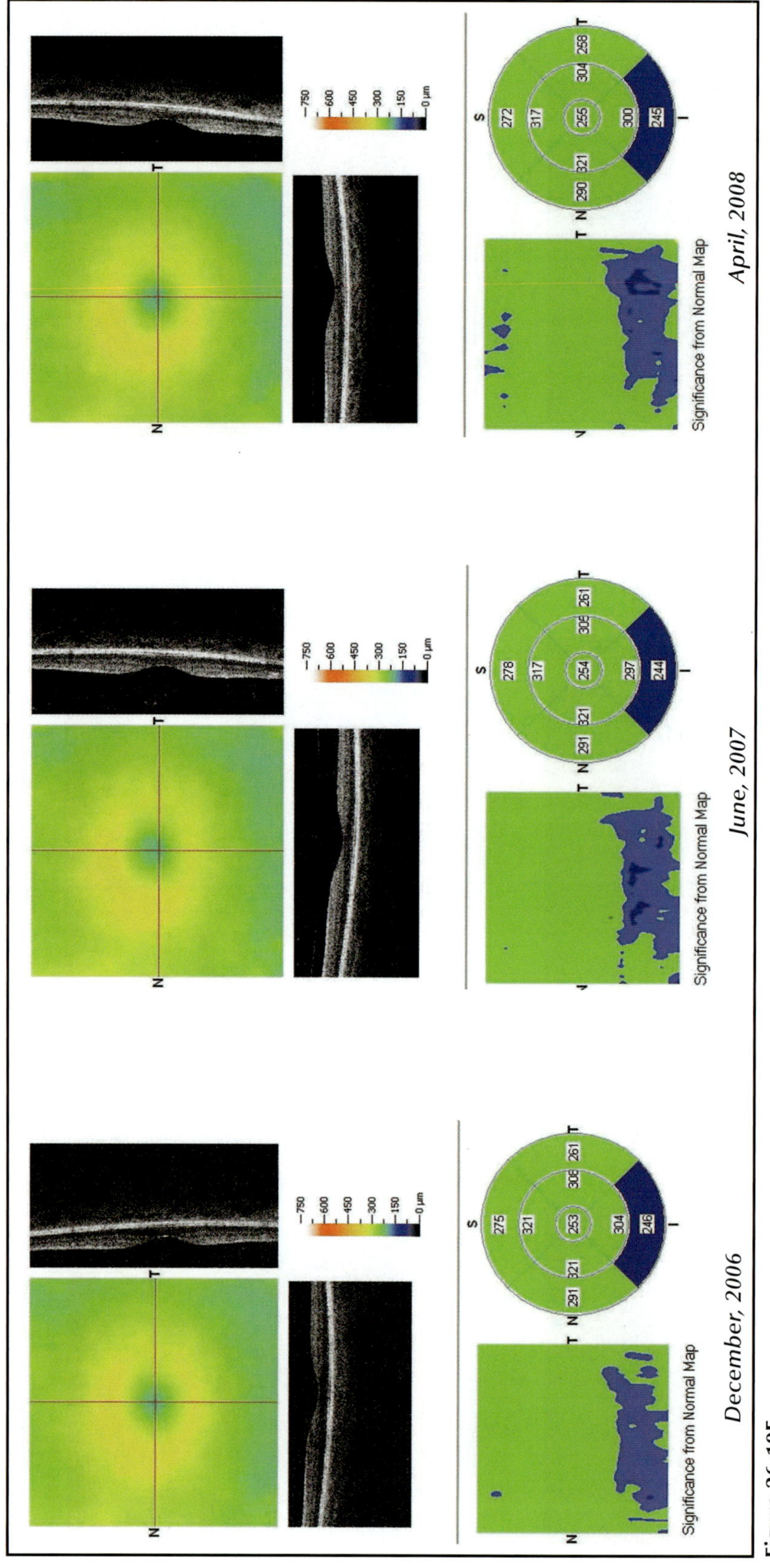

Figure 26-10E.

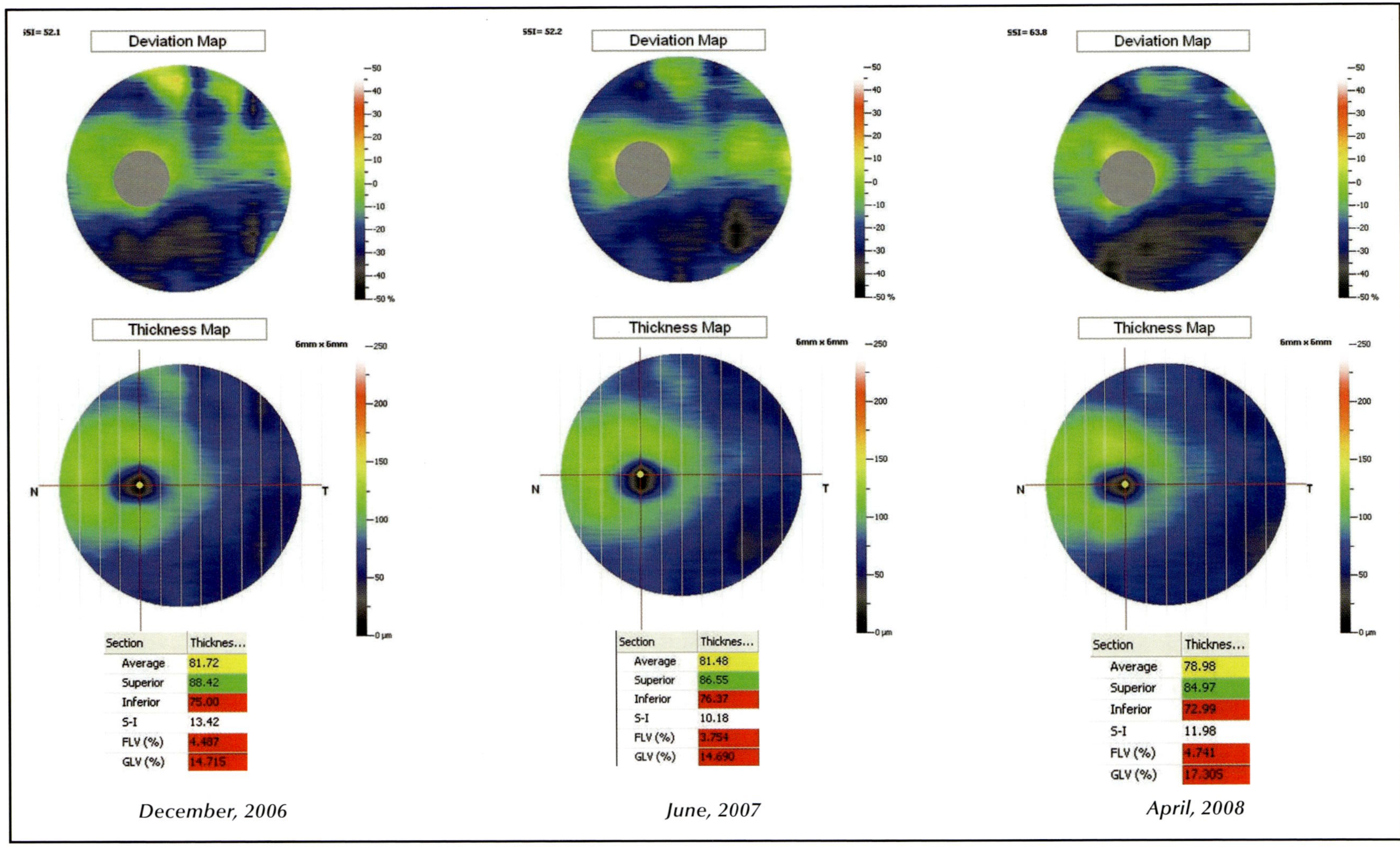

Figure 26-10F.

Detection of Glaucomatous Progression Using Spectral-Domain Optical Coherence Tomography

Gianmarco Vizzeri, MD; Linda M. Zangwill, PhD; and Robert N. Weinreb, MD

INTRODUCTION

Glaucoma is a slowly progressing optic neuropathy characterized by the loss of retinal ganglion cells and their axons. For this reason, the detection of glaucomatous progression is a critical aspect of glaucoma management. The identification of glaucomatous changes, such as progressive thinning of the retinal nerve fiber layer (NFL), not only can help clinicians in confirming the initial diagnosis but, more importantly, can alert them that further treatment may be required to prevent visual impairment due to glaucoma.

At present, however, clinicians attempting to establish the occurrence of glaucomatous progression face numerous challenges. First, because risk factors associated with glaucomatous progression are not well established, it is difficult to identify patients at increased risk for disease progression that may require closer follow-up. Second, the task is complicated by the fact that changes occurring at the level of the disc and the retinal NFL are often subtle and difficult to detect on clinical examination. Third, to better target treatment, not only should one know whether progression has occurred but also how fast the change is occurring.

Imaging techniques used to objectively assess the optic nerve head (ONH) and the retinal NFL, such as OCT, provide clinicians with a promising tool to monitor disease progression. It is essential that clinicians understand the strengths and limitations of these instruments so that they are interpreted appropriately for patient management decisions.

TIME-DOMAIN OPTICAL COHERENCE TOMOGRAPHY TO DETECT GLAUCOMATOUS PROGRESSION

OCT, by providing noninvasive quantitative assessment of ocular microstructures,[1] has shown good reproducibility,[2-5] together with the ability to detect glaucomatous structural damage in numerous studies.[6-17] However, there are few reports documenting the ability of Stratus OCT (Carl Zeiss Meditec, Dublin, CA), the only time-domain OCT device available on the market, for detecting glaucomatous progression.[18,19] This may in part be due to the fact that progression software for the Stratus OCT has only recently become available for the retinal NFL circle scan to be placed around the optic disc to quantify the retinal NFL thickness. However, progression software is not yet available for the ONH scan, consisting of 6 radial scan lines for optic disc evaluation. A previous investigation using a prototype time-domain OCT instrument with higher variability than the commercial Stratus OCT device suggests that change in retinal NFL thickness can be documented with time-domain OCT.[18] A more recent investigation has shown that time-domain OCT has the potential for providing the rate of retinal NFL thickness change for clinical use.[19]

However, there are several theoretical limitations to time-domain OCT that may weaken its ability to detect progression.[20-29] First, time-domain OCT lacks a scan registration feature to ensure that the scan and subsequent measurements are obtained at the same location dur-

Huang D, Duker JS, Fujimoto JG, Lumbroso B, Schuman JS, Weinreb RN.
Imaging the Eye from Front to Back with RTVue Fourier-Domain Optical Coherence Tomography (pp 245-252).

ing follow-up examinations.[24-27] For example, to obtain retinal NFL thickness estimates at 3.4 mm around the optic disc, the instrument currently relies on the operator to consistently center the circle scan on the optic disc in follow-up scans with the help of a landmark feature (ie, placing a landmark on a branching vessel or at the disc margin to be used as reference for follow-up scans). Second, numerous studies have shown that scan quality and improper alignment can significantly affect the variability of time-domain OCT measurements and limit the ability to detect "true" change over time.[27-29] In addition, cut-offs for which a decrease in average retinal NFL thickness can be attributed to disease progression and not to signal strength variability have not yet been clearly established.

Spectral-Domain Optical Coherence Tomography to Detect Glaucomatous Progression

Spectral-domain OCT has the potential to overcome most of the limitations of time-domain OCT technology. Spectral-domain OCT technology currently is capable of an acquisition speed of up to 29,000 A-scans per second, compared to 400 A-scans per second with time-domain OCT. The increased acquisition speed results in a significant reduction in motion artefacts that are known to decrease scan resolution. In addition, more scans to be recorded per second allow for less data interpolation compared to time-domain OCT. With the current software analysis, a cube scan of 6 mm by 6 mm that contains more than 40,000 scans can be acquired in less than 2 seconds. This process results in a more complete assessment of the ONH and its structures.

Moreover, spectral-domain OCT offers a higher resolution than time-domain OCT. The current resolution of spectral-domain OCT is 5 µm, compared to approximately 10 µm axial resolution of time-domain OCT in tissue.[30-33] Recent studies have shown that the reproducibility of spectral-domain OCT retinal NFL thickness measurements is excellent and similar in healthy eyes and eyes of glaucoma patients. In addition, retinal NFL thickness measurements taken with time-domain OCT and spectral-domain OCT were found to be highly correlated, although measurements from the 2 instruments may not be interchangeable.[34,35] In part, due to the increased resolution, it is possible that spectral-domain OCT devices may incorporate more robust detection algorithms that are less influenced by changes in the scans' signal-to-noise ratio.[36] These enhancements result in more consistent measurements over time.

To overcome the lack of scan registration with previous OCT technology, the recently introduced RTVue spectral-domain OCT employs an optic disc scan registration feature based on automated disc contour line drawing and blood vessel alignment to ensure that the retinal NFL thickness is measured at the same location around the optic disc over time. First, using a 3D scan, the software automatically delineates the optic disc margin to be placed on all subsequent scans. The process is completed based on software algorithms that automatically detect the end of the retinal pigment epithelium or Bruch's membrane at the disc margin and allow calculation of the retinal NFL thickness along with the optic disc parameters (eg, cup area and rim area). Second, a series of scans are automatically aligned and registered to each other based on matching the blood vessel pattern on the *en face* OCT image from one scan to the next. These automated steps used to perfect scan registration allow the instrument to be more operator independent, compared to previous OCT devices, potentially eliminating one of the possible variance components affecting the long-term variability of time-domain OCT measurements.

However, the failure of an algorithm to accurately detect the end of the retinal pigment epithelium or Bruch's membrane is always possible. Therefore, the user should always verify accuracy, particularly as the software allows the operator to modify the disc margin as desired. In addition, because RTVue lacks an eye tracking system, scan distortions or imperfect blood vessel alignment due to incorrect post-processing may occur as a result of eye movements. As a consequence, the ability of the software algorithms to accurately detect retinal NFL thickness changes over time may be impaired. Finally, RTVue should be capable of identifying significant change that is greater than the variability of the measurements, and this calculation is not available with the current software version.

Nevertheless, these enhanced features suggest that spectral-domain OCT may be even more capable of monitoring disease progression than Stratus OCT. Following is a description on how RTVue progression software can allow detection of glaucomatous structural changes over time.

RTVue Progression Software

RTVue (software version 4.0) offers 4 unique scanning protocols of analysis designed for glaucoma evaluation and for detecting disease progression:

1. The 3D disc
2. The ONH
3. The ganglion cell complex (GCC)
4. The retinal NFL 3.45

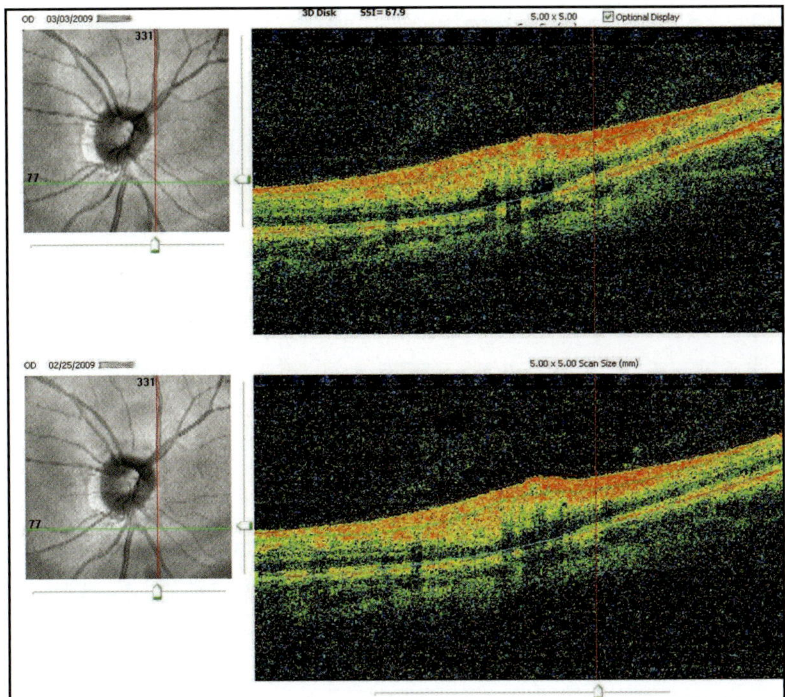

Figure 27-1. 3D disc glaucoma progression printout report of a healthy eye. The 2 *en face* OCT images can be visualized simultaneously. Each A-scan and B-scan is visually compared between scans by moving the green and red lines horizontally and vertically, respectively, on the *en face* OCT images.

(Each of these protocols has been previously described in detail elsewhere in this book.)

In brief, while the 3D disk provides a 3D image (4 x 4 mm) of the optic disc and the surrounding area, the ONH is used to quantitatively assess the retinal NFL thickness and the ONH features, such as disc area and rim area. The GCC employs a unique algorithm for the evaluation of the GCC in the macular region. Finally, the retinal NFL 3.45 uses a circle scan with a standardized diameter of 3.45 mm centered on the optic disc to obtain retinal NFL thickness measurements, similar to previous OCT technology.

With the current software version, a specific progression analysis feature is available for all 4 protocols. In order for the scans to be processed by the software, at least 1 baseline and 1 follow-up scan must be acquired.

The current RTVue progression software allows direct visual comparison of the scans obtained at baseline and different follow-ups and automatically quantifies the amount of change present at the last follow-up visit as compared to the baseline. For the average retinal NFL thickness and other important parameters, such as GCC, for example, the software also provides a graph displaying the change in retinal NFL thickness over a series of visits.

In this chapter, a case example consisting of a healthy eye of a healthy subject imaged 4 times over a 1-month period is used to describe the progression report available for each of the 4 protocols of analysis.

3D DISC GLAUCOMA PROGRESSION REPORT

The 3D disc protocol is composed of 101 B-scans each composed of 512 A-scans, all acquired in 2.2 seconds. The resulting scan provides a 3D image of the optic disc and surrounding area. The *en face* image generated by this scanning protocol is then used to automatically draw the contour line describing the disc margin that is required to generate optic disc parameters and to ensure scan registration at follow-up visits.

The 3D disc progression software allows visual comparison of sets of scans by simultaneously visualizing 2 *en face* OCT images taken at different times. A-scan and B-scan from the baseline and the follow-up image can be visually compared by scrolling horizontally or vertically along the *en face* OCT image. However, no quantitative assessment of changes over time is available with this protocol (Figure 27-1).

OPTIC NERVE HEAD MAP (NHM4) GLAUCOMA PROGRESSION REPORT

The ONH is composed of 12 radial scans of 3.4 mm length (452 A-scans each) and 6 concentric ring

Figure 27-2. ONH map (NHM4) glaucoma progression printout report of a healthy eye. All baseline and up to 3 follow-up nerve head maps can be visualized simultaneously. The different TSNIT plots also can be visually compared. The relative change in parameters' measurements between baseline and the last follow-up is shown. Note: the change in microns shown for the parameters described may still be within the test-retest variability of the measurements. Significant change is not calculated with the current software version.

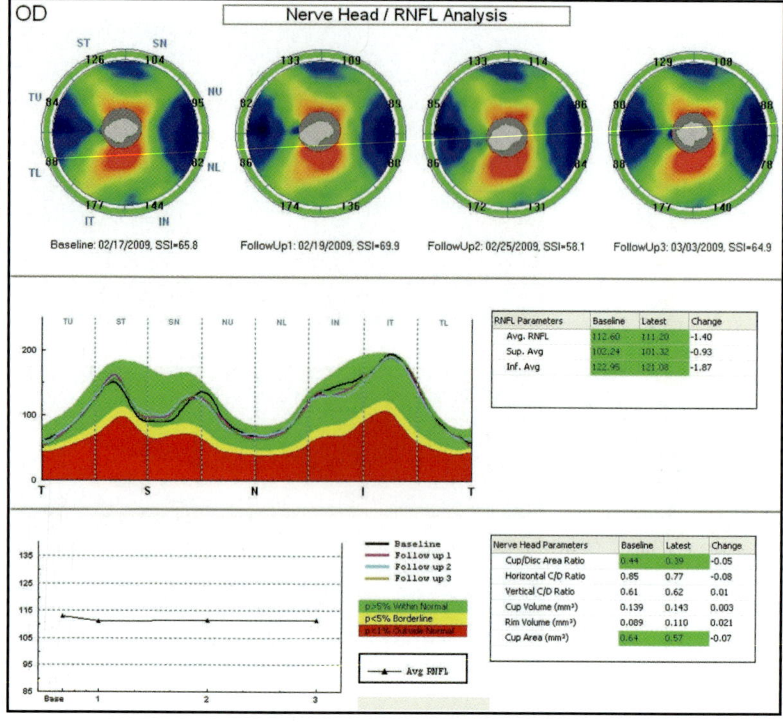

RNFL Parameters	Baseline	Latest	Change
Avg. RNFL	112.60	111.20	-1.40
Sup. Avg	102.24	101.32	-0.93
Inf. Avg	122.95	121.08	-1.87

Nerve Head Parameters	Baseline	Latest	Change
Cup/Disc Area Ratio	0.44	0.39	-0.05
Horizontal C/D Ratio	0.85	0.77	-0.08
Vertical C/D Ratio	0.61	0.62	0.01
Cup Volume (mm³)	0.139	0.143	0.003
Rim Volume (mm³)	0.089	0.110	0.021
Cup Area (mm²)	0.64	0.57	-0.07

scans ranging from 2.5 to 4.0 mm diameters (587 or 775 A-scans each), all centered on the optic disc using an automatic contour line previously drawn on the *en face* OCT image to ensure scan registration. This scan configuration provides 9510 A-scans in 0.39 seconds. A polar retinal NFL thickness map and various parameters to describe optic disc features are provided.

The ONH map progression report can display up to 4 scans at a time (1 baseline and 3 follow-up scans) separately for each eye. The retinal NFL thickness maps in pseudo colors along with sectoral retinal NFL thickness measurements from the 3.45 mm circle scan (colored in green for within normal limits, in yellow for borderline, or in red for outside normal limits based on comparisons with results from the instrument normative database) are displayed for each scan. Below each scan are also the date the scan was recorded and the signal strength index (SSI) expressed on a scale from 1 to 100. This information allows for a visual comparison of the retinal NFL thickness maps and measurements across visits. The report also contains a TSNIT (temporal, superior, nasal, inferior, temporal) graph to display the double hump retinal NFL thickness profile at 3.45 mm around the optic disc for all scans. A "fit each value" graph for the average retinal NFL thickness to describe changes in retinal NFL thickness values from baseline is also shown. In addition, for each retinal NFL parameter and nerve head parameter evaluated, the amount of change between the baseline scan the last follow-up scan is automatically calculated (eg, for the parameter average retinal NFL thickness, the baseline and last follow-up scans are 112.60 and 111.20 μm, respectively, and the change is -1.40 μm; Figure 27-2).

GANGLION CELL COMPLEX GLAUCOMA PROGRESSION REPORT

The GCC protocol consists of a macula grid that can be used for the assessment of the macular GCC thickness. The GCC is composed of 3 layers: the retinal NFL layer made up of the ganglion cell axons, the ganglion cell layer made up of the ganglion cell bodies, and the inner plexiform layer made up of the ganglion cell dendrites. Taking advantage of the higher scan speed, RTVue can generate a macular map by evenly sampling the macular region over a 7 mm square area. The center of the GCC protocol is shifted 0.75 mm temporally to improve sampling of the temporal periphery. The GCC pattern consists of 14928 A-scans from 1 horizontal line and 15 vertical lines with 0.5 mm intervals. The scan time for the GCC pattern is 0.6 seconds.

The GCC progression report can display up to 4 scans at a time (1 baseline and 3 follow-up scans) separately for each eye. The printout can be divided into 4 sections. The first 3 sections provide the *GCC thickness map* (color-coded thickness map representing the thickness of the GCC of the macular region), the *deviation map* (color-coded map showing the percent loss from normal as determined by the instrument normative database), and the *significance map* (color-coded map showing regions where the change from normal reaches statistical significance, with green, yellow, and red representing values within the normal range, borderline with a p value of <5% and outside normal limits with a p value of <1%, respectively) for each scan to allow visual comparison across

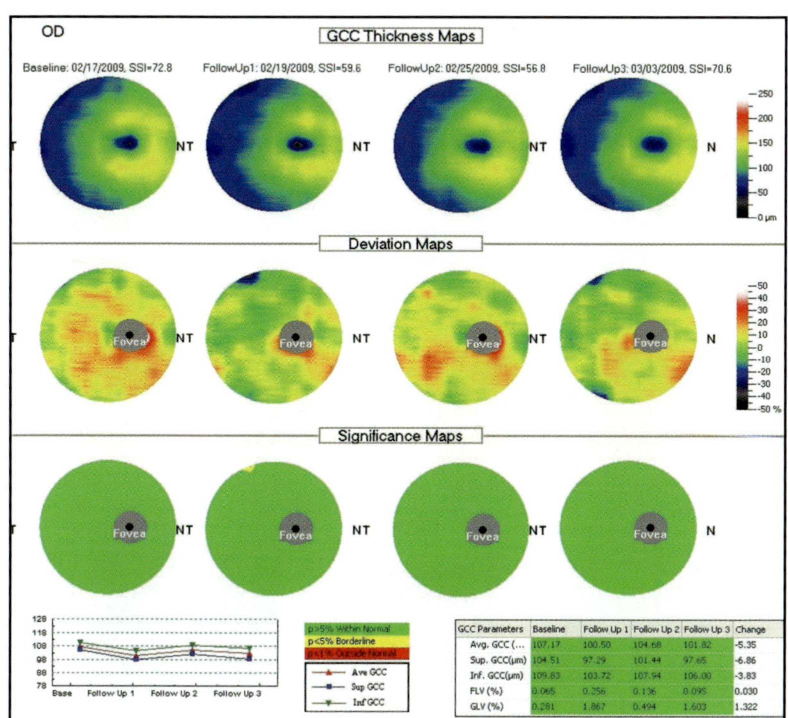

Figure 27-3. GCC glaucoma progression printout report of a healthy eye. All baseline and up to 3 follow-up GCC thickness, deviation, and significance maps can be visualized simultaneously. The change in microns between GCC thickness measured at baseline and at the last follow-up is shown. Note: the change in microns shown for the parameters described may still be within the test-retest variability of the measurements. Significant change is not calculated with the current software version.

visits. The fourth section contains a "fit each value" graph for the average, superior, and inferior GCC (ie, the GCC thickness values for the full retina and for the superior and inferior hemiretinas, respectively, with the exclusion of the circle mask in the fovea). This section also shows a table of the results for baseline scan and each follow-up scan, along with the change in microns calculated between the baseline and the last follow-up scan (Figure 27-3).

RETINAL NERVE FIBER LAYER GLAUCOMA PROGRESSION REPORT

The retinal NFL 3.45 protocol assesses the peripapillary retinal NFL thickness by placing a 3.45 circle scan around the optic disc, similar to previous time-domain OCT technology. The retinal NFL 3.45 is the average of 4 circle scans, each composed of 1024 A-scans, for a total of 4096 data points, all acquired in 0.15 seconds.

The retinal NFL progression printout can be divided into 3 sections. In the first section, sectoral retinal NFL thickness results and SSI values at baseline and each subsequent visit are reported. In the second section, the TSNIT graph is displayed for visual comparison of the retinal NFL double hump pattern across scans. The third section contains a fit each value graph for the average, superior, and inferior retinal NFL thickness, and a table showing the results for baseline scan and each follow-up scan along with the change in microns calculated between the baseline and the last follow-up scan (Figure 27-4).

CONCLUSION

When transitioning to a new and more advanced technology, such as spectral-domain OCT for retina and optic disc imaging, evidence is required to confirm that the new instrument can offer clinicians a significant advantage over previous technologies. With an automated scan registration and specific software designed for glaucoma progression, RTVue has the potential to assist clinicians in the difficult task of detecting glaucomatous structural changes at the level of the optic disc, the retinal NFL thickness, and the entire GCC. Numerous studies are underway to evaluate how spectral-domain OCT can benefit the management of glaucoma. In addition, software is improving rapidly, so it is likely that more powerful tools for monitoring glaucoma progression using RTVue will be available in the near future. In particular, one critical aspect that remains to be addressed relates to measurements' variability. In fact, for any progression algorithm to be valuable, it must identify significant change that is greater than the variability of the measurements. Regardless of the method used to detect structural change, careful clinical examination and visual function testing also should be considered when making patient management decisions.

REFERENCES

1. Huang D, Swanson EA, Lin CP, et al. Optical coherence tomography. *Science.* 1991;254:1178-1181.

Figure 27-4. Retinal NFL glaucoma progression printout report of a healthy eye. All baseline and up to 3 follow-up retinal NFL maps can be visualized simultaneously. The different TSNIT plots also can be visually compared. The relative change in parameters' measurements between baseline and the last follow-up is shown. Note: the change in microns shown for the parameters described may still be within the test-retest variability of the measurements. Significant change is not calculated with the current software version.

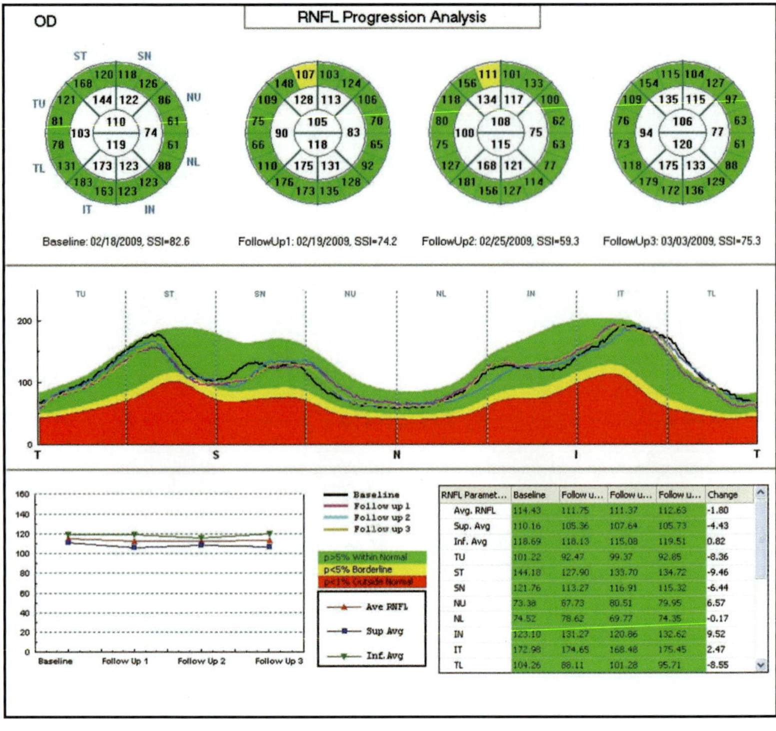

2. Budenz DL, Chang RT, Huang X, et al. Reproducibility of retinal nerve fiber thickness measurements using the Stratus OCT in normal and glaucomatous eyes. *Invest Ophthalmol Vis Sci.* 2005;46:2440-2443.

3. Paunescu LA, Schuman JS, Price LL, et al. Reproducibility of nerve fiber layer thickness, macular thickness, and optic nerve head measurements using Stratus OCT. *Invest Ophthalmol Vis Sci.* 2004;45:1716-1724.

4. Budenz DL, Fredette MJ, Feuer WJ, et al. Reproducibility of peripapillary retinal nerve fiber thickness measurements with Stratus OCT in glaucomatous eyes. *Ophthalmology.* 2008;115:661-666.

5. Carpineto P, Ciancaglini M, Aharrh-Gnama A, et al. Custom measurement of retinal nerve fiber layer thickness using Stratus OCT in normal eyes. *Eur J Ophthalmol.* 2005;15:360-366.

6. Schuman JS, Hee MR, Puliafito CA, et al. Quantification of nerve fiber layer thickness in normal and glaucomatous eyes using optical coherence tomography. *Arch Ophthalmol.* 1995;113:586-596.

7. Budenz DL, Michael A, Chang RT, et al. Sensitivity and specificity of the Stratus OCT for perimetric glaucoma. *Ophthalmology.* 2005;112:3-9.

8. Medeiros FA, Zangwill LM, Bowd C, et al. Evaluation of retinal nerve fiber layer, optic nerve head, and macular thickness measurements for glaucoma detection using optical coherence tomography. *Am J Ophthalmol.* 2005;139:44-55.

9. Brusini P, Salvetat ML, Zeppieri M, et al. Comparison between GDx VCC scanning laser polarimetry and Stratus optical coherence tomography in the diagnosis of chronic glaucoma. *Acta Ophthalmol Scand.* 2006;84:650-655.

10. Sihota R, Sony P, Gupta V et al. Diagnostic capability of optical coherence tomography in evaluating the degree of glaucomatous retinal nerve fiber damage. *Invest Ophthalmol Vis Sci.* 2006;47:2006-2010.

11. Cheng HY, Huang ML. Discrimination between normal and glaucomatous eyes using Stratus optical coherence tomography in Taiwan Chinese subjects. *Graefes Arch Clin Exp Ophthalmol.* 2005;243:894-902.

12. Medeiros FA, Zangwill LM, Bowd C, Weinreb RN. Comparison of the GDx VCC scanning laser polarimeter, HRT II confocal scanning laser ophthalmoscope, and Stratus optical coherence tomography for the detection of glaucoma. *Arch Ophthalmol.* 2004;122:827-837.

13. Jeoung JW, Park KH, Kim TW, et al. Diagnostic ability of optical coherence tomography with a normative database to detect localized retinal nerve fiber layer defects. *Ophthalmology.* 2005;112:2157-2163.

14. Bowd C, Zangwill LM, Medeiros FA, et al. Structure-function relationships using confocal scanning laser ophthalmoscopy, optical coherence tomography, and scanning laser polarimetry. *Invest Ophthalmol Vis Sci.* 2006;47:2889-2895.

15. Hoffmann EM, Medeiros FA, Sample PA, et al. Relationship between patterns of visual field loss and retinal nerve fiber layer thickness measurements. *Am J Ophthalmol.* 2006;141:463-471.

16. Shah NN, Bowd C, Medeiros FA, et al. Combining structural and functional testing for detection of glaucoma. *Ophthalmology.* 2006;113:1593-1602.

17. Zangwill LM, Bowd C. Retinal nerve fiber layer analysis in the diagnosis of glaucoma. *Curr Opin Ophthalmol.* 2006;17:120-31.

18. Wollstein G, Schuman JS, Price LL, et al. Optical coherence tomography longitudinal evaluation of retinal nerve fiber layer thickness in glaucoma. *Arch Ophthalmol.* 2005;123:464-470.

19. Leung CK, Cheung CY, Liu S, et al. *Evaluation of glaucoma progression by measuring the rate of change of retinal nerve fiber layer thickness.* ARVO Abstract E-2571, 2009.

20. Savini G, Zanini M, Barboni P. Influence of pupil size and cataract on retinal nerve fiber layer thickness measurements by Stratus OCT. *J Glaucoma.* 2006;15:336-340.

21. van Velthoven ME, van der Linden MH, de Smet MD, et al. Influence of cataract on optical coherence tomography image quality and retinal thickness. *Br J Ophthalmol.* 2006;90:1259-1262.

22. Stein DM, Wollstein G, Ishikawa H, et al. Effect of corneal drying on optical coherence tomography. *Ophthalmology.* 2006;113:985-991.

23. Kok PH, Van Dijk HW, Van den Berg TJ, et al. A model for the effect of disturbances in the optical media on the OCT image quality. *Invest Ophthalmol Vis Sci.* 2009;50:787-792.

24. Vizzeri G, Bowd C, Medeiros FA, et al. Effect of improper scan alignment on retinal nerve fiber layer thickness measurements using Stratus optical coherence tomography. *J Glaucoma.* 2008;17:341-349.

25. Gabriele ML, Ishikawa H, Wollstein G, et al. Optical coherence tomography scan circle location and mean retinal nerve fiber layer measurement variability. *Invest Ophthalmol Vis Sci.* 2008;49:2315-2321.

26. Vizzeri G, Bowd C, Medeiros FA, et al. Scan tracking coordinates for improved centering of Stratus OCT scan pattern. *J Glaucoma.* 2009;18:81-87.

27. Vizzeri G, Bowd C, Medeiros FA, et al. Effect of signal strength and improper alignment on the variability of Stratus optical coherence tomography retinal nerve fiber layer thickness measurements. *Am J Ophthalmol.* 2009;148:249-255.e1.

28. Wu Z, Vazeen M, Varma R, et al. Factors associated with variability in retinal nerve fiber layer thickness measurements obtained by optical coherence tomography. *Ophthalmology.* 2007;114:1505-1512.

29. Cheung CY, Leung CK, Lin D, et al. Relationship between retinal nerve fiber layer measurement and signal strength in optical coherence tomography. *Ophthalmology.* 2008;115:1347-1351.e1-2.

30. van Velthoven ME, Faber DJ, Verbraak FD, et al. Recent developments in optical coherence tomography for imaging the retina. *Progr Retin Eye Res.* 2007;26:57-77.

31. Chen TC, Cense B, Pierce MC, et al. Spectral domain optical coherence tomography: ultra-high speed, ultra-high resolution ophthalmic imaging. *Arch Ophthalmol.* 2005;123:1715-1720.

32. Wojtkowski M, Leitgeb R, Kowalczyk A, et al. In vivo human retinal imaging by Fourier domain optical coherence tomography. *J Biom Opt.* 2002;7:457-463.

33. Nassif N, Cense B, Hyle Park B, et al. In vivo human retinal imaging by ultrahigh-speed spectral domain optical coherence tomography. *Optic Letters.* 2004;29:480-482.

34. Vizzeri G, Weinreb RN, Gonzalez-Garcia AO, et al. Agreement between spectral-domain and time-domain OCT for measuring retinal NFL thickness. *Br J Ophthalmol.* 2009;93:775-781.

35. González-García AO, Vizzeri G, Bowd C, et al. Reproducibility of RTVue retinal nerve fiber layer thickness and optic disc measurements and agreement with Stratus optical coherence tomography measurements. *Am J Ophthalmol.* 2009;147:1067-1074.e1.

36. Johannes F, de Boer, Cense B, et al. Improved signal-to-noise ratio in spectral-domain compared with time-domain optical coherence tomography. *Optic Letters.* 2003;28:2067-2069.

28 Nonglaucomatous Optic Neuropathies

Divya Aggarwal, MD; Ou Tan, PhD; and
Alfredo A. Sadun, MD, PhD

The optic nerve is about 50 mm in length and extends from the back of the eye posteriorly to the brain. It traverses through the eye wall (1 mm), orbit (25 mm), optic canal (9 mm), and cranium (16 mm), and partially decussates with the other optic nerve to form the optic chiasm. Any insult along the entire course of the optic nerve can result in damage to the fibers of the optic nerve (Figure 28-1) and retrograde degeneration to the retinal ganglion cell.

OCT is an objective method of assessing the optic nerve head (ONH) and peripapillary region. It provides quantitative measurements of the optic cup and disc and retinal nerve fiber layer (NFL) thickness, which are helpful in assessing optic disc edema and atrophy. It is useful in delineating diagnosis at early stages of disease. Its approximation to histological analysis may also provide insights into the pathogenesis of the diseases of the optic nerve.

NONANTERIOR ISCHEMIC OPTIC NEUROPATHY

A 51-year-old woman complained of painless sudden onset blurred vision in the left eye 6 months before presentation. She described a visual field defect. She did not have a history of headache, scalp tenderness, jaw claudication, fatigue, malaise, loss of weight, fever, or joint pains.

On examination, her best-corrected visual acuities (BCVAs) were 20/20 and 20/25. Optic nerve function studies showed very severe impairment OS; however, she had good color vision. She could discern 12 out of 12 American

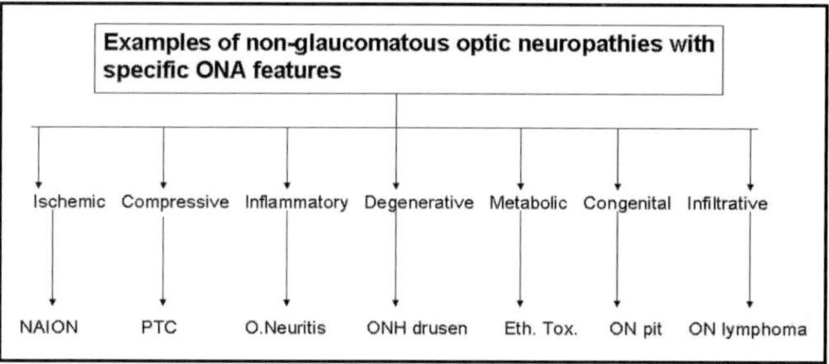

Figure 28-1. ONA: optic nerve atrophy, NAION: nonanterior ischemic optic neuropathy, PTC: pseudotumor cerebri, O. Neuritis: optic neuritis, ONH drusen: optic nerve head drusen, Eth. Tox.: ethambutol toxicity, ON pit: optic nerve pit, ON lymphoma: optic nerve lymphoma.

Optical (AO) test plates in either eye, although her brightness sense was down in the left eye to 85% of the right. There was an afferent pupillary defect (APD) OS. Amsler grid testing was full and normal in both eyes. Fundus examination showed 0.0 cup-to-disc ratio in both the eyes and superimposed mild (1+) optic atrophy in the left eye. Humphrey visual field 30-2 was normal in the right eye, but in the left eye there was an inferior nasal quadrantanopia that respected the horizontal and extended over the vertical midline (Figure 28-2, OS), and hence classical for nonanterior ischemic optic neuropathy (NAION).

Blood tests revealed normal erythrocyte sedimentation rate, C-reactive protein, angiotensin converting enzyme level and negative anti-nuclear antibody (ANA) panel, antiphospholipid, anticardiolipin and anti-beta-2 glycoprotein-1 antibodies. Rheumatoid factor was found to be mildly high.

The diagnosis of NAION was made.

Huang D, Duker JS, Fujimoto JG, Lumbroso B, Schuman JS, Weinreb RN.
Imaging the Eye from Front to Back with RTVue Fourier-Domain Optical Coherence Tomography (pp 253-260).

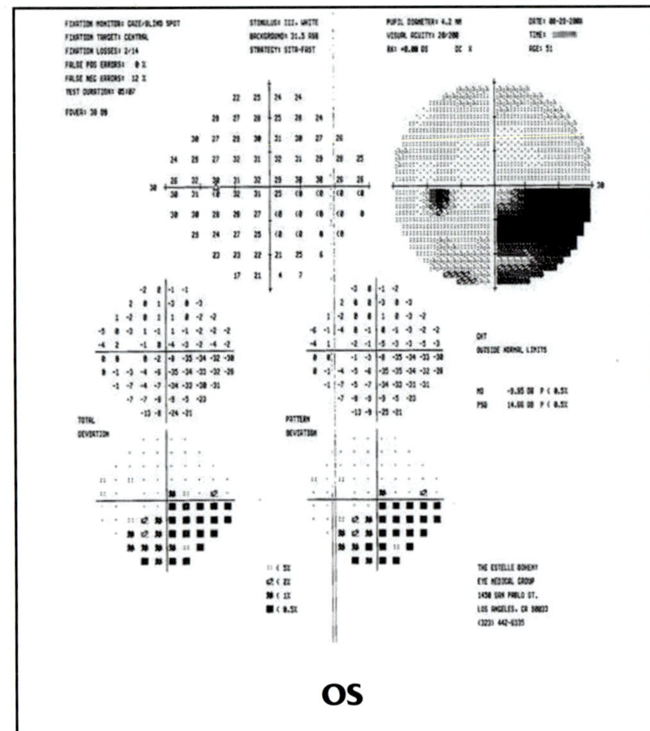

Figure 28-2. OS: inferior nasal quadrantic defect in NAION obeying the horizontal meridian.

Optical Coherence Tomography

OCT of the ONH scan showed crowding of the optic disc in both eyes. It also showed elevation of the ONH in the superior nasal region in the right eye (Figure 28-3, OD) and thinning of the NFL in the superior hemisphere in the left eye (Figure 28-3, OS). The peripapillary retinal nerve fiber layer (ppNFL) scan showed normal thickness, except for mild disc swelling in the superonasal quadrant in the right eye (Figure 28-4, OD). The ppNFL scan in the left eye showed NFL thinning in the superotemporal quadrant (ST and TU) corresponding to the inferonasal quadrant defect in the visual field (Figure 28-4, OS), and optic atrophy was observed on fundus examination.

PSEUDOTUMOR CEREBRI

A 37-year-old obese woman was referred to the neuro-ophthalmology service at Doheny Eye Institute with the chief complaint of visual distortion in the left eye of 1 year's duration. She described the visual distortion as heat coming off the pavement. This distortion got worse when she got up from supine position (transient obscurations of vision typical of papilledema). She also complained of headaches which got worse when lying down. She had no history of neurological symptoms like numbness, weakness, tingling, incontinence, etc.

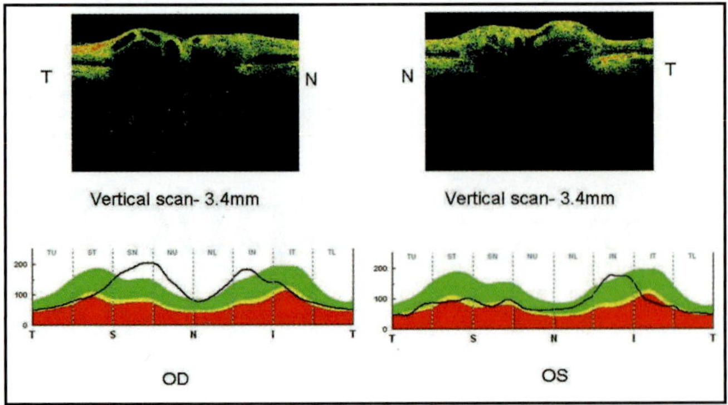

Figure 28-3. ONH scan, OD (upper left): vertical line scan (D = 3.4 mm) shows the crowding of optic disc (bottom left), retinal NFL thickness profile shows elevation of the optic disc in the superior nasal region. OS (upper right): vertical line scan (D = 3.4 mm) shows the crowding of optic disc (bottom right), retinal NFL thickness profile shows thinning of NFL in the superior hemisphere. Crowding of optic discs in both OD and OS cause cup-to-disc ratio of 0.

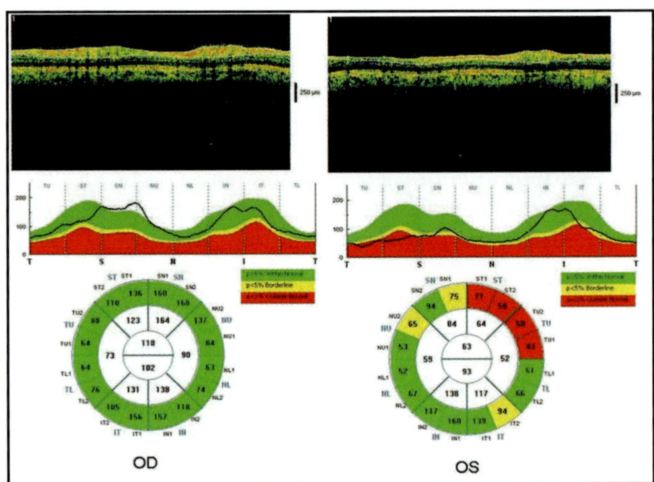

Figure 28-4. Retinal NFL 3.45 scan in NAION. OD: mild retinal NFL swelling in the superior nasal region. OS: retinal NFL thinning in the superior temporal quadrant.

On examination, visual acuities were 20/20 in both eyes. Optic nerve function studies were all normal including color vision, brightness sense, and pupils. Amsler grid testing was full and normal in both eyes. Fundus examination showed some disc edema and hyperemia in both eyes (Figure 28-5, OD and OS). Visual field examination revealed a normal visual field in the right eye (Figure 28-6, OD) and an enlarged blind spot in the left eye (Figure 28-6, OS).

Magnetic resonance imaging (MRI) of the brain was normal. Blood tests revealed a positive ANA. She was clinically diagnosed with pseudotumor cerebri (PTC) and advised to undergo lumbar puncture and weight loss.

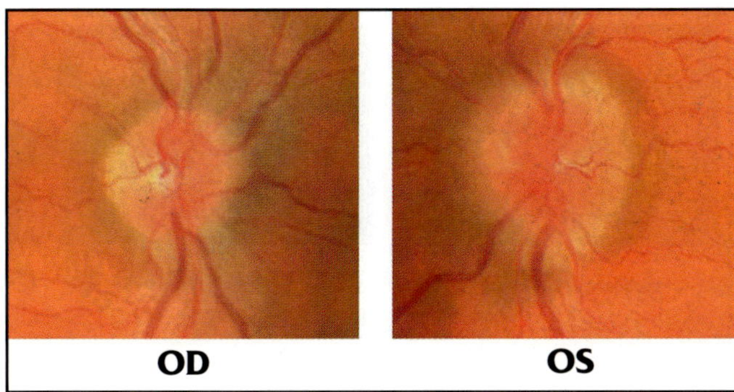

Figure 28-5. Fundus photograph. OD: shows disc edema worse nasally. OS: shows diffuse optic disc edema.

Optical Coherence Tomography

A 3D scan through the ONH demonstrated optic disc edema in both eyes (Figure 28-7, OD and OS). In the right eye, the edema was greatest supero- and inferonasally around the optic disc margins (Figure 28-7, OD). In the left eye, optic disc edema also extended to involve the temporal quadrants (Figure 28-7, OS). The cup-to-disc ratio was 0.01 in the left eye, whereas it was completely obliterated in the left eye with the cup-to-disc ratio being 0.0. The average retinal NFL thickness was 144.68 μm in the right eye and 175.40 μm in the left eye. Ganglion cell complex (GCC) scan though the macula was normal.

Optic Neuritis

A 39-year-old woman complained of a subacute loss of vision in the left eye that progressed over 3 to 4 days. This was associated with retrobulbar pain exacerbated by eye movements. She also complained of double vision of 1 week's duration and the left eye wandering out.

On examination, visual acuities were 20/20 and 20/400 respectively. Optic nerve function was impaired OS. She could discern 10/10 AO test plates OD, but none OS. The brightness sense was down in the left eye to 20% of OD. She had 2+ APD OS. Amsler grid testing was normal OD, but she had a large central scotoma OS.

Fundus examination was normal OD and showed a significant amount of temporal pallor OS (Figure 28-8).

She had 6 D of exotropia in all fields of gaze and full ductions and vergences, however, she showed a mild internuclear ophthalmoplegia on her left. Humphrey visual field testing 30-2 was normal in the right eye and showed generalized depression throughout the central field (Figure 28-9). MRI of the brain and orbit was normal without any lesions or plaques, but showed some enhancement of the left optic nerve.

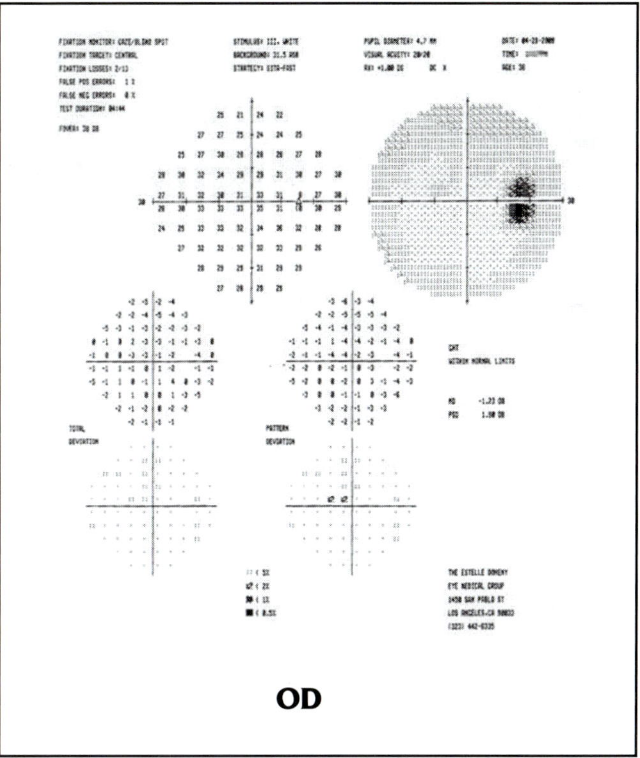

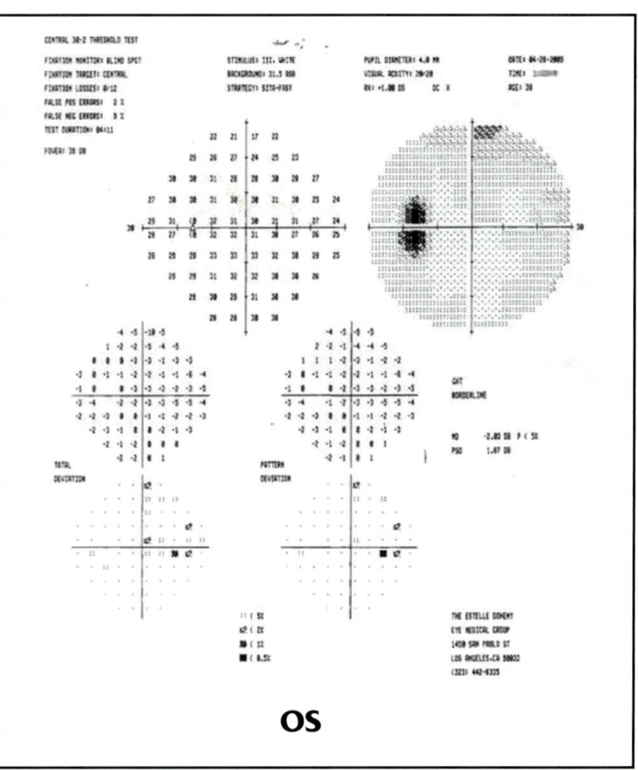

Figure 28-6. OD: normal visual field. OS: enlarged blind spot in PTC.

Thus, this was an episode of retrobulbar optic neuritis diagnosed clinically and confirmed on MRI and accompanied by a resolving internuclear ophthalmoplegia in the left eye.

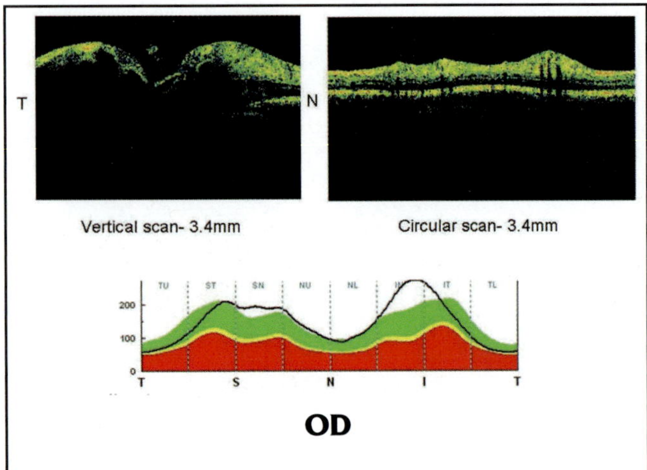

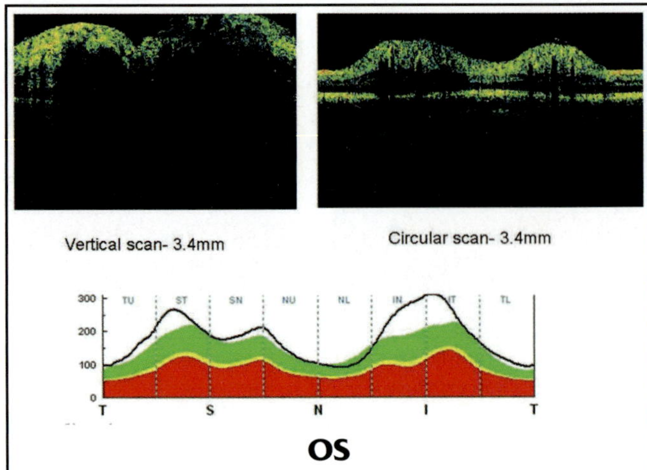

Figure 28-7. OD: ONH vertical and circular scans showing mild optic disc edema superonasally and moderate disc edema inferonasally with average retinal NFL thickness being 144.68 µm. OS: ONH vertical and circular scans showing optic disc edema in both the superior and inferior hemispheres with average retinal NFL thickness being 175.40 µm. Note the obliteration of the cup.

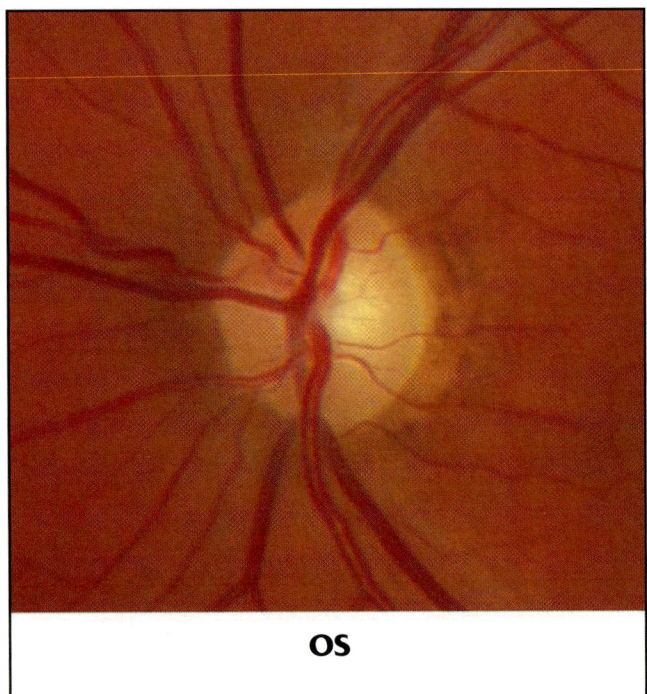

Figure 28-8. Fundus photograph. OS: showing temporal pallor in optic neuritis.

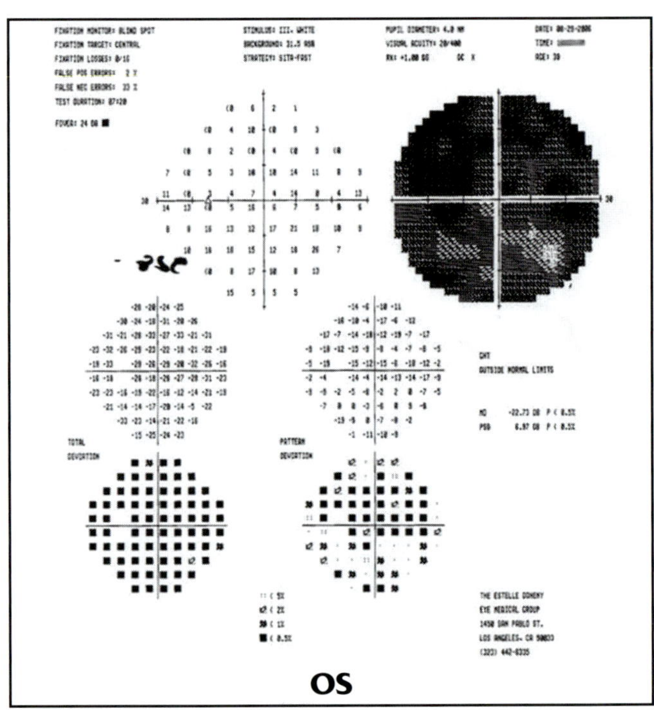

Figure 28-9. OS: generalized depression of visual field in optic neuritis.

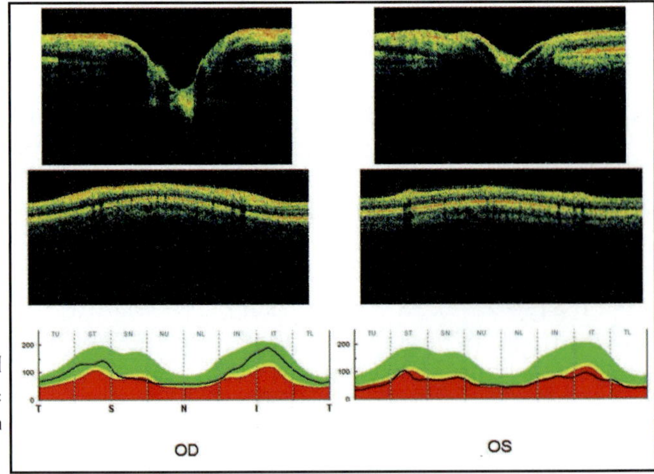

Figure 28-10. Upper: vertical line scan (D = 3.4 mm) from ONH scan. Middle: circular scan (D = 3.45 mm) from ONH scan. Bottom: retinal NFL thickness profile. OD: normal. OS: NFL is thinned in all quadrants.

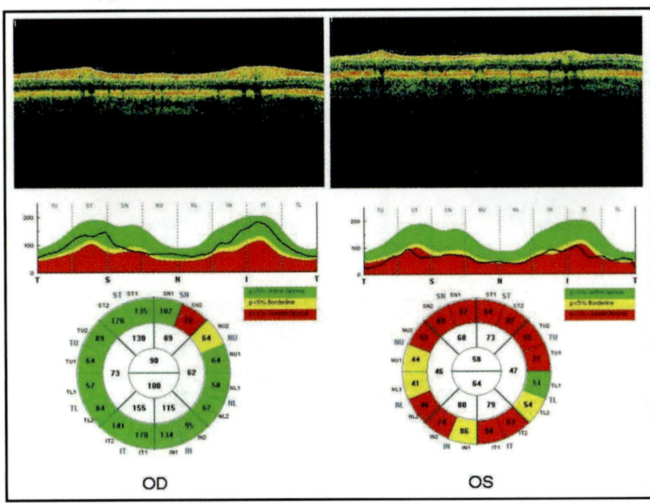

Figure 28-11. OD: ppNFL scan, NFL thinned in superior nasal (SN) and nasal upper (NU) region. OS: diffuse thinning of NFL, with average thickness being 61 μm.

Optical Coherence Tomography

The 3D ONH scan showed diffuse thinning of the NFL thickness in the left eye (Figure 28-10, OS) as compared to the right (Figure 28-10, OD). The ppNFL scan showed some thinning of the ppNFL in the superior nasal region in the right eye (Figure 28-11, OD), whereas the left eye showed diffuse loss of ppNFL (Figure 28-11, OS).

Optic Nerve Head Drusen

A 66-year-old woman came in to the clinic with the complaint of a dark film in front of the right eye for 2 months of gradual onset. She described blurred vision each morning that became worse during the day.

On examination, BCVAs were 20/100 and 20/40 respectively. Optic nerve function studies were abnormal bilaterally with right eye being worse. She saw 0 out of 8 AO test plates OD, but 2 out of 8 OS. Her brightness sense was down in the right eye to 40% of the left eye. She had an APD on the right. Amsler grid testing showed areas of decreased sensation superiorly OD. Fundus examination showed significant ONH drusen throughout the right optic disc and in the temporal and superior regions of the left disc. Humphrey visual field examination showed generalized depression in both eyes with the right visual field being the worse (Figure 28-12, OD and OS). MRI of the brain was normal. She was prescribed latanoprost for lowering the intraocular pressure.

Optical Coherence Tomography

The ppNFL scan in the right eye showed an irregular NFL in the right eye around the optic disc corresponding

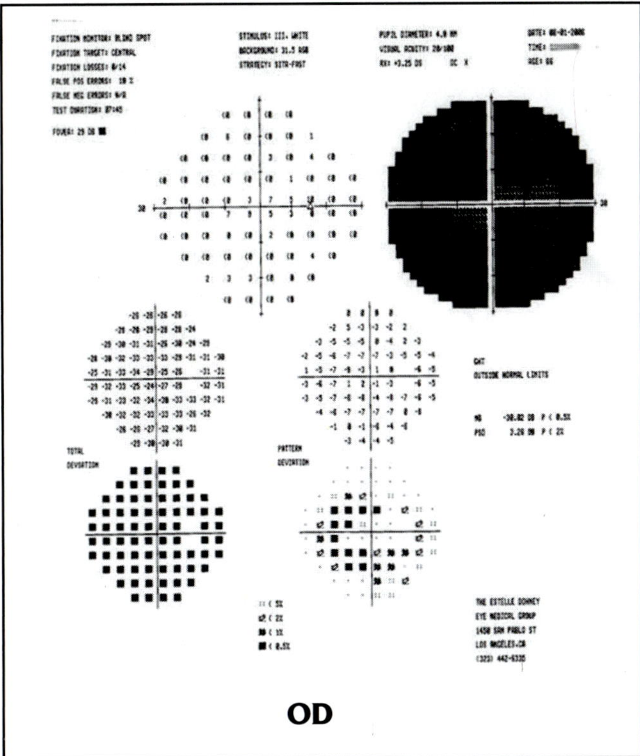

OD

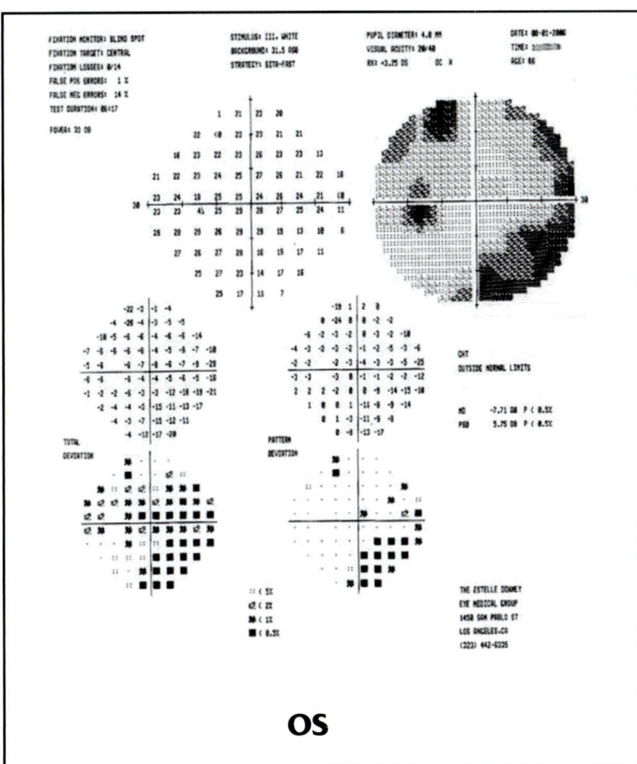

OS

Figure 28-12. OD: ONH drusen visual field showing diffuse loss. OS: ONH drusen visual field showing loss predominantly in the inferior nasal quadrant.

to the drusen with more loss temporally (Figure 28-13, OD). Drusen were also seen in the left optic disc (Figure 28-13, OS). GCC scan of both eyes showed thinning of

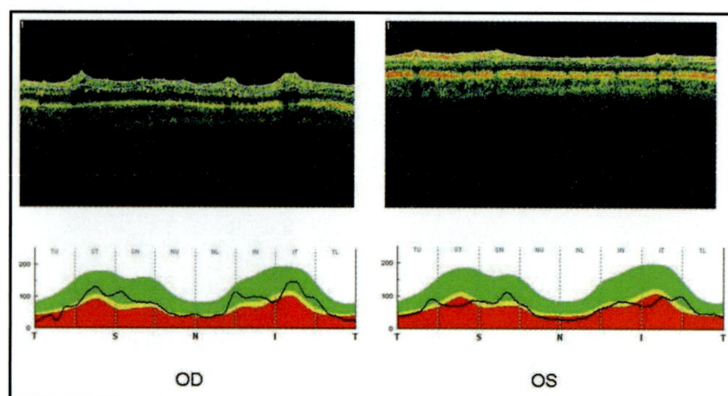

Figure 28-13. Upper: circular scan from retinal NFL 3.45 scan. OD, OS: ONH drusen.

the macula due to retinal ganglion cells loss from retrograde axonal degeneration secondary to ONH drusen (Figure 28-14, OD and OS).

METABOLIC OPTIC NEUROPATHY— ETHAMBUTOL TOXICITY

A 64-year-old female complained of painless central loss of vision in both eyes that worsened over a few days and then stabilized for 3 months until presentation. She also noted some color vision loss, along with general diminution of vision. Her past medical history revealed that she was prescribed ethambutol for an MAC infection over a 4-month period, but this was discontinued 4 days after the visual loss.

On examination, her BCVAs were 20/200 in both eyes. Color vision was compromised. She could see 12 out of 15 plates in the right eye and she could see only 9 out of 15 plates in the left eye. Brightness sense was normal and there was no APD. Fundus examination showed slight hyperemia of the optic disc nasally in both eyes (Figure 28-15, OD and OS). Humphrey visual fields showed bilateral centrocecal defects and some nasal arcuate defects in right eye (Figure 28-16, OD and OS).

Optical Coherence Tomography

The ppNFL scan showed bilateral symmetrical loss of NFL in the inferior temporal region (Figure 28-17, OD and OS) whereas the remaining NFL looked robust. GCC scan showed inferior perifoveal (Figure 28-18, OD) and mild perifoveal (Figure 28-18, OS) thinning secondary to ethambutol toxicity.

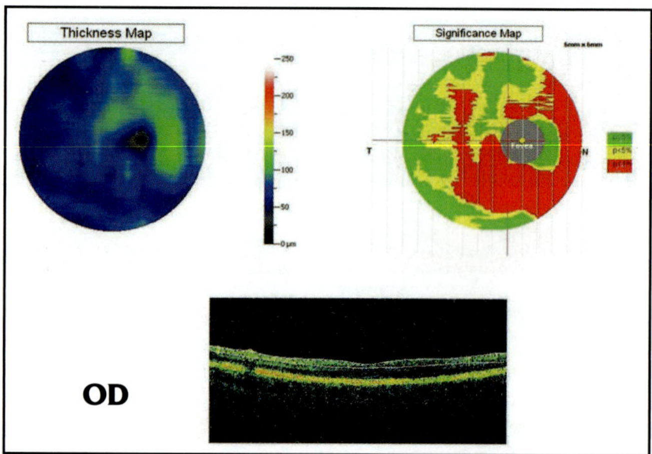

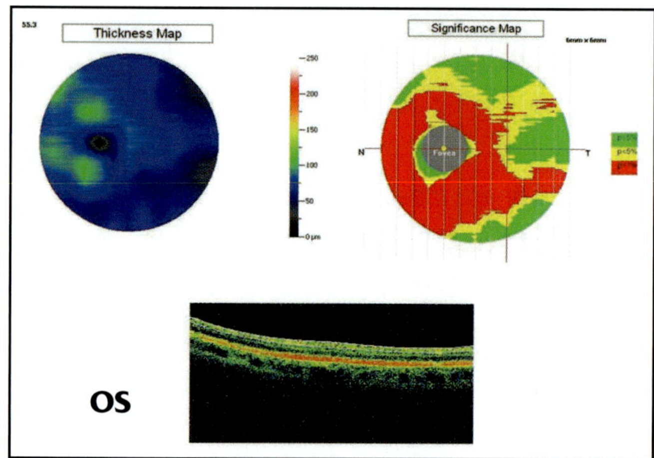

Figure 28-14. OD, OS: GCC scan. Upper left: GCC thickness map. Upper right: significant loss map. Bottom: OCT image of the vertical line marked in red on the significant loss map. Thin macula due to retinal ganglion cell loss from retrograde axonal degeneration secondary to ONH drusen.

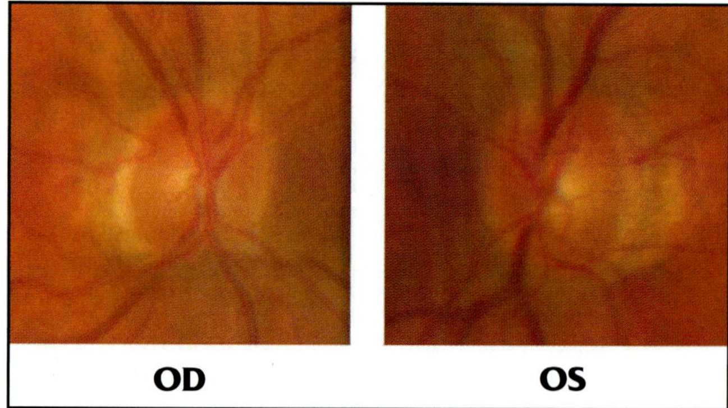

Figure 28-15. OD, OS: fundus photograph showing slight hyperemia nasally and mild temporal pallor more evident OS.

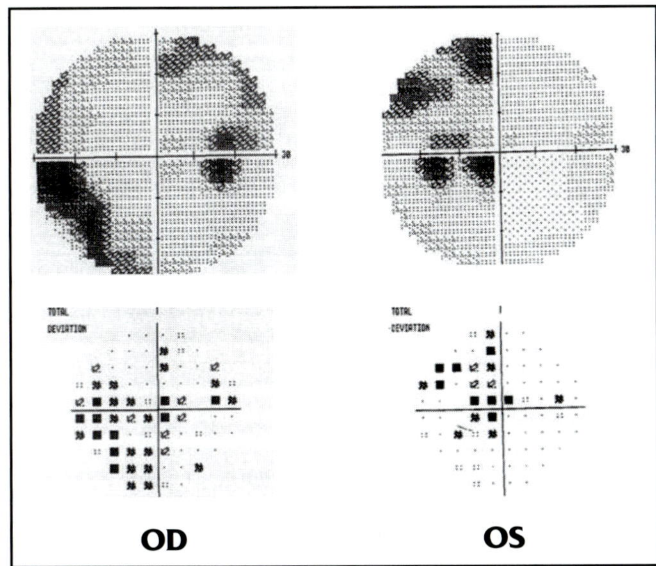

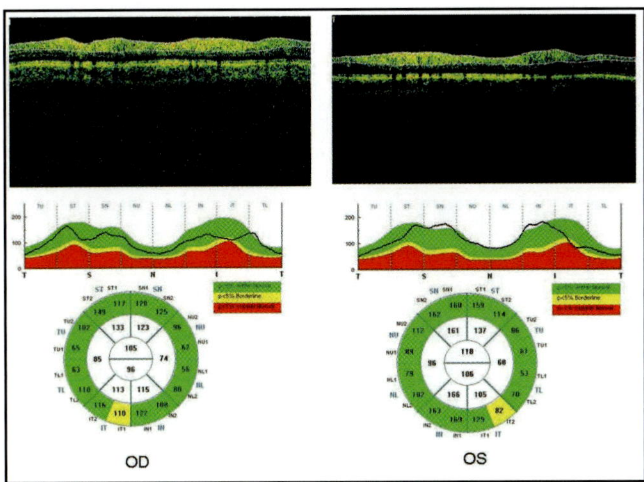

Figure 28-17. OD, OS: ppNFL, inferotemporal loss bilaterally.

Figure 28-16. OD: inferonasal visual field defect. OS: centrocecal visual field defect.

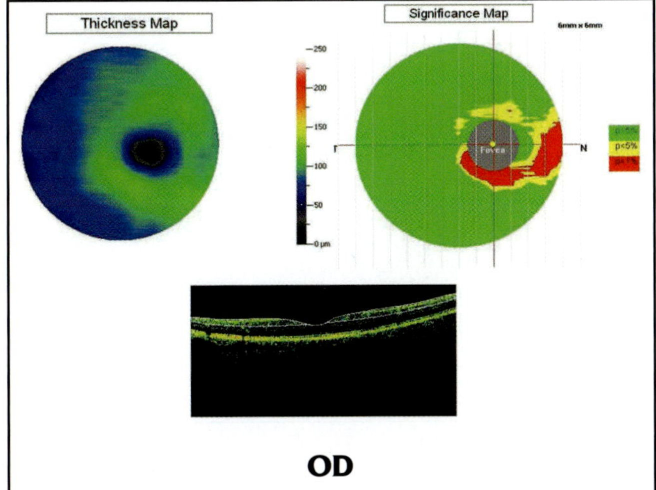

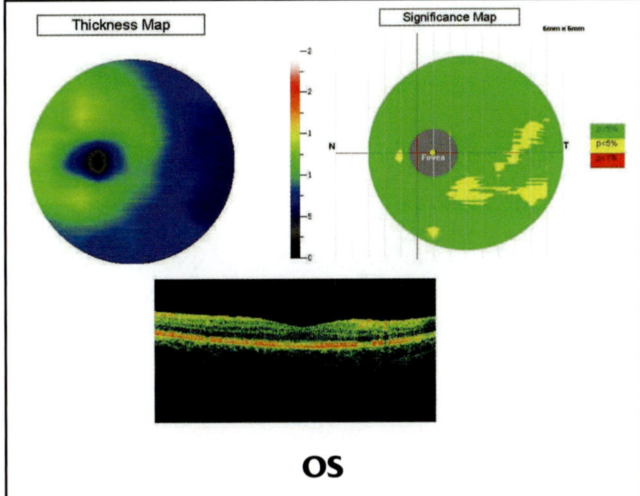

Figure 28-18. OD: GCC scan, inferior perifoveal thinning secondary to ethambutol toxicity. OS: minimal perifoveal thinning secondary to ethambutol toxicity.

Financial and Proprietary Disclosures

Dr. Divya Aggarwal has not disclosed any relevant financial relationships.

Dr. Perry S. Binder is a medical monitor for Abbott Medical Optics; a consultant for Acufocus Inc, Optovue Inc, Aqucsys Inc, Ophthonix Inc, Allergan Inc; an owner of Outcomes Analysis Software Inc; and an investor.

Dr. Leonardo C. Castro has not disclosed any relevant financial relationships.

Mr. William D. Dilworth is employed by Optovue Inc.

Dr. Jay S. Duker has not disclosed any relevant financial relationships.

Dr. Amani A. Fawzi receives grant support from Optovue Inc.

Ms. Lindsey S. Folio has no financial or proprietary interest in the materials presented herein.

Dr. James G. Fujimoto is a scientific advisor and has stock options in Optovue Inc. He receives royalties for intellectual property licensed by the Massachusetts Institute of Technology to Carl Zeiss Meditec Inc.

Dr. Fumi Gomi has no financial or proprietary interest in the materials presented herein.

Dr. Tarek S. Hemeida has no financial or proprietary interest in the materials presented herein.

Dr. Martin Heur has no financial or proprietary interest in the materials presented herein.

Dr. David Huang is a consultant and speaker for Optovue Inc and receives stock options, research grants, patent royalties, and travel support from the company. He also receives patent royalty from the Massachusetts Institute of Technology based on OCT technology licensed to Carl Zeiss Meditec Inc.

Dr. Yutaka Imamura has not disclosed any relevant financial relationships.

Dr. Alessandro Invernizzi has no financial or proprietary interest in the materials presented herein.

Dr. Hiroshi Ishikawa receives royalties on image processing algorithms licensed to Bioptigen Inc.

Mr. Larry Kagemann has no financial or proprietary interest in the materials presented herein.

Dr. Yan Li receives research support and travel support from Optovue Inc.

Dr. Bruno Lumbroso has not disclosed any relevant financial relationships.

Dr. Sheila Mahdaviani has no financial or proprietary interest in the materials presented herein.

Dr. Farnaz Memarzadeh has not disclosed any relevant financial relationships.

Dr. Nobuyuki Ohguro has not disclosed any relevant financial relationships.

Dr. Marco Pellegrini has no financial or proprietary interest in the materials presented herein.

Dr. Marco Rispoli has no financial or proprietary interest in the materials presented herein.

Dr. Alfredo A. Sadun has not disclosed any relevant financial relationships.

Dr. Camila Haydée Rosas Salaroli has no financial or proprietary interest in the materials presented herein.

Dr. Maria Cristina Savastano has not disclosed any relevant financial relationships.

Dr. Kaori Sayanagi has not disclosed any relevant financial relationships.

Dr. Joel S. Schuman receives royalties for intellectual property owned by MIT and Massachusetts Eye and Ear Infirmary and licensed to Carl Zeiss Meditec Inc and for intellectual property owned by University of Pittsburgh and licensed to Bioptigen Inc.

Dr. Giovanni Staurenghi has not disclosed any relevant financial relationships.

Dr. Ou Tan receives grant support and has a patent with Optovue Inc.

Dr. Maolong Tang receives research and grant support from Optovue Inc.

Dr. Stephen L. Trokel has not disclosed any relevant financial relationships.

Dr. Motokazu Tsujikawa has not disclosed any relevant financial relationships.

Dr. Gianmarco Vizzeri has no financial or proprietary interest in the materials presented herein.

Dr. Yimin Wang has one patent licensed to Optovue Inc.

Dr. Robert N. Weinreb is a consultant with Optovue Inc and Carl Zeiss Meditec Inc. He receives research support (instrumentation) from Optovue Inc, Carl Zeiss Meditec Inc, Topcon, and Heidelberg Engineering.

Dr. Andre J. Witkin has not disclosed any relevant financial relationships.

Dr. Gadi Wollstein receives royalties on intellectual properties licensed by the University of Pittsburgh to Bioptigen Inc and research support from Optovue Inc.

Dr. Linda M. Zangwill receives research support (equipment) and service agreement from Optovue Inc.

Index